AF292486

Plastic Design and Second-Order Analysis of Steel Frames

W.F. Chen   I. Sohal

Plastic Design and Second-Order Analysis of Steel Frames

With 312 illustrations

Includes two diskettes

Springer Science+Business Media, LLC

W.F. Chen
Department of Structural Engineering
Civil Engineering Bldg.
Purdue University
West Lafayette, IN 47907
USA

I. Sohal
Department of Civil Engineering
Rutgers University
Piscataway, NJ 08855
USA

Library of Congress Cataloging-in-Publication Data
Chen, Wai-Fah, 1936–
      Plastic design and second-order analysis of steel frames / W.F.
   Chen, I. Sohal.
          p. cm.
      Includes bibliographical references and index.
      ISBN 978-1-4613-8430-4
      1. Structural frames--Design. 2. Plastic analysis (Engineering)
   3. Steel, Structural. I. Sohal, I. II. Title.
   TA660.F7C44 1994
   624.1'821--dc20   94-11614

Printed on acid-free paper.

Production managed by Karen Phillips; manufacturing supervised by Gail Simon.
Typeset by Asco Trade Typesetting Ltd., Hong Kong.

9 8 7 6 5 4 3 2 1

ISBN 978-1-4613-8430-4   ISBN 978-1-4613-8428-1 (eBook)
DOI 10.1007/978-1-4613-8428-1

To our families

Preface

This book grew out of lectures which the senior author gave for a number of years to graduate students of structural engineering at Purdue University. Its primary purpose is to present the basic concept and methods of analysis of plastic theory, show how to use the theory in practical frame design, and discuss how the practical design rules in the AISC-LRFD specifications are related to theoretical considerations. These include the effect of axial load and shear force on plastic moment capacity, frame, member and local instability, and the significance of connection detailing in plastic design. Emphasis upon these and other design problems commences in Chapter 2 ("Plastic Hinges") and continues in Chapter 4 ("Equilibrium Method") and in Chapter 5 ("Work Method") where the design examples are given and calculations are made as complete as possible. The methods described in the first six chapters are suitable for hand calculations. Chapter 7 presents a computer-based method for the first-order plastic hinge-by-hinge analysis for frame design. The computer program FOPA developed and provided in this chapter can be used by students to check their homework problems given at the end of each chapter in a direct manner.

The advent of personal computers, particularly in the computing and graphics performance of engineering workstations, has made more sophisticated methods of analysis feasible in design practice. While the use of first-order analysis for elastic or plastic design is still the norm of engineering practice, a new generation of codes has emerged which recommends the second-order theory as the preferred method of analysis (AISC-LRFD, 1993). The advantage of using second-order theory for design practice is that the effect of lateral deflections of a structure under loading upon the overall geometry can be accounted for in a direct and more accurate manner. The result is more realistic and economical design. For this reason, Chapter 8 provides a compact and convenient summary of the second-order plastic hinge-by-hinge analysis methods suitable for computer application. This chapter also attempts to take stock of where the structural engineering profession stands with regard to direct analysis of inelastic strength and stability for frame design, and where it might be going. Included with the text are two diskettes containing two computer programs: one for Chapter 7 (FOPA) and the other for Chapter 8 (PHINGE). Both the menu-driven, user-friendly programs capable of tracing every plastic hinge formation throughout the entire range of loading up to plastic collapse (Chapter 7) or stability failure (Chapter 8).

In writing this book, we have endeavored to present the plastic methods in as simple a manner as possible. It also serves as an introduction to the second-order theory for inelastic frame design. Attention is directed to both analysis and design, and emphasis is placed on the physical significance of the various calculations involved. The book is aimed squarely for students of structural engineering who are familiar with the processes of elastic analysis and design of building frames. The first six chapters present the fundamental concepts, theorems, and the plastic methods of analysis and design; numerous examples suitable for hand calculation are included for illustration, and suitable problems provided at the end of each chapter for the student. The last two chapters are concerned specifically with the computer-based analysis methods for frame design. Here, for second-order inelastic analysis (Chapter 8) only an introduction to this quite difficult subject is given. It combines the structural stability theory with the plastic theory described in this book. The AISC-LRFD provisions for the use of plastic theory in practical design are the basis for the solution of various practical building frame design problems developed in the book.

The two computer programs were developed by Dr. M. Abdel-Ghaffar (Chapter 7) and Dr. R.J.Y. Liew (Chapter 8) as a part of their Ph.D. thesis work in the School of Civil Engineering at Purdue University for the research project entitled "Second-Order Inelastic Analysis for Frame Design" sponsored by the National Science Foundation (Dr. Ken Chong, Program Director).

Dr. Sohal wishes to thank his Department Chairman Yong S. Chae, for reducing the teaching load for a few semesters; Dean Ellis H. Dill, for his encouragement; and his teachers, students, colleagues, friends, computer and administrative staff, laboratory technicians, AAUP, research and sponsored programs personnel, secretaries, neighbors and the family, for their technical and personal support. Particularly, he is indebted to Dr. Pritam and Rupinder Dhillon, Dr. Rakesh and Madhu Kapania, Ms. Jessica K. Dembski, Drs. J. Wiesenfeld, L.S. Beedle, M. Shinozuka, T.V. Galambos, D.R. Sherman, R. Bjorhovde, S.C. Goel, S. Sridharan, S.T. Mau, J.T.P. Yao, J.T. Gaunt, V.J. Meyers, A.F. Grandt, Jr., T.Y. Yang, R.H. Lee, Donald White, K.C. Sinha, K.L. Bhanot, M.S. Ghuman, A.F. Saleeb, Eric M. Lui, Y. Ohtani, W.O. McCarron, Susan Pritchard, M. Taheri, Andrew J. Hinkle, Bernard Stahl, Nipen Saha, Ers. Robinder S. Sandhu, Gopal Gupta, Harjinder Singh, Gursharan Wason, Pushpinder Singh, Bhagwan D. Garg, Ajay Garg, Tarsem Lal Dhall, Daljit Mand and Dr. Harjit Bhatia, for their informal academic and personal support. Special thanks are due to Drs. Eiki Yamaguchi, Lian Duan, W.S. King, Mrs. Liping Cai, Messrs, John Sayler, Rajesh Mankani,

Edward Gray and Seeth Ramakrishnan, who contributed to the examples and solution of the problems in this book, through their home works. He will also like to thank Messrs. Ashish Patel, Ghassan Habib, Peter Tardy, Jae Chung, Gwo-Gong Huang, Shay Burrows, Mohammed El-Hawwat, Young Cho, David Brill, Nadeem Syed, Satvinder Singh, Mark Palus, Michael Steiner, David Stanger, Drs. Anil Khajuria, James Stewart, Ahmed Ezeldin, Zheng Zang, Benxian Chen, Luis Aguiar, Ms. Kristi Latimer, Ms. Gargi Shah and several others; who participated in useful discussions on the text, during their structural analysis and design classes at Rutgers University.

West Lafayette, IN   W.F. Chen
                                                            I.S. Sohal

Contents

1
Basic Concepts

1.1 Plastic Design vs. Elastic Design

The plastic design of steel structures has several advantages over the elastic design, of which the most important are simplified procedures, savings in the cost, and more realistic representation of the actual behavior of steel structures [1.1]. These advantages are due to the fact that the plastic design fully uses the important property of steel called *ductility*. This chapter will focus on the effects of ductility of the steel on the behavior of steel structures and show the benefits of the plastic methods that are derived from this property. To demonstrate the benefits of ductility, we will present two examples: first a hot-rolled section with residual stresses and second a plate with a hole. For both examples, the material is idealized to have an elastic–perfectly plastic stress-strain behavior as shown in Fig. 1.4.

1.1.1 Redistribution of Stresses in a Hot-Rolled Section with Residual Stresses

A hot-rolled wide flange section with residual stresses is shown in Fig. 1.1(a). These residual stresses are in self-equilibrium. If we apply an axial compressive load to this section, the section will yield first at the *elastic limit* load $P_y = (\sigma_y - \sigma_{rc})A$, in which σ_y is the yield stress of steel, σ_{rc} is the maximum compressive residual stress induced as a result of manufacturing process, and A is the cross-sectional area. At this load P_y, the stress distribution is shown in Fig. 1.1(b). However, since the steel is ductile, it can take a load higher than P_y. At a higher load, the section calls upon its less-stressed elastic portions to carry the increase in the load while the yielded portions remain at the yield stress level σ_y as shown in Fig. 1.1(c). At the plastic limit state, the stress distribution is shown in Fig. 1.1(d) and the load corresponding to this state has the value $P_p = \sigma_y A$ in which P_p is called the *plastic limit load*.

From this example, it is obvious that when we compare this plastic limit load with the elastic limit load, the plastic design/analysis is *simpler* because

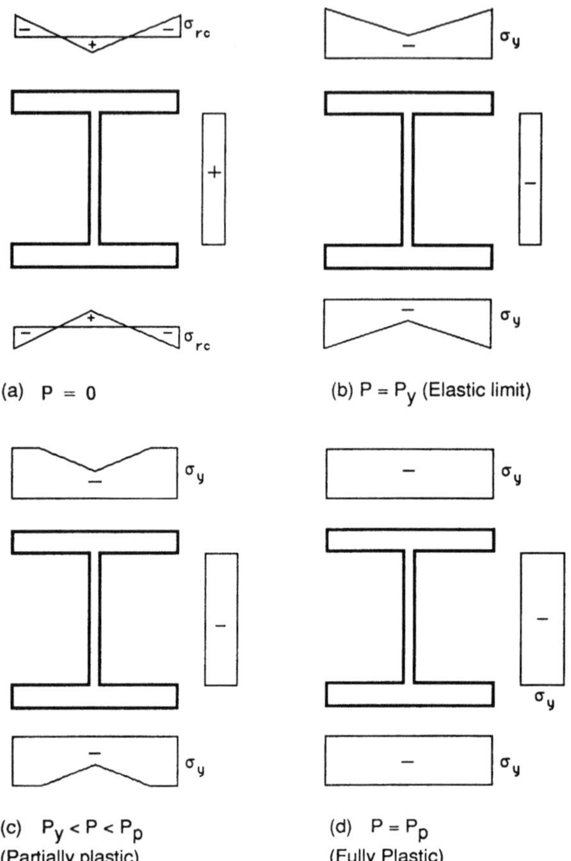

FIGURE 1.1. Stresses in a rolled I-section: (a) residual stresses; (b) stresses at initial yielding (or elastic limit load); and (c) elastic-plastic stresses; and (d) stresses at plastic limit load.

the residual stress (parallel to the direction of applied load) has no influence on the computation of the plastic limit load P_p. It is *economical* because a given section takes a higher load on the basis of the *plastic method* than that of the *elastic method*, i.e., $P_p > P_y$. Since the plastic design considers the ultimate limit state, it represents a more realistic estimation of the maximum load-carrying capacity of an actual structure.

1.1.2 Redistribution of Stresses in a Plate with a Hole

Another simple example is a plate with a hole, subjected to a tensile load as shown in Fig. 1.2(a). In the elastic range, due to *stress concentration* at the hole, the stresses are not uniform as shown in Fig. 1.2(a). So, the plate yields first at the elastic limit load $P_y = \sigma_y A/K$, in which A is the net area of cross section of the plate excluding the area of the hole, and K is the *stress concentration factor*. Again, the plate will take higher loads by using its property of ductility and redistributing stresses to its less-stressed elastic portions as

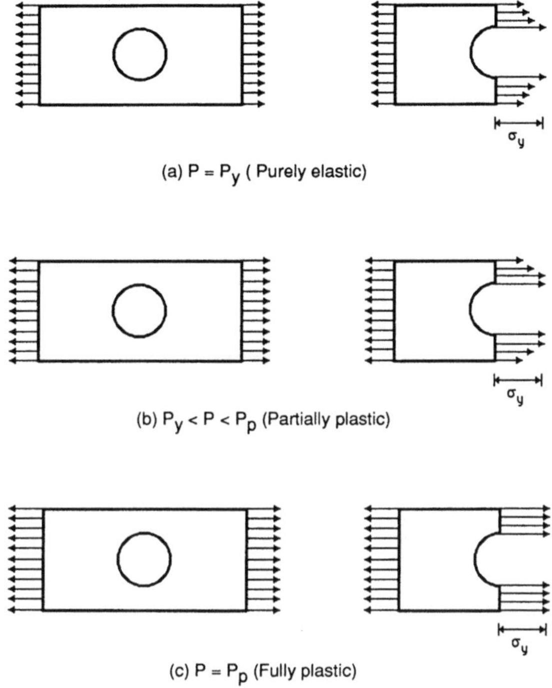

FIGURE 1.2. Stresses in a tension member with a hole: (a) stresses at initial yielding (elastic limit load); (b) elastic-plastic stresses; and (c) stresses at plastic limit load.

shown in Fig. 1.2(b). At the plastic limit load, the stress distribution becomes uniform as shown in Fig. 1.2(c), and the load corresponding to this limit state is $P_p = \sigma_y A$. Here, as in the first example, the advantages of the plastic design over the elastic design are obvious.

Here, as in the elastic design, factors such as buckling, fatigue, and deflection limitations will require special consideration.

1.2 The Ductility of Steel

As mentioned previously, the plastic design has several advantages over the elastic design because it fully uses the important property of steel, namely, *ductility*, which may be defined here as the ability of a material to undergo large deformation without much loss in its strength. Herein, we will discuss the ductile behavior of steel and the redistribution of forces/moments as an important benefit of the ductile behavior of steel.

1.2.1 Stress-Strain Relationship of Steel

The ability of structural steel to deform plastically at and above yield point is illustrated graphically in Fig. 1.3. Note that after the *elastic limit* is reached,

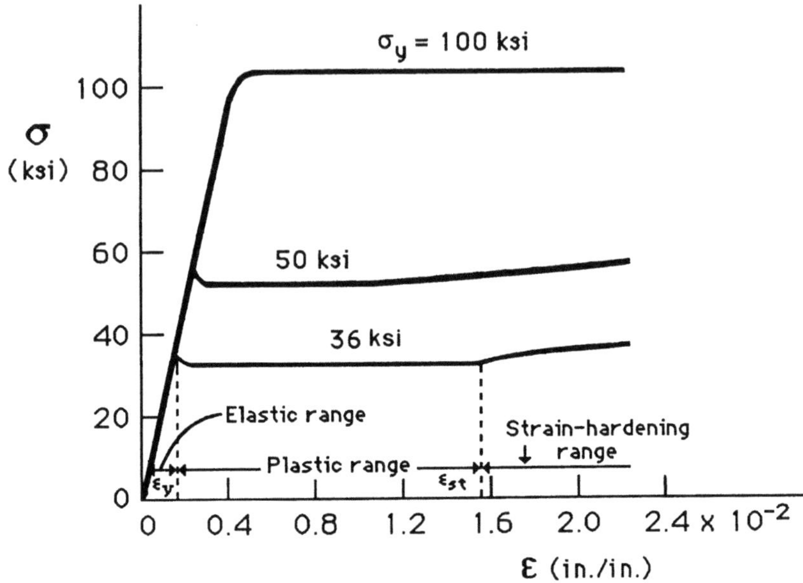

FIGURE 1.3. Stress-strain curve of various steels.

elongations of up to 15 times the elastic limit strain ε_y take place with no significant increase or decrease in stress, thus showing the ductile behavior of steel. After that, strain hardening commences at ε_{st}, and further deformation can take place only with some increase in stress.

Generally, steels with higher strength, such as 100 ksi, have relatively lower ductility. So, to ensure adequate ductility for plastic analysis and design, AISC-LRFD requires that the following specifications be satisfied: LRFD A5.1 (page 6–31)—the steel must exhibit a plastic plateau on the stress-strain curve; consequently, $F_y \leq 65$ ksi must be used.

For simplicity, the stress-strain curves of steel may be idealized by two straight lines as shown in Fig. 1.4. Up to the yield stress level, the material is elastic. After the yield stress has been reached, the strain is assumed to increase without further increase or decrease of the stress. This is known as the *elastic–perfectly plastic* idealization of the material behavior.

1.2.2 Redistribution of Forces in a Three-Bar Structure

Consider the simple three-bar structure shown in Fig. 1.5. This structure is statically indeterminate since the internal bar forces cannot be determined uniquely by statics alone. In order to determine forces in this structure, we need to consider not only the equilibrium condition, but also the compatibility condition and the stress-strain relationship of the steel. The equilibrium

FIGURE 1.4. Elastic–perfectly plastic idealization for the stress-strain relationship of steel.

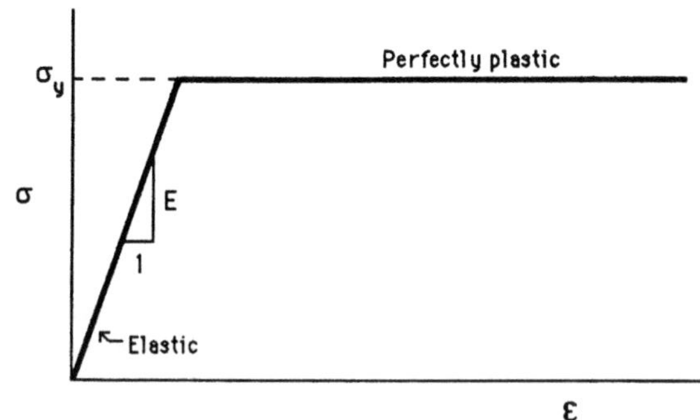

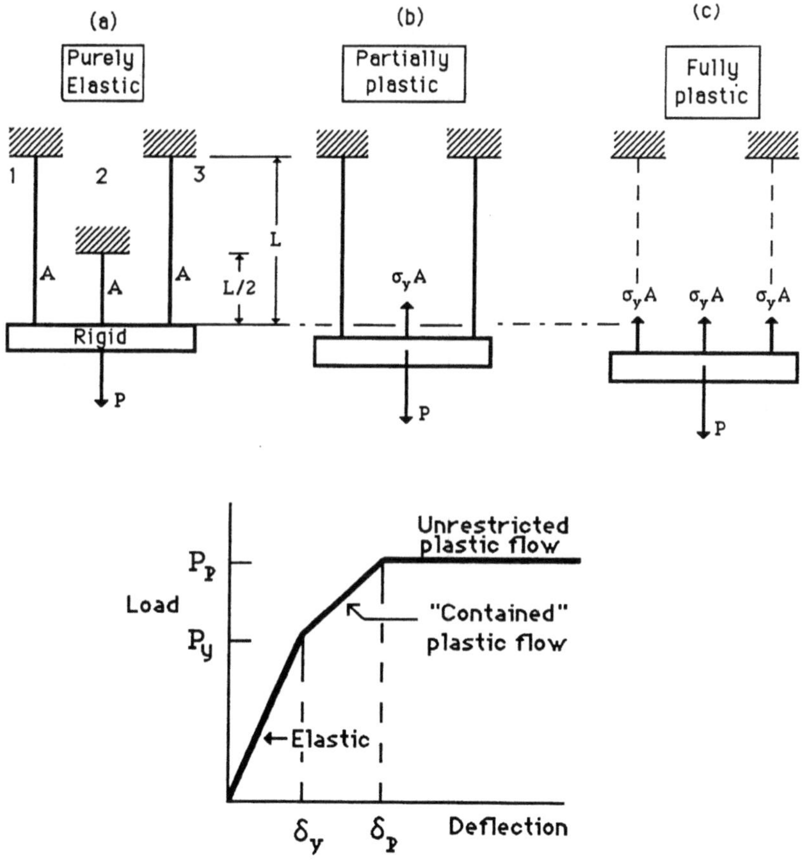

(d) Load-deflection relationship

FIGURE 1.5. Benefits of ductility in a three-bar structure due to force redistribution: (a) Purely elastic; (b) partially plastic; (c) fully plastic; and (d) load-deflection relationship.

condition of the three bars in Fig. 1.5(a) is

$$2T_1 + T_2 = P \tag{1.2.1}$$

where T_1 is the force in Bars 1 and 3, and T_2 is the force in Bar 2. The compatibility condition of the three bars is that the displacement of Bars 1 and 3, $\delta_1 = \delta_3$, must be equal to that of Bar 2, δ_2. Using the elastic stress-strain relation, $\delta_1 = T_1 L_1/AE$ and $\delta_2 = T_2 L_2/AE$ with $L_1 = L_3 = L$ and $L_2 = L/2$ where L_1, L_2, and L_3 are the lengths of Bars 1, 2, and 3, respectively, and A is the area of each of the Bars 1, 2, and 3, we obtain

$$\frac{T_1 L_1}{AE} = \frac{T_2 L_2}{AE} \tag{1.2.2}$$

or

$$T_1 = \frac{T_2}{2}. \tag{1.2.3}$$

From Eqs. (1.2.1) and (1.2.3), we find

$$T_2 = \frac{P}{2}. \tag{1.2.4}$$

Since the force in Bar 2 is greater than that in Bar 1, Bar 2 will yield first. Therefore, the load at which the structure will first yield ($P = P_y$) may be determined by substituting $T_2 = \sigma_y A$ in Eq. (1.2.4). Thus,

$$P_y = 2T_2 = 2\sigma_y A. \tag{1.2.5}$$

The corresponding displacement at this yield load is equal to the yield displacement of Bar 2. Using the elastic stress-strain relationship $\varepsilon_y = \sigma_y/E$, we have

$$\delta_y = \varepsilon_y L_2 = \frac{\sigma_y L}{2E}. \tag{1.2.6}$$

After the yielding of Bar 2, the structure reduces to a two-bar structure with a constant force equal to $\sigma_y A$ in Bar 2 [Fig. 1.5(b)]. The structure is now statically determinate. This two-bar structure can carry further loading until the outer two bars also yield at the plastic limit load P_p [Fig. 1.5(c)] given by

$$P_p = 3\sigma_y A. \tag{1.2.7}$$

Notice how easily one can compute the ultimate load, that is, the sum of the yield loads of each of the three bars. Unlike the elastic analysis, the compatibility condition is not required in the determination of the plastic limit load. This process of successive yielding of bars causing change of forces among the bars in a structure as the load is increased is known as *force redistribution*.

The corresponding load-deflection relationship of the three-bar structure is shown in Fig. 1.5(d). The load reaches the ultimate load (or the plastic limit load P_p) at a deflection δ_p given by

$$\delta_p = \varepsilon_y L_1 = \frac{\sigma_y L}{E} \tag{1.2.8}$$

and beyond δ_p, the deflections increase without limit while the load remains constant at P_p.

1.2.3 Plastification and Moment Redistribution in Beams

The stress redistribution in a beam is similar to that in the plate with a hole and that in the three-bar structure. The beam can be visualized as being made of many horizontal bars or fibers, some of which are in tension while others are in compression. At the *yield moment M_y*, only the extreme fibers yield. At moment higher than M_y, yielding spreads to the interior fibers too. At the *plastic limit moment M_p*, all fibers are yielded. The process of successive yielding of fibers causing change in stresses carried by the fibers as bending moment is increased is called *plastification*.

For statically indeterminate beams and frames, the benefits of ductility are even higher than that of simple beams and bars. In these beams and frames, the plastic limit load will be much higher than the initial yield load because of the two processes, namely, *plastification* and *redistribution*.

In statically indeterminate frames, the moment diagram has more than one peak moment. As the loads are applied to and increased in such a beam or frame, the cross section at the greatest peak moment will reach the yield moment first. As the loads are further increased, this cross section goes through the plastification process and a zone of yielding (called *plastic zone*) is formed around this cross section. As the loads are further increased, the moment at the yielded or plastic zone remains almost the same and the additional loads on the beam or frame are now taken by its less stressed sections, thus changing the distribution of moments among various cross sections with peak moments. This process of *moment redistribution* continues until plastic zones are formed at other cross sections with peak moments. The beam or frame will eventually fail when a sufficient number of these yielded zones are developed to transform the beam or frame into a *failure mechanism*.

1.3 Moment-Curvature Relationship

The basic information required in any calculations for plastic behavior and strength of framed structures is the relationship between the value of the applied bending moment M and the angle of relative rotation θ of the ends of a beam segment [Fig. 1.6(b)]. The gain in moment-carrying capacity of a beam due to plastification depends on this *moment-curvature* relationship, which in turn depends to a large extent on the shape of the cross section. Herein, the moment-curvature relationship of a beam with rectangular cross

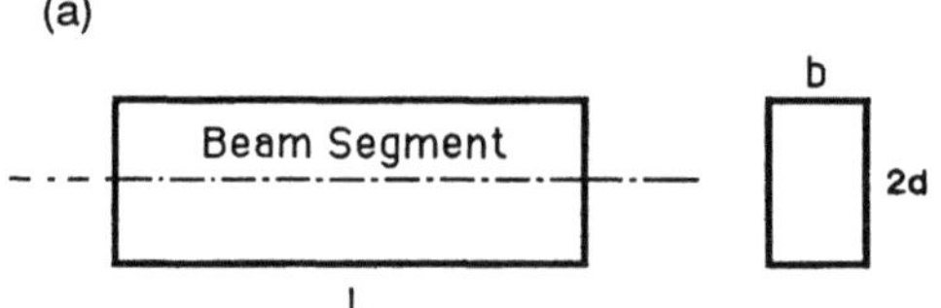

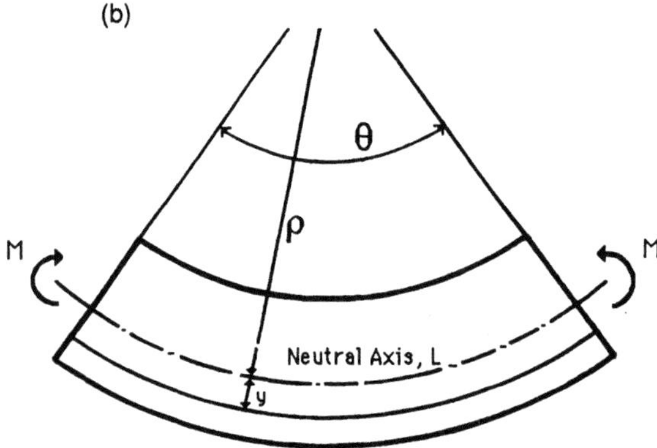

FIGURE 1.6. Bending of a rectangular beam segment.

section in the elastic and elastic-plastic regimes is first derived. Next, full plastic moment M_p and shape factor f are described. Then, the moment-curvature curves of various cross sections are presented and discussed.

1.3.1 Elastic Regime

Consider a rectangular beam segment of elastic–perfectly plastic material having length L, width b, and depth $2d$ as shown in Fig. 1.6(a). When this beam segment is subjected to bending moment M at its ends, it will bend into an arc of radius ρ as shown in Fig. 1.6(b). The central angle θ is related to the radius of curvature ρ by

$$\theta = \frac{L}{\rho}. \tag{1.3.1}$$

The curvature Φ, the relative rotation of two sections at a unit distance apart, can be expressed as

$$\Phi = \frac{\theta}{L} = \frac{1}{L}\left(\frac{L}{\rho}\right) = \frac{1}{\rho}. \tag{1.3.2}$$

Assume that after bending the plane section remains plane and the transverse fibers remain normal to the deflected axis, i.e., shear deformation is negligible. Thus, the length of a longitudinal fiber at a distance y from the neutral axis is $(\rho + y)\theta$, and the axial strain in the fiber is proportional to the distnace y from the neutral axis as [Fig. 1.6(b)]

$$\varepsilon = \frac{(\rho + y)\theta - L}{L} = \Phi y. \tag{1.3.3}$$

The *moment-curvature relationship* can now be obtained by combining the compatibility Eq. (1.3.3) with the following stress-strain relationships (1.3.4) and equilibrium equations (1.3.5) and (1.3.6). The idealized stress-strain relationship, as shown in Fig. 1.4, can be written as

$$\sigma = E\varepsilon \quad (\varepsilon < \varepsilon_y) \tag{1.3.4a}$$

$$\sigma = \sigma_y \quad (\varepsilon \geq \varepsilon_y) \tag{1.3.4b}$$

in which ε_y is the yield axial strain and E is Young's modulus.

The two equilibrium equations required in the derivation of the moment-curvature relationship of a segment are

$$P = \int_A \sigma \, dA = 0 \tag{1.3.5}$$

$$M = \int_A \sigma y \, dA. \tag{1.3.6}$$

Equation (1.3.5) is used to locate the neutral axis of the section. The neutral axis for a rectangular section, due to symmetry, is at the centroid of the section in the elastic and elastic-plastic regimes. Equation (1.3.6) is used to obtain the moment-carrying capacity of a section from a known stress distribution.

When all fibers of the segment are in the elastic regime ($\varepsilon < \varepsilon_y$), the stress distribution in the section is linear, as shown in Fig. 1.7(a). The moment-curvature relationship in this regime is likewise linear and can be obtained by substituting σ in Eq. (1.3.6) from Eq. (1.3.4a) and ε in the resulting equation from Eq. (1.3.3) as follows

$$M = E \int_A \varepsilon y \, dA = E\Phi \int_A y^2 \, dA = EI\Phi \tag{1.3.7}$$

in which I is the moment of inertia of the section.

As the moment is increased, the axial strain in the fibers increases. The segment begins to yield when the axial strain in the extreme fibers reaches the yield strain ε_y. The curvature corresponding to this initial yielding can be written from Eq. (1.3.3) as [Fig. 1.7(b)]

$$\Phi_y = \frac{\varepsilon_y}{d} \tag{1.3.8}$$

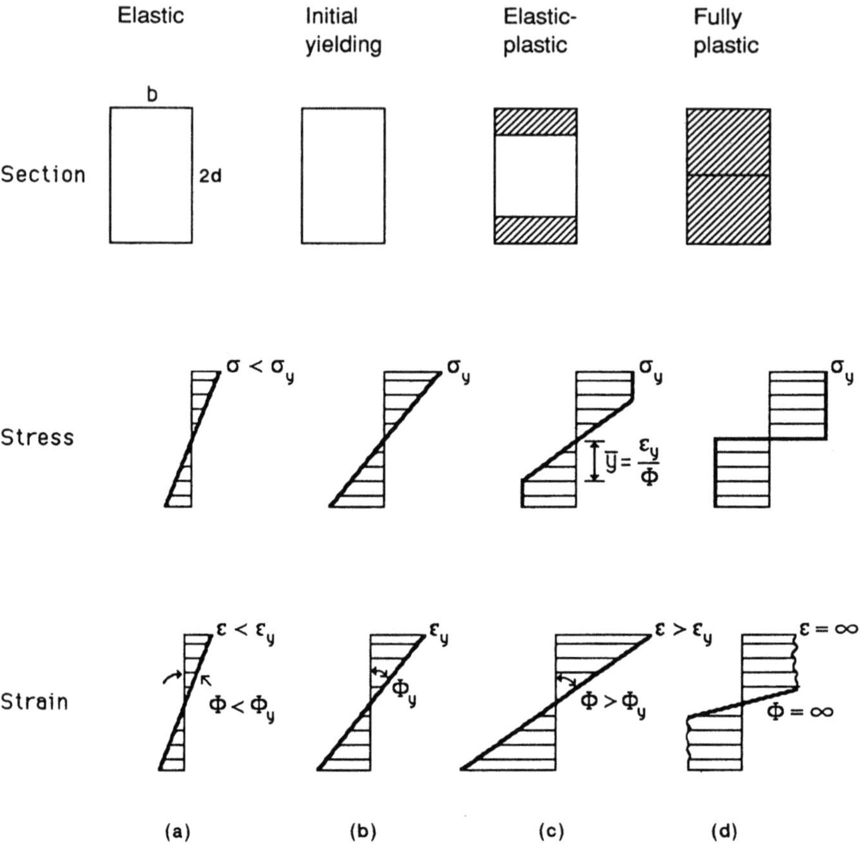

FIGURE 1.7. Stress and deformation states of rectangular section under pure bending.

in which $2d$ is the depth of the rectangular section. The moment capacity at this initial yield state can be obtained from Eq. (1.3.7) as

$$M_y = EI\Phi_y = E\varepsilon_y \frac{I}{d} = \sigma_y \left(\frac{2bd^2}{3} \right) = \sigma_y S \tag{1.3.9}$$

where b is the width of the rectangular section and S is the *elastic section modulus* of the section. For later comparisons, scales in Fig. 1.7 for stress and strain profiles have been chosen such that, up to yield point, stress and strain are represented by identical horizontal distances on the diagrams, i.e., σ_y in stress diagram is shown equal to maximum ε_y in the strain diagram in Fig. 1.7(b).

1.3.2 Elastic-Plastic Regime

A further increase in the moment results in the plastification of the section. Here, as in the three-bar example in Section 1.2, the yielded rod/fibers continue to carry the constant yield stress σ_y while the less stressed interior elastic fibers take additional stresses induced by the increase in the moment as shown in Fig. 1.7(c). Here, the strain at the outermost fibers of the beam has been doubled. This is possible only by doubling the curvature of the bent beam segment. No increase in maximum stress accompanied this increase in strain; yield stress σ_y has penetrated one-half the distance in toward the neutral axis. Further increases in strain of the beam outer fibers will result in a corresponding increase in the beam curvature Φ, but will only produce a further penetration of the constant yield stress σ_y in the beam. This process of plastification continues until all fibers are yielded as shown in Fig. 1.7(d). During the plastification process, the section is in the elastic-plastic—or partially elastic and partially plastic—regime as shown in Fig. 1.7(c). The boundary between the elastic and plastic portions is given by

$$\bar{y} = \frac{\varepsilon_y}{\Phi} \tag{1.3.10}$$

in which $\bar{y}$ is defined more clearly in Fig. 1.8. The moment-carrying capacity of the section in this elastic-plastic regime is obtained from Eq. (1.3.6) by substituting $\sigma = E\varepsilon = E\Phi y$ for the elastic portion and $\sigma = \sigma_y$ for the plastic portion as follows

$$M = \int_{A_e} E\Phi y^2 \, dA + \int_{A_p} \sigma_y y \, dA \tag{1.3.11}$$

in which A_e and A_p are, respectively, the area of elastic and plastic portions of the section. By substituting $dA = b \, dy$, M can be written as

$$M = 2 \int_0^{\bar{y}} E\Phi y^2 b \, dy + 2 \int_{\bar{y}}^d \sigma_y y b \, dy \tag{1.3.12}$$

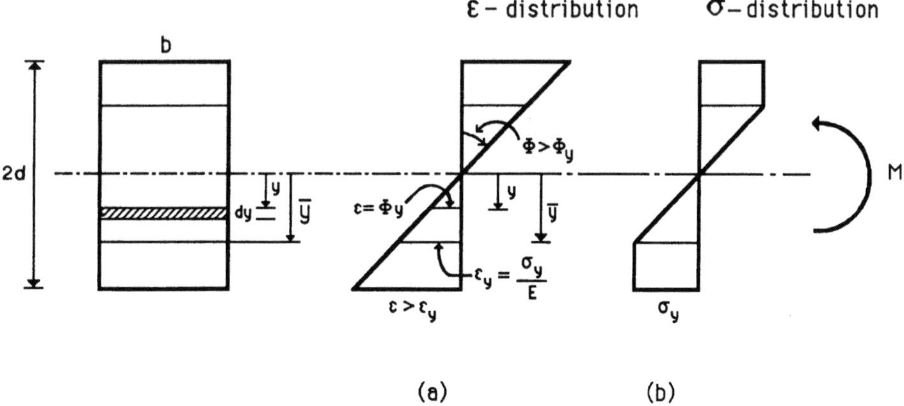

FIGURE 1.8. Elastic-plastic strains and stresses in rectangular beam under pure bending.

or

$$M = 2E\Phi b\frac{\bar{y}^3}{3} + 2\sigma_y\frac{b}{2}(d^2 - \bar{y}^2). \tag{1.3.13}$$

Substitution of $\bar{y}$ in Eq. (1.3.13) from Eq. (1.3.10) and of ε_y in the resulting equation from Eq. (1.3.8) along with $\sigma_y = E\varepsilon_y$, results in the following equation for the elastic-plastic moment capacity of the rectangular section

$$M = \frac{3}{2}M_y\left[1 - \frac{1}{3}\left(\frac{\Phi_y}{\Phi}\right)^2\right] \tag{1.3.14}$$

in which M_y is the *yield moment* of the section given by Eq. (1.3.9). Note that Φ_y/Φ in Eq. (1.3.14) is equal to $\bar{y}/d$. The moment-curvature relationship (1.3.14) is shown in Fig. 1.9. The full plastic moment capacity M_p corresponds to $\Phi \to \infty$, or $\Phi_y/\Phi \to 0$, i.e., $M_p = 1.5M_y$ (or $1.5\sigma_y S$).

1.3.3 Full Plastic Moment and Shape Factor

The moment-capacity in the elastic-plastic regime can also be expressed in terms of this full plastic moment M_p by substituting Φ in Eq. (1.3.11) from Eq. (1.3.10) as follows

$$M = \int_{A_e} E\frac{\varepsilon_y}{\bar{y}}y^2\,dA + \int_{A_p} \sigma_y y\,dA \tag{1.3.15}$$

or

$$M = \frac{\sigma_y}{\bar{y}}\int_{A_e} y^2\,dA + \sigma_y\int_{A_p} y\,dA \tag{1.3.16}$$

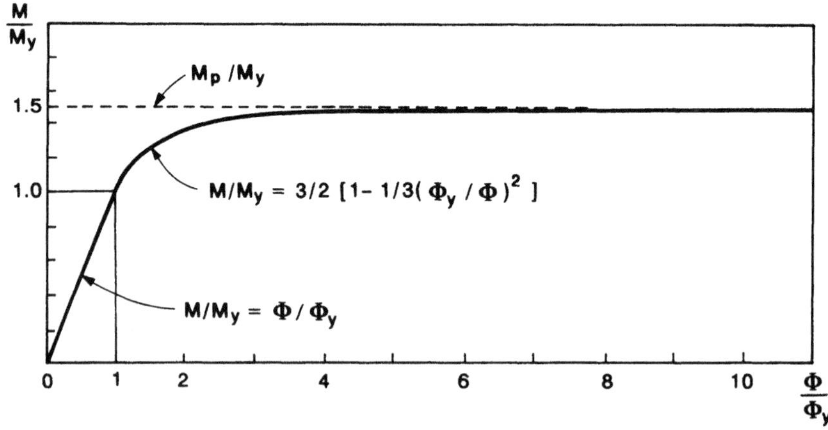

FIGURE 1.9. Nondimensionalized moment-curvature relationship of a rectangular beam.

or

$$M = \frac{\sigma_y}{\bar{y}} \int_{A_e} y^2 \, dA + \sigma_y \int_{A-A_e} y \, dA \qquad (1.3.17)$$

or

$$M = \frac{\sigma_y}{\bar{y}} \int_{A_e} y^2 \, dA + \sigma_y \int_{A} y \, dA - \sigma_y \int_{A_e} y \, dA \qquad (1.3.18)$$

or

$$M = \sigma_y S_e + \sigma_y Z - \sigma_y Z_e \qquad (1.3.19)$$

in which the second term on the right-hand side of Eq. (1.3.19) $\sigma_y Z = M_p$ is known as the full plastic moment and is equal to the moment of the stresses in the section at the fully plastic state as shown in Fig. 1.7(d); Z is known as the *plastic section modulus* and is equal to the first moment of the area of the whole cross section; S_e is the elastic section modulus of the elastic portion of the section; and Z_e is the plastic section modulus of the elastic portion of the section. Note that Eq. (1.3.19) can also be written directly from Fig. 1.10 by the method of superposition. When the section is fully plastic, the elastic portion disappears and S_e and Z_e in Eq. (1.3.19) reduce to zero. Thus, the *full plastic moment* M_p has the general form

$$M_p = \sigma_y Z. \qquad (1.3.20)$$

This is the maximum bending strength of the section, which is equal to the numerical sum of the moments of the fully plastic stress profile areas above and below the neutral axis, taken about that axis. The plastic moment value M_p is the basis for plastic design. The ratio of the plastic moment M_p to the yield moment M_y represents the amount of reserve strength due to the ductility of the material leading to plastification and is a function of the cross-sectional form or shape. This ratio, called *shape factor*, is a good indicator of

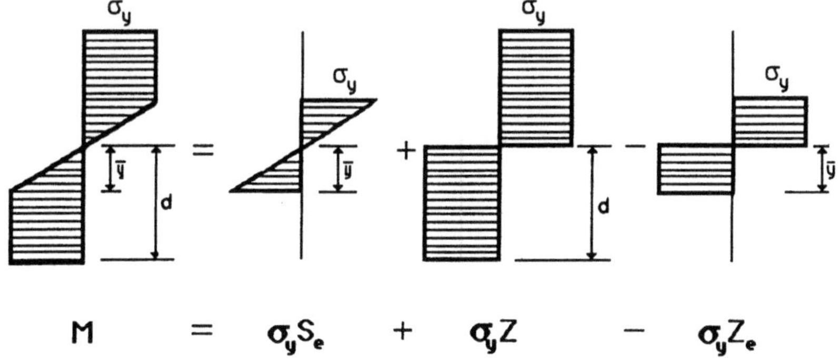

FIGURE 1.10. Moment capacity of a section in the elastic-plastic regime using the method of superposition.

TABLE 1.1. Shape factor for several cross sections

Shape of cross section	Shape factor
Isosceles triangle	2.32
Diamond	2.0
Round bar	1.7
Rectangle	1.5
Circular tube	1.27
Wide flange	1.14
Idealized I-section	1.0

the potential of a given section to gain strength by its plastification process. The shape factor is defined as

$$f = \frac{M_p}{M_y} = \frac{\sigma_y Z}{\sigma_y S} = \frac{Z}{S}. \tag{1.3.21}$$

For a rectangular section, we have $Z = bd^2$ and $S = 2bd^2/3$, and thus the shape factor has the value

$$f = \frac{M_p}{M_y} = \frac{Z}{S} = \frac{bd^2}{2bd^2/3} = 1.5. \tag{1.3.22}$$

The shape factors for some other cross-sectional shapes are given in Table 1.1. The shape factor is higher for sections with mass concentrated near the centroid of the section and lower for sections with mass concentrated away from the centroid. For rolled I-beams and wide flange shapes bent about their strong axis, the shape factor f varies between 1.10 and 1.18. For most W shapes, it is very close to the value of 1.12.

The moment-curvature curves for other sections listed in Table 1.1 can be derived in a manner similar to that of a rectangular section described here. The moment-curvature curves of diamond, rectangular, W14 × 426, W21 × 109, and idealized I-sections are shown in Fig. 1.11. The idealized curves (elastic–perfectly plastic) for these sections are shown by dashed lines in this figure.

1.3.4 Discussion of Moment-Curvature Curves

The moment-curvature relationship of the rectangular section, Eq. (1.3.7) for the elastic regime and Eq. (1.3.14) for the elastic-plastic regime, was plotted previously in Fig. 1.9. As the curvature is increased, the moment capacity approaches rapidly to the full *plastic moment* M_p. Note that the full plastic moment M_p is higher than the yield or elastic limit moment M_y by 50%. Theoretically, the full plastic moment capacity M_p will not be reached until the curvature approaches infinity. Practically, 99% of the plastic moment capacity is attained at a curvature equal to only four times the *yield curvature*

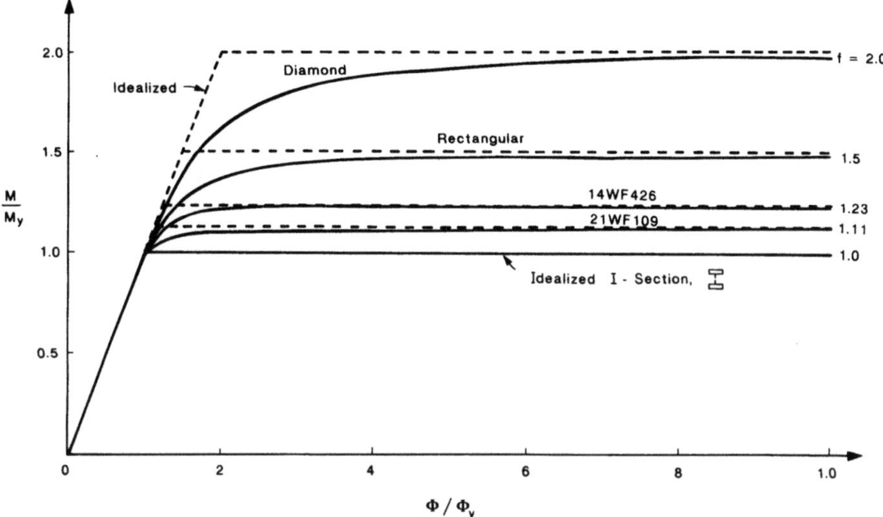

FIGURE 1.11. Actual and idealized moment-curvature curves.

Φ_y. However, it is expected that the moment capacity of an actual section will reach full plastic moment at an early curvature and will have a higher value than the full plastic moment at a larger curvature because extreme fibers will enter into the strain-hardening regime that has been neglected in the present derivation.

1.4 Flexure of a Fixed-Ended Beam with Uniformly Distributed Load

As discussed in the previous sections, the load-carrying capacity of steel structures obtained by plastic analysis is higher than that by elastic analysis based on the first yielding of the material in the structures. In statically determinate structures, only material plastification at the critical section contributes to this higher load-carrying capacity. While in statically indeterminate structures, both material plastification at the critical sections and moment redistribution among these sections contribute to the higher load-carrying capacity. Herein, we shall first illustrate this point through the study of the bending behavior of a fixed-ended beam subjected to a distributed lateral load as shown in Fig. 1.12(a). Next, the hinge-by-hinge method and the required rotation capacity, as applied to the present example, are described. Finally, some comments are made about the process of plastification and moment redistribution of the present example.

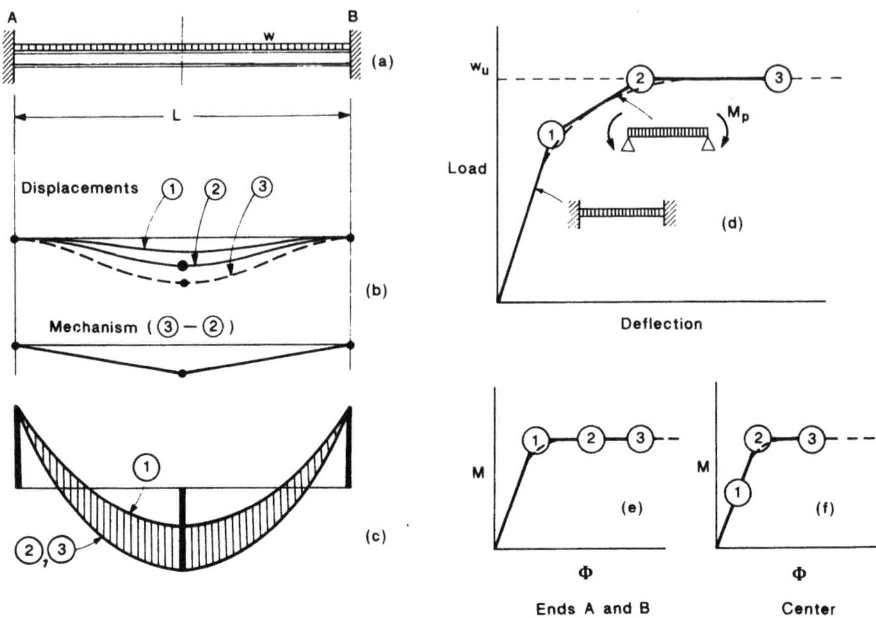

FIGURE 1.12. Redistribution of moment in a fixed-ended beam with uniformly distributed load.

1.4.1 Behavior with Actual and Idealized M-Φ Relationships

In the elastic range, the deflected shape and bending moment diagram of this beam can be obtained by solving the governing equilibrium and compatibility equations. The results are shown by Curve 1, respectively, in Parts (b) and (c) of Fig. 1.12. Note that there are three peaks in the moment diagram. The moment at the end peaks is $wL^2/12$, and at the center peak it is $wL^2/24$, in which w is the intensity of the distributed load and L is the length of the beam. As the load is increased, the moment at the greater peaks (end peaks) reaches the yield moment M_y first, and the moment at the smaller (central) peak reaches only half the yield moment. In the elastic range, the moment at all other sections increases proportionately, thus maintaining the same shape of the moment diagram. With a further increase of the load, the plastification of some sections near the end peaks starts, and zones of yielding begin to form near these peaks. The bending stiffness (or the slope of the moment-curvature curve) of these yielded zones is smaller compared to that of the elastic zones (or the initial slope of the moment-curvature curve). This results in a different shape of the bending moment diagram with a higher rate of increase in moment in the central elastic portion than that at the plastic end portions of the beam. In other words, there is a continuous redistribution of moments along the length of the beam during loading. As the load is further increased, the moment at the end peaks will approach the plastic moment, while the moment at the central peak will reach the yield moment. The beam will fail when the bending moments at all three peaks are equal to the plastic moment (Curve 2), thus forming a *failure mechanism* [see Curve 3, Fig. 1.12(b)].

The process of moment redistribution can be seen more clearly when we carry out the plastic analysis of the fixed-ended beam by employing the idealized moment-curvature relationship (the elastic–perfectly plastic type shown by the dashed lines in Fig. 1.11). Figure 1.12(d) shows the load-deflection results of the three-stage loading. In the elastic range, the midspan bending moment remains half that at the ends as the load is increased, until the beam reaches the limit of its elastic behavior. Stage 1 in Fig. 1.12(d) corresponds to the load when the moments at the ends have just reached the full plastic moment of the section. As the load w is further increased, the beam enters into the elastic-plastic range. In this range, plastic hinges form at the fixed ends, permitting them to rotate with constant moment capacity. Stage 2 corresponds to the theoretical plastic limit load when the moment at the center has also reached the full plastic moment capacity of the section. Since sufficient plastic hinges have now formed, a failure mechanism has developed and no further loading can be supported beyond this stage. Stage 3 corresponds to an arbitrary deformation obtained by a continued deformation beyond Stage 2. The deflected shapes and the moment diagrams at these three loading stages are shown in Fig. 1.12(b) and (c), respectively. The load-deflection curve corresponding to the actual moment-curvature relationship of the beam is shown as the dashed curve in Figs. 1.12(d). The moment-curvature responses at the ends and center are shown in Figs. 1.12(e) and (f), respectively. At Stage 1, the moment at the ends has just reached M_p [Fig. 1.12(e)], while at the center of the beam the moment is only half of the plastic moment [Fig. 1.12(f)]. As the load is increased beyond this stage, the sections at two ends rotate with a constant plastic moment capacity (*plastic hinge action*). The beam now behaves as a simply supported beam with constant end moments equal to M_p. At Stage 2, the moment at center also reaches the plastic moment. This corresponds to the maximum load-carrying capacity state of the beam. Beyond Stage 2, the beam continues to deform as a rigid body under constant load.

1.4.2 The Hinge-by-Hinge Method and the Required Rotation Capacity

It is evident from the load-deflection curve shown in Fig. 1.12(d) that the formation of each plastic hinge removes one degree of indeterminacy from the structure, and the subsequent load-deflection relationship correspons to a new and simpler structure. For example, in the elastic range, the deflection under the given load can be determined by an elastic analysis with fixed ends. The load-deflection curve between Stages 1 and 2 can be determined by an elastic analysis with simply supported ends. The method of finding deflection by the elastic analysis of a new structure after the formation of a plastic hinge is known as the "hinge-by-hinge method." This method will be further elaborated in Example 1.8.2. The computer-based hinge-by-hinge analysis, together with its computer program for framed structures, will be presented in Chapter 7.

Note that the fixed-ended beam in this example can go through moment redistribution and develop a higher load-carrying capacity only if the end sections have adequate *rotation capacity*, i.e., the section can rotate the required amount without any significant loss in moment capacity. For example, for the fixed-ended beam of Fig. 1.12(a) to reach Stage 2, the end sections of the beam must be able to rotate by an angle of $M_p L/6EI$. This required rotation capacity for plastic design must be provided by the ductility of the material. This is not a problem for ductile steels, but is can be a problem for more brittle materials such as reinforced concrete.

1.4.3 *Some Comments on Plastification and the Moment-Redistribution Process*

Note that the process of plastification and moment redistribution in the present example beam, in principle, is similar to that of force redistribution in the previous example of a three-bar structure. In the case of the three-bar structure, when the middle bar yielded, the force in that bar remained constant while forces in Bars 1 and 3 continued to increase. The ultimate load was reached when all three bars became plastic. Similarly, during the plastification in beam cross sections, when the extreme fibers yielded, the stresses in those fibers remained constant while stresses in the intermediate fibers continued to increase. The full plastic moment capacity was reached when all fibers in the cross section were plastic. Thus, the process of *plastification* results in a successive yielding of fibers in the cross section of a member as the bending moment is increased and the yielded zone spreads.

Similarly, during the moment redistribution in the fixed-ended beam, when the moments at end sections reached plastic moment capacity, they remained constant, while moments at other sections continued to increase. The ultimate load was reached when the moment at the central section also reached the plastic moment capacity. As a result, a failure mechanism was developed. Thus, the process of *moment redistribution* results in a successive formation of plastic hinges so that less-stressed portions of a structure will carry increased moments.

1.5 Margin of Safety in Plastic Design with Load Factor

In the *allowable stress design*, the safety is achieved by using an allowable stress that is obtained by applying a factor of safety to the stress level assumed to represent failure. In the plastic design, the safety is achieved by using a factored load obtained by multiplying the given service loads by a *load factor* λ. The application of factor of safety to loads is better since the uncertainty associated with loads is higher than that associated with resistances. The load factor by definition is

$$\lambda = \frac{\text{limit load}}{\text{working load}}. \tag{1.5.1}$$

This factor is obtained by first considering the margin of safety of a simply supported beam with uniformly distributed lateral load w designed by the plastic method and then calibrating it against the allowable stress method as laid down in current regulations. By substituting the ultimate load or the plastic limit load w_p and the working load w_a in Eq. (1.5.1) in terms of the *plastic moment* $M_p = w_p L^2/8$ and the *allowable moment* $M_a = w_a L^2/8$, the load factor λ for a simply supported beam can be determined as

$$\lambda = \frac{w_p}{w_a} = \frac{8M_p/L^2}{8M_a/L^2} = \frac{M_p}{M_a} \tag{1.5.2}$$

in which L is the length of a simply supported beam. By replacing M_p and M_a in terms of stresses, λ can be written as

$$\lambda = \frac{\sigma_y Z}{\sigma_a S} = \frac{\sigma_y}{\sigma_a} f. \tag{1.5.3}$$

In building design, the *allowable working stress* σ_a for compact beams (beams having width-to-thickness ratios of their flanges and webs less than those specified by AISC specifications) with continuous lateral support is equal to 0.66 of the yield stress of the steel section. By substituting $\sigma_a = 0.66\sigma_y$ in Eq. (1.5.3), the load factor λ is found to be

$$\lambda = \frac{\sigma_y}{0.66\sigma_y} f = 1.52 f. \tag{1.5.4}$$

The *shape factor f* for wide-flange beams varies from 1.10 to 1.18 with an average value of 1.134 and a mode of 1.12, and for wide flange columns the shape factor varies from 1.10 to 1.23, with an average of 1.137 and a mode of 1.115. Using the shape factor equal to the mode of 1.12, we find the load factor λ for the plastic design of steel structures as

$$\lambda = 1.52 \times 1.12 = 1.70. \tag{1.5.5}$$

Thus, in the United States, a load factor of 1.70 is being used for the plastic design of steel structures under gravity loads. Note that this load factor is also consistent with the usual factor of safety of 1.67 for most structural members, used in the allowable stress design specifications.

For the case of gravity load in combination with wind or earthquake forces, allowable-stress design specifications permit a one-third increase in computed stresses. So, to be consistent with ASD specification, the load factor λ for the plastic design will be reduced by 25% to

$$\lambda = \frac{1.70}{1.33} = 1.30. \tag{1.5.6}$$

This smaller load factor is justified by the fact that the probability of simultaneous occurrence of all these load effects is unlikely.

The values of load factors recommended for plastically designed steel structures elsewhere in the world are listed in Table 1.2.

Since plastic analysis enables the designer to calculate the maximum load that the structure is capable of supporting in a direct manner, the corresponding working load is determined by dividing this load by a load factor.

TABLE 1.2. Load factors for plastic design in various countries

Country (1)	Assumed shape factor (2)	Dead load + live load (3)	Dead load + live load + wind or earthquake forces (4)	Number of load factors (5)
		(a)		
USA	1.12	1.70	1.30	2
Australia	1.15	1.75	1.40	2
Canada	1.12	1.70	1.30	2
Germany	——	$1.71f$	$1.50f$	2
India	1.15	1.85	1.40	2
Mexico	1.12	1.70	1.30	2
Sweden	——	1.57	1.34	2
United Kingdom	1.15	1.75 (portal frames)	1.40	3
		(b) Multiple Load Factors		
Japan	——	$1.2D + 2.1 (L + S)$ or $1.4 (D + L + S)$ (normal condition) $(D + L) + 1.5E$ or $(D + L + nS) + 1.5E$ (under earthquake) $(D + L) + 1.5W$ or $(D + L + nS) + 1.5W$ (under typhoon)		6

The following symbols are used:
D = dead load
L = live load
E = earthquake load
f = shape factor
S = maximum snow load
W = wind force

n	a period of snowdrifts
0	less than one month
0.5	one month
1.0	three months

Thus, in plastic design, if the required working load is multiplied by 1.70, the fixed-ended beam designed for this increase (factored) load will have the same factor of safety 1.70 against plastic collapse as the simple beam. In allowable stress design, however, the safety factor against plastic collapse is 1.70 for the simple beam case, but it increases to $1.7 \times 4/3 = 2.27$ for the case of fixed-end beam.

The load factors recommended for all structures including redundant structures are the same as those given earlier. Since redundant structures such as the fixed-ended beam can develop higher load-carrying capacity through redistribution of moments, such structures designed by the plastic method will be more economical compared to those when designed by the allowable stress method. For example, a 30-foot-long fixed-ended beam with continuous lateral support and 1 kip/ft uniformly distributed load, designed by the allowable stress method will require a W16 $\times$ 26 A36 section. However, if it is designed by the plastic method, it will require only W14 $\times$ 22 A36 secton, thus resulting in an over 15% saving. Note that the W14 $\times$ 22 beam will remain elastic at working load and the deflection at this load will be about 1.23 times greater than that of the W16 $\times$ 26 beam designed by the allowable stress method. Thus, at working loads, plastically designed structures generally deflect slightly more than a similar continuous structure designed to a working stress limitation. When required, deflections at working load can be computed by means of the usual elastic methods. An estimate of the working load deflection can generally be obtained more easily by dividing the computed deflection at plastic collapse load by the load factor. This will be described in Chapter 6. No more rules are provided to govern deflections in plastically designed structures; they are subject to the same limitations as those governing working stress designs.

1.6 A Brief Historical Account of Plastic Design

Plastic concepts were applied to the design of building frames as early as 1914 when Kazinczy [1.2] of Hungary published results of his tests on clamped girders. He suggested the concept of a *plastic hinge*. In 1927, Maier-Leibnitz [1.3] of Germany conducted experiments on continuous beams and showed that the ultimate capacity of continuous beams is not affected by settlement of their supports. During the late 1930s and afterwards, Baker and his associates [1.4] in Great Britain continued tests on steel structures and formulated design rules to use the plastic reserve strength. In the 1940s, significant progress was made at Brown University in the United States in the theory of plastic analysis of structures [1.5].

In the late 1940s and during the 1950s, full-scale tests were conducted at Lehigh University in the United States by Beedle and his associates [1.6–1.8] to study the plastic behavior of large steel frames. These studies focused on the verification of plastic analysis and design methods and the conditions that must be satisfied to avoid secondary failure modes such as column buckling, local buckling, fatigue, and fracture.

The plastic design method was approved as an alternative design method by the AISC in 1958 and is now treated as a separate chapter of the updated Allowable Stress Design Specification issued in 1989 [1.9]. A guide and commentary for plastic design of steel was published by the ASCE in 1971 [1.10]. Recently, the AISC (1993) included plastic design as a part of the general limit states design specifications known as the load and resistance factor design (LRFD) specification [1.11].

The rapid advancements in computer hardware and software have motivated researchers and engineers in recent years to make more sophisticated inelastic analysis techniques practical for direct engineering design. To this end, attempts have been made to modify the *plastic hinge method* to include both strength and stability considerations and to make it consistent with the current LRFD specifications [1.12]. In this way it will account for both inelastic redistribution of forces and member and system stability in a direct manner. This development, known as *advanced inelastic analysis*, will be described in Chapter 8.

1.7 Current and Future Design Philosophies

Current and future design philosophies can be classified as: allowable stress design, plastic design with load factor, LRFD with elastic analysis, plastic design with LRFD, and design with advanced inelastic analysis. A brief description of these design philosophies is given in the following.

1.7.1 Allowable Stress Design

In the *allowable stress design* (ASD), it is ensured that the stresses in a structure under working or service loads do not exceed some predesignated allowable values. These allowable values are usually obtained by dividing the yield stress or ultimate stress of the material by a factor of safety. The general format for an allowable stress design is thus

$$\frac{R_n}{\text{F.S.}} \geq \sum_{i=1}^{m} Q_{ni} \qquad (1.7.1)$$

where R_n = nominal resistance of the structural member expressed in unit of stress; Q_n = nominal working or service stresses computed under working load conditions; F.S. = factor of safety (e.g., 1.5 for beams, 1.67 for tension members, 1.92 for long columns); i = type of load (i.e., dead load, live load, wind load); and m = number of load types.

The left-hand side of Eq. (1.7.1) represents the allowable stress of the structural member or component under a given loading condition (for example, tension, compression, bending, or shear). The right-hand side of the equation represents the combined stress produced by various load combinations (for example, dead load, live load, or wind load). Formulas for the allowable stresses for various types of structural members under various types of loadings are specified in the AISC specification [1.9]. A satisfactory design is reached when the stresses in the member computed using a first-order elastic analysis under working load conditions do not exceed their allowable values. The effects of secon-order moments and inelasticity are considered in an indirect manner. One should realize that in the allowable stress design, the factor of safety is applied only to the resistance term, and safety is evaluated at the service load. Thus, ASD is characterized by the use of unfactored "working" loads in conjunction with a single factor of safety applied to the resistance. Because of the greater variability and unpredictability of the live load and other loads in comparison with dead load, a uniform reliability is not possible with ASD.

1.7.2 Plastic Design with Load Factor

In the *plastic design (PD) with load factor*, it is ensured that the factored load combinations or their effects do not exceed the maximum plastic strength of the structure or component. It has the format

$$R_n \geq \lambda \sum_{i=1}^{m} Q_{ni} \qquad (1.7.2)$$

where R_n = nominal plastic strength of the structure or component; Q_n = nominal load or load effect (e.g., axial force, shear force, bending moment); i = type of load (D = dead load, L = live load, W = wind load); λ = load factor [e.g., 1.70 for (D + L), 1.30 for (D + L + W)]; and m = number of load types.

Note that in this method of design, safety is incorporated only in the load term and is evaluated at the ultimate (plastic strength) limit sate. Applying a factor of safety to the load term is more appropriate because uncertainty associated with loads is higher than that associated with resistances. The method is superior than the allowable stress design approach in the sense that it considers redistribution of forces in beams and frames in a more direct manner. The effects of secon-order moments are considered indirectly. Since only a single factor of safety (called load factor) is applied to all loads, a uniform reliability cannot be fully achieved with PD.

1.7.3 Load and Resistance Factor Design with Elastic Analysis

In the *load and resistance factor design*, it is ensured that the factored load effects do not exceed the factored nominal resistance of the structural member or component. Here, we have two safety factors. One is applied to the

loads, the other to the resistance of the material. This is more realistic because both loads and resistances have different uncertainties. Thus, the load
and resistance factor design has the format

$$\phi R_n \geq \sum_{i=1}^{m} \gamma_i Q_{ni} \tag{1.7.3}$$

where R_n = nominal resistance of the structural member; Q_n = nominal load
effect (e.g., axial force, shear force, bending moment); ϕ = resistance factor
(≤ 1.0) (e.g., 0.9 for beams, 0.85 for columns); i = type of load (e.g., D = dead
load, L = live load, S = snow load); γ_i = load factor (usually > 1.0) corresponding to Q_{ni} (e.g., 1.4D and 1.2D + 1.6L + 0.5S are the factored load
combinations recommended by LRFD); and m = number of load types.

Note that LRFD uses separate factors for each load and can therefore
reflect the degree of uncertainty of different loads and combinations of loads.
As a result, a more uniform reliability can be achieved.

In the 1993 LRFD specification [1.11], the resistance factors were developed mainly through calibration with ASD [1.13], whereas the load factors were developed based on a statistical analysis [1.14–1.15]. A satisfactory
design is one in which the probability of exceeding a limit state of the structural member (for example, yielding, fracture, or buckling) is minimal. Based
on the first-order second-moment probabilistic analysis [1.16], the safety of
the structural member is measured by a *reliability* or *safety index* [1.11]
definded as

$$\beta = \frac{\ln(\overline{R}_n/\overline{Q}_n)}{\sqrt{V_R^2 + V_Q^2}} \tag{1.7.4}$$

where $\overline{R}$ = mean resistance; $\overline{Q}$ = mean load effect; V_R = coefficient of variation of resistance = $\sigma_R/\overline{R}$; and V_Q = coefficient of variation of load effect =
$\sigma_Q/\overline{Q}$ in which σ is the standard deviation.

The physical interpretation of the *reliability index* β is shown in Fig. 1.13.
The shaded area in the figure represents the probability in which $\ln(R/Q) < 0$,
i.e., the probability that the resistance will be smaller than the load effect,
indicating that a limit state has been exceeded. The larger the value of β, the
smaller the shaded area, so that it becomes more improbable that a limit
state will be exceeded. Thus, the magnitude of β reflects the safety of the
member. In the development of the current LRFD specification [1.11], the
target values of β were selected as 3.0 for members and 4.5 for connectors
under dead plus live and/or snow loading. A higher value of β for connectors
ensures that the connections are designed to be stronger than their adjoining
members.

In this design method, the designer has an option to either carry out the
second-order elastic analysis for directly computing load effects (moments
and forces) in the members or carry out the first-order elastic analysis and

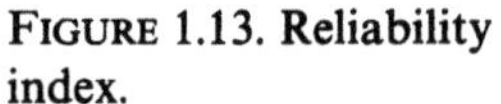

FIGURE 1.13. Reliability
index.

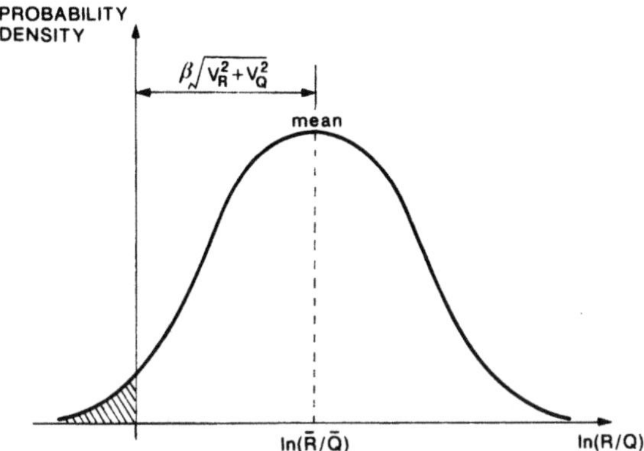

estimate the second-order effects by using the moment amplification factors B_1 and B_2 provided in the specifications [1.11]. The effects of inelasticity are considered indirectly. The method is more appropriate for tall buildings in which secon-order effects are more pronounced than the inelasticity effects.

1.7.4 *Plastic Design with LRFD*

This method combines the advantages of plastic design with load factor (Section 1.7.2) and load and resistance factor design with elastic analysis (Section 1.7.3). The format for this design method is the same as that given by Eq. (1.7.3). In this method, it is ensured that the member forces caused by factored loads and determined by plastic analysis are less than the resistances (mainly moments) reduced by the ϕ-factors. The plastic method will be illustrated in detail in Chapters 4 and 5. This design method has the following characteristics.

1. It considers the redistribution of first-order forces/moments in structures in a direct manner.
2. The second-order moments can be estimated by using the amplification factors B_1 and B_2.
3. It is more appropriate for the design of low-rise buildings for which the effects of inelasticity are more pronounced than the effects of instability.
4. It gives a more realistic representation of the actual behavior of structures, and it is simple to use.

Plastic design with LRFD is a method for proportioning structures so that no *strength limit state* is exceeded when the structure is subjected to all appropriate factored load combinations. Here, as in PD with load factor, strength limit states are the basis for design and are related to safety and load-carrying capacity (e.g., the limit state of plastic moment and buckling).

1.7.5 Design with Advanced Inelastic Analysis

Currently, inelastic analysis is addressed by design specifications under the category of *plastic analysis/design*. The distinguishing feature of the plastic method of analysis/design is that it accounts for inelastic force redistribution in the calculation of load effects. Given the calculated load effects for a particular member, the specifications provide design equations, which the member forces must not violate if the member is to be deemed adequate. However, advancements in computer hardware, particularly in the computing and graphics performance of engineering workstations, are making more sophisticated methods of analysis feasible in design practice. These more sophisticated analysis techniques hold the promise of more realistic prediction of load effects and overall frame performance, and therefore in certain cases, greater economy and more uniform safety. If the significant behavioral effects are considered properly in these more sophisticated methods, separate checks of member design equations become unnecessary. Any method of second-order inelastic analysis involving the direct consideration of both member strength and stability effects such that separate member capacity checks are not needed is referred to here as an *advanced inelastic analysis*. At present, AS4100 [1.17] is the only design specification that explicitly allows the designer to disregard member capacity checks if an advanced inelastic analysis is employed.

There are three different inelastic analysis approaches currently available with respect to their use in the design of planar frames. These advanced methods are referred to here as the *plastic-zone*, the *rigid-plastic hinge*, and the *elastic-plastic hinge* approaches. As implied by the name, the elastic-plastic hinge approach involves the modeling of inelastic behavior through "zero-length" plastic hinges that remain elastic until the plastic cross-section strength of the member is reached. The rigid-plastic hinge approach also represents the material yielding effects through a plastic hinge model, but the effects of elastic deformations in the structure are neglected. The plastic-zone approach involves the explicit modeling of the distribution of plasticity throughout the structural members.

Because the members are modeled as elastic elements between plastic hinge locations in the elastic-plastic hinge approach, this method generally overestimates the strength and stiffness of the actual structure. However, a number of research studies have demonstrated that for practically designed frames, elastic-plastic hinge approaches predict essentially the same system strength and stability as a more refined (and more expensive) plastic-zone analysis [1.18]. In other words, a second-order elastic-plastic hinge analysis may in many cases satisfy the requirements for advanced inelastic analysis.

At present the elastic-plastic hinge method is still in its active development stage. Attempts are being aimed (a) to make the *plastic hinge method* work in engineering practice using workstations and (b) to make the design process consistent with design codes and specifications. A detailed description of this new development is given in Chapter 8.

1.8 Examples

To further illustrate the benefits of ductility-induced redistribution of forces and moments in a structural system, we present three additional examples in this section. The first one deals with the redistribution of forces, and the second and third deal with the redistribution of moments.

Example 1.8.1. The three bars of the symmetric truss shown in Fig. 1.14 have equal cross-sectional area A and are made of steel with yield stress σ_y and Young's modulus E:

  i. Plot the load-deflection (P-Δ) relationship of the truss.
 ii. Plot the relationships between member forces and the total force P.
iii. Determine the ultimate load and the plastic reserve strength beyond the elastic limit.

Solution: The first step toward obtaining the load-deflection relationship is to determine the member forces in terms of the total applied force P. The equilibrium of the truss in the vertical direction has the simple form for the symmetric structure

$$T_2 + 2T_1 \cos 45° = P \tag{1.8.1}$$

in which T_1 and T_2 are the forces in Bars 1 and 2, respectively. Note that forces in Bars 1 and 3 are equal. Since there are two unknowns and only one equilibrium equation, the structure is statically indeterminate by one degree and one more equation is required to determine T_1 and T_2. The second equation has different forms depending on whether the stress state in each bar is in the elastic or plastic regime. This is given in the following.

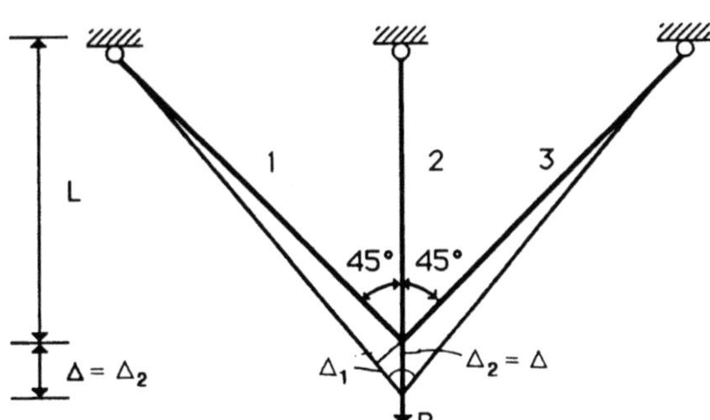

FIGURE 1.14. Three-bar symmetric truss.

Elastic Regime: Referring to Fig. 1.14, the compatibility condition at the joint is

$$\Delta_1 = \Delta_2 \cos 45° \tag{1.8.2}$$

in which Δ_1 and Δ_2 are the extensions of Bars 1 and 2, respectively. In the elastic regime, the extensions Δ_1 and Δ_2 can be expressed in terms of T_1 and T_2 by the elastic stress-strain relationships

$$\Delta_1 = \frac{T_1 L}{AE \cos 45°} \tag{1.8.3}$$

and

$$\Delta_2 = \frac{T_2 L}{AE}. \tag{1.8.4}$$

Substitution of Δ_1 and Δ_2 in Eq. (1.8.2) results in

$$T_1 = \frac{T_2}{2}. \tag{1.8.5}$$

Solution of Eqs. (1.8.1) and (1.8.5) gives

$$T_1 = \frac{P}{2 + \sqrt{2}} \tag{1.8.6}$$

and

$$T_2 = \frac{2P}{2 + \sqrt{2}}. \tag{1.8.7}$$

The load-deflection relationship in this regime can now be obtained from Eq. (1.8.4) as

$$\Delta = \Delta_2 = \frac{2P}{2 + \sqrt{2}} \frac{L}{AE}. \tag{1.8.8}$$

Since T_2 is greater than $T_1 = T_3$, Bar 2 will yield first. The *elastic limit load P_y* is determined by equating stress in Bar 2 to the yield stress σ_y.

$$\sigma_2 = \frac{T_2}{A} = \frac{2P_y}{(2 + \sqrt{2})A} = \sigma_y \tag{1.8.9}$$

or

$$P_y = \frac{2 + \sqrt{2}}{2} A\sigma_y. \tag{1.8.10}$$

The corresponding *elastic limit deflection Δ_y* of the truss at P_y is

$$\Delta_y = \Delta_{2y} = \varepsilon_y L = \frac{\sigma_y}{E} L. \tag{1.8.11}$$

Elastic-Plastic Regime: When the load P is increased beyond the elastic limit load P_y, the truss enters into the elastic-plastic regime. Here, Bar 2 is in the yield state while Bars 1 and 3 remain in the elastic state. The compatibility equation (1.8.2) is now replaced by the yield condition

$$T_2 = A\sigma_y. \tag{1.8.12}$$

Solution of Eqs. (1.8.1) and (1.8.12) gives

$$T_1 = \frac{P - A\sigma_y}{\sqrt{2}}. \tag{1.8.13}$$

Since Bars 1 and 3 are still elastic, the load-deflection relationship in this regime can be obtained from Eq. (1.8.3) as

$$\Delta = \frac{\Delta_1}{\cos 45°} = \frac{(P - A\sigma_y)2L}{\sqrt{2}\,AE}. \tag{1.8.14}$$

Plastic Regime: In the fully plastic regime, all three bars yield, and the yield condition for the three bars is

$$T_1 = T_2 = T_3 = A\sigma_y. \tag{1.8.15}$$

Solution of Eqs. (1.8.1) and (1.8.15) gives the *plastic limit load* or the *plastic collapse load* as

$$P_p = A\sigma_y(1 + \sqrt{2}) = \sqrt{2}P_y. \tag{1.8.16}$$

Note that at the limit load, the deflection will be unrestricted in this regime as shown in Fig. 1.15(a). However, the deflection at the onset of this limit state can be determined from Eq. (1.8.2) as

$$\Delta_p = \frac{\Delta_{1y}}{\cos 45°} = \frac{2\sigma_y L}{E} = 2\Delta_y. \tag{1.8.17}$$

The percentage of the plastic reserve strength beyond the elastic limit load for the three-bar truss is

$$\frac{P_p - P_y}{P_y} = \frac{P_y(\sqrt{2} - 1)}{P_y} = 41.4\%. \tag{1.8.18}$$

The load-deflection relationship and the relationship between member forces and the total load P are plotted, respectively, in parts (a) and (b) of Fig. 1.15. It is seen that the truss can carry an ultimate load in excess of P_y by 41.4%, but only at the expense of a larger deflection. However, the total displacement under the limit load P_p is only twice that under P_y. In fact, the real truss will carry loads in excess of P_p with the actual displacement at P_p less than Δ_p because of strain-hardening for real materials.

Example 1.8.2. Use the hinge-by-hinge method and plot the load-deflection relationship of a fixed-ended beam with a concentrated lateral load at one-

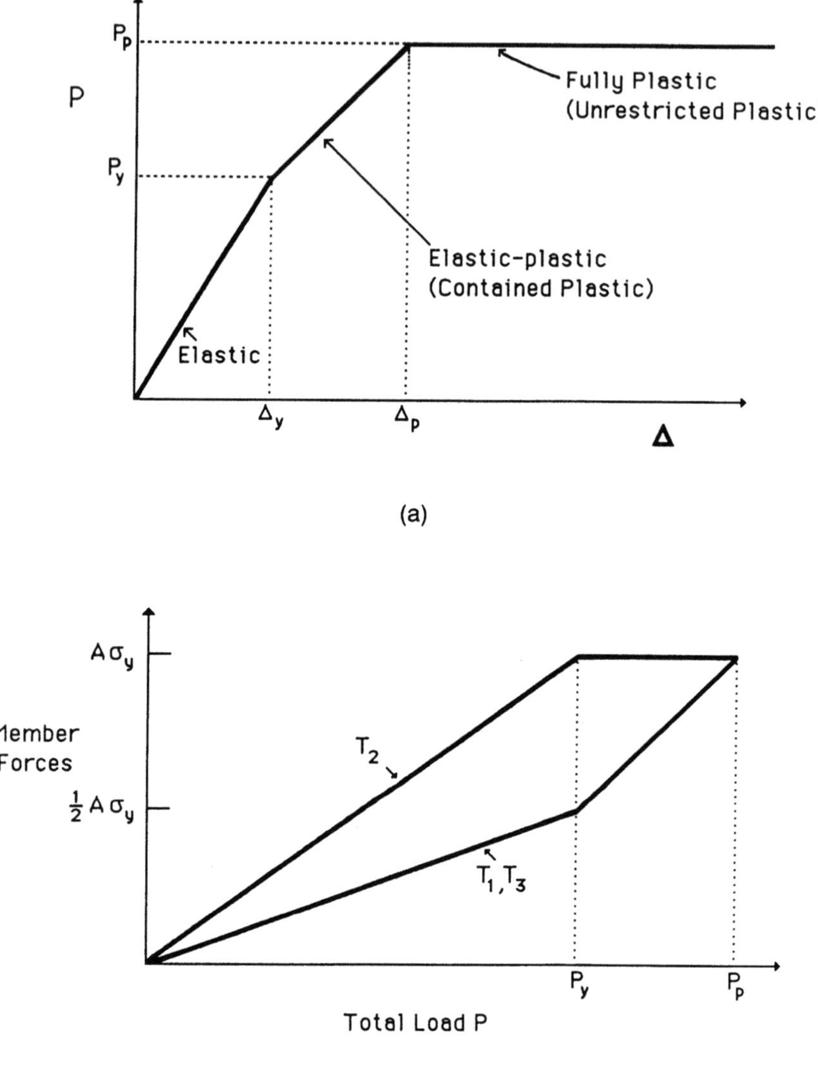

FIGURE 1.15. The force redistribution in a fixed-ended beam: (a) load-deflection relationship and (b) member forces versus total load relationship.

third point as shown in Fig. 1.16(a). Determine the amount of plastic reserve strength contributed by the process of moment redistribution. Assume that the bending stiffness of the beam is *EI*.

Solution: As described in Section 1.4, the hinge-by-hinge method consists of a series of elastic analyses. It begins with an elastic analysis of the original beam. When the maximum moment in the beam reaches the plastic moment

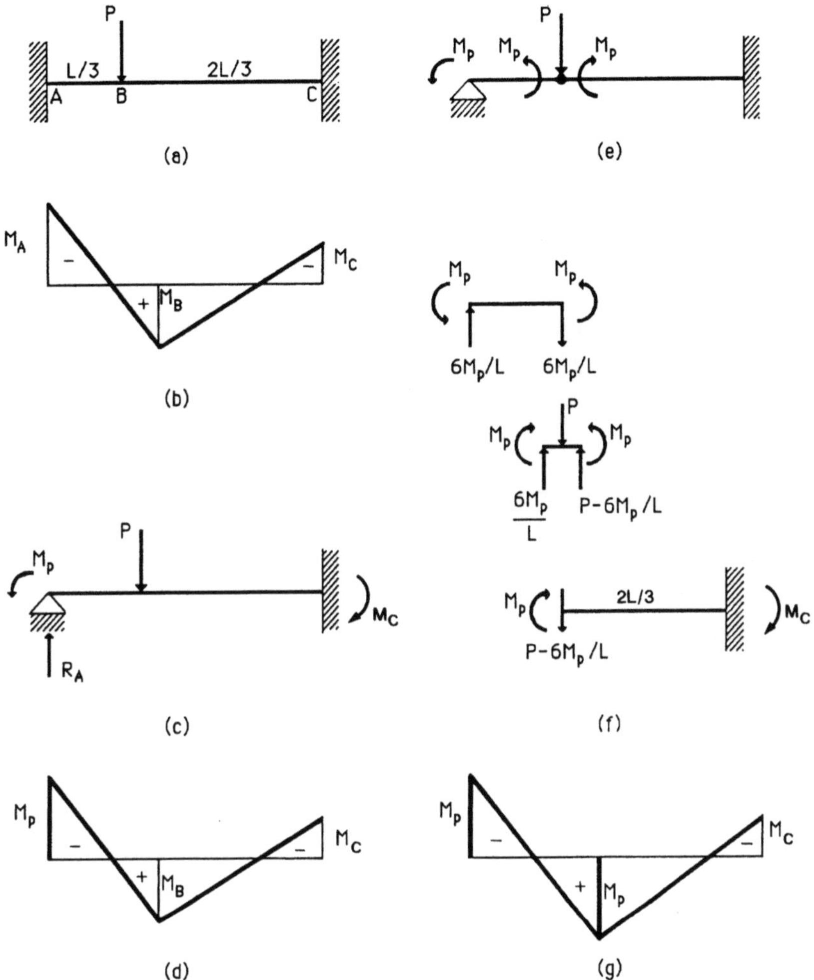

FIGURE 1.16. The changes in moment diagrams through hinge-by-hinge analysis of a fixed-ended beam.

M_p, a hinge with the plastic moment capacity M_p is inserted at the point of maximum moment (called *plastic hinge*), and an elastic analysis is performed on the resulting simpler structure. This process is continued until a failure mechanism is formed. Since the load-deflection relationship is linear for each of these elastic analyses, it is necessary to compute the load and deflection only when a new plastic hinge is introduced to the beam.

Elastic Analysis of the Original Beam: The moment distribution for this indeterminate beam is shown in Fig. 1.16(b) with (see, for example, AISC steel manual [1.9])

$$M_A = -\frac{4}{27}PL \qquad (1.8.19)$$

$$M_B = \frac{8}{81}PL \qquad (1.8.20)$$

and

$$M_C = -\frac{2}{27}PL. \qquad (1.8.21)$$

Since the maximum moment occurs at A, a plastic hinge will form first at A when $|M_A| = M_p$. The *first hinge load* $P = P_1$ is then

$$P_1 = \frac{27}{4}\frac{M_p}{L} = 6.75\frac{M_p}{L}. \qquad (1.8.22)$$

The *deflection* at B corresponding to this first hinge load is

$$\Delta_1 = \frac{2}{81}\frac{M_pL^2}{EI} = 0.0247\frac{M_pL^2}{EI}. \qquad (1.8.23)$$

Elastic Analysis After the Formation of Plastic Hinge at A: The moments at B and C for the indeterminate beam shown in Fig. 1.16(c) are:

$$M_B = -\frac{M_p}{2} + \frac{14}{81}PL \qquad (1.8.24)$$

and

$$M_C = \frac{M_p}{2} - \frac{4}{27}PL. \qquad (1.8.25)$$

Since M_B has a larger numerical value [Fig. 1.16(d)], the next plastic hinge will form at B, when M_B is equal to M_p. The *second hinge load* $P = P_2$ is then

$$P_2 = \frac{243}{28}\frac{M_p}{L} = 8.67\frac{M_p}{L}. \qquad (1.8.26)$$

The moment M_C at C corresponding to this load is

$$M_C = -\frac{11}{14}M_p = -0.785M_p. \qquad (1.8.27)$$

The *deflection* at B corresponding to the second hinge load has the value

$$\Delta_2 = 0.0423\frac{M_pL^2}{EI}. \qquad (1.8.28)$$

Elasitc Analysis After the Formation of Hinges at A and B: The beam is now statically determinate [Fig. 1.16(c)] and its internal forces and moments are shown in Fig. 1.16(f). The moment at C [Fig. 1.16(g)] can be expressed as

$$M_C = \left(P - \frac{6M_p}{L}\right)\left(\frac{2}{3}L\right) - M_p. \qquad (1.8.29)$$

The next plastic hinge will therefore form at C when $|M_C| = M_p$. The *third hinge load* $P = P_3$ is then

$$P_3 = \frac{9M_p}{L}. \qquad (1.8.30)$$

The *deflection* at B corresponding to the third hinge load is the deflection of the elastic cantilever beam of length $2L/3$ under the vertical load $(P_3 - 6M_p/L = 3M_p/L)$ combined with the end moment M_p as

$$\Delta_3 = -\frac{M_p}{2EI}\left(\frac{2}{3}L\right)^2 + \frac{(3M_p/L)}{3EI}\left(\frac{2}{3}L\right)^3$$

or

$$\Delta_3 = 0.0741 \frac{M_p L^2}{EI}. \qquad (1.8.31)$$

At $P = P_3$, a *failure mechanism* has formed and the deflection will become unrestricted. Note that the *plastic limit load* $(P_p = P_3)$ can be determined directly from Fig. 1.16(g) with $M_c = M_p$. There is no need to know the order of hinge formation for the direct calculation of the limit load. The load-deflection relationship is plotted in Fig. 1.17.

The percentage of plastic reserve strength for the beam beyond the elastic limit through the moment redistribution is

$$\frac{9M_p/L - 6.75M_p/L}{6.75M_p/L} = 33.3\%.$$

Example 1.8.3. Plot the load-deflection relationship of a fixed-ended beam with a concentrated load acted at the midspan. Determine the contribution of the moment redistribution to the plastic reserve strength, if any.

Solution: The moment diagram for the given beam is shown in Fig. 1.18(a) with

$$|M_A| = M_B = |M_C| = \frac{PL}{8}. \qquad (1.8.32)$$

Since the moments at A, B, and C are all equal, they reach the value M_p simultaneously, thus forming the three hinges at the same time. The load corresponding to this state has the value

$$P_1 = \frac{8M_p}{L}.$$

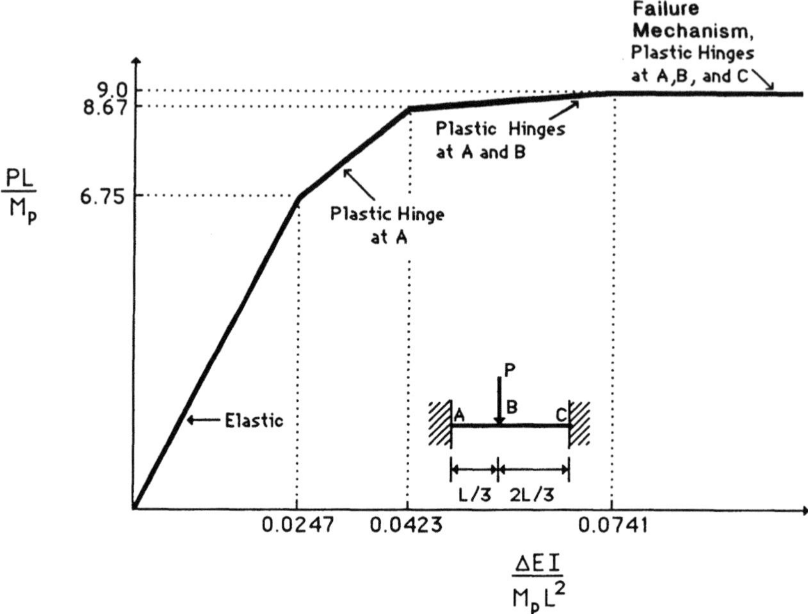

FIGURE 1.17. The load-deflection relationship reflecting the moment redistribution in a fixed-ended beam.

The deflection at C corresponding to this load can be determined directly from the elastic limit solution in usual manner

$$\Delta_1 = 0.0417 \frac{M_p L^2}{EI}.$$

Note that at this load a failure mechanism has developed and the deflection will become unlimited as shown in Fig. 1.18(b).

This example shows that in the present case, contribution through moment redistribution to the plastic reserve strength of the indeterminate beam cannot be realized.

1.9 Summary

From the preceding discussions of simple examples, it is seen that one of the major advantages of plastic analysis and design is its relative simplicity. To solve the elastic problem, it was necessary to formulate an equilibrium equation and a compatibility equation. Thus, the problem involved solution of two simultaneous linear algebraic equations. However, the solution of the plastic problem was much simplified by substitution of the yield stress or the

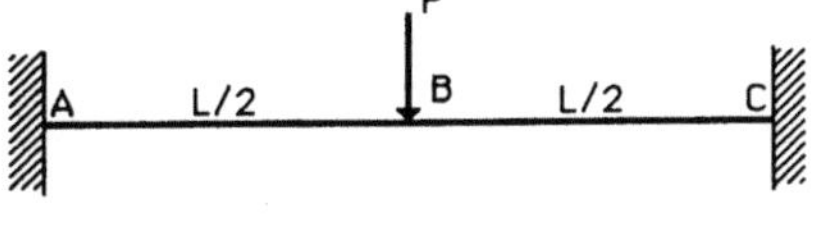

FIGURE 1.18. No contribution to the plastic reserve strength due to moment redistribution.

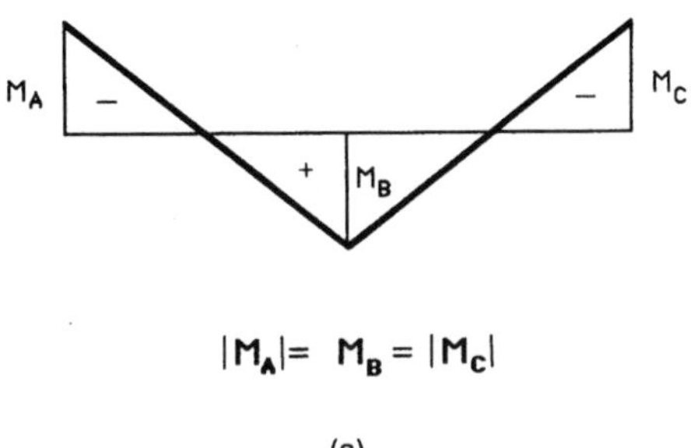

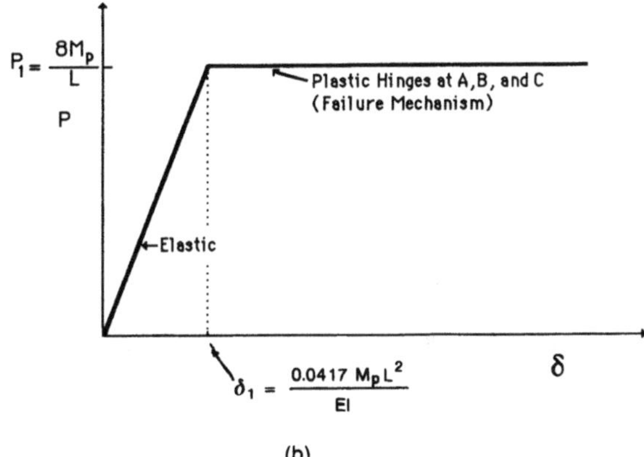

plastic moment into the equation of equilibrium. Furthermore, the maximum load-carrying capacity computed by the plastic analysis based on an idealized plastic material has real significance as a measure of ultimate strength of a real structure, and the load at which the limit state is reached can usually be determined in a direct and simple manner. On the other hand, the elastic limit load computed by the elastic analysis based on a linear elastic material is relatively meaningless as a measure of strength, although it may be appropriate for the computation of stresses and strains under working loads.

The general description of plastic analysis applies to any structural material with sufficient *ductility*. It is, of course, particularly appropriate to mild steel with its sharply defined yield point and large strain value before the beginning of strain hardening. Thus, the *plastic reserve strength* cannot be fully realized for structures made of brittle materials that will crack or soften under relatively small strains, or for structures made of slender bars that will buckle in compression within either the elastic or the plastic range. For bridge-type structures, repeated loadings might fatigue the material, and the plastic analysis may not be appropriate for this type of application.

The major portion of this book will be concerned with the plastic analysis of steel-framed structures. In general, if the analysis problem can be solved, design can always be achieved by an inverse trial-and-error procedure. However, some direct approach to the design problem based on the *equilibrium method and work method* will be presented in Chapters 4 and 5, repectively. In the following chapters the basic plastic theory and the methods of plastic analysis and design will be set forth and the necessary secondary design factors such as the details of connecting joints, the deflection limits, and the overall and local buckling requirements for compression members will be treated in accordance with the 1993 AISC load and resistance factor design specifications [1.11] or the plastic design chapter as a part of the 1989 AISC allowable stress design specifications [1.9]. Since the serviceability provisions (e.g., deflections, drift, and vibration) of ASD and LRFD are similar, members controlled by serviceability criteria are not affected by the choice of design method. They are subject to the same limitations and checking procedures as those used in allowable stress designs. The reader who is interested in the early experimental verifications of the *"simple plastic theory"* as applied to engineering practice may find the 1971 ASCE manual on "Plastic Design in Steel" interesting and informative [1.10].

References

1.1. Drucker, D.C., "Plastic Design Methods—Advantages and Limitations," *Trans. of Society of Naval Architects and Marine Engineers*, Vol. 65, pp. 172–190, 1957.

1.2. Kazinczy, G., "Kiserletek Befalazott Tartokkal" ("Experiments with Clamped Girders"), *Betonszemle*, 2(4), p. 68; 2(5), p. 83; 2(6), p. 101, 1914.

1.3. Maier-Leibnitz, H., "Contribution to the Problem of Ultimate Carrying Capacity of Simple and Continuous Beams of Structural Steel and Timber," *Die Bautechnik*, 1(6), 1927.

1.4. Baker, J.F., "A Review of Recent Investigations into the Behavior of Steel Frames in the Plastic Range," *J. Inst. Civil Eng.*, London, 31(3), pp. 185–240, 1949.

1.5. Symonds, P.S., and Neal, B.G., "Recent Progress in the Plastic Methods of Structural Analysis," *J. Franklin Inst.*, 252, pp. 383–407 and 469–492, 1951.

1.6. Johnston, B.G., Yang, C.H., and Beedle, L.S., "An Evaluation of Plastic Analysis as Applied to Structural Design," *Welding J.*, 32(5), p. 224-s, 1953.

1.7. Beedle, L.S., Thurlimann, B., and Ketter, R.L., *Plastic Design in Structural Steel*, Lehigh University, Bethlehem, Pa., and American Institute of Steel Construction, New York, 1955.

1.8. Beedle, L.S., *Plastic Design of Steel Frames,* John Wiley and Sons, New York 1958.

1.9. *Manual of Steel Construction—Allowable Stress Design,* ninth edition, AISC, Chicago, July 1989.

1.10. *Plastic Design in Steel—A Guide and Commentary,* Manual 41, American Society of Civil Engineers, New York, 1971.

1.11. *Load and Resistance Factor Design Specification for Structural Steel Buildings,* AISC, Chicago, Dec. 1, 1993.

1.12. White, D.W., Liew, J.Y.R., and Chen, W.F., "Second-Order Inelastic Analysis for Frame Design," A report submitted to SSRC Task Group 29 on Recent Research and the Perceived State-of-the-Art, CE-STR-91-12, School of Civil Engineering, Purdue University, West Lafayette, IN, 115 pp.

1.13. Ravindra, M.K., and Galambos, T.V., "Load and Resistance Factor Design for Steel," *J. Structural Division,* ASCE, 104(ST9), pp. 1337–1354, Sept. 1978.

1.14. Ellingwood, B., MacGregor, J.G., Galambos, T.V., and Cornell, C.A., "Probability-Based Load Criteria, Load Factors and Load Combinations," *J. Structural Division,* ASCE, 108(ST5), pp. 978–997, May 1982.

1.15. *Building Code Requirements for Minimum Design Loads in Buildings and Other Structures,* ANSI A58.1-1982, American National Standards Institute, New York, 1982.

1.16. Ang, A.H-S., and Cornell, C.A., "Reliability Bases of Structural Safety and Design," *J. Structural Division,* ASCE, 100(ST9), pp. 1755–1769, Sept. 1974.

1.17. Australian Limit States Specification, AS4100, Australian Institute of Steel Construction, Oct. 1990.

1.18. Vogel, U., "Calibrating Frames," *Stahlbau,* 54, pp. 295–301, 1985.

Problems

1.1. A member made of A36 steel has a maximum compressive residual stress of 10 ksi and maximum tensile residual stress of 15 ksi. If the area of cross section of the member is 2 in^2, determine the yield load and plastic limit load for the member when it is subjected to

(a) compressive axial load ($P_{cy} = 52$ kips, $P_{cp} = 72$ kips).

(b) tensile axial load ($P_{ty} = 42$ kips, $P_{tp} = 72$ kips).

1.2. Determine the yield and plastic limit loads for the A36 steel plate shown in Fig. P1.2. Assume that the radius r of the fillet is 0.5 inch ($K = 1.8$). ($P_y = 37.5$ kips, $P_p = 67.5$ kips).

1.3. A rigid cross-beam is supported by three equally spaced tension rods. The two outside rods are of length L and area A and the center rod is of length $2L$ and area A. What maximum load P_p will the cross-beam support if the load is applied at the center? Draw the load-deflection curve. Compute P_y, δ_y, and δ_p ($P_p = 3A\sigma_y, P_y = (2.5)A\sigma_y, \delta_p = 2L\varepsilon_y, \delta_y = L\varepsilon_y$).

1.4. By assuming that the horizontal bar of the structure shown in Fig. P1.4 is rigid, determine

(a) yield load $P = P_y$ [$P_y = (5/3)A\sigma_y$].

(b) end deflection of the bar at yield load ($\delta = L\varepsilon_y$).

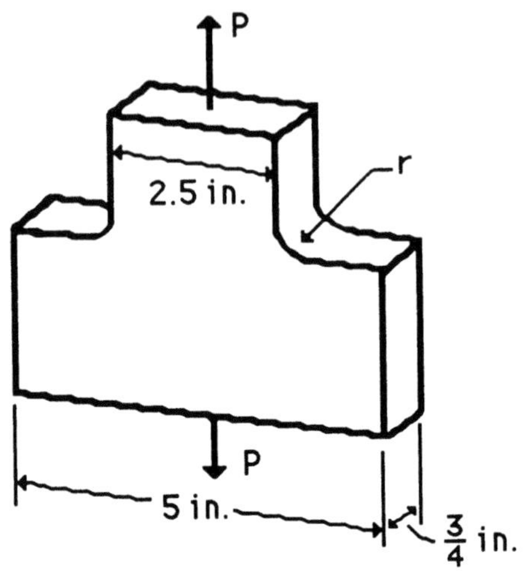

FIGURE P1.2

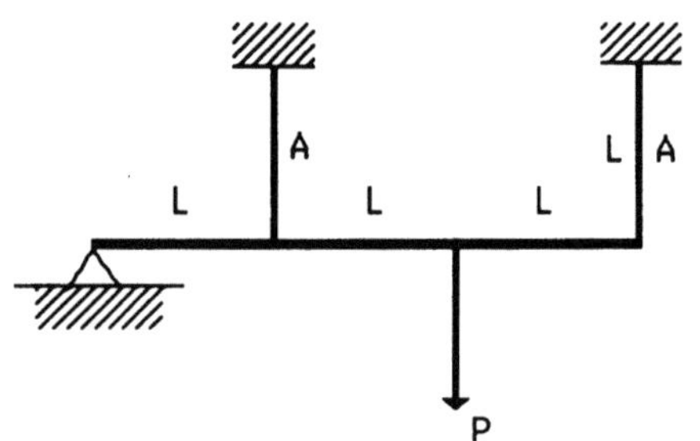

FIGURE P1.4

(c) plastic limit load $P = P_p$ $(P_p = 2A\sigma_y)$.

(d) end deflection of the bar when load starts to increase above plastic limit load. Assume $\varepsilon_{st} = 10\varepsilon_y$ $(\delta = 10L\varepsilon_y)$.

1.5. Each cable of the structure shown in Fig. P1.5 has a cross-sectional area of 0.2 in² and is made of A36 steel. By assuming horizontal bar to be rigid, determine

(a) yield load $P = P_y$ $[P_y = 1.25A\sigma_y]$.

(b) deflection of C at yield load $(\delta = L\varepsilon_y)$.

(c) plastic limit load $P = P_p$ $[P_p = 1.5A\sigma_y]$.

(d) deflection of C when load P starts to increase above the plastic limit load. Assume $\varepsilon_{st} = 12\varepsilon_y$ $(\delta_c = 12L\varepsilon_y)$.

1.6. The three-bar truss shown in Fig. 1.14 is subjected to horizontal load Q rather than the vertical load. Determine the plastic limit load Q_p $(Q_p = \sqrt{2} A\sigma_y)$.

1.7. The three-bar structure shown in Fig. 1.5 is subjected to the vertical load P half way between Bars 1 and 2, rather than directly under Bar 2. Determine the value of P at elastic and plastic limit $(P_y = P_p = 2A\sigma_y)$.

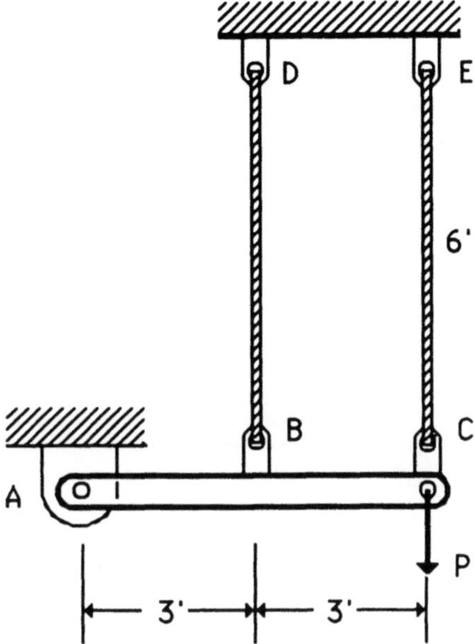

FIGURE P1.5

1.8. What is the plastic limit load P_p of the three-bar truss shown in Fig. 1.14, if Bars 1 and 3 are mild steel and Bar 2 is high-strength steel with a yield stress of three times of mild steel. Also, compute Δ_p. Assume the same Young's modulus for both steels $[P_p = (\sqrt{2} + 3)A\sigma_{y1}, \Delta_p = 3L\varepsilon_{y1})]$.

1.9. Compute the plastic limit load P_p for the structure shown in Fig. 1.5, if Bar 1 is rigid, Bar 2 is mild steel, and Bar 3 is high-strength steel with a yield stress of twice that of mild steel. What is the load at first yield? Also, compute δ_y and δ_p at Bar 3 $(P_y = 3A\sigma_{y2}, P_p = 5A\sigma_{y2}, \delta_{y3} = L\varepsilon_y, \delta_{p3} = 2L\varepsilon_y)$.

1.10. The three-bar truss shown in Fig. 1.14 is subjected to the following loading path: The vertical load P is first increased from zero to the elastic limit load P_y and then held constant at P_y, while a horizontal load H is applied and increased to the collapse state. Determine the maximum value of H $(H_p = A\sigma_y/\sqrt{2})$?

1.11. What size cross-beam is required to support the maximum load $P_p = 3\sigma_y A$ of the structure shown in Fig. 1.5, if the cross-beam of length L is not rigid but made of the same material as the bars $(Z = 1/2AL)$?

1.12. Using the hinge-by-hinge method, plot the load-midspan deflection relationship of the beams shown in Fig. P1.12. Determine the plastic reserve strength contributed by the moment redistribution process. Assume that the bending stiffness of the beams is EI.

1.13. For mild steel, there exists an upper yield stress σ_y^u and a lower yield stress σ_y at the same yield strain ε_y. Find the corresponding initial yield moment

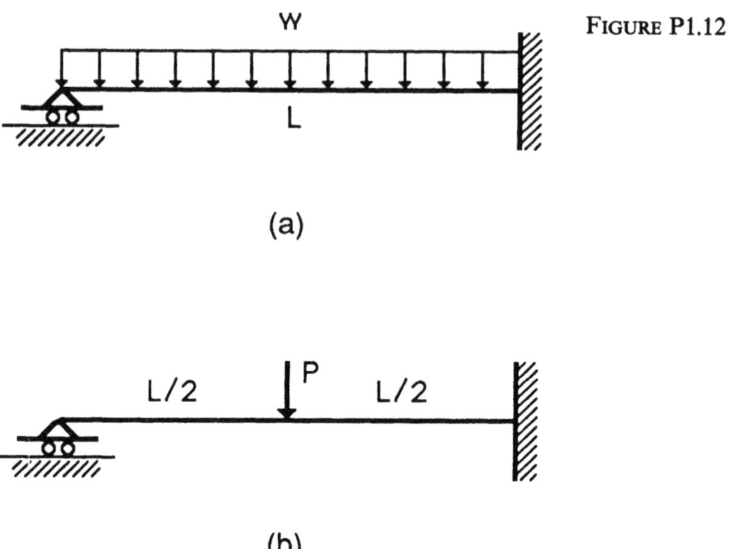

FIGURE P1.12

M_y, fully plastic moment M_p, and shape factor f of a rectangular section $(M_y = \sigma_y^u S, M_p = \sigma_y Z, f = 1.5\sigma_y/\sigma_y^u)$.

1.14. To attain the fully plastic moment condition, the corresponding curvature Φ will theoretically have to approach infinity. For all practical purposes, the fully plastic moment capacity can be approximately reached when the curvature is less than four times that of the initial yield curvature Φ_y. Show this fact for an actual beam of rectangular section $(M = 0.98M_p)$.

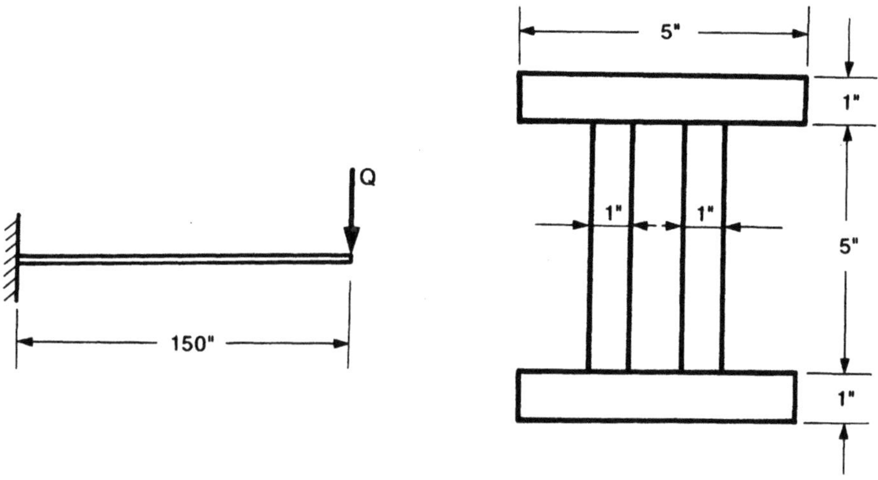

FIGURE P1.15

1.15. A cantilever beam with a double-web section shown in Fig. P1.15 was designed to resist a concentrated load of 5 kips under allowable stress conditions ($\sigma_a = 20$ ksi).
   a. Determine the factor of safety against plastic collapse.
   b. If an additional load of 20 kips is to be added to the free end of the beam, design the required cover plates. (Use 1″ thick plates and A36 steel.)

2
The Plastic Hinge

2.1 Introduction

The plastic hinge concept was introduced in Chapter 1 when the "hinge-by-hinge" method was applied to solve the fixed-ended beam problems. In this chapter, we shall elaborate the concept of the *plastic hinge* and *plastic moment*. The methods of computing the plastic moment and the methods of using plastic moment in the design of a cross section will be presented first. The plastic moment capacity of a section is significantly affected by factors such as axial and shear forces on the section, and the *compactness* of the section. The effects of these factors on the plastic moment capacity of a section will then be described. Since the strength of a connection of a given member to the adjoining members may govern the ultimate moment-carrying capacity of the member, methods of estimating the strength of a given connection and methods of designing the connection under a given loading will be presented in the later part of this chapter.

2.2 Moment-Curvature Relationship and Plastic Hinge Length

To illustrate the concept of plastic hinge, it is instructive to look at the elastic-plastic behavior of a simple structure. Consider a simply supported beam with an I-shaped cross section, subjected to a concentrated load at the midspan (Fig. 2.1). The behavior of the beam mainly depends on the moment-curvature (M-Φ) relationship of its cross section. The M-Φ relationship of an I-section will thus be derived in the following.

2.2.1 Moment-Curvature Relationship of I-Section

Here, as for a rectangular section in Section 1.3, the M-Φ relationship of an I-shaped section (Fig. 2.2a) is derived based on the usual assumptions: (1) the

FIGURE 2.1. A simply supported beam of an I-shaped section with a concentrated load at midspan.

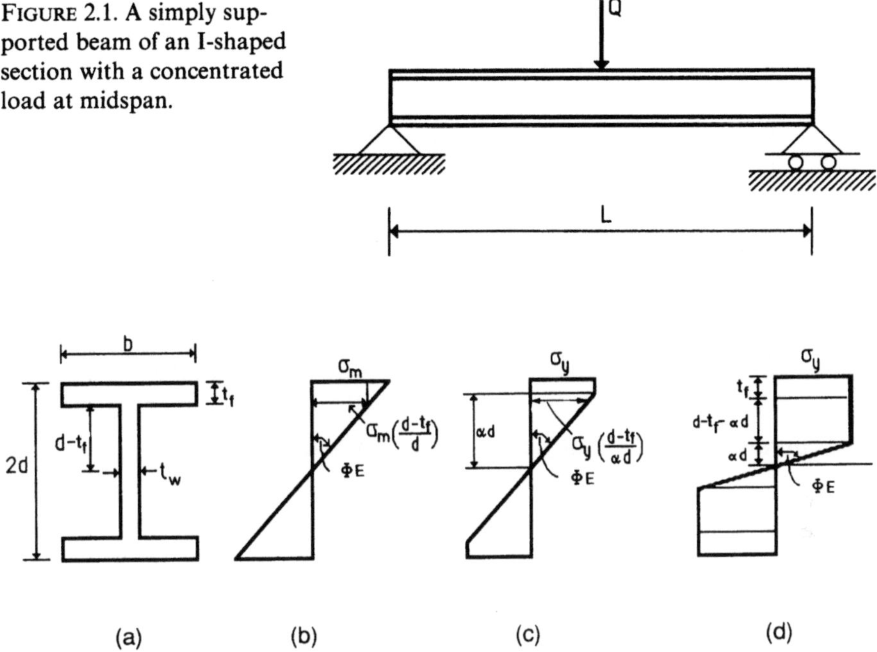

FIGURE 2.2. Elastic-plastic stress distributions in an I-section.

plane section remains plane after bending of the section, (2) the elastic–perfectly plastic stress-strain relationship of the material, and (3) the equilibrium conditions. The moment-curvature relationship of an I-section can be divided into three regimes: (1) the entire section is elastic; (2) when web is elastic and flanges are partially plastic; and (3) web is partially plastic and flanges are fully plastic.

Regime I: Elastic—In the elastic regime, the stress distribution will be linear throughout the cross section as shown in Fig. 2.2(b). The moment resistance M_R of the section can be obtained by summing up the moments of the stresses shown in Fig. 2.2(b) as

$$M_R = \frac{1}{2}\left[\sigma_m\left(\frac{d-t_f}{d}\right)\right](d-t_f)t_w\left[\frac{4}{3}(d-t_f)\right]$$
$$+ \frac{1}{2}\left[\sigma_m - \sigma_m\left(\frac{d-t_f}{d}\right)\right]t_f b\left(2d - \frac{2}{3}t_f\right) + \left[\sigma_m\left(\frac{d-t_f}{d}\right)\right]t_f b(2d-t_f)$$

$$(2.2.1)$$

in which σ_m is the maximum stress in the section and d, t_f, t_w, and b are dimensions of the I-section as shown in Fig. 2.2(a). The first terms in the right-hand side of Eq. (2.2.1) comprise the moment due to the linear elastic stress in the web, the second term is the moment due to the triangular stress $[\sigma_m - \sigma_m(d - t_f)/d]$ in the flanges, and the last term is the moment due to the uniform stress $[\sigma_m(d - t_f)/d]$ in the flanges. Equation (2.2.1) can be simplified to

$$M_R = \frac{2}{3}\sigma_m(d - t_f)^3\left(\frac{t_w}{d}\right) + \sigma_m\left(\frac{t_f^2}{d}\right)b\left(d - \frac{t_f}{3}\right) + \sigma_m(d - t_f)\left(\frac{b}{d}\right)(2d - t_f)t_f.$$

$$(2.2.2)$$

The flanges begin to yield when σ_m is equal to σ_y. The moment resistance at this stage is called *yield moment* M_y, given by

$$M_y = \frac{2}{3}\sigma_y(d - t_f)^3\left(\frac{t_w}{d}\right) + \sigma_y\left(\frac{t_f^2}{d}\right)b\left(d - \frac{t_f}{3}\right) + \sigma_y(d - t_f)\left(\frac{b}{d}\right)(2d - t_f)t_f.$$

$$(2.2.3)$$

The curvature at the initial yield moment is given by

$$\Phi_y = \frac{\varepsilon_y}{d} = \frac{\sigma_y}{Ed}.$$

$$(2.2.4)$$

Regime II: Flange Partially Plastic—The stress distribution in this regime is shown in Fig. 2.2(c) and the moment resistance corresponding to this stress distribution can be written as

$$M_R = \frac{1}{2}\left[\sigma_y\left(\frac{d - t_f}{\alpha d}\right)\right](d - t_f)t_w\left[\frac{4}{3}(d - t_f)\right]$$

$$+ \frac{1}{2}\left[\sigma_y - \sigma_y\left(\frac{d - t_f}{\alpha d}\right)\right][t_f - d(1 - \alpha)]b$$

$$\left\{2(d - t_f) + \frac{4}{3}[t_f - d(1 - \alpha)]\right\}$$

$$+ \left[\sigma_y\left(\frac{d - t_f}{\alpha d}\right)\right][t_f - d(1 - \alpha)]b\{2(d - t_f) + [t_f - d(1 - \alpha)]\}$$

$$+ \sigma_y[d(1 - \alpha)]b[2d - d(1 - \alpha)]$$

$$(2.2.5)$$

in which the first term is the moment due to the elastic stress in the web, the second and third terms are the moments due to the elastic stress in the flanges, and the last term is the moment due to the yield stress in the flanges

$$\alpha = \frac{\varepsilon_y}{\Phi d} = \frac{\sigma_y}{E\Phi d}.$$

$$(2.2.6)$$

Simplification of Eq. (2.2.5) gives

$$M_R = \frac{2}{3}\sigma_y(d - t_f)^3\left(\frac{t_w}{\alpha d}\right) + \sigma_y b(t_f - d + \alpha d)^2\left(\frac{2}{3}\alpha d + \frac{1}{3}d - \frac{1}{3}t_f\right)\left(\frac{1}{\alpha d}\right)$$

$$+ \sigma_y b\left(\frac{d - t_f}{\alpha d}\right)(t_f - d + \alpha d)(d - t_f + \alpha d) + \sigma_y bd^2(1 - \alpha^2). \quad (2.2.7)$$

Regime III: Web is Partially Yielded—The stress distribution in this regime is shown in Fig. 2.2(d), and the moment resistance corresponding to this stress distribution can be written as

$$M_R = \sigma_y t_f b(2d - t_f) + \sigma_y(d - t_f - \alpha d)t_w[2\alpha d - (d - t_f - \alpha d)]$$

$$+ \frac{1}{2}\sigma_y(\alpha d)t_w\left(\frac{4}{3}\alpha d\right) \tag{2.2.8}$$

in which the first term is the moment due to the yield stress in the flanges, and the second and third terms are the moments due to the yield stress and elastic stress in the web, respectively. Simplification of Eq. (2.2.8) leads to

$$M_R = \sigma_y b t_f(2d - t_f) + \sigma_y t_w(d - t_f - \alpha d)(d - t_f + \alpha d) + \frac{2}{3}\sigma_y\alpha^2 d^2 t_w. \tag{2.2.9}$$

The section will become fully plastic when α reduces to zero. The moment resistance corresponding to this fully plastic state can be obtained by substituting $\alpha = 0$ in Eq. (2.2.9)

$$M_p = \sigma_y b t_f(2d - t_f) + \sigma_y t_w(d - t_f)^2. \tag{2.2.10}$$

The M-Φ relationship of other shapes of sections can be obtained in a similar manner. Figure 2.3 shows the M-Φ relationship of W8 × 31 plotted by substituting appropriate values of d, t_f, b and t_w in Eqs. (2.2.2), (2.2.7), and (2.2.9).

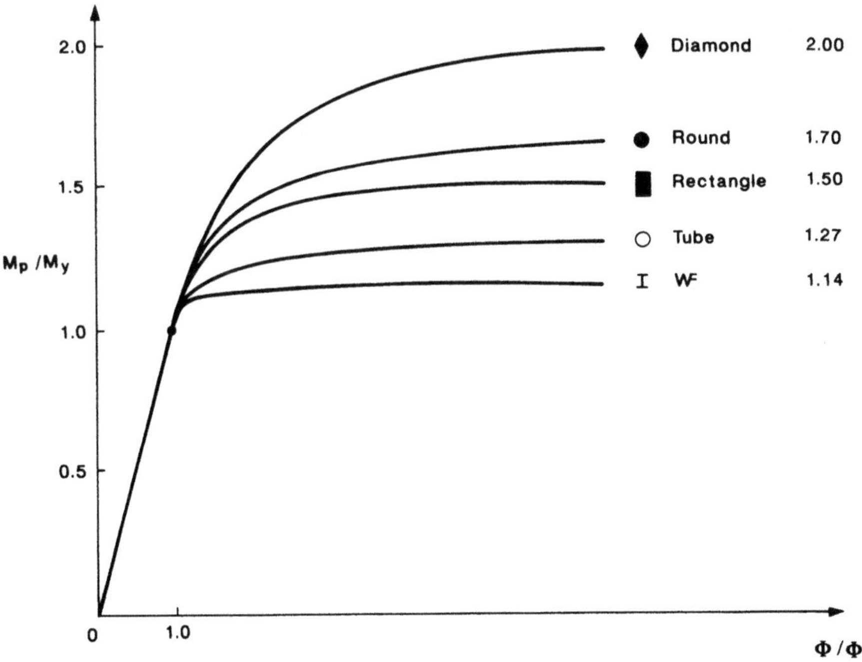

FIGURE 2.3. Moment-curvature curves of various sections.

The $M - \Phi$ curves of rectangular and some other sections have also been included here for comparison.

2.2.2 Plastic Hinge Length

Now, we return to the simply supported beam shown in Fig. 2.1. The beam will remain elastic when Q is less than $Q_y = 4M_y/L$. When $Q = Q_y$, the extreme fibers of the section at midspan begin to yield. When the load Q is increased beyond Q_y, the maximum moment at midspan and the moments at the sections near the midspan exceed the yield moment M_y [Fig. 2.4(b)], thus spreading the yielding over a length of the beam like that shown in Fig. 2.4(c). The spreading of the yielded zone continues until the maximum moment at midspan reaches M_p [Fig. 2.4(d)]. At this state the entire section at midspan is yielded, and this yielded zone spreads out over a length called *plastic hinge length* [Fig. 2.4(e)]. The increase in beam curvature corresponding to the development of the full plastic moment M_p at midspan does not produce a sharp kink in the beam. Nevertheless, it is sufficient to simulate the effect of a hinge. The location at which the value of M_p is reached in a structure is called *plastic hinge*. The actual extent and shape of spread of plasticity in the beam depend on the moment diagram. For design purposes, however, we shall assume each plastic hinge action takes place only at a single section of the beam. The plastic hinge length or the yield length and the actual extent of yielding for a simply supported beam of rectangular section with a concentrated load at midspan are determined in the forthcoming.

Referring to Fig. 2.4(e), the distance between section C where yielding has just begun and the end support A can be obtained by equating the moment

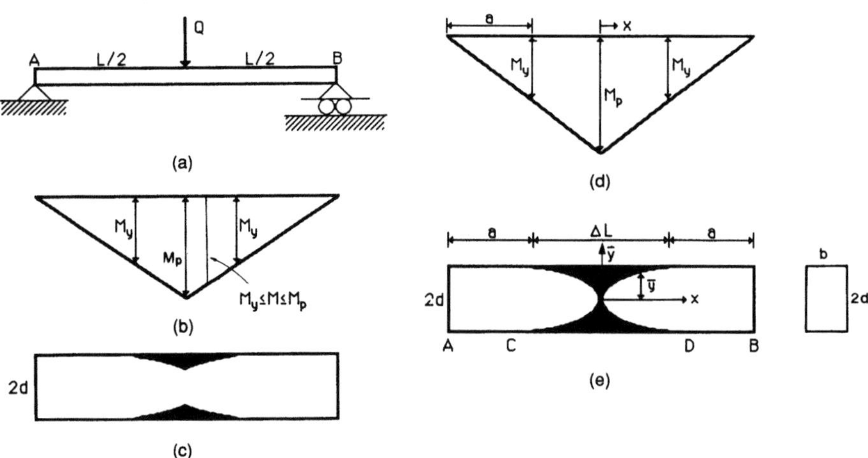

FIGURE 2.4. Hinge length of a simply supported beam with a concentrated load at midspan.

at C to the yield moment as

$$M_C = \frac{Q_p}{2}a = M_y \tag{2.2.11}$$

where Q_p is the plastic limit load. Equation (2.2.11) leads to

$$a = \frac{2M_y}{Q_p}. \tag{2.2.12}$$

Substitution of $Q_p = 4M_p/L$ in Eq. (2.2.12) gives

$$a = \frac{L}{2f} \tag{2.2.13}$$

where f is the *shape factor* of the section (see Section 1.3.3 and Table 1.1) and L is the length of the beam. From Fig. 2.4(e), the hinge length ΔL can be found as

$$\Delta L = L - 2a = L\left(1 - \frac{1}{f}\right). \tag{2.2.14}$$

This is the hinge length for a simply supported beam with a concentrated load at midspan. This length will be different for other boundary and loading conditions. As noted previously, the hinge length depends on the shape of moment diagram, the length of the beam, and the shape factor of the section. The distribution of yielded zone within the hinge length also depends on the shape of the moment-curvature curve of the section. The plastic zone distribution can be obtained by equating the moment within the hinge length to the M-Φ expression in the elastic-plastic regime. For example, for the rectangular section, the moment M_R in the elastic-plastic regime has the value [Eqs. (1.3.14), (1.3.10), and (1.3.8)]:

$$M_R(\bar{y}) = M_p\left[1 - \frac{1}{3}\left(\frac{\bar{y}}{d}\right)^2\right]. \tag{2.2.15}$$

The moment distribution for the given beam is [Fig. 2.4(d)]

$$M(x) = M_p\left(\frac{L - 2x}{L}\right) \tag{2.2.16}$$

in which x is the distance from the midspan. By equating right-hand sides of Eqs. (2.2.15) and (2.2.16), $\bar{y}$ can be written as

$$\bar{y} = d\sqrt{\frac{6x}{L}} \tag{2.2.17}$$

in which $\bar{y}$ defines the boundary between the elastic and plastic regions. The distribution of the yielded portion for an I-shaped or any other shaped section can be obtained in a similar manner. Since for an I-section, there are two expressions for the M-Φ curve in the elastic-plastic regime, the yielded zone is also defined by two expressions.

2.2.3 Plastic Hinge Idealization

The use of exact nonlinear moment-curvature curve (Fig. 2.3) in the analysis of steel structures beyond elastic regime requires an iterative process for a solution. However, this iteration process can be eliminated and the solution procedure drastically simplified by using the moment-curvature relationship idealized by two straight lines represented by

$$M = EI\Phi, (0 < \Phi < \Phi_p) \tag{2.2.18}$$

and

$$M = M_p, (\Phi > \Phi_p) \tag{2.2.19}$$

where

$$\Phi_p = M_p/EI. \tag{2.2.20}$$

Equations (2.2.18) and (2.2.19) are plotted in Fig. 2.5. The relationship is elastic up to moment M_p and plastic thereafter. In this idealization, all the plastic rotation is assumed to occur at the plastic hinge and the length of the plastic hinge is assumed to be zero.

This idealization results in a considerable simplification of the analysis procedure without making significant compromise in the accuracy of the computed plastic limit load. It results in a series of piecewise linear load-deflection relationships while the exact load-deflection relationship based on the exact elastic-plastic M-Φ curve is a smooth curve bounded by the piecewise linear relationship. For example, the idealized load-deflection relationship for a beam shown in Fig. 2.1 is linearly elastic up to the lateral load Q_p and thereafter, a plastic hinge is formed at the midspan, and the beam behaves as if it were hinged at the midspan but with a restraining moment M_p.

To show the influence of loading and boundary conditions on the hinge length, two examples of determining the plastic hinge length are presented in the following.

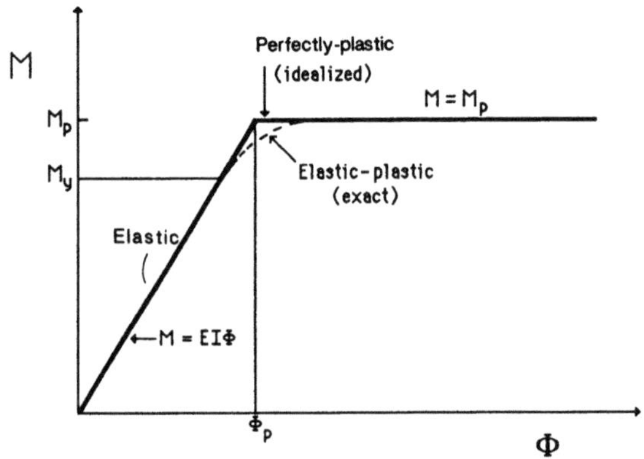

FIGURE 2.5. The idealized and the actual moment-curvature relationships.

FIGURE 2.6. Hinge length of a simply supported I-beam with uniformly distributed load.

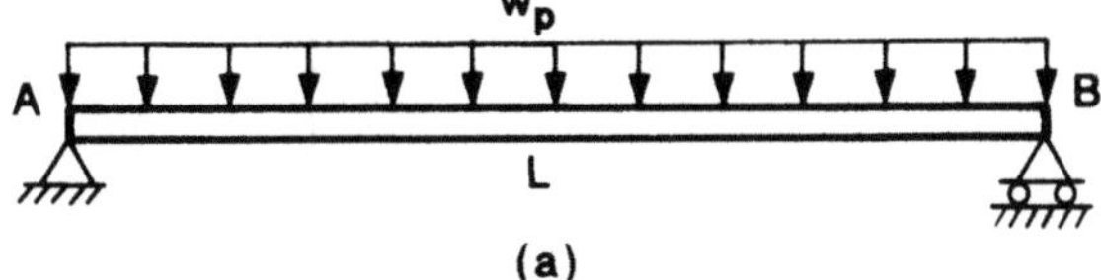

(a)

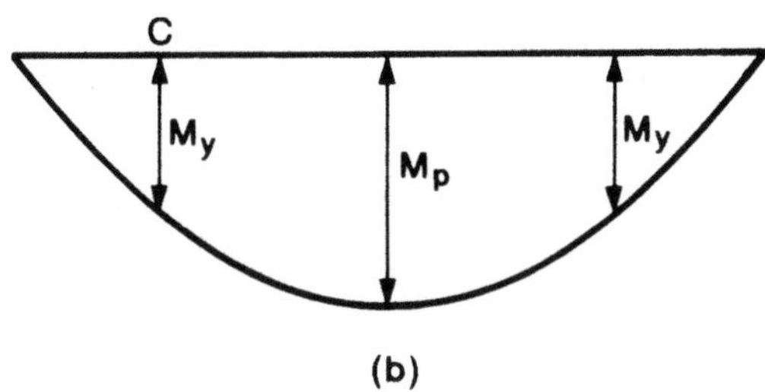

(b)

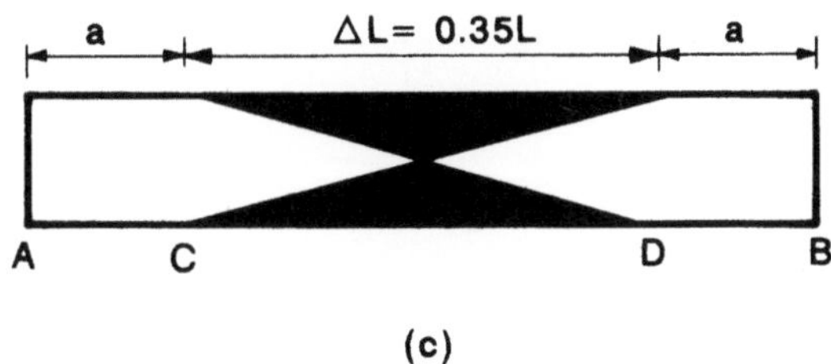

(c)

Example 2.2.1. Determine the plastic hinge length of a simply supported beam with uniformly distributed load as shown in Fig. 2.6.

Solution: The distance between section C, where yielding just began, and the end support A [Fig. 2.6(c)] is obtained by equating the moment at C to the yield moment M_y as

$$\frac{w_p L}{2} a - w_p \frac{a^2}{2} = M_y \tag{2.2.21}$$

in which w_p is the distributed load at which the moment at midspan reaches M_p and it has the value

$$w_p = \frac{8 M_p}{L^2}. \tag{2.2.22}$$

By substituting Eq. (2.2.22) in Eq. (2.2.21) and solving the resulting equation for a, the distance a is found to be

$$a = \frac{L}{2}[1 - \sqrt{1 - 1/f}]. \tag{2.2.23}$$

Thus, from Fig. 2.6(c), the plastic hinge length ΔL is determined as

$$\Delta L = L - 2a = L\sqrt{1 - 1/f}. \tag{2.2.24}$$

With the shape factor $f = 1.14$, median for wide flange sections, ΔL comes out to be $0.35L$. Note that for a simply supported beam, the plastic hinge length for the distributed load case is higher than that for the concentrated load case (Eq. 2.2.14).

Example 2.2.2. Determine the plastic hinge lengths at the ends and midspan of a fixed-ended beam with a concentrated load at the midspan as shown in Fig. 2.7.

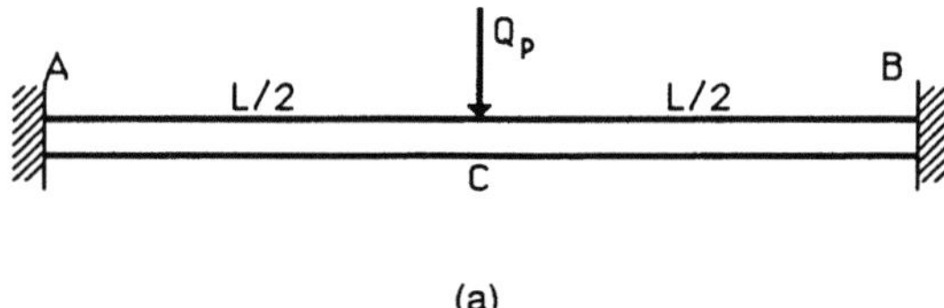

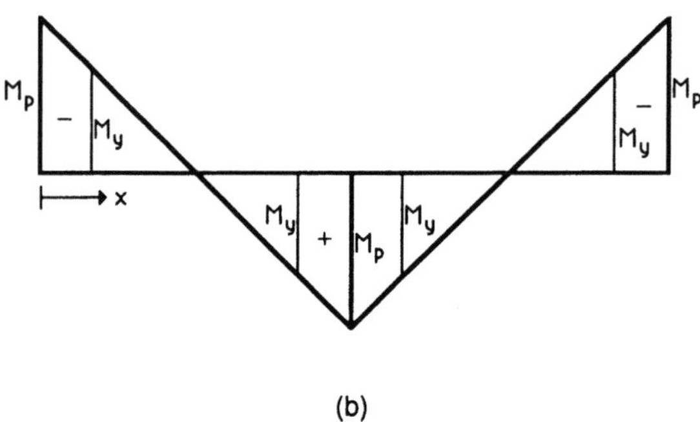

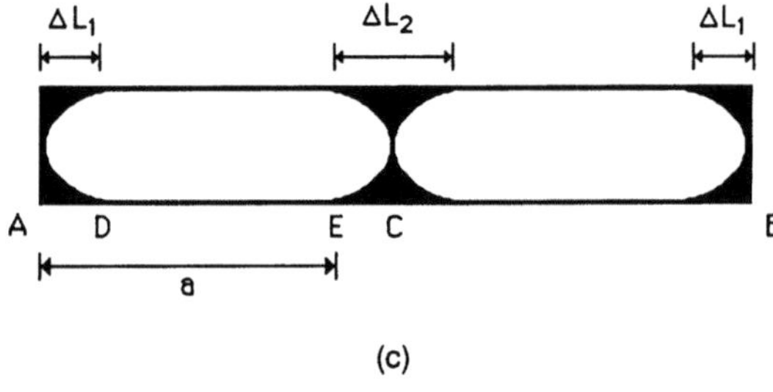

FIGURE 2.7. Hinge lengths of a fixed-ended beam with concentrated load at midspan.

Solution: Referring to the moment diagram at the collapse state [Fig. 2.7(b)], the bending moment at a distance x from end A is given by

$$M_x = \frac{Q_p}{2} x - M_p \qquad (2.2.25)$$

in which Q_p is the plastic limit load when moments at the ends and midspan all reach M_p and can be written as

$$Q_p = \frac{8M_p}{L}. \qquad (2.2.26)$$

Hinge Length at Ends: ΔL_1 in Fig. 2.7(c) can be obtained by equating the moment at D to $-M_y$

$$M_D = \frac{Q_p}{2} \Delta L_1 - M_p = -M_y. \qquad (2.2.27)$$

By substituting Q_p from Eq. (2.2.26) into Eq. (2.2.27) and solving the resulting equation, ΔL_1 is found to be

$$\Delta L_1 = \frac{L}{4}(1 - 1/f). \qquad (2.2.28)$$

Hinge Length at the Midspan: Location of point E in Fig. 2.7(c) can be obtained by equating the bending moment at E to M_y as

$$M_E = \frac{Q_p}{2} a - M_p = M_y. \qquad (2.2.29)$$

Substituting Q_p from Eq. (2.2.26) into Eq. (2.2.29) and solving for a, we have

$$a = \frac{L}{4}[1 + 1/f]. \qquad (2.2.30)$$

Thus, from Fig. 2.7(c), the hinge length ΔL_2 is found to be

$$\Delta L_2 = L - 2a = \frac{L}{2}[1 - 1/f]. \qquad (2.2.31)$$

2.3 The Full Plastic Moment

The plastic analysis is drastically simplified by using the idealized moment-curvature relationship shown in Fig. 2.5. In this simplification, the bending curvature beyond the *elastic limit curvature* Φ_p increases indefinitely with the constant moment capacity M_p. This limit moment capacity M_p is known as the *full plastic moment*. The adjective *full* means that all fibers of the section are plastic.

A knowledge of the value of the full plastic moment capacity of a section is very important in the plastic analysis and design. For example, if full plastic moments of various members of a frame are known, then the plastic limit load of the frame can be determined quickly. similarly, the design of a frame requires an assignment of certain minimum values of full plastic moment to its members to carry the factored loads. For steel sections, M_p depends on the yield stress of the steel and the geometry of the section. The calculation of M_p of a section can be summarized in the following two steps.

First, the *plastic neutral axis* is located. Like the elastic neutral axis, the plastic neutral axis is determined by considering equilibrium of forces in the axial direction, i.e., $\sum F_x = 0$. Since at the fully plastic state, the stress is equal to the yield stress over the entire section (both in compression and in tension), the calculation of plastic neutral axis is simpler than that of the elastic neutral axis for which a consideration of varying stress over the cross section is required. In fact, if the entire section is made of only one type of steel (same yield stress over the section), the plastic neutral axis can be determined by simply dividing the cross-sectional area into two equal parts. However, if a section is made of more than one type of steel (sometimes a standard section is modified by the addition of coverplates of a different type of steel), the plastic neutral axis must be determined by considering the axial equilibrium condition.

Second, the full plastic moment capacity is determined by summing the moments of the forces resulting from stresses in the section. The computations of plastic moments of several shapes of cross section are illustrated in the following example.

Example 2.3.1. Determine the full plastic moment capacity of the rectangular section, wide flange section, solid circular section, hollow thin-walled circular section, and triangular section as shown in Fig. 2.8.

Solution: (a) Rectangular section: Since the section is made of only one material, the plastic neutral axis (PNA) divides the section into two equal parts [Fig. 2.8(a)], i.e., the PNA is at a distance d from the top of the section. So, we have

$$\text{compressive force} = \text{tensile force} = bd\sigma_y$$

and the corresponding lever arm is d. Thus the full plastic moment is

$$M_p = (bd\sigma_y)d = bd^2\sigma_y. \tag{2.3.1}$$

(b) Wide Flange Section: The PNA divides the section into two equal parts, i.e., it is at a distance d from the top of the section [Fig. 2.8(b)]. For this section, the full plastic moment can be assumed to consist of two parts: one due to the couple formed by flange forces and the other due to the couple formed by web forces. The forces and the associated lever arms are given by

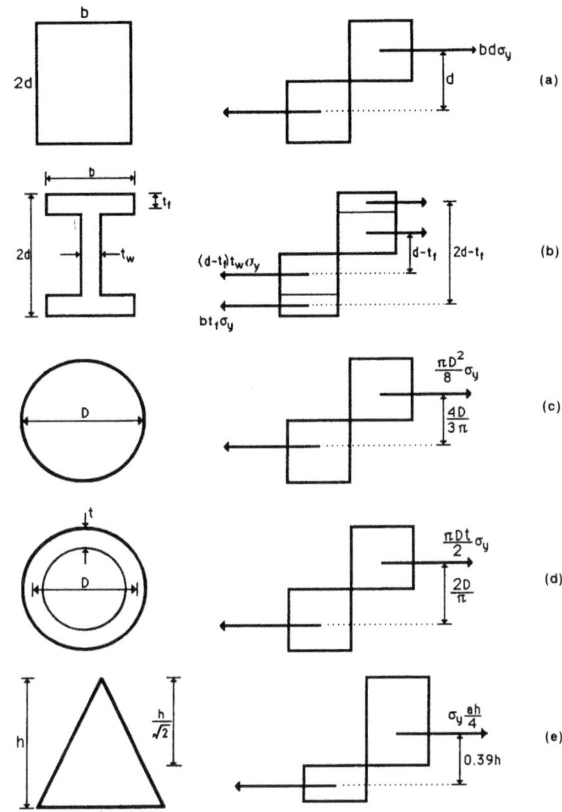

FIGURE 2.8. Computations of the fully plastic moment capacity of several shapes of cross section.

Flange Forces $= bt_f\sigma_y$
Lever Arm $= 2d - t_f$
Web Forces $= (d - t_f)t_w\sigma_y$

lever arm $= d - t_f$.

So the full plastic moment turns out to be

$$M_p = bt_f\sigma_y(2d - t_f) + (d - t_f)^2 t_w\sigma_y. \tag{2.3.2}$$

(c) Solid Circular Section: The PNA of the section passes through the center of the circle [Fig. 2.8(c)]. So, the compressive force = tensile force equation is given by

$$C = T = \frac{\pi D^2}{8}\sigma_y.$$

The lever arm for these two forces is two times the distance of the centroid of half-circle from the center of circle, i.e., $4D/3\pi$. So, the full plastic moment is

$$M_p = \left(\frac{\pi D^2}{8}\sigma_y\right)\left(\frac{4D}{3\pi}\right) = \frac{D^3}{6}\sigma_y. \tag{2.3.3}$$

(d) Hollow Thin-Walled Circular Section: The PNA passes through the centroid of the section [Fig. 2.8(d)]. The compressive force = tensile force equation is given by

$$C = T = \frac{\pi Dt}{2}\sigma_y.$$

The lever arm associated with these two forces is two times the distance of the centroid of the half of the hollow circular section from the centroid of the section, i.e., $2D/\pi$. So the full plastic moment is

$$M_p = \left(\frac{\pi Dt}{2}\sigma_y\right)\left(\frac{2D}{\pi}\right) = D^2 t\sigma_y. \tag{2.3.4}$$

(e) Triangular Section: The PNA divides the triangular area into two equal parts. So, the distance of the PNA from the top is $h/\sqrt{2}$ [Fig. 2.8(e)]. The compressive force = tensile force equation is

$$C = T = \frac{ah}{4}\sigma_y.$$

The lever arm corresponding to these two forces is $0.39h$. So, the full plastic moment is

$$M_p = \left(\sigma_y\frac{ah}{4}\right)(0.39h) = 0.098\sigma_y ah. \tag{2.3.5}$$

2.4 Design of a Cross Section

In the plastic design of steel structures, a certain value of M_p is assigned to each member of the structure. For design purposes, it is convenient to write M_p in the form

$$M_p = Z\sigma_y \tag{2.4.1}$$

in which Z is called the *plastic section modulus* and it depends solely on the geometry of the cross section. It can be determined simply by computing the first moment of area of the section about its plastic neutral axis. The values of Z for all hot-rolled sections are given in the AISC *Manual of Steel Construction.*

Very often, the standard hot-rolled sections are modified by either the addition of cover plates or the curtailment of flanges. For these cases, the values of Z are not given in the manual and they have to be calculated.

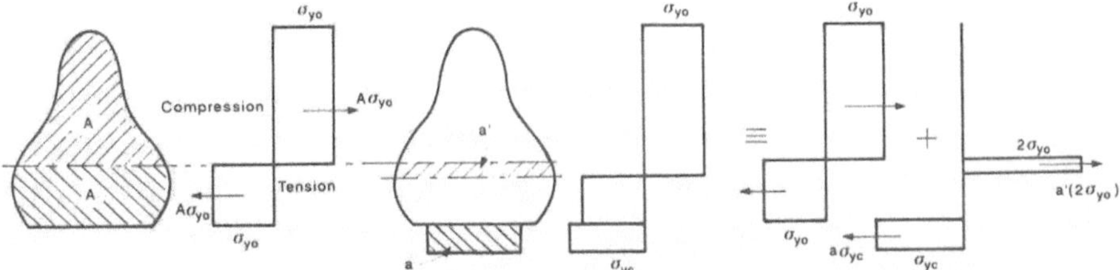

FIGURE 2.9. Plastic modulus of a section with a cover plate.

However, these calculations can be shortened by modifying the value of Z for the standard section given in the manual.

Consider the example of an arbitrary section with area $2A$ and one axis of symmetry. Assume that it is strengthened by the addition of a single cover plate of area a, as shown in Fig. 2.9. Further assume that the yield strengths of the original section and the cover plate are σ_{y0} and σ_{yc}, respectively. In the fully plastic state, the total force on the cover plate will be $a\sigma_{yc}$. Therefore, to maintain equilibrium in the axial direction, it is clear from the diagram that the plastic neutral axis must shift downward from its original position so that an area

$$a' = \frac{a\sigma_{yc}}{2\sigma_{y0}} \tag{2.4.2}$$

of the original cross section moves from the tension side to the compression side of the axis. The resulting fully plastic stress distribution may be considered as the sum of two parts: the stress distribution in the original section and the change in stress distribution introduced by the addition of the cover plate as shown in Fig. 2.9.

The full plastic moment of the modified section can now be determined by summing the full plastic moment of the original section and the moment contributed by the cover plate. The contribution of the cover plate is equal to the moment caused by a couple formed by the cover plate force $a\sigma_{yc}$ and a force due to the fictitious stress $2\sigma_{y0}$ acting on area a' that has been transferred from tension to compression, as shown in Fig. 2.9.

Example 2.4.1. A member of a frame is supposed to have the plastic moment capacity of 300 kip-ft. Select the lightest wide flange section for this member. Use A36 steel.

Solution: The required plastic section modulus is

$$Z = \frac{300 \times 12}{36} = 100 \text{ in.}^3$$

The following sections satisfy the requirement of minimum plastic modulus

(from the AISC manual)

$$W16 \times 57 \qquad Z_x = 105 \text{ in}^3$$
$$W18 \times 50 \qquad Z_x = 101 \text{ in}^3$$
$$W21 \times 50 \qquad Z_x = 110 \text{ in}^3$$
$$W24 \times 55 \qquad Z_x = 134 \text{ in.}^3$$

Use W21 × 50.

Example 2.4.2. A W21 × 50 section of A36 steel is modified by the addition of a single cover plate. Determine the full plastic moment capacity of the modified section if

(a) A36, 8″ × 3/4″ plate is used as a cover plate.
(b) Grade 50, 8″ × 3/4″ plate is used as a cover plate.

Solution: The dimensions of W21 × 50, taken from the AISC manual, are shown in Fig. 2.10(a). Also from the manual

$$Z_x = 110 \text{ in.}^3$$

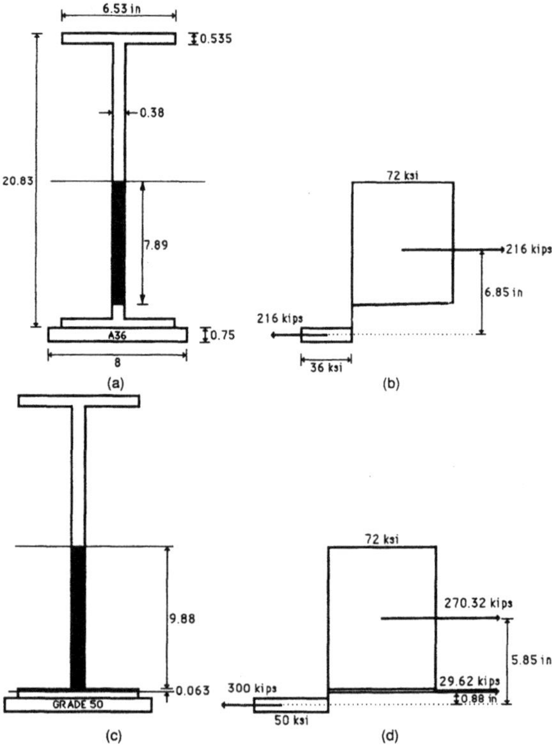

FIGURE 2.10. Computations of plastic moment capacity of W21 × 50 section with a cover plate.

(a) A36 Steel Cover Plate: Area of the original section that must be transferred from the tension to the compression side

$$a' = \frac{8 \times 0.75}{2} = 3 \text{ in.}^2$$

The distance by which the neutral axis should be shifted is

$$y = \frac{3}{0.38} = 7.89 \text{ in.}$$

The plastic moment of the modified section is now equal to the sum of the full plastic moment of the original section and the moment caused by a couple formed by the cover plate force of 6×36 kips and the web force of 3×72 kips as shown in Fig. 2.10(b).

The lever arm for the couple is

$$d = \frac{20.83}{2} + \frac{3}{8} - \frac{7.89}{2} = 6.85 \text{ in.}$$

Thus, the full plastic moment for the modified section is

$$M_p = 110 \times 36 + 3 \times 72 \times 6.85 = 5439.6 \text{ kip-in.}$$

(b) Grade 50 Steel Cover Plate: Area of the original section that must be transferred from the tension to the compression side is

$$a' = \frac{(8 \times 0.75)}{2} \times \frac{50}{36} = 4.17 \text{ in.}^2$$

Since this is greater than half the web area ($[20.83/2 - 0.535]0.38 = 3.75$), a part of the flange area must also be transferred. For an extremely accurate answer, a graphical method should be used to include the curved portion at the junction of flange and web. However, sufficient accuracy can be achieved by simply treating the section as three rectangles, Fig. 2.10(c). The shaded portion in Fig. 2.10(c) shows the area that must be transferred from tension to compression. So, the additional moment is contributed by two couples: one couple formed by the web force and a part of the cover plate force and the other formed by the flange force and the remaining part of the cover plate force.

$$\text{web force} = \left(\frac{20.83}{2} - 0.535\right)0.38 \times 72 = 270.32 \text{ kips.}$$

The distance between the web force and the cover plate force is

$$d_1 = \frac{20.83}{2} + \frac{3}{8} - \frac{1}{2}\left(\frac{20.83}{2} - 0.535\right) = 5.85 \text{ in.}$$

Flange force $= 0.063 \times 6.53 \times 72 = 29.62$ kips.

The distance between the flange force and the cover plate force is

$$d_2 = \frac{3}{8} + 0.535 - \frac{0.063}{2} = 0.88 \text{ in.}$$

$$M_p = 110 \times 36 + 270.32 \times 5.85 + 29.62 \times 0.88 = 5,567.44 \text{ kip-in.}$$

2.5 Effect of Axial Load

The application of an axial compression to a cross section results in a uniform compressive stress over the section. The addition of a bending moment to this axial compression produces a linear variation of elastic stress across the section as shown in Fig. 2.11. Further increases of bending moment, with the axial compression remaining constant, eventually cause yielding on the compression face of the section, followed by yielding on the tension face, and eventual yielding of the entire cross section. During this process, the neutral axis initially lies outside the section for very small values of bending moment and it shifts progressively toward the final position in the section in the fully plastic state.

The full plastic moment capacity of a section in the presence of axial compression can be determined from the two usual equilibrium conditions

$$P = \int_A \sigma \, dA \tag{2.5.1}$$

$$M = \int_A \sigma y \, dA. \tag{2.5.2}$$

The presence of axial compression reduces the full plastic moment capacity of a section. This reduced moment capacity is designated M_{pc}. The extent of this reduction is dependent on the magnitude of the axial load. Using Eqs. (2.5.1) and (2.5.2), the *reduced moment capacities M_{pc}* for rectangular, circular tubular, wide flange, and T-sections are obtained in the following.

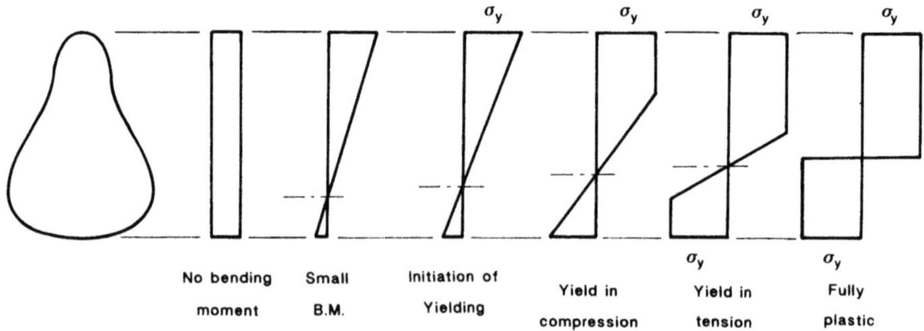

FIGURE 2.11. Stress distributions under combined action of bending and axial compression.

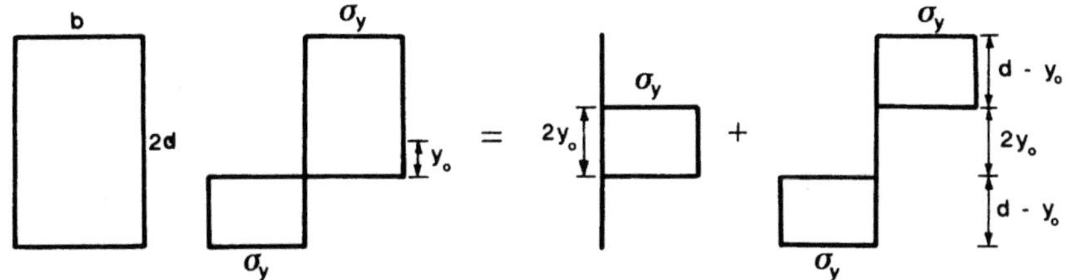

FIGURE 2.12. The reduced plastic moment capacity M_{pc} of a rectangular section considering the influence of axial compression.

2.5.1 Rectangular Section

The full plastic stress diagram consists of two portions as shown in Fig. 2.12. The axial load P is assumed to be supported entirely by a centrally located portion of the total cross-sectional area stressed to the yield point in compression, and the bending moment to be resisted by the top and bottom portions stressed to full yielding in tension in the bottom portion and in compression in the upper portion.

The extent of central portion $2y_0$ can be determined from Eq. (2.5.1) as

$$y_0 = \frac{P}{2\sigma_y b}, \tag{2.5.3}$$

and the reduced plastic bending strength can be expressed in terms of y_0 from Eq. (2.5.2) as

$$M_{pc} = \sigma_y b(d^2 - y_0^2). \tag{2.5.4}$$

By substituting Eq. (2.5.3) into Eq. (2.5.4) and by noting that $2\sigma_y bd = \sigma_y A = P_y$ and $M_p = \sigma_y bd^2$, the nondimensionalized expression for the reduced plastic moment capacity of a rectangular section can be written as

$$\frac{M_{pc}}{M_p} = 1 - \left(\frac{P}{P_y}\right)^2. \tag{2.5.5}$$

Compared with the stress diagram shown in Fig. 2.8(a) for the case where there was no axial load, it is seen that the bending resistance indicated in Fig. 2.12 is reduced by an amount equal to the moment of the central stress area used to carry the axial load.

2.5.2 Circular Tubular Section

Like a rectangular section, the value of M_{pc} for circular tubular section can be expressed as

$$\frac{M_{pc}}{M_p} = \cos\left(\frac{\pi}{2}\frac{P}{P_y}\right). \tag{2.5.6}$$

2.5.3 *Wide-Flange Section Bending About Strong Axis*

Under low values of axial compression, the plastic neutral axis for wide-flange sections bending about a strong axis will be in the web, while for high values of axial compression, the plastic neutral axis will be in a flange. The resulting equations for M_{pc} for these two cases are given in the following. Note that in all the previous derivations, for simplicity we have taken the depth of a cross section to be $2d$. Herein and from here on, we shall follow the usual notation and take the full depth of a wide-flange section to be d instead of $2d$ as shown, among others, in the inset of Fig. 2.13.

For a neutral axis in a web $[0 \leq P/P_y \leq 1/(1 + 2bt_f/t_w d_w)]$

$$\frac{M_{pc}}{M_p} = 1 - \frac{\left(\dfrac{P}{P_y}\right)^2 \left(1 + \dfrac{2bt_f}{t_w d_w}\right)^2}{\left(1 + \dfrac{4bt_f d_f}{t_w d_w^2}\right)}. \tag{2.5.7}$$

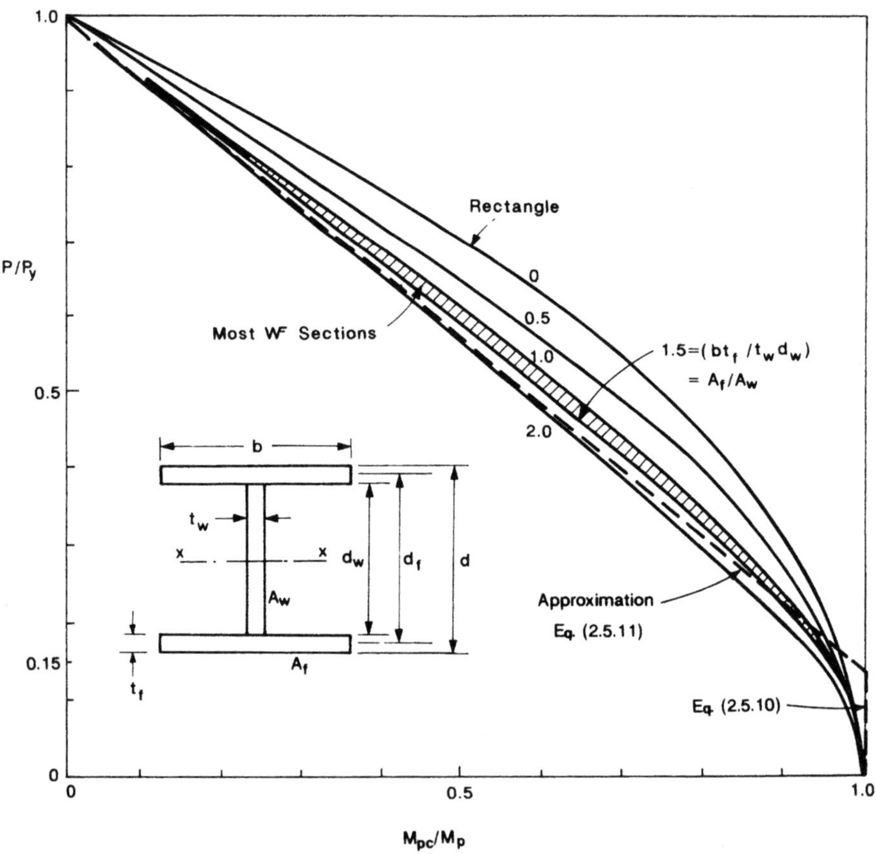

FIGURE 2.13. Strength interaction curves for wide-flange sections bending about a strong axis.

For a neutral axis in a flange $[1/(1 + 2bt_f/t_w d_w) \le P/P_y < 1.0]$

$$\frac{M_{pc}}{M_p} = \frac{2\left(\dfrac{l}{d_w}\right)\left(1 - \dfrac{P}{P_y}\right)\left(1 + \dfrac{2bt_f}{t_w d_w}\right)}{1 + \left(1 + \dfrac{d}{d_w}\right)\dfrac{2bt_f}{t_w d_w}} \tag{2.5.8}$$

where l is the lever arm of the couple formed by the tensile and compressive forces in the flanges and is given by

$$l = d - t_f\left(1 + \frac{t_w d_w}{2bt_f}\right)\left(1 - \frac{P}{P_y}\right) \tag{2.5.9}$$

and b, t_f, t_w, and d_w are dimensions of wide-flange section as shown in the inset of Fig. 2.13. Note that the maximum and minimum values of lever arm l are, respectively, d and d_f. The interaction curves plotted between moment M_{pc} and axial force P for $bt_f/t_w d_w = A_f/A_w = 0.5, 1, 1.5$, and 2.0 are shown in Fig. 2.13 in which $A_f = bt_f =$ area of one flange and $A_w = t_w d_w =$ area of web. These curves are plotted by assuming that $l/d_w = d/d_w$ and d_f/d_w are about the same for all shapes used as columns and by approximating these values as 1.10 and 1.05, respectively. The shaded area in Fig. 2.13 shows graphically the extent of variation that would result by applying such an expression to all of the rolled shapes likely to be used in the plastic design.

For design purposes, the interaction in Eqs. (2.5.7) and (2.5.8) can be approximated by the following two simple equations

For $0 \le P \le 0.15P_y$,

$$M_{pc} = M_p \tag{2.5.10}$$

and for $0.15P_y \le P \le P_y$

$$M_{pc} = 1.18\left(1 - \frac{P}{P_y}\right)M_p. \tag{2.5.11}$$

Note that this approximation is somewhat conservative for most shapes, except in the region where P/P_y is amall. Even here, the maximum error is less than 5% (dotted line, Fig. 2.13).

2.5.4 Wide-Flange Section Bending About Weak Axis

For weak-axis bending, the plastic neutral axis, depending on the value of axial load, may fall in the web or the flanges. The resulting M_{pc} equations for these two cases are given in the following.

For neutral axis in web $[0 \le P/P_y \le (d/d_w)(1/(1 + 2bt_f/t_w d_w))]$

$$\frac{M_{pc}}{M_p} = 1 - \left(\frac{t_w}{b}\right)\left(\frac{d_w}{d}\right)\left[\frac{\left(1 + \dfrac{2bt_f}{t_w d_w}\right)^2}{\dfrac{2bt_f}{t_w d_w} + \dfrac{t_w}{b}}\right]\left(\frac{P}{P_y}\right)^2. \tag{2.5.12}$$

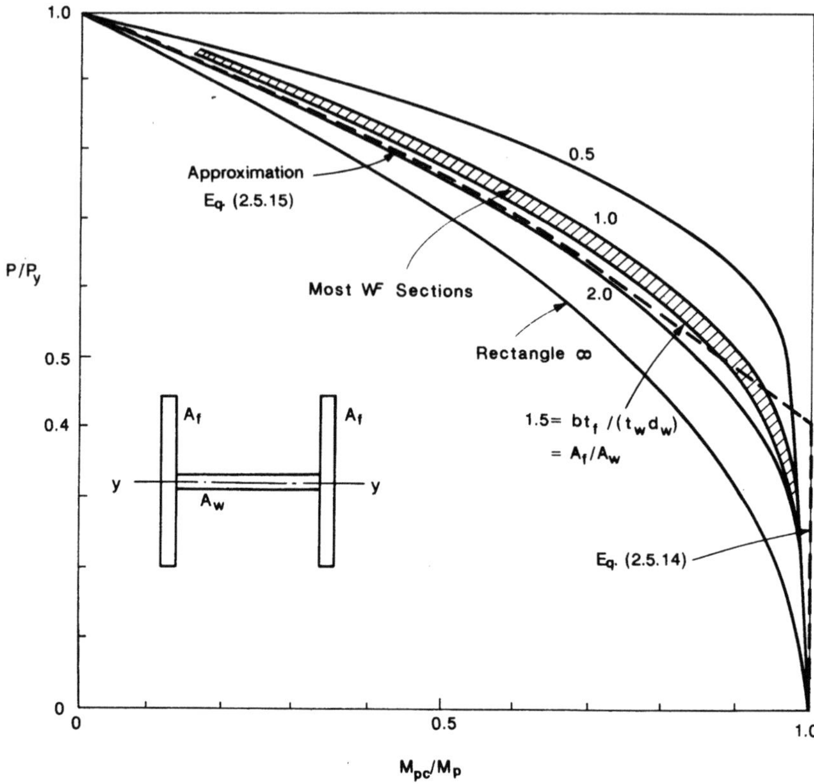

FIGURE 2.14. Strength interaction curves for wide-flange sections bending about a weak axis.

For neutral axis in flanges $[(d/d_w)(1/(1 + 2bt_f/t_w d_w)) \leq P/P_y < 1]$

$$\frac{M_{pc}}{M_p} = \left(1 - \frac{P}{P_y}\right)\left[\frac{\left(1 + \frac{2bt_f}{t_w d_w}\right)^2}{\left(\frac{2bt_f}{t_w d_w}\right)\left(\frac{2bt_f}{t_w d_w} + \frac{t_w}{b}\right)}\right]\left[\frac{2}{\left(1 + \frac{t_w d_w}{2bt_f}\right)} - \left(1 - \frac{P}{P_y}\right)\right].$$

$$(2.5.13)$$

The interaction curves plotted between moment and axial force for $bt_f/t_w d_w = A_f/A_w = 0.5, 1, 1.5,$ and 2.0 are shown in Fig. 2.14. These curves are plotted by assuming that t_w/b and d/d_w are about the same for all shapes used as columns and approximating these values as 0.04 and 1.10, respectively.

For design purposes, interaction Eqs. (2.5.12) and (2.5.13) can be approximated by the following two equations (dotted line, Fig. 2.14).

For $0 \leq P \leq 0.4P_y$

$$M_{pc} = M_p.$$

$$(2.5.14)$$

For $0.4P_y \leq P \leq P_y$

$$M_{pc} = 1.19\left[1 - \left(\frac{P}{P_y}\right)^2\right]M_p. \tag{2.5.15}$$

Note that the influence of the axial compressive load P on the plastic bending strength of columns may have to be further reduced below the value M_{pc} to guard against premature buckling of columns. This is provided for in Chapter H of the LRFD rules on beam-column design where columns in frames are classified as either a sidesway prevented case or sidesway permitted case.

2.5.5 T-Sections

Under the combined action of bending and axial compression, the line of action of an axial load may significantly affect the calculation of plastic moment capacity of a section. Mostly, this line of action is assumed to pass through the centroid of the section. For doubly symmetric sections such as rectangular, circular, and wide-flange sections, the plastic neutral axis (PNA) under pure bending moment passes through the centroids of the sections. The moment capacity of such sections, therefore, can be determined by summing up the moments about the centroidal/plastic neutral axis.

For monosymmetric sections such as T-sections, PNA under pure moment does not pass through the centroids of the sections. If the line of action of axial load passes through the original plastic neutral axis, then the full plastic moment is always reduced by the axial load. However, if the line of action of axial load passes through the centroid, then axial load will generate an additional bending moment equal to the moment of a couple formed by the applied axial load acting through the centroid and the internal axial resistance acting through the plastic neutral axis. If the sign of this additional moment is the same as that of the applied moment, the reduced plastic moment capacity M_{pc} of the section will artificially be greater than M_p for some cases. The following example demonstrates this fact.

Example 2.5.1. Determine M_{pc} of a T-section shown in Fig. 2.15 by assuming that

(a) the axial load is acting through the original plastic neutral axis under pure moment.
(b) the axial load is acting through the centroidal axis.

Solution: (a) Axial Load Through Original PNA: Under pure bending, the PNA for the given T-section is at the junction of the two rectangles. In the presence of axial compressive load, the PNA shifts by a distance y_2 as shown in Fig. 2.15. The distances y_1 and y_2 in Fig. 2.15(a) can be determined from the axial equilibrium condition in Fig. 2.15(b) as

$$y_1 B\sigma_y + y_2 T\sigma_y = P \tag{2.5.16}$$

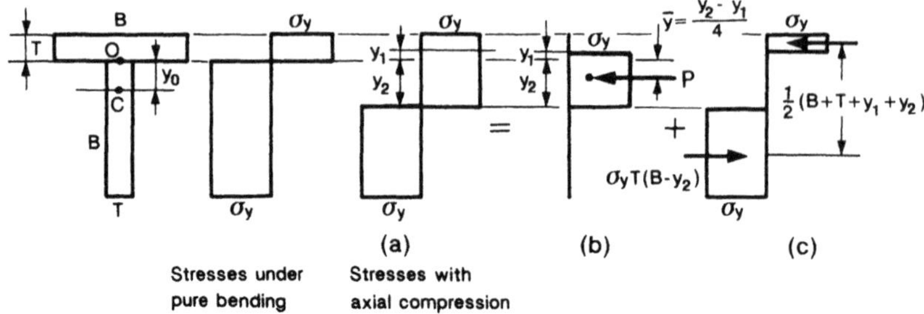

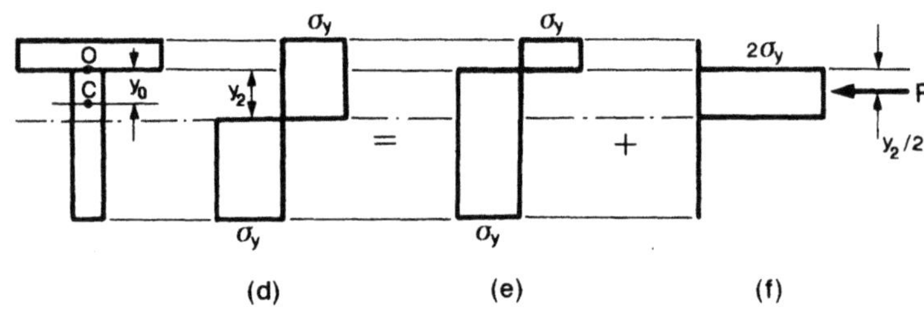

FIGURE 2.15. Computations of the reduced moment capacity M_{pc} for a T-section.

and the pure bending condition in Fig. 2.15(c) as

$$\sigma_y T(B - y_2) = \sigma_y B(T - y_1) \tag{2.5.17}$$

or

$$y_1 B = y_2 T.$$

Solving these two equations, we obtain

$$y_1 = \frac{P}{2\sigma_y B} \tag{2.5.18}$$

$$y_2 = \frac{P}{2\sigma_y T}. \tag{2.5.19}$$

By summing up moments about PNA, the reduced moment capacity M_{pc} is found to be

$$M_{pc} = \sigma_y T(B - y_2)\frac{1}{2}(B + T + y_1 + y_2) - P\bar{y}. \tag{2.5.20}$$

Substituting y_1 and y_2 from Eqs. (2.5.18) and (2.5.19), respectively, using $P_y = 2\,BT\sigma_y$ and $\bar{y} = (y_2 - y_1)/4$ and simplifying, M_{pc} can be written as

$$M_{pc} = \frac{\sigma_y BT}{2}(B + T)\left[1 - \left(\frac{P}{P_y}\right)^2 \frac{2B}{B + T}\right] \tag{2.5.21}$$

or

$$M_{pc} = M_p \left[1 - \left(\frac{P}{P_y}\right)^2 \frac{2B}{B+T} \right] \tag{2.5.22}$$

in which $M_p = [(\sigma_y BT/2)(B+T)]$ is the full plastic moment when axial load is equal to zero.

(b) Axial Load Through Centroid: The distance y_0, between the centroid and the PNA under pure bending, can be expressed as

$$y_0 = \frac{-BT\dfrac{T}{2} + BT\dfrac{B}{2}}{2BT} = \frac{B-T}{4}. \tag{2.5.23}$$

By taking moments about the centroid of section, M_{pc} is found to be

$$M_{pc} = M_p \left[1 - \left(\frac{P}{P_y}\right)^2 \frac{2B}{B+T} \right] + Py_0. \tag{2.5.24}$$

By substituting y_0 from Eq. (2.5.23) and using $P_y = 2BT\,\sigma_y$, M_{pc} can be expressed as

$$M_{pc} = M_p \left[1 - \left(\frac{P}{P_y}\right)^2 \frac{2B}{B+T} + \frac{P}{P_y}\left(\frac{B-T}{B+T}\right) \right]. \tag{2.5.25}$$

Using the superposition of the stress diagram shown in Figs. 2.15(d) to (f), Equations (2.5.22) and (2.5.25) can be derived in a simple and direct manner.

2.6 Effect of Shear Force

The shear force combined with bending moment results in a two-dimensional stress system in a section, which makes consideration of the effect of shear force on M_p much more complex than that of axial force only. Axial force and bending moment both result in longitudinal stresses that can be superimposed directly. So, for the combined bending and axial force, we were able to obtain the exact solution of the problem in a simple manner by using the equilibrium equations, the kinematic assumption of plane section remains plane, and the yield condition $\sigma = \sigma_y$. For the combined action of shear force and bending moment, the exact solution of the governing equations is often intractable for most cases and recourse must be made through approximations and simplifications for practical solutions by using the following equilibrium equations

$$M = \int_A \sigma y\, dA \tag{2.6.1}$$

$$V = \int_A \tau\, dA \tag{2.6.2}$$

and the *von Mises yield condition* [2.1]

$$\sigma^2 + 3\tau^2 \leq \sigma_y^2 \tag{2.6.3}$$

in which σ and τ are, respectively, the normal and shear stresses at a point in the beam section at a distance y from the neutral axis.

These approximate solutions satisfying only the equilibrium equations and the yield condition but not the kinematic condition are always lower than the exact solution. This will be proved in Chapter 3. The highest solution satisfying Eqs. (2.6.1–2.6.3) will therefore be the best and closest to the exact solution. The lower-bound solutions for the moment-carrying capacity considering the effect of shear force for rectangular and wide-flange sections are presented in the following.

2.6.1 *Rectangular Section*

Consider an element of a beam with rectangular cross section [Fig. 2.16(a)]. The elastic solution of the beam section under the combined bending and shear has the following stress distribution [Fig. 2.16(b)]

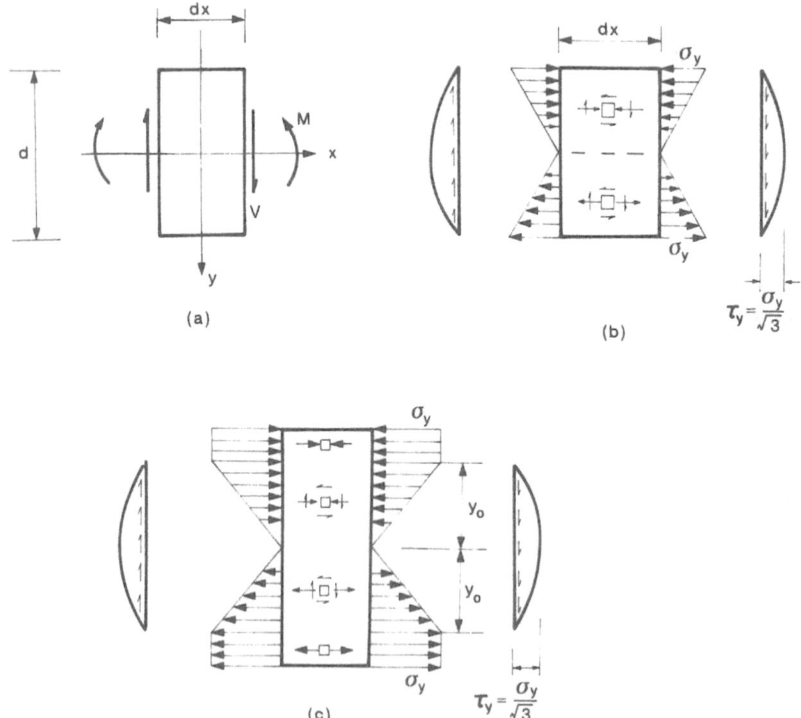

FIGURE 2.16. Assumed stress distributions in rectangular cross section under combined bending and shear loads.

$$\sigma = \sigma_y \left(\frac{2y}{d}\right) \tag{2.6.4}$$

and

$$\tau = \frac{\sigma_y}{\sqrt{3}} \left[1 - \left(\frac{2y}{d}\right)^2\right]^{1/2} \tag{2.6.5}$$

in which d is the depth of the beam. It can easily be shown that the yield condition (2.6.3) is not violated over the entire section. The top and bottom fibers are in the yield state of simple compression $-\sigma_y$ and simple tension $+\sigma_y$, respectively, and the center fiber is in the yield state of pure shear, $\tau_y = \sigma_y/\sqrt{3}$, according to the von Mises yield condition (2.6.3). The rest of the fibers are in the elastic state. By using this elastic stress distribution, the lower-bound solution for the reduced bending moment M_{ps} considering the effect of shear force can be obtained as

$$M_{ps} = \frac{1}{6}\sigma_y bd^2 = \frac{2}{3}M_p \tag{2.6.6a}$$

$$V = \frac{2}{3}\frac{\sigma_y}{\sqrt{3}}bd. \tag{2.6.6b}$$

This lower-bound solution can be improved by assuming a better stress distribution with more fibers yielded as shown in Fig. 2.16(c). M_{ps} and shear force V corresponding to this distribution can be written as

$$M_{ps} = M_p \left[1 - \frac{1}{3}\left(\frac{2y_0}{d}\right)^2\right] \tag{2.6.7}$$

$$V = \frac{4}{3}\tau_y by_0 \tag{2.6.8}$$

in which b is the width of the beam and $\tau_y = \sigma_y/\sqrt{3}$ is the yield stress of the material in pure shear. By eliminating y_0 from Eqs. (2.6.7) and (2.6.8), M_{ps} can be expressed as

$$\frac{M_{ps}}{M_p} = 1 - \frac{3}{4}\left(\frac{V}{V_p}\right)^2 \tag{2.6.9}$$

in which $V_p = bd\sigma_y/\sqrt{3}$ is the maximum shear force capacity of rectangular section in the absence of moment.

The lower-bound solution can be further improved by first assuming a general stress distribution and then using the maximization process to obtain the best distributions [2.2]. However, the process and the resulting equations are too much involved from a practical point of view. For design purposes, the following interaction equation proposed by Drucker [2.3] can be used as a good approximation to the exact solution

$$\frac{M_{ps}}{M_p} = 1 - \left(\frac{V}{V_p}\right)^4. \tag{2.6.10}$$

Interaction equation (2.6.10) is plotted in Fig. 2.17.

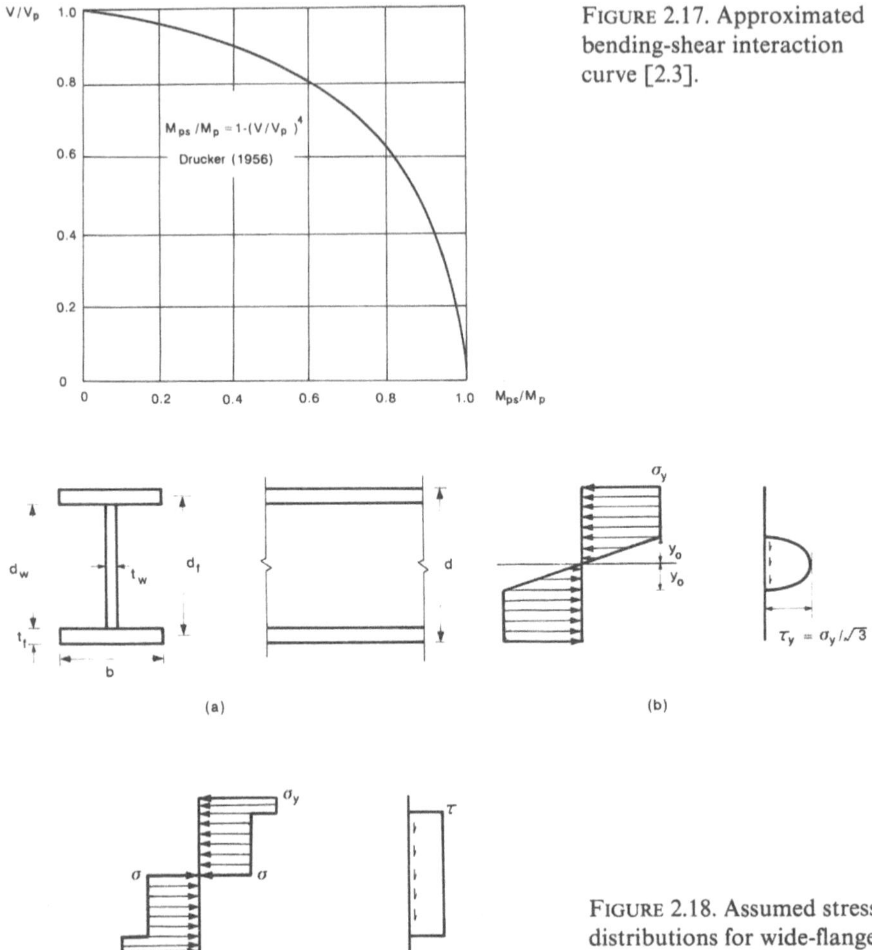

FIGURE 2.17. Approximated bending-shear interaction curve [2.3].

FIGURE 2.18. Assumed stress distributions for wide-flange sections under combined bending and shear force.

2.6.2 *Wide-Flange Section*

Consider the beam element with wide-flange section as shown in Fig. 2.18(a). Assume that the flanges and part of the web yield under normal stress and the stress distribution in the remaining portion of the web is parabolic for shear stress and linear for normal stress [Fig. 2.18(b)]. These assumed stress distributions satisfy the yield criterion (2.6.3). The resultant moment M_{ps} and shear force V corresponding to this stress distribution are

$$M_{ps} = M_p - \frac{1}{3}\sigma_y y_0^2 t_w \tag{2.6.11}$$

$$V = \frac{4}{3}\frac{\sigma_y}{\sqrt{3}}t_w y_0. \tag{2.6.12}$$

By eliminating y_0 from Eqs. (2.6.11) and (2.6.12), the reduced moment M_{ps} can be expressed as

$$\frac{M_{ps}}{M_p} = 1 - \frac{\frac{3}{4}\left(\frac{V}{V_p}\right)^2}{1 + \frac{4bt_f d_f}{t_w d_w^2}} \quad \text{for } V \leq \frac{2}{3}V_p \tag{2.6.13}$$

in which b, t_f, d_f, and t_w are dimensions as shown in Fig. 2.18(a) and

$$V_p = t_w d_w \tau_y = t_w d_w \frac{\sigma_y}{\sqrt{3}} \tag{2.6.14}$$

is the maximum shear capacity of the web.

Expression (2.6.13) for M_{ps} can be improved by assuming that the flanges will carry only the normal stresses and the web will carry uniform normal and shear stresses as shown in Fig. 2.18(c). With this assumption, M_{ps} can be written as

$$M_{ps} = \sigma_y bt_f d_f + \frac{1}{4}\sigma t_w d_w^2 \tag{2.6.15}$$

where σ is the uniform normal stress in the web and can be expressed in terms of shear force V by using $\tau = V/t_w d_w$ in the von Mises yield criterion, Eq. (2.6.3) as

$$\frac{\sigma}{\sigma_y} = \sqrt{1 - \left(\frac{\sqrt{3}\,\tau}{\sigma_y}\right)^2} = \sqrt{1 - \left(\frac{V}{V_p}\right)^2}. \tag{2.6.16}$$

By substituting σ from Eq. (2.6.16) in Eq. (2.6.15), M_{ps} can be expressed as

$$\frac{M_{ps}}{M_p} = \frac{1 + \frac{1}{4}\frac{t_w d_w^2}{bt_f d_f}\sqrt{1 - \left(\frac{V}{V_p}\right)^2}}{1 + \frac{1}{4}\frac{t_w d_w^2}{bt_f d_f}}. \tag{2.6.17}$$

Though the stress distribution in Fig. 2.18(c) does not fully satisfy the equilibrium conditions at the web-flange junction (the shear stress has a jump at this junction), it provides a reasonably good theoretical estimate of the effects of shear force on the plastic moment capacity of a wide-flange section.

The effect of shear force on the full moment capacity of members in a practical frame is generally negligible. Because in frames, high shear and moment occur in localized zones where strain hardening of material will set in quickly, thus, in most cases permitting the moment to reach or exceed the full plastic value. Therefore, in actual design of frames, as far as shear is concerned, the full plastic moment M_p may be used in design, provided that the total transverse shear force V on the section at ultimate loading is no more than V_p given by Eq. (2.6.14).

Example 2.6.1. Determine the percentage of reduction in the plastic modulus of W14 × 82 of A36 steel when a shear force of 100 kips is to accompany the bending moment.

Solution: From Fig. 2.18(c), the reduced plastic moment capacity of a wide-flange section in the presence of shear can be written as

$$M_{ps} = M_f + M_{ws} \tag{2.6.18}$$

in which M_f is the moment capacity contributed by the flanges and M_{ws} is the moment capacity of the web in the presence of shear force and can be written as

$$M_{ws} = M_w \frac{\sigma}{\sigma_y} \tag{2.6.19}$$

where M_w is the moment capacity of the web when shear force is absent and σ is the normal stress in the web reduced by the presence of shear force. Thus, Eq. (2.6.18) can be expressed as

$$M_{ps} = M_f + M_w - M_w + M_{ws} \tag{2.6.20}$$

or

$$M_{ps} = M_p - M_w \left(1 - \frac{\sigma}{\sigma_y} \right). \tag{2.6.21}$$

Dividing both sides by σ_y, Eq. (2.6.21) can be rewritten as

$$Z_{ps} = Z - Z_w \left(1 - \frac{\sigma}{\sigma_y} \right)$$

in which Z_{ps} is the plastic modulus of a wide-flange section reduced for the presence of shear force and Z_w is the plastic modulus of the web of a wide-flange section.

From the AISC manual, the following properties of W14 × 82 section can be noted

$$Z = 139 \text{ in}^3$$

$$d_w = d - 2t_f = 14.31 - 2 \times 0.855 = 12.6 \text{ in.}$$

$$t_w = 0.510 \text{ in.}$$

$$Z_w = t_w \frac{d_w^2}{4} = 0.51 \times \frac{(12.6)^2}{4} = 20.24 \text{ in.}^3$$

The shear stress in the web is

$$\tau = \frac{V}{d_w t_w} = \frac{100}{12.6 \times 0.51} = 15.56 \text{ ksi}$$

from the von Mises yield criterion (2.6.3), the normal stress reduced for the presence of shear stress is

$$\sigma = \sqrt{\sigma_y^2 - 3\tau^2} = \sqrt{(36)^2 - 3(15.56)^2} = 23.87 \text{ ksi.}$$

Therefore, we have

$$Z_{ps} = 139 - 20.24\left(1 - \frac{23.87}{36}\right) = 132.18 \text{ in}^3$$

$$\% \text{ reduction} = \frac{139 - 132.18}{139} = 4.9\%.$$

2.7 Effect of Combined Axial and Shear Force

In a manner similar to that used for considering the individual effects of axial and shear force, the effects of combined axial and shear on M_p can be considered.

For a rectangular cross section, Near [2.4] has suggested the following approximate interaction equation for the combined bending, axial load, and shear force

$$\frac{M}{M_p} + \left(\frac{P}{P_y}\right)^2 + \frac{\left(\dfrac{V}{V_p}\right)^4}{1 - \left(\dfrac{P}{P_y}\right)^2} = 1. \tag{2.7.1}$$

Equation (2.7.1) is a good approximation of the exact interaction relation. This relation reduces to Drucker's approximation (2.6.10) for the special case of $P = 0$ and it reduces to the exact Eq. (2.5.5) for the special case of $V = 0$. Over the full range of values of M/M_p, P/P_y, and V/V_p, the error never exceeds 5%.

For wide-flange sections, the effects of combined axial and shear force on M_p can be considered by assuming a stress distribution shown in Fig. 2.19. By using this stress distribution, the reduced plastic moment capacity of the section is found to be

$$M = \sigma_y b t_f d_f + \sigma t_w\left(\frac{d_w^2}{4} - y_0^2\right)$$

in which σ is the normal stress in the web and can be expressed in terms of shear force V by using $\tau = V/t_w d_w$ in the von Mises yield criterion (2.6.3) and y_0 can be related to the axial load P by the relation $P = 2t_w y_0 \sigma$.

2.8 Compactness

In the plastic analysis of steel structures, it is tacitly assumed that the moment capacity of a section will remain at the level of the plastic moment until a sufficient number of plastic hinges are developed to transform the structure

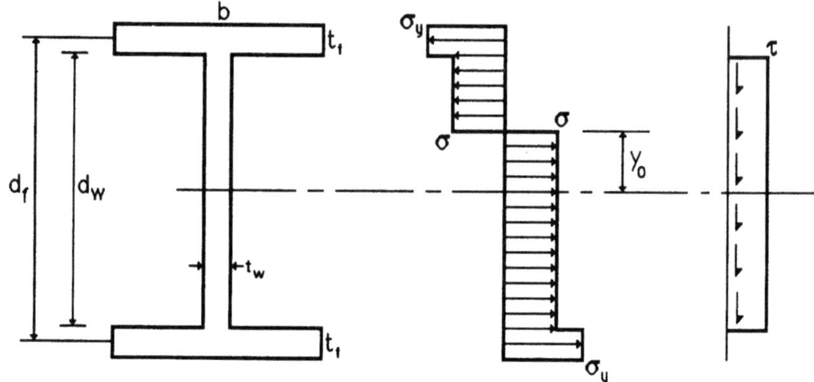

FIGURE 2.19. Assumed stress distributions under combined bending, axial compression, and shear force for wide-flange sections.

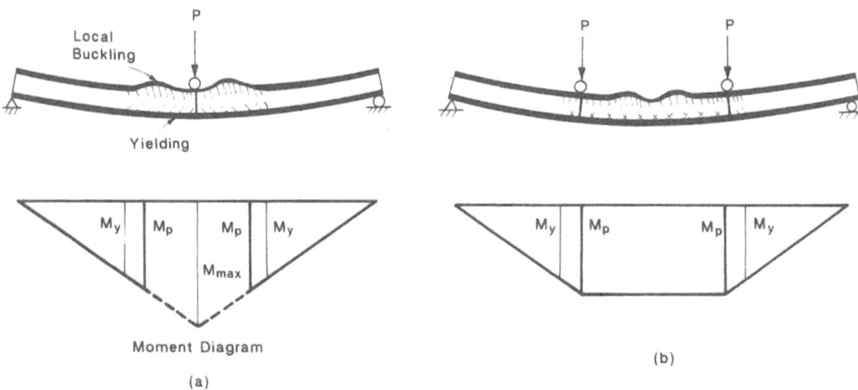

FIGURE 2.20. Local buckling of flanges of beams.

into a failure mechanism. However, if the section is made of thin plate elements, premature local buckling of some of its element may occur and the section may not be able to attain the value of plastic moment or after attaining the plastic moment, it may not be able to sustain the plastic moment up to the desired rotation capacity. For example, if the width-to-thickness ratios of the flanges and web are too high, they may buckle locally, thus violating the basic assumptions underlying the plastic analysis. Figure 2.20 shows possible local buckling modes of compression flanges of two beams: one under moment gradient and the other under uniform moment.

2.8.1 LRFD Definitions of Compact, Non-Compact and Slender Sections

To ensure proper *compactness* of section for various purposes, the LRFD specification has defined two sets of limiting width-to-thickness ratios

TABLE 2.1. Limiting values of width-to-thickness ratio ($\lambda = b/t$) to avoid premature local buckling

Type of element	$k_{\min}$	λ_p (compact)	λ_r (noncompact)
Unstiffened			
Single angles	0.425	NA	$76/\sqrt{F_y}$
Flanges of I-shaped rolled beams and channel section in flexure	0.7	$65/\sqrt{F_y}$	$141/\sqrt{F_y - 10}$
Flanges of I-shaped hydrid or welded beams in flexure	0.7	$65/\sqrt{F_{yf}}$	$162/\sqrt{(F_{yf} - 16.5)/k_c}$
Flanges of channels or I-shapes in pure compression	0.7	NA	$95/\sqrt{F_y}$
Stems of tees	1.277	NA	$127/\sqrt{F_y}$
Stiffened			
Uniform thickness flanges of tubular sections and flange cover plates	4.7	$190/\sqrt{F_y}$	$238/\sqrt{F_y}$
Webs in flexural compression	——	$640/\sqrt{F_y}$	$970/\sqrt{F_y}$
Performated cover plates	6.97	NA	$317/\sqrt{F_y}$
All other uniformly stressed stiffened elements, i.e., supproted along two edges	5.0	NA	$253/\sqrt{F_y}$
Circular hollow sections, D/t, in axial compression, in flexure	——	$2070/F_y$	$3300/F_y$, $8970/F_y$

F_{yf} = yield stress of the flange, ksi.
h = clear distance between flanges when welds are used.
$k_c = 4/\sqrt{h/t_w}$ but not less than $0.35 \leq k_c \leq 0.763$.
D = diameter.

($\lambda = b/t$): λ_p and λ_r (Table 2.1). On the basis of these limiting values, the LRFD has divided the steel sections into three categories: *compact* sections, *noncompact* sections, and *slender* sections. If the width-to-thickness ratios of all elements of a section are less than λ_p, then, the section is a compact section. Compact sections are capable of developing a fully plastic stress distribution (plastic moment), and they have a rotation capacity of approximately 3 times the yield rotation capacity before the possible occurrence of the local buckling. If the width-to-thickness ratios of all elements of a section are less than λ_r and the width-to-thickness ratio of one or more elements of a section is greater than λ_p, then, the section is a noncompact section. Noncompact sections can develop the yield stress in compression elements before local buckling occurs. However, these sections may not resist the local buckling at the strain levels required to develop the fully plastic stress distribution. If the width-to-thickness ratio of one or more elements of a section is greater than λ_r, then the section is a slender section. These sections may develop local buckling elastically before the yield stress is achieved. For determining the width-to-thickness ratio (b/t) of a plate element in a thin-walled section, the width b is explicitly defined in Fig. 2.21. Further discussion of Table 2.1 is given in the forthcoming.

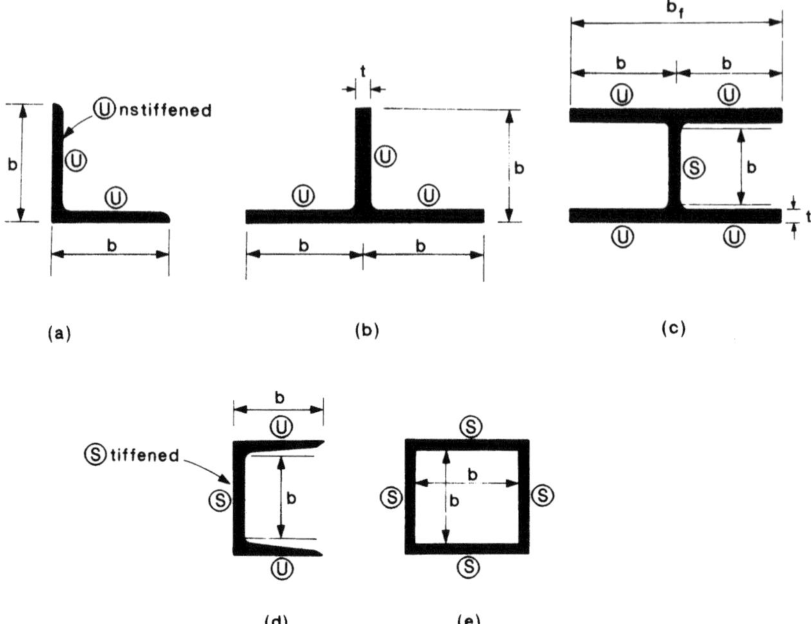

FIGURE 2.21. Width of stiffened and unstiffened elements for computing width-to-thickness ratio b/t.

2.8.2 Relation of λ_p and λ_r with Buckling Strength of Plate Elements

The recommended values of λ_p and λ_r are based on the buckling strength of plate element with various boundary conditions. With the formal mathematics, the elastic buckling strength of an axially loaded steel plate element can be developed as

$$F_{cr} = \frac{k\pi^2 E}{12(1 - v^2)\left(\dfrac{b}{t}\right)^2} = 26{,}210\,\frac{k}{\left(\dfrac{b}{t}\right)^2} \qquad (2.8.1)$$

in which v is the Poisson ratio of the material, b/t is the width-to-thickness ratio of the plate, and k is the buckling coefficient. For elements found in most structural members (high aspect ratio), the value of k depends mainly on the boundary conditions (or edge conditions) of the plate element.

On the basis of boundary conditions, the plate elements in actual structural members can conveniently be divided into two categories: *unstiffened* elements and *stiffened* elements. Unstiffened elements are supported only along one of the edges parallel to the axial stress, for example, legs of single angles, flanges of wide-flange sections, and stems of tees. Stiffened elements are supported along both the edges parallel to the axial stress, for example, flanges of square and rectangular tubular sections, perforated cover plates, and webs of wide-flange and channel sections. The recommended values of the buckling coefficient k for several types of thin elements are listed in Table 2.1.

A plate element can develop full yield stress without occurrence of buckling only if its width-to-thickness ratio does not exceed a certain value. For a plate without residual stresses, this value can be obtained by rearranging Eq. (2.8.1)

$$\frac{b}{t} = 162 \sqrt{\frac{k}{F_y}} \qquad (2.8.2)$$

where F_y is yield strength of steel in ksi.

2.8.3 LRFD Recommended Values of λ_p and λ_r

To attain the full yielding, the elements with residual stresses have to undergo larger strains than those for the elements without residual stresses. Thus, the LRFD recommendations for the limiting b/t denoted as λ_r are based on the requirement that is more strict than Eq. (2.8.2) and is approximately given by

$$\lambda_r = 113 \sqrt{\frac{k}{F_y}}. \qquad (2.8.3)$$

In order to attain fully plastic stress distribution over the entire section, the plate elements have to undergo even higher strains. Thus, λ_p in LRFD is based on even more strict requirements approximately given by the following:

for unstiffened elements

$$\lambda_p = 75 \sqrt{\frac{k}{F_y}} \qquad (2.8.4)$$

and for stiffened elements

$$\lambda_p = 94 \sqrt{\frac{k}{F_y}}. \qquad (2.8.5)$$

The values of λ_r and λ_p recommended by LRFD [1.11] in Table B5.1 can approximately be obtained by using an appropriate value of k in Eqs. (2.8.3) to (2.8.5). These value of λ_r and λ_p are given herein in Table 2.1. Since webs of sections under flexural compression are partially under tension, the recommended values of λ_r and λ_p for these elements are much higher than those for the uniformly stressed stiffened elements (Table 2.1). For webs under a combined flexural and axial compression, λ_p is recommended as the following:

for $P/(\phi_b P_y) \leq 0.125$

$$\lambda_p = \frac{640}{\sqrt{F_y}}\left(1 - 2.75\frac{P}{\phi_b P_y}\right) \qquad (2.8.6)$$

and for $P/(\phi_b P_y) > 0.125$

$$\lambda_p = \frac{191}{\sqrt{F_y}}\left[2.33 - \frac{P}{\phi_b P_y}\right] \geq \frac{253}{\sqrt{F_y}} \qquad (2.8.7)$$

in which P is the required axial load capacity, P_y is the yield axial load, and ϕ_b is the resistance factor for bending, i.e., 0.9. Note that the limiting b/t values denoted as λ_p are to ensure that the member can be designed by the plastic analysis methods. However, in areas of high *seismicity*, sections must also be able to develop higher ductility/rotation capacity (7 to 9 times the elastic rotation). Therefore, for such areas, the limiting values of λ_p should be further reduced as recommended by LRFD. For flanges of I-shaped and channel sections, λ_p recommended by LRFD in seismic area is

$$\lambda_p = \frac{52}{\sqrt{F_y}}. \qquad (2.8.8)$$

For webs under a combined flexural and axial compression, λ_p recommended by LRFD in seismic area is the following:

for $P/(\phi_b P_y) \leq 0.125$

$$\lambda_p = \frac{520}{\sqrt{F_y}}\left[1 - 1.54\frac{P}{\phi_b P_y}\right] \qquad (2.8.9)$$

for $P/(\phi_b P_y) \geq 0.125$

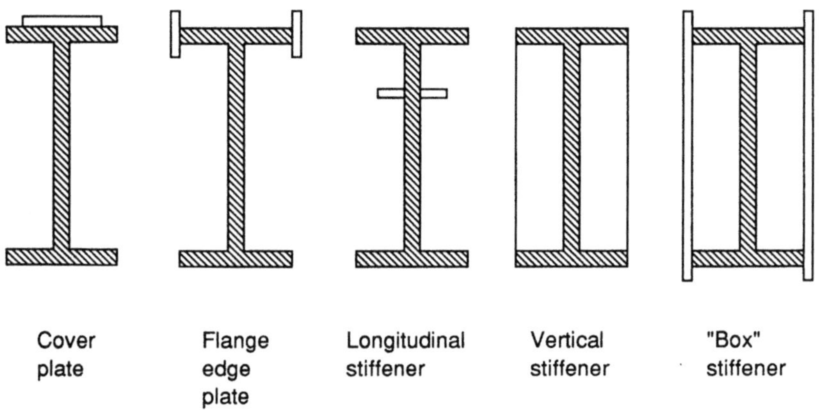

FIGURE 2.22. Methods of stiffening wide-flange shapes to prevent local buckling of compression elements.

$$\lambda_p = \frac{191}{\sqrt{F_y}}\left[2.33 - \frac{P}{\phi_b P_y}\right] \geq \frac{253}{\sqrt{F_y}}. \tag{2.8.10}$$

The sections that do not meet width-to-thickness requirements may be strengthened/stiffened in the region of the plastic hinge. Figure 2.22 suggests some methods by which this may be accomplished.

2.9 Connections

A structural frame will be able to reach its plastic limit load only if its connections and members are capable of developing and subsequently maintaining the required plastic moment up to the desired rotation capacity. The various types of connections that are encountered in steel framed structures are designated in Fig. 2.23. These include corner connections (straight and haunched), beam-to-column connections (interior, top, and side), beam-to-beam connections, splices, column anchorages, and miscellaneous connections (purlins, girts, and bracing).

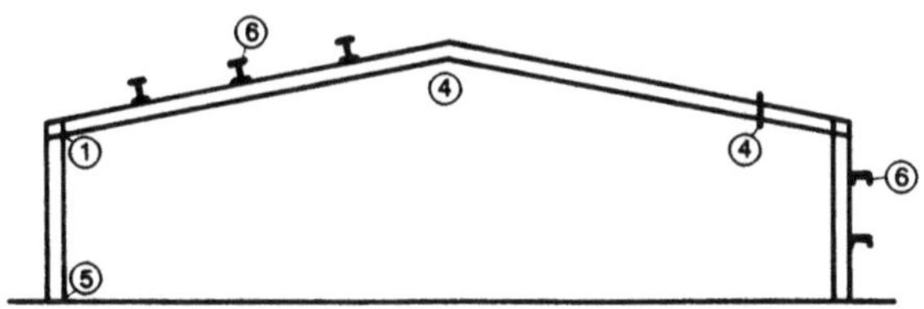

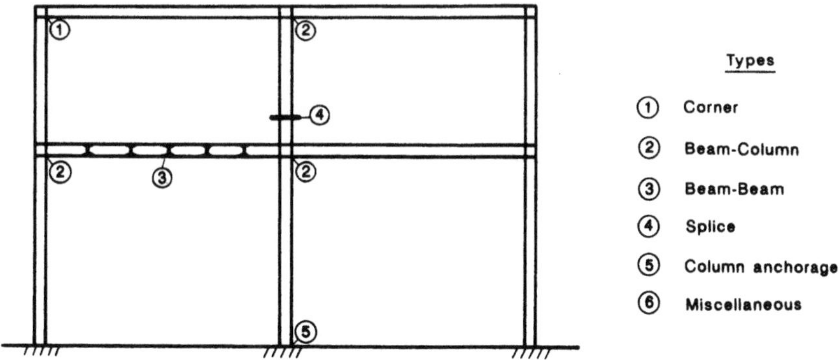

FIGURE 2.23. Types of connections in frames: 1 corner, 2 beam-column, 3 beam-beam, 4 splice, 5 column anchorage, and 6 miscellaneous.

Herein, the primary attention will be focused on the straight corner and interior beam-to-column connections under both balanced and unbalanced conditions. The basic analysis procedure discussed here is also applicable to other types of connections. The analysis will be performed to achieve the minimum thicknesses for the parts of the connections that have to transfer the load from one member to other members. The analysis will use plastic stress distributions that satisfy equilibrium and yield criterion (2.6.3). Since the stress distributions that satisfy only the equilibrium and yield criterion but not the kinematic condition are lower bounds, the procedure will lead to conservative estimates of the thicknesses of plate elements required to transfer a given load.

2.9.1 Requirements for Connections

The principal requirements for connections used in a plastically designed frame are:

1. sufficient strength.
2. adequate rotation capacity.
3. adequate overall stiffness in the elastic range.
4. economical fabrication and ease of erection.

The connection must have sufficient strength so that the full plastic moment M_p of the connecting members can be developed (i.e., the weaker of the two members). In addition to strength, these connections must have the capacity to rotate while sustaining the plastic moment to permit redistribution of moments so that plastic hinges can subsequently form at other critical locations resulting in the formation of a mechanism. The connection must also exhibit overall elastic stiffness under working load to maintain the relative positions of all structural components so that excessive drift of the frame will not occur. Finally, fabrication and erection of connections should be easy and economical because minor material and labor savings in a connection can considerably reduce the cost of steel structures where connections are repeated many times.

Figure 2.24(a) shows the moment-rotation behavior of four connection tests under symmetric loading. Connection A is considered to be properly designed and detailed since it is strong enough to carry the plastic moment of the connecting beam and allow the beam to rotate inelastically through a large angle. Connection B is not acceptable because it does not have enough rotation capacity even though it can carry the plastic moment of the beam. Connection C exhibits a large rotation capacity but does not have enough strength. Connection D is the worst of all since it does not have the required moment and rotation capacities that are imperative in plastically designed frames.

Figure 2.25 shows the failure modes of a fully welded connection that met the criteria for sufficient strength and rotation capacity in plastic design.

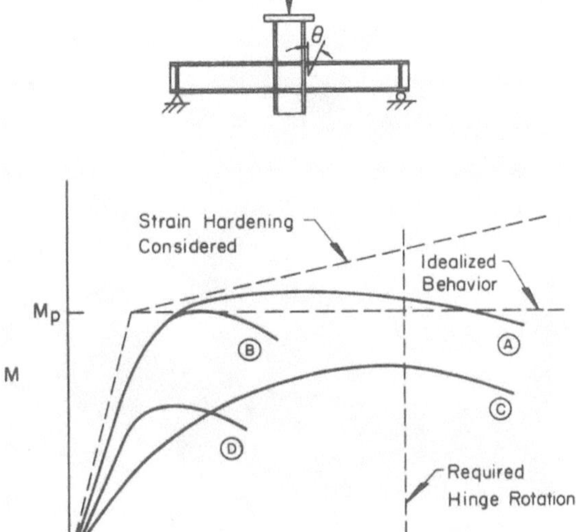

FIGURE 2.24. Requirements of moment-rotation behavior for connections.

Failure of the specimen was due primarily to a combination of excessive buckling of the column web and fracture of the column flange material around the weld at the tension flange. Since no premature weld failure and buckling occurred, the current provisions for welding and connection detailing are adequate [2.5].

In the following sections, these provisions of (1) strength, (2) stiffness, and (3) rotation capacity will be discussed in light of the behavior and design of *corner* and *interior connections*. Obviously, among other factors, extra connecting materials must be kept to a minimum for overall economy. Both *unstiffened* and *stiffened* connections will be considered.

2.9.2 Corner Connections

Without Stiffeners: Consider a typical straight corner connection without stiffeners, as sketched in Fig. 2.26(a). Assume that the horizontal beam continues through the knee. The moment, axial force, and shear acting on the connection are shown in Fig. 2.26(b). The free body diagrams of the parts of the corner are shown in Fig. 2.26(c). The tensile force in the outer flange of the beam is transferred as shear in the web along line AB. In the same manner, the tensile force in the outer flange of the column is transferred through the end plate as shear to the web of the beam along line AD. Note that the tensile stress in the outer flange of the beam and the end plate will vary from σ_y at point B (or D) to zero at the external corner A.

A

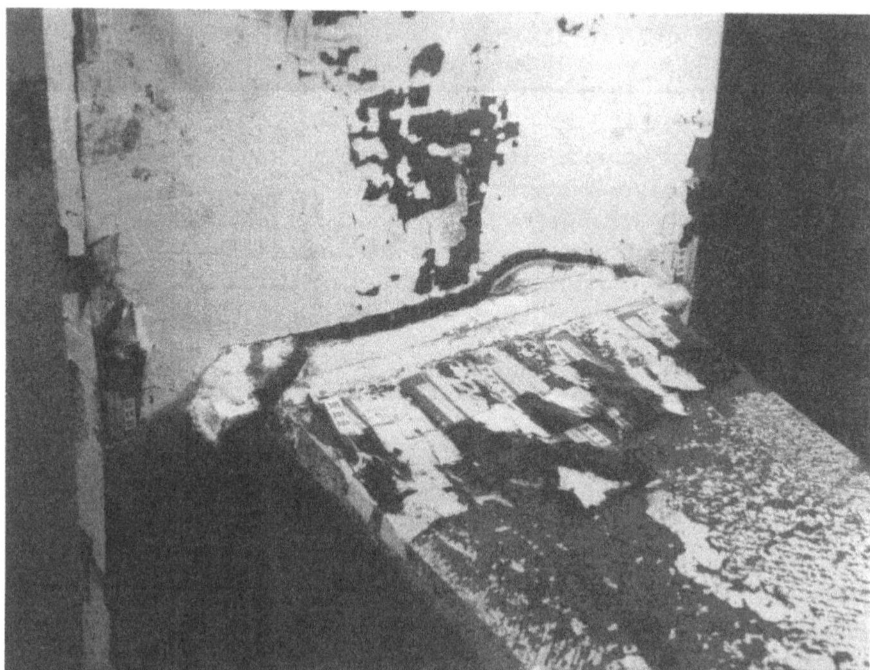

B

C

FIGURE 2.25. Failure modes of an interior beam-to-column connection under symmetrical loadings: **(a)** overall view of a failure mode of a properly designed and detailed moment connection (connection A defined in Fig. 2.24); **(b)** fracture failure mode of the column flange material around the weld at the tension flange; and **(c)** buckling failure mode of stiffeners in the compression zone of the column web.

The inner flange of the beam carries two forces: the shear of the column, and the flange force due to bending and thrust in the beam. These two forces are transferred as shear to the corner along line DC [Fig. 2.26(c)]. Similarly, the two forces on the inner flange of the column are transferred as shear in the web along line BC. Note that the shear forces in the web panel tend to deform it as shown in Fig. 2.26(d).

The minimum required thickness t_w of the web panel ABCD can be obtained by considering the equilibrium of horizontal forces on the portion of the outer flange between A and B and by assuming that (i) the beam flanges carry the moment M_p and the beam web carries the shear; (ii) the axial force in the beam is negligible; and (iii) the distribution of shear stress in the panel web along line AB is uniform [Fig. 2.26(c)]. Thus, the flange force T can be expressed in terms of M_p and yield stress τ_{yb} as

$$T = \frac{M_p}{d_b} = \tau_{yb} t_w d_{col} \tag{2.9.1}$$

in which $d_b = d_{bm} - t_{fb}$ is the lever arm corresponding to the center-to-center beam flange forces, d_{col} is the depth of the column and t_w is the thickness of the panel web. Equation (2.9.1) can be solved for t_w to obtain the required thickness of the panel web

$$t_w = \frac{M_p}{\tau_{yb} d_b d_{col}}. \tag{2.9.2}$$

The shear stress at yield in Eq. (2.9.2) can be taken from the von Mises yield criterion as $\tau_{yb} = \sigma_{yb}/\sqrt{3} = 0.577\sigma_{yb}$. However, LRFD uses $\tau_y = 0.6\sigma_y$. Introducing resistance factor ϕ_v, we have

$$t_w = \frac{M_p}{\phi_v \tau_{yb} d_b d_{col}} \tag{2.9.3}$$

where ϕ_v has the value 0.9.

With Stiffeners: When the thickness of the panel web is less than the required thickness t_w, the web can be reinforced either by using a doubler plate or by a symmetrical pair of diagonal stiffeners whose cross-sectional area is sufficient to transmit that portion of the shear in excess of the web capacity. Thus, when a doubler plate is used as reinforcement, its thickness should be such that the total thickness of the web is equal to or more than the minimum required as given by Eq. (2.9.3). Welds should be arranged at the edges of doubler plates so as to transmit shear stress directly to the boundary of plate stiffeners and flanges. When diagonal stiffeners are added as reinforcement, the required strength of the pair of diagonal stiffeners [Fig. 2.26(e)] can be obtained by considering equilibrium of horizontal forces.

FIGURE 2.26. Analysis and
design of straight corner
connections.

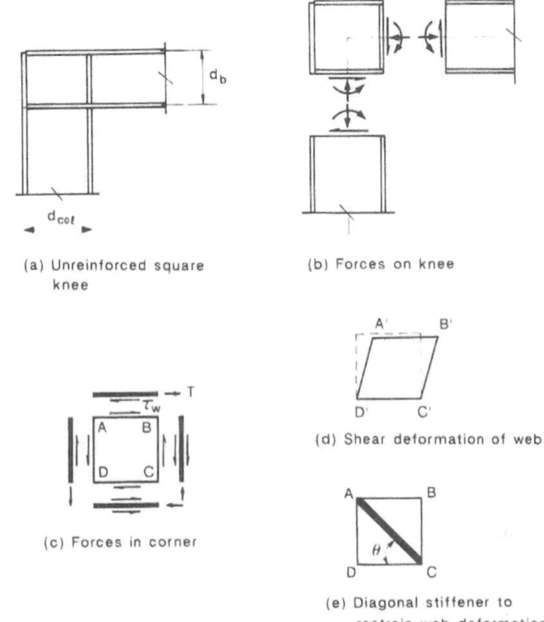

$$T = \frac{M_p}{d_b} = \phi_v \tau_{yb} t_{wb} d_{col} + T_s \cos\theta \qquad (2.9.4)$$

in which T is the flange force, t_{wb} is the thickness of beam web, T_s is the required strength of the area of a symmetrical pair of stiffeners and θ is the angle of diagonal stiffener with the horizontal, i.e., $\tan\theta = (d_{bm} - 2t_{fb})/(d_{col} - 2t_{fc})$ as shown in Fig. 2.26(e). From Eq. (2.9.4), T_s can be determined as

$$T_s = \frac{1}{\cos\theta}\left[\frac{M_p}{d_b} - \frac{\phi_v t_{wb} d_{col}\sigma_y}{\sqrt{3}}\right]. \qquad (2.9.5)$$

Stiffeners should be welded for their full area across their ends and continuously fillet welded to column web. To avoid overall buckling, the design of stiffeners must also satisfy the following LRFD requirements (Section K1.9).

The stiffeners shall be designed as axially compressed members (columns) in accordance with the requirements of Sec. E2 with an effective length equal to 0.75 l_s, a cross section composed of two stiffeners and a strip of the web having a width of $25t_w$ at interior stiffeners and $12t_w$ at the ends of members, where l_s is the length of stiffeners.

Design of compressive members and background of column strength equations, in Section E2 of LRFD, are presented in details in Section 4.6.2. Herein, we will only apply these equations to check the overall buckling of the web stiffeners designed in Examples (2.9.1) and (2.9.2).

To avoid local buckling of the stiffeners, the stiffeners of A36 steel must satisfy the criteria (Table 2.1)

$$\frac{b_s}{t_s} \leq \frac{65}{\sqrt{F_y}} \tag{2.9.6}$$

in which b_s and t_s are, respectively, the width and thickness of each stiffener, and F_y is the yield stress of the stiffener in ksi.

2.9.3 Balanced Interior Beam-to-Column Connections

Without Stiffeners: A sketch of a typical interior beam-to-column connection is shown in Fig. 2.27(a). The forces introduced in the beam flanges by moment are transferred to the column flanges and then to the column web [Fig. 2.27(b)]. If the thickness of the column web is insufficient, then it can fail either by yielding and/or buckling due to the beam compression flange force or fracture due to the beam tension flange force. If the thickness of the column flange is insufficient, the tensile force in the beam flange tends to pull the outstanding column flanges outward, resulting in a possible initiation of fracture at the junction of column flanges and web, as shown in Fig. 2.28.

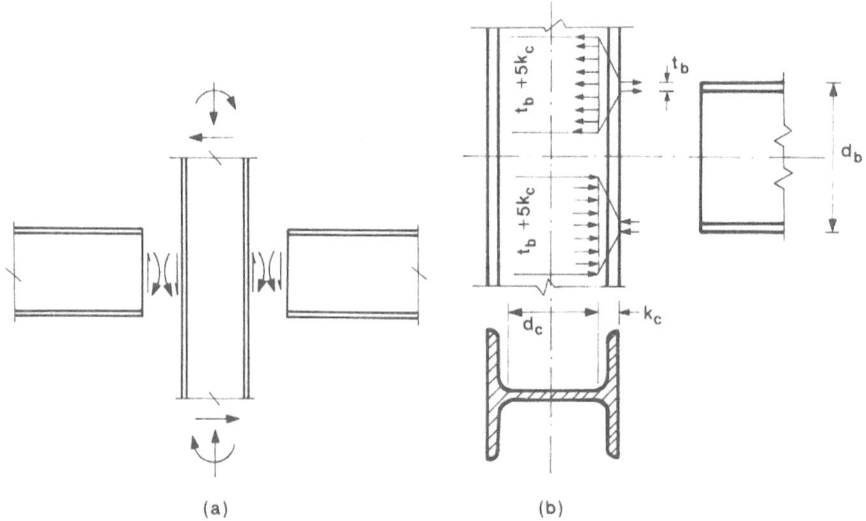

FIGURE 2.27. Yielding in an interior beam-to-column connection.

Column Web Yielding

The stresses in the column web resulting from the pair of concentrated beam flange forces spread out as they penetrate deeper into the column web. Tests of these connections show that stresses in the column web can be estimated by assuming a slope of 2.5 : 1 from the point of contact to the column k-line [Fig. 2.27(b)]. The minimum thickness of the column web required to prevent yielding under the applied concentrated force of tensile or compressive flange of the beam can therefore be determined simply by

$$R = \sigma_{yc} t_{wc}(t_{fb} + 5k_c) \geq T_{fb} \tag{2.9.7}$$

in which σ_{yc} is the yield stress of the column, T_{fb} is the beam flange force, t_{wc} is the thickness of column web, t_{fb} is the thickness of beam flange, and k_c is the column k-line as shown in Fig. 2.27(b). Note that Eq. (2.9.7) is the same as Eq. (K1-2) of LRFD and ϕ corresponding to this case is 1.

Column Web Buckling

To avoid possible buckling of the column web, the buckling strength of the web as given by Eq. K1.8 of LRFD must also be checked against the concentrated force from the beam flange as follows

$$R = \frac{4100\phi t_{wc}^3 \sqrt{\sigma_{yc}}}{d_c} \geq T_{fb} \tag{2.9.8}$$

in which ϕ is the resistance factor and has the value 0.9, d_c is clear depth of the column and is shown in Fig. 2.27(b), and σ_{yc} is in ksi.

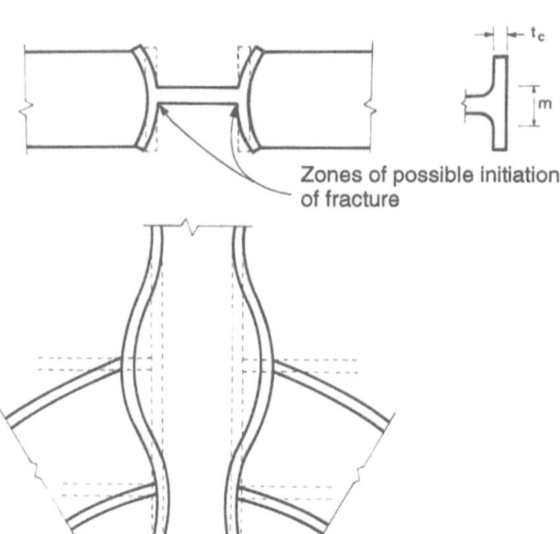

FIGURE 2.28. Column flange bending in tension region of interior beam-to-column connection.

Column Flange Bending

The bending of column flanges is a much more complex problem because the flanges bend in two directions, both longitudinal and transverse to the axis of the column, as sketched in Fig. 2.28. In order to avoid the excessive bending due to the forces from the beam flange, LRFD Eq. K1-1 requires that the thickness of the flange of the column must satisfy the following equation

$$\phi R = \phi 6.25 t_{fc}^2 \sigma_{yc} \geq T_{fb} \tag{2.9.9}$$

in which ϕ has the value 0.9, and t_{fc} is in in. Equation (2.9.9) is obtained by using the *yield line theory* for plates and some simplifications about the relative dimensions of beams and columns.

With Stiffeners: If the thickness of the column web is insufficient to avoid yielding and/or buckling, the web can be reinforced by a pair of stiffeners. Then the beam flange force is resisted jointly by the column web and stiffeners. The strength T_s of the stiffeners required to prevent column web yielding or buckling can be determined from the simple equilibrium condition

$$T_s = T_{fb} - P_{max} \tag{2.9.10}$$

in which P_{max} is the capacity of the web without stiffeners, and it is taken as the smaller of:

the yield strength of the column web [LRFD Eq. (K1-2)]

$$P_{max} = \sigma_{yc} t_{wc}(t_{fb} + 5k_c) \tag{2.9.11}$$

or the buckling strength of the column web [LRFD Eq. (K1-8)]

$$P_{max} = \frac{4100\phi t_{wc}^3 \sqrt{\sigma_{yc}}}{d_c} \tag{2.9.12}$$

in which ϕ is the resistance factor and has the value 0.9, σ_{yc} is in ksi, t_{wc} is in inches and $d_c = d_{col} - 2k_c$ is the clear depth of the column and is also in inches.

If the thickness of the column flanges is less than that from Eq. (2.9.9), stiffeners must be used to provide support to column flanges. The dimensions of all the stiffeners should satisfy local and overall buckling checks. Stiffeners should be welded for full strength across their ends, in contact with the inner face of flanges to which the supported members are framed, and the welding connecting them to the web should be strong enough to transmit the net stress applied to the web by the supporting members. The types of stiffeners that are commonly used to reinforce the column web and to prevent the bending of column flanges are shown in Fig. 2.29.

2.9.4 Unbalanced Interior Beam-to-Column Connections

When the moments in the two beams at an interior connection differ by a large amount, they may cause large shears in the column web. As a result, a

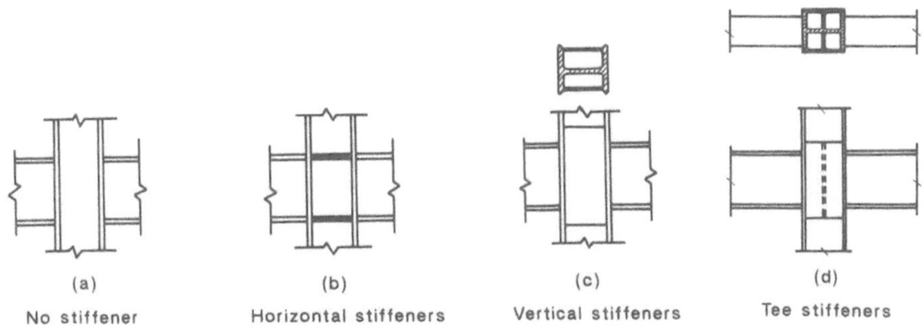

FIGURE 2.29. Stiffeners for interior beam-to-column connections.

column web may deform in the same manner as in a corner connection. Figure 2.30(a) shows shears and moments acting on a typical unbalanced interior beam-to-column connection and Fig. 2.30(b) shows a free body diagram of the forces acting on the top flange stiffener AB. The forces are V, the horizontal shear present in the column above the connection and two tensile flange forces, T_1 and T_2. The net result of these forces must be resisted by a shear stress τ acting on an area of column web equal to $t_{wc}d_{col}$. Thus

$$\phi_v \tau t_{wc} d_{col} = \frac{M_2}{d_{b2}} - \frac{M_1}{d_{b1}} - V. \tag{2.9.13}$$

Using τ equal to the shear yield strength τ_{yc}, the required web thickness to resist shear can be obtained as

$$t_{wc} = \frac{1}{\phi_v \tau_{yc} d_{col}} \left[\frac{M_2}{d_{b2}} - \frac{M_1}{d_{b1}} - V \right]. \tag{2.9.14}$$

If the actual web thickness of the column is less than that given by Eq. (2.9.14), diagonal stiffners or web doubler plates, similar to those discussed in Section 2.9.2, should be provided to carry the excess shearing force. The connection should also be checked for all the failure modes discussed in Section 2.9.3.

Example 2.9.1. Design the corner connection shown in Fig. 2.31. Assume that the yield strength of both beam and column is 36 ksi.

Solution: For W14 × 34 beam, the following properties are noted from the AISC manual

$$Z = 54.6 \text{ in}^3, \; t_f = 0.455 \text{ in.}, \; t_w = 0.285 \text{ in.}, \; k_c = 1 \text{ in.}$$

$$d_b = d - t_f = 13.98 - 0.455 = 13.525 \text{ in.}$$

in which the more accurate center-to-center distance between the two beam flanges is used for the calculation of the flange force. The plastic moment capacity of beam is

$$M_p = \sigma_{yb} Z = 36 \times 54.6 = 1965.6 \text{ kip-in.}$$

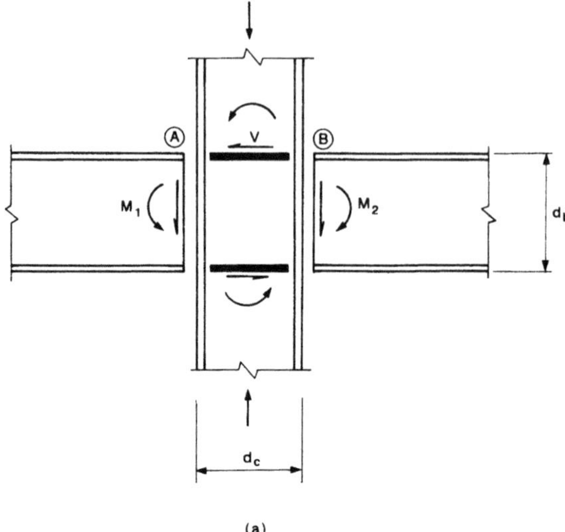

FIGURE 2.30. Forces acted on an unbalanced interior beam-to-column connection.

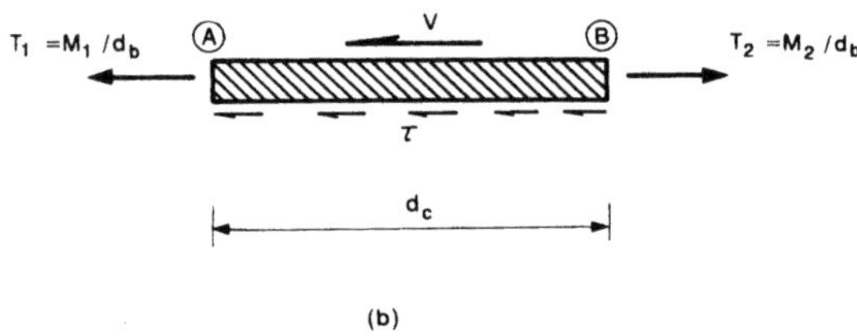

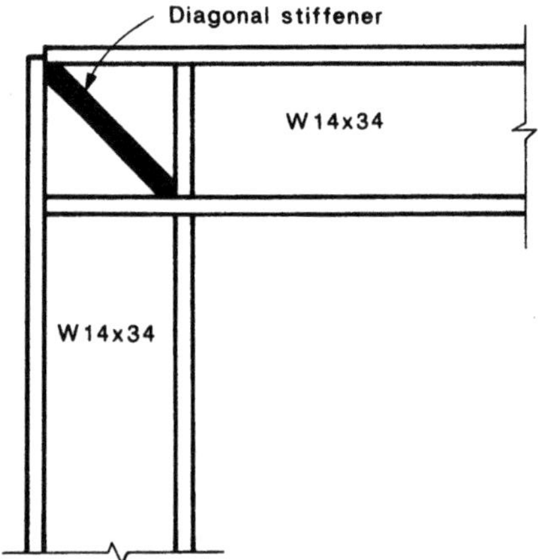

FIGURE 2.31. Design of a pair of diagonal stiffener in a straight corner connection.

The force in each flange of the beam is

$$T = \frac{M_p}{d_b} = \frac{1{,}965.6}{13.525} = 145.33 \text{ kips.}$$

The shear resistance provided by the column web has the value

$$T_w = \phi_v \tau_{yc} t_{wc} d_{col} = 0.9(0.6 \times 36)(0.285)(13.98) = 77.45 \text{ kips.}$$

The required strength of stiffeners is

$$T_s \cos \pi/4 = T - T_w = 145.33 - 77.45 = 67.88 \text{ kips}$$

or

$$T_s = 96 \text{ kips.}$$

Using a pair of A36 stiffeners, the required area of each stiffener is

$$A_s = \frac{T_s}{2\phi_c \sigma_y} = \frac{96}{2 \times 0.85 \times 36} = 1.57 \text{ in.}^2$$

Try a 1/2-inch-thick plate. The required width of stiffener is

$$b_s = \frac{1.57}{1/2} = 3.14 \text{ in.}$$

Try a 3 1/4-inch-wide plate.

Check Buckling of the Stiffeners in the Plane of the Web

Assume that the stiffeners are welded at its ends only. The compressive strength of each of the stiffeners against their buckling in the plane of the web can be determined by using the LRFD column strength equation.

The length of the stiffeners is

$$l_s = \frac{d - 2t_f}{\cos \theta} = (13.98 - 2 \times 0.455)\sqrt{2} = 18.48 \text{ in.}$$

Area, moment of inertia, radius of gyration, effective slenderness ratio, and λ_c of each of the stiffeners are calculated as

$$A_s = b_s t_s = 3.25 \times 0.5 = 1.625 \text{ in.}^2$$

$$I_s = \frac{1}{12}(3.25)(0.5)^3 = 0.0339 \text{ in.}^4$$

$$r_s = \sqrt{\frac{I_s}{A_s}} = \sqrt{\frac{0.0339}{1.625}} = 0.144 \text{ in.}$$

$$\left(\frac{KL}{r}\right) = \frac{0.75 \times 18.48}{0.144} = 96.25$$

$$\lambda_c = \frac{1}{\pi}\frac{KL}{r}\sqrt{\frac{F_y}{E}} = \frac{1}{\pi}(96.25)\sqrt{\frac{36}{29{,}000}} = 1.079.$$

Note that the effective length factor K for the stiffener is taken as 0.75 as per Section K1.9 of LRFD. Since $\lambda_c < 1.5$, the in-plane buckling strength of each stiffener is

$$\phi_c P_n = \phi_c A_s 0.658^{\lambda_c^2} F_y$$

or

$$\phi_c P_n = 0.85 \times 1.625 \times 0.658^{(1.079)^2} \times 36 = 30.55 \text{ kips}$$

$$(\phi_c P_n = 30.55 \text{ kips}) < \left(\frac{T_s}{2} = 48 \text{ kips}\right), \quad \text{not okay.}$$

The strength of stiffeners can be increased either by increasing the size of the stiffeners or by welding the stiffeners to the web panel throughout their length. Here, we assume the latter so that the stiffeners shall not buckle in the plane of web.

Check Out-of-Plane Buckling of the Stiffeners

To check out-of-plane buckling, we shall evaluate the compressive strength of a combined stiffener and web assembly as per Section K1.8 of LRFD. The area of the assembly is

$$A_s = 2 \times b_s \times t_s = 2 \times 3.25 \times \frac{1}{2} = 3.25 \text{ in}^2.$$

Note that the contribution of the web to the assembly area for this corner connection is neglected. Now, moment of inertia, radius of gyration, effective slenderness ratio, and λ_c of the assembly of the two stiffeners are calculated as

$$I_s = \frac{\frac{1}{2}(2 \times 3.25 + 0.285)^3}{12} = 13.01 \text{ in.}^4$$

$$r_s = \sqrt{\frac{I_s}{A_s}} = \sqrt{\frac{13.01}{3.25}} = 2 \text{ in.}$$

$$\left(\frac{KL}{r}\right)_s = 0.75 \frac{18.48}{2} = 6.93$$

$$\lambda_c = \frac{1}{\pi}\left(\frac{KL}{r}\right)\sqrt{\frac{F_y}{E}} = \frac{6.93}{\pi}\sqrt{\frac{36}{29{,}000}} = 0.078.$$

Since $\lambda_c < 1.5$, the compressive strength of the assembly is checked by using the LRFD column strength equation

$$\phi_c P_n = \phi_c A_s 0.658^{\lambda_c^2} F_y$$

or

$$\phi_c P_n = 0.85 \times 3.25 \times 0.658^{(0.078)^2} \times 36 = 99.2 \text{ kips}$$

$$(\phi_c P_n = 99.2 \text{ kips}) > (T_s = 96 \text{ kips}) \quad \text{okay.}$$

FIGURE 2.32. Design of a pair of horizontal stiffeners in an interior beam-to-column connection.

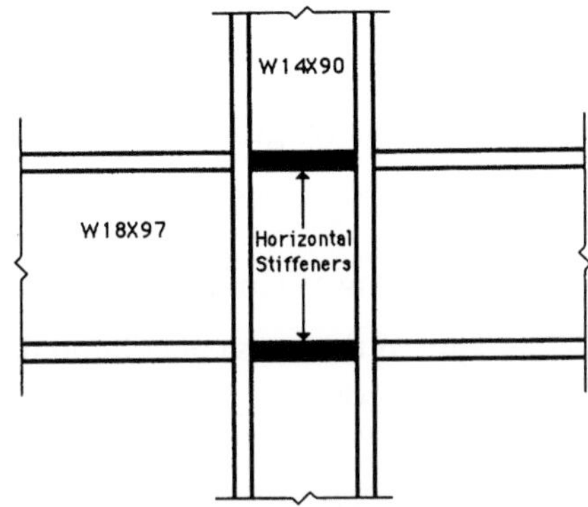

Check Local Buckling of the Stiffeners

$$\frac{b_s}{t_s} = \frac{3.25}{1/2} = 6.5 < \frac{65}{\sqrt{F_y}} = 10.8 \quad \text{okay.}$$

Use a pair of A36-3 1/4 × 1/2″-inch diagonal stiffeners.

Example 2.9.2. Design the interior beam-to-column connection shown in Fig. 2.32. Assume that both beam and column are made of A36 steel.

Solution: From the AISC manual, the following properties of W18 × 97 beam and W14 × 90 column are noted.

Beam	Column
W18 × 97	W14 × 90
$t_{fb} = 0.87$ in.	$t_{wc} = 0.44$ in.
$b_f = 11.145$ in.	$k_c = 11/8$ in.
$d_b = 18.59$ in.	$d_c = 45/4$ in.
$Z_{x,b} = 211$ in.^{3}	$t_{fc} = 0.71$ in.
	$d_{col} = 14.02$ in.

Check Yielding and Buckling of Column Web

The maximum beam flange force is

$$T_{fb} = \frac{M_{p,b}}{d_b - t_{fb}} = \frac{211 \times 36}{18.59 - 0.87} = 428.7 \text{ kips.}$$

The resistance of column web against yielding has the value

$$T_w = \sigma_{yc} t_{wc}(t_{fb} + 5k_c)$$

$$= 36 \times 0.44\left(0.87 + 5 \times \frac{11}{8}\right) = 122.7 \text{ kips} < 428.7 \quad \text{not okay.}$$

The resistance of column web against buckling, Eq. (2.9.12)

$$T_w = \frac{4100 \times 0.9 \times (0.44)^3 \sqrt{36}}{14.02 - 2(11/8)} = 167.6 \text{ kips.}$$

The required strength of stiffeners is

$$T_s = 428.7 - 122.7 = 306 \text{ kips.}$$

Check Bending of Column Flanges

The required thickness of column flange is determined from Eq. (2.9.9) in the rearranged form as

$$t_{fc} = 0.4\sqrt{\frac{T_{fb}}{0.9\sigma_{yc}}} = 0.4\sqrt{\frac{428.7}{0.9 \times 36}} = 1.46 \text{ in.} > 0.71 \text{ in.}   \text{not okay.}$$

Select Dimensions of Stiffeners

Stiffeners are needed to reinforce the column web against yielding and to strengthen the column flanges against excessive distortion. Providing a pair of horizontal stiffeners both in compression and tensile zones, and using A36 steel, the required area of each stiffener is

$$A_s = \frac{T_s}{2\phi_c\sigma_{ys}} = \frac{306}{2 \times 0.85 \times 36} = 5 \text{ in.}^2$$

Try 5- × 1-inch plate.

Check Buckling of Stiffeners in the Plane of Column Web

Assume that the stiffeners are not welded to the column web. Then the compressive strength of the stiffeners against their buckling in the plane of the column web is determined from the LRFD column strength equation as follows:

$$\text{length of stiffener } l_s = d - 2t_{fc} = 14.02 - 2 \times 0.71 = 12.6 \text{ in.}$$

Area, moment of inertia, radius of gyration, effective slenderness ratio, and λ_c of each of the stiffeners are calculated as

$$A_s = b_s t_s = 5 \times 1 = 5 \text{ in.}^2$$

$$I_s = \frac{1}{12}(5)(1)^3 = 0.417 \text{ in.}^4$$

$$r_s = \sqrt{\frac{0.417}{5}} = 0.289 \text{ in.}$$

$$\left(\frac{KL}{r}\right) = \frac{0.75 \times 12.6}{0.289} = 32.7$$

$$\lambda_c = \frac{1}{\pi}\left(\frac{KL}{r}\right)\sqrt{\frac{F_y}{E}} = \frac{1}{\pi}(32.7)\sqrt{\frac{36}{29,000}} = 0.367.$$

Note that $K = 0.75$ is taken as recommended in Section K1.8 of LRFD. Since $\lambda_c < 1.5$, the buckling strength of each stiffener is

$$\phi_c P_n = \phi_c A_s 0.658^{\lambda_c^2} F_y$$

or

$$\phi_c P_n = 0.85 \times 5 \times 0.658^{(0.367)^2} \times 36 = 144.6 \text{ kips}$$

$$(\phi_c P_n = 144.6 \text{ kips}) < \left(\frac{T_s}{2} = 153 \text{ kips}\right) \quad \text{not okay.}$$

The strength of stiffeners may be increased by either welding the stiffeners to the column web or increasing the size of the stiffeners. Here we increase the stiffener size to 5 1/2- × 1-inch. As λ_c remains the same, the increased strength of the stiffeners becomes

$$\phi_c P_n = 0.85 \times 5.5 \times 0.658^{(0.367)^2} \times 36 = 159.1 \text{ kips}$$

$$\phi_c P_n > \frac{T_s}{2} = 153 \text{ kips}, \quad \text{okay.}$$

Provide a pair of A36-5 1/2- × 1-inch stiffeners in both compression and tension zones.

2.10 Examples

Herein, we present examples of calculating the full plastic moment of a given section with or without the presence of axial load and shear force. The section also includes the examples dealing with plastic analysis and section design of given determinate beams. More examples can be found in the book by Baker and Heyman [2.6], among others.

Example 2.10.1. Calculate the plastic section modulus Z, the elastic section modulus S, and the shape factor f for the following sections with dimensions shown in Fig. 2.33.

(a) an I-section bending about strong axis [Fig. 2.33(a)].
(b) an I-section bending about weak axis [Fig. 2.33(a)].
(c) an I-section with a cover plate [Fig. 2.33(b)].
(d) a T-section [Fig. 2.33(c)].
(e) a square shaft with keyway [Fig. 2.33(d)].
(f) a thin-walled tubular section [Fig. 2.33(e)].
(g) an isosceles triangle [Fig. 2.33(f)].

Solution: The plastic section modulus Z is determined by calculating the first moment of area about the plastic neutral axis while the moment of inertia I

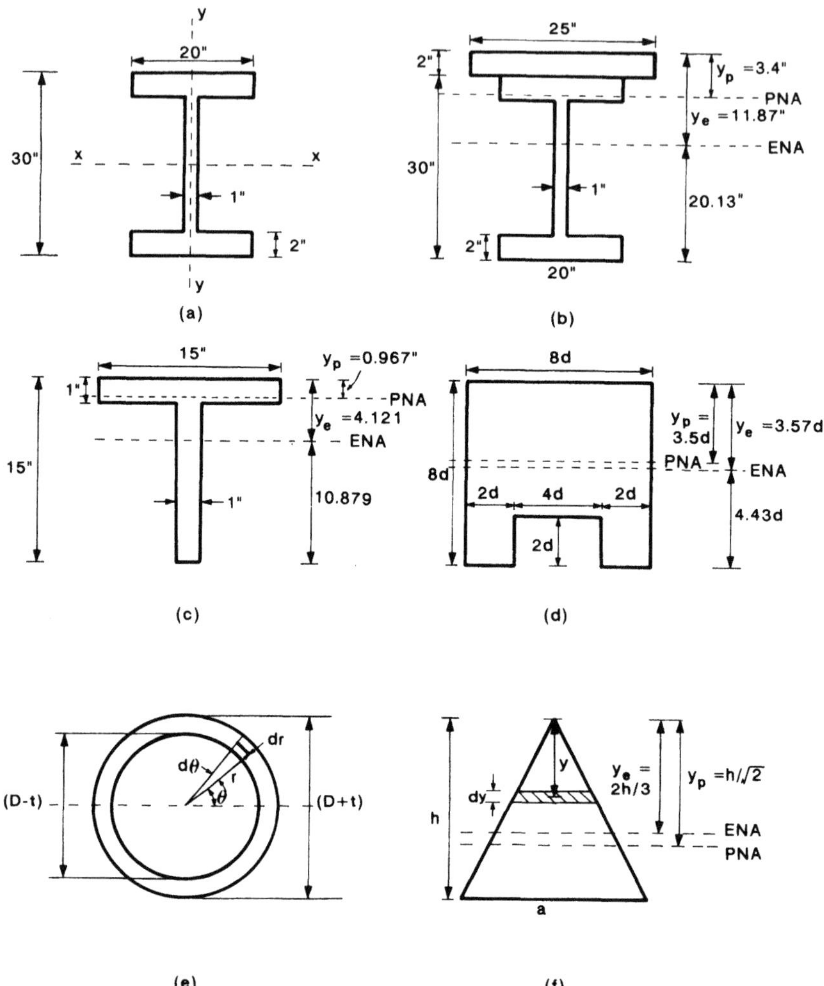

FIGURE 2.33. Computation of plastic and elastic moduli and shape factors of various sections.

is determined by calculating the second moment of area about the elastic neutral axis.

a) I-Section Bending About Strong Axis: Both elastic and plastic neutral axes pass through the centroid of the section as shown by the dashed line x-x in Fig. 2.33(a). Thus, we have

$$Z = 2[(1 \times 13) \times 6.5] + 2[(20 \times 2) \times 14] = 1289 \text{ in.}^3$$

$$I = \frac{1}{12}[1 \times (26)^3] + 2\left[\left(\frac{1}{12}\right)(20)(2)^3\right] + 2[(20)(2)(14)^2] = 17{,}171 \text{ in.}^4$$

$$S = \frac{I}{c} = \frac{17,171}{15} = 1145 \text{ in.}^3$$

$$f = \frac{Z}{S} = \frac{1289}{1145} = 1.126.$$

b) I-Section Bending About Weak Axis: Both elastic and plastic neutral axes are the same as shown by the dashed line y-y in Fig. 2.33(a). Thus, we have

$$Z = 2(0.5 \times 26) \times 0.25 + 4[(2 \times 10) \times 5] = 406.5 \text{ in.}^3$$

$$I = \frac{1}{12}[26 \times (1)^3] + 2\left[\left(\frac{1}{12}\right)(2)(20)^3\right] = 2669 \text{ in.}^4$$

$$S = \frac{I}{c} = \frac{2669}{10} = 267 \text{ in.}^3$$

$$f = \frac{Z}{S} = \frac{406.5}{267} = 1.522.$$

c) I-Section with a Cover Plate: Assume that the cover plate material is of the same yield stress as the original section.

Plastic Modulus: Area of the built-up section [Fig. 2.33(b)] is

$$A = 2 \times (20 \times 2) + 26 \times 1 + 25 \times 2 = 156 \text{ in.}^2$$

The plastic neutral axis will divide the area of cross section into two equal halves. Since the area of cover plate is less than $A/2$ which is less than the area of cover plate and top flange, the PNA is in the top flange and the distance y_p of the PNA from the top of the section can be obtained by

$$20(y_p - 2) + 50 = \frac{A}{2} = \frac{156}{2} = 78$$

$$y_p = 3.4 \text{ in.}$$

By calculating the first moment of area about the neutral axis, Z is obtained as

$$Z = (25 \times 2)(2.4) + (20 \times 1.4)(0.7)$$
$$+ (20 \times 2)(27.6) + (26 \times 1)(13.6) + (20 \times 0.6)(0.3) = 1601 \text{ in.}^3$$

Elastic Modulus and Shape Factor: The distance y_e of the elastic neutral axis (ENA) from the top of the section can be obtained by

$$y_e = \frac{(25 \times 2) \times 1 + (20 \times 2) \times 3 + (26 \times 1) \times 17 + (20 \times 2) \times 31}{156}$$

$$= 11.87 \text{ in.}$$

Thus, we have

$$I = \frac{1}{12} \times 25 \times (2)^3 + (25 \times 2)(10.87)^2 + \frac{1}{12} \times 20 \times (2)^3 + (20 \times 2)(8.87)^2$$

$$+ \frac{1}{12} \times 1 \times (26)^3 + (26 \times 1)(5.13)^2 + \frac{1}{12} \times 20 \times (2)^3 + (20 \times 2)(19.3)^2$$

or

$$I = 25{,}885 \text{ in.}^4$$

$$S = \frac{I}{c} = \frac{25{,}885}{20.13} = 1286 \text{ in.}^3$$

$$f = \frac{Z}{S} = \frac{1601}{1286} = 1.245.$$

d) T-Section:

Plastic Modulus: The area of the T-section shown in Fig. 2.33(c) is

$$A = 15 \times 1 + 14 \times 1 = 29 \text{ in.}^2$$

Since $A/2$ is less than the area of the flange, the PNA is in the flange and the distance y_p of the PNA from the top of the section can be obtained by

$$15y_p = \frac{A}{2} = \frac{29}{2} = 14.5,$$

which gives

$$y_p = 0.967 \text{ in.}$$

Thus, we obtain

$$Z = (15 \times 0.967)\left(\frac{0.967}{2}\right) + (15 \times 0.033)\left(\frac{0.033}{2}\right) + (14 \times 1) \times 7.033$$

$$= 105.5 \text{ in.}^3$$

Elastic Modulus and Shape Factor: The distance y_e of the ENA from the top of the section can be obtained as

$$y_e = \frac{(15 \times 1) \times 0.5 + (14 \times 1) \times 8}{29} = 4.121 \text{ in.}$$

Thus, we obtain

$$I = \frac{1}{12} \times 15 \times (1)^3 + (15 \times 1) \times (3.621)^2 + \frac{1}{12} \times 1 \times (14)^3$$

$$+ (14 \times 1) \times (3.879)^2 = 637 \text{ in.}^4$$

$$S = \frac{I}{c} = \frac{637}{10.879} = 58.6 \text{ in.}^3$$

$$f = \frac{105.5}{58.6} = 1.8.$$

e) Square Shaft with Keyway:

Plastic Modulus: The area of cross section of the shaft is

$$A = 8d \times 8d - 2d \times 4d = 56d^2.$$

The distance y_p of the PNA from the top of the section can be computed from

$$8d \times y_p = \frac{A}{2} = 28d^2$$

$$y_p = 3.5d.$$

Thus, we obtain

$$Z = (8d \times 3.5d)(1.75d) + (8d \times 4.5d)(2.25d) - (2d \times 4d) \times 3.5d = 102d^3.$$

Elastic Modulus and Shape Factor: The distance y_e of the elastic neutral axis from the top of the section can be obtained as

$$y_e = \frac{(8d \times 8d)(4d) - (4d \times 2d)(7d)}{56d^2} = 3.57d.$$

Thus, we have

$$I = \frac{1}{12}(8d)(8d)^3 + (8d \times 8d)(4d - 3.57d)^2$$

$$- \frac{1}{12}(4d)(2d)^3 - (4d \times 2d)(7d - 3.57d)^2 = 256d^4$$

$$S = \frac{I}{c} = \frac{256d^4}{8d - 3.57d} = 57.8d^3$$

$$f = \frac{Z}{S} = \frac{102d^3}{57.8d^3} = 1.765.$$

f) Thin-Walled Tubular Section: Both elastic and plastic neutral axes pass through the centroid of the section. The first moment of area about this axis is

$$Z = 4 \int_0^{\pi/2} \int_{(D-t)/2}^{(D+t)/2} (r\,dr\,d\theta) r \sin\theta$$

or

$$Z = \frac{4}{3}\left[\left(\frac{D+t}{2}\right)^3 - \left(\frac{D-t}{2}\right)^3\right] \int_0^{\pi/2} \sin\theta\,d\theta$$

or

$$Z = tD^2 + \frac{1}{3}t^3 \approx D^2 t.$$

The second moment of area about the neutral axis is

$$I = 4 \int_0^{\pi/2} \int_{(D-t)/2}^{(D+t)/2} (r\,dr\,d\theta)(r\sin\theta)^2$$

or

$$I = \left[\left(\frac{D+t}{2}\right)^4 - \left(\frac{D-t}{2}\right)^4 \right] \int_0^{\pi/2} \sin^2 d\theta$$

or

$$I = \frac{\pi}{64}[8D^3 t + 8Dt^3] \approx \frac{\pi}{8}D^3 t.$$

Thus, we have

$$S = \frac{I}{c} = \frac{\dfrac{\pi}{8}D^3 t}{\dfrac{D+t}{2}} \approx \frac{\pi}{4}D^2 t$$

$$f = \frac{Z}{S} = \frac{D^2 t}{\dfrac{\pi}{4}D^2 t} = 1.273.$$

g) Isosceles Triangle:

Plastic Modulus:

$$\text{area of isosceles triangle } A = \frac{1}{2}ah$$

The distance y_p of the PNA from the top of the section is determined from

$$\frac{1}{2}\left(a\frac{y_p}{h}\right)y_p = \frac{A}{2} = \frac{ah}{4},$$

which gives

$$y_p = \frac{h}{\sqrt{2}}.$$

Thus, the plastic modulus Z can be expressed as [Fig. 2.33(f)]

$$Z = \int_0^{y_p} \left(\frac{ay}{h}dy\right)(y_p - y) + \int_{y_p}^{h} \left(\frac{ay}{h}dy\right)(y - y_p)$$

$$= \frac{a}{h}\left[y_p\frac{y_p^2}{2} - \frac{y_p^3}{3} \right] + \frac{a}{h}\left[\frac{1}{3}(h^3 - y_p^3) - \frac{1}{2}y_p(h^2 - y_p^2) \right].$$

By substituting $y_p = h/\sqrt{2}$, Z can be computed as

$$Z = 0.0976ah^2.$$

Elastic Modulus and Shape Factor: The elastic neutral axis is at a distance of $y_e = (2/3)h$ from the top of the section. The second moment of inertia about this axis can be expressed as

$$I = \int_0^h \left(a\frac{y}{h}dy\right)\left(\frac{2h}{3} - y\right)^2$$

or

$$I = \frac{a}{h}\int_0^h \left(y^3 - \frac{4}{3}hy^2 + \frac{4}{9}h^2y\right)dy = \frac{ah^3}{36}.$$

Thus, we have

$$S = \frac{I}{c} = \frac{ah^3/36}{(2h/3)} = \frac{ah^2}{24}$$

$$f = \frac{Z}{S} = \frac{0.0976ah^2}{ah^2/24} = 2.342.$$

Note that I is obtained by using one integral for the entire section with limits from 0 to h, while two separate integrals—one each for the portions above and below the neutral axis—are used for obtaining Z. This is because the integral for I, second moment of area, is a function of square of the distance of the area from the neutral axis and therefore the sign of this distance/lever arm ($[2h/3 - y]$ for the portion above the neutral axis and $[y - 2h/3]$ for the portion below the neutral axis) does not affect the results. While the integral for the plastic modulus Z, first moment of area, is a function of only the lever arm and not its square, the sign of the lever arm is important and if Z is obtained by using one integral with limits from 0 to h as for I, the value of Z will come out to be $0.0202\ ah^2$, which is wrong.

Example 2.10.2. The dimensions of a T-section are as shown in Fig. 2.34. The yield stress of the material in compression is $1.5\ \sigma_y$ and the tensile yield stress

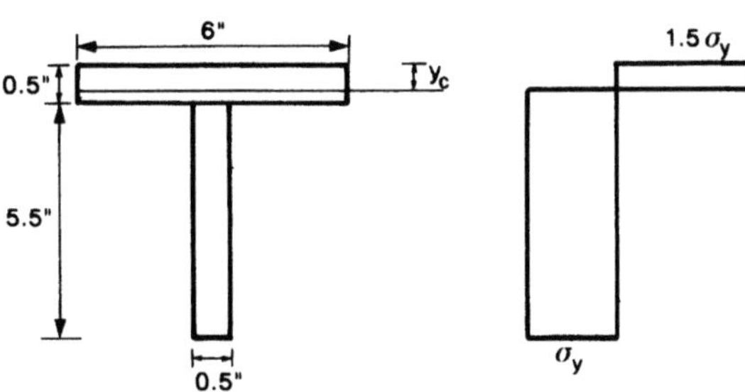

FIGURE 2.34. Calculations of plastic moment of a T-section with different yield stress in tension and compression.

is σ_y. For bending about the strong axis, the elastic modulus is given as 4.61 in.^{3}. Find (a) the full plastic moment when the tip of the web is in tension and (b) the corresponding value of the shape factor.

Solution: (a) When the section is made of the same material and the yield stress in tension and compression is the same, the plastic neutral axis divides the section into two equal parts. Otherwise, the PNA must be determined from the basic equilibrium condition that compressive force must be equal to tensile force, i.e., see Fig. 2.34:

$$6 \times y_c \times 1.5\sigma_y = 5.5 \times 0.5 \times \sigma_y + (0.5 - y_c) \times 6 \times \sigma_y$$

which gives

$$y_c = \frac{5.75}{15} = 0.383 \text{ in.}$$

Thus, we have

$$M_p = 6 \times y_c \times \frac{y_c}{2} \times 1.5\sigma_y + 5.5 \times 0.5\left(\frac{1}{2} \times 5.5 + 0.5 - y_c\right)\sigma_y$$

$$+ 6 \times \frac{(0.5 - y_c)^2}{2} \times \sigma_y$$

or

$$M_p = 8.59\sigma_y.$$

(b)

$$f = \frac{M_p}{M_y} = \frac{8.59\sigma_y}{4.61\sigma_y} = 1.86.$$

Example 2.10.3. Design an I-section, 50 inches deep and 20 inches wide, made up of rectangles with a plastic modulus of 2449 in.^{3} about the strong axis and 411.5 in.^{3} about the weak axis.

Solution: The dimensions t and T of the required section (Fig. 2.35) are unknown but can be determined by equating the strong- and weak-axis plastic moduli in terms of t and T to the required plastic moduli as

$$Z_x = 2\left[(25 - T)t\frac{(25 - T)}{2}\right] + 2\left[(20)(T)\left(25 - \frac{T}{2}\right)\right] = 2449 \text{ in.}^3 \quad (2.10.1)$$

$$Z_y = 2(50 - 2T)\left(\frac{t}{2}\right)\left(\frac{t}{4}\right) + 4[(T)(10)(5)] = 411.5 \text{ in.}^3 \quad (2.10.2)$$

The solution of Eqs. (2.10.1) and (2.10.2) for the unknowns t and T leads to

$$t = 1 \text{ in.;} \quad T = 2 \text{ in.}$$

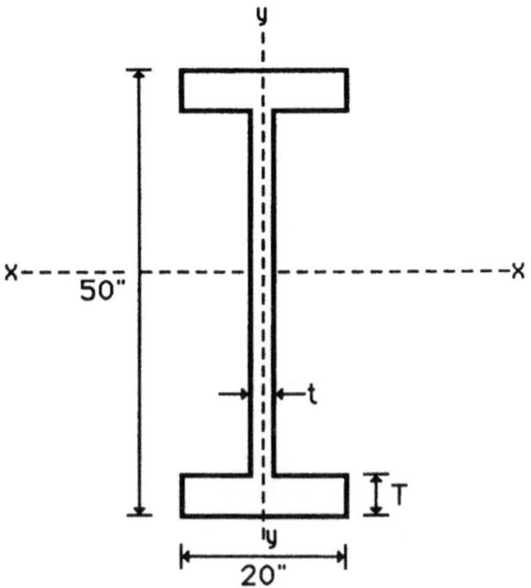

FIGURE 2.35. Design of an I-section for a given set of plastic modulus.

Example 2.10.4. A beam having the cross section shown in Fig. 2.36 is made of steel with 36-ksi yield stress. Calculate the percentage of reductions in the full plastic moments about both principal axes due to axial loads of (a) 381.6 kips and (b) 1908 kips. How do your results compare with those from interaction Eqs. (2.5.10), (2.5.11), (2.5.14) and (2.5.15).

Solution:

(a) $P = 381.6$ **kips: Strong Axis:** Since $P = 381.6$ kips is less than the axial load capacity of web $= (26 \times 1) \times 36 = 936$ kips, the plastic neutral axis is in the web. The distance y_0 of shift of the neutral axis from its original position can be determined from [Fig. 2.36(a)]

$$P = 2y_0(1)\sigma_y = 2(y_0)(1)(36) = 381.6$$

which gives

$$y_0 = 5.3 \text{ in.}$$

The reduced plastic moment capacity is obtained by subtracting the contribution of the portion of the cross section carrying axial load from the original plastic moment of the section as

$$M_{pc} = M_{px} - \sigma_y(1)y_0^2$$

in which $M_{px} = \sigma_y Z_x$, and Z_x from Example 2.10.1(a) is 1289 in.[3] Thus, we obtain

$$M_{pc} = 36 \times 1{,}289 - 36 \times 1 \times (5.3)^2 = 46{,}404 - 1011 = 45{,}393 \text{ kip-in.}$$

The percentage of reduction in the plastic modulus is

$$\frac{1011}{46{,}404} = 2.2\%.$$

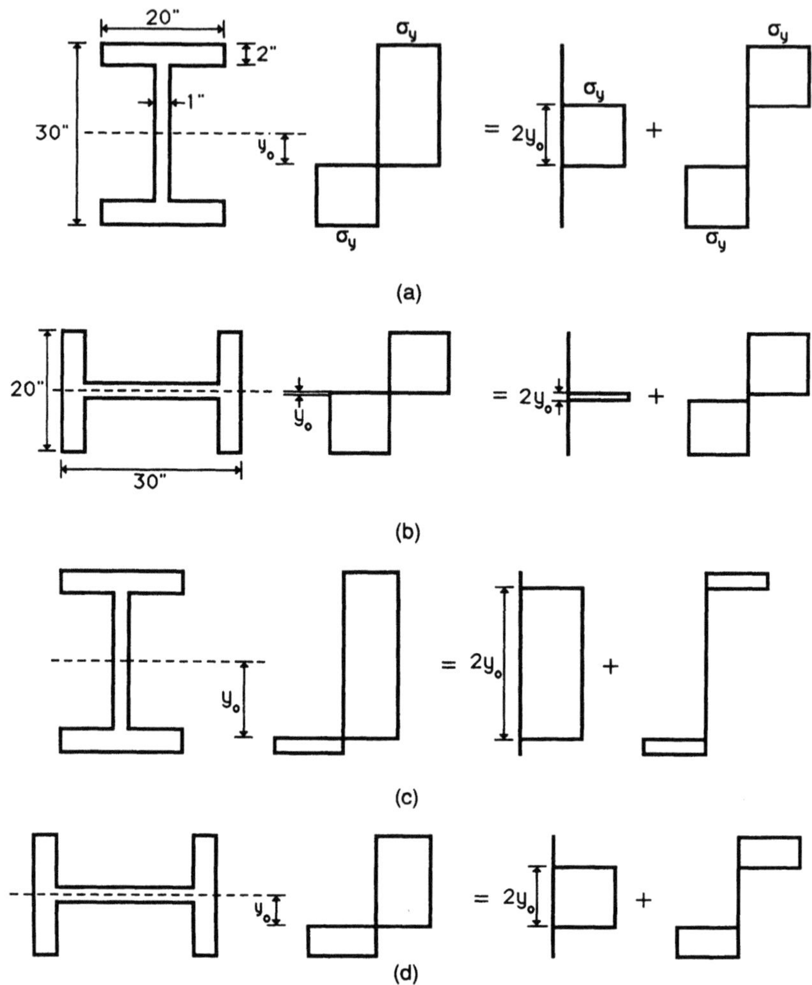

FIGURE 2.36. Computations for the effect of axial loading on M_p of an I-section.

Note that since $P/P_y = 381.6/[(2 \times 20 \times 2 + 26 \times 1)36] = 0.1 < 0.15$, the interaction Eq. (2.5.10) permits no reduction in the full plastic moment capacity.

Weak Axis: For the case of weak-axis bending, the plastic neutral axis is also in the web. The distance y_0 of shift of the neutral axis from its original position can be determined from [Fig. 2.36(b)]

$$P = 2y_0(30)\sigma_y = 2(y_0)(30)(36) = 381.6 \text{ kips}$$

which gives

$$y_0 = 0.177 \text{ in.}$$

Thus, we obtain

$$M_{pc} = M_{py} - \sigma_y(30)(y_0)^2$$

in which $M_{py} = \sigma_y Z_y$. Since Z_y from Example 2.10.1(b) is 406.5 in.^{3}, we obtain

$$M_{pc} = 36 \times 406.5 - 36 \times 30 \times (0.177)^2$$

or

$$M_{pc} = 14{,}634 - 34 = 14{,}600 \text{ kip-in.}$$

The percentage of reduction in the plastic modulus is

$$\frac{34}{14{,}634} = 0.23\%.$$

Note that since $P/P_y = 0.1$ is less than $0.4P_y$, the interaction Eq. (2.5.14) permits no reduction in the full plastic modulus of the section.

(b) $P = 1908$ **kips: Strong Axis:** Since $P = 1908$ kips is greater than the axial capacity of web $= 936$ kips, the neutral axis is in the flange. The distance y_0 of shift of the neutral axis from its original position can be determined from [Fig. 2.36(c)]

$$P = 2\sigma_y[13 \times 1 + (y_0 - 13) \times 20] = 2 \times 36[20y_0 - 247] = 1908 \text{ kips}$$

which gives

$$y_0 = 13.68 \text{ in.}$$

The reduced plastic moment can be computed by taking the moment of stresses [Fig. 2.36(c)] about the neutral axis as

$$M_{pc} = 2\sigma_y(20)(15 - y_0)\left(y_0 + \frac{15 - y_0}{2}\right) = \sigma_y(20)[(15)^2 - y_0^2]$$

or

$$M_{pc} = (36)(20)[(15)^2 - (13.68)^2] = 27{,}357 \text{ kip-in.}$$

From part (a), we have $M_{px} = 46{,}404$ kip-in.

The percentage of reduction in the plastic modulus is thus

$$\frac{46{,}404 - 27{,}357}{46{,}404} = 41\%.$$

M_{pc} from Eq. (2.5.11) is

$$M_{pc} = 1.18\left(1 - \frac{P}{P_y}\right)M_p = 1.18\left[1 - \frac{1908}{(36)[2(20 \times 2) + (26 \times 1)]}\right]M_p$$

$$= 0.59M_p.$$

The percentage of reduction in the plastic modulus using the approximate interaction equation

$$\frac{M_p - 0.59M_p}{M_p} = 41\%.$$

The approximate interaction equation gives the same reduction as that given by exact calculations.

Weak Axis: For the case of weak axis bending, the neutral axis is also in the flanges. The distance y_0 of shift of the neutral axis from the original position can be determined from [Fig. 2.36(d)]

$$P = 2\sigma_y[26 \times 0.5 + 2y_0 \times 2] = 2 \times 36[4y_0 + 13] = 1908 \text{ kips}$$

which gives

$$y_0 = 3.38 \text{ in.}$$

Thus, we obtain

$$M_{pc} = 4\sigma_y(2)(10 - y_0)\left(y_0 + \frac{10 - y_0}{2}\right) = 4 \times 36[(10)^2 - (3.38)^2]$$

$$M_{pc} = 12{,}755 \text{ kip-in.}$$

From part (a), we have $M_{py} = 14{,}634$ kip-in. So the percentage of reduction due to axial load is

$$\frac{14{,}634 - 12{,}755}{14{,}634} = 12.8\%.$$

Since $P/P_y = 0.50 > 0.4$, M_{pc} from the approximate interaction Eq. (2.5.15) is

$$M_{pc} = 1.19\left[1 - \left(\frac{P}{P_y}\right)^2\right]M_p$$

or

$$M_{pc} = 1.19\left[1 - \left(\frac{1{,}908}{36 \times 106}\right)^2\right]M_p = 0.893M_p.$$

The percentage of reduction from approximate Eq. (2.5.15) is

$$\frac{M_p - 0.893M_p}{M_p} = 10.7\%.$$

Example 2.10.5. Flange cover plates 25×2.5 inches are added to the section in Example 2.10.4.

(a) Find the increase in the full plastic moment capacity about the strong axis, and

(b) calculate the reduction in the plastic moment due to an axial load of 1200 kips.

Solution: (a) Increase in the Full Plastic Moment: From Example 2.10.4, the original M_p of the sections is

$$M_{po} = 1289 \times 36 = 46{,}404 \text{ kip-in.}$$

The area of each cover plate is

$$A_p = 25 \times 2.5 = 62.5 \text{ in.}^2$$

The value of M_p of the section with cover plates is

$$M_p = M_{po} + 2\sigma_y A_p \left[15 + \frac{2.5}{2} \right]$$

or

$$M_p = 46{,}404 + 2 \times 36 \times 62.5 \times 16.25 = 119{,}529 \text{ kip-in.}$$

The increase in moment capacity due to the addition of cover plates is

$$M_p - M_{po} = 119{,}529 - 46{,}404 = 73{,}125 \text{ kip-in.}$$

(b) Reduction in the Moment Capacity Due to Axial Load: Since (axial load capacity of web $= 26 \times 1 \times 36 = 936$ kips) $< (P = 1200$ kips$) <$ (axial load capacity of web and two flanges $= 936 + 2 \times 36 \times 20 \times 2 = 3816$ kips), the plastic neutral axis in the presence of an axial load falls in the bottom flange as shown in Fig. 2.37. The shift of distance y_0 of the neutral axis from its original position is

$$P = 936 + 2 \times 36 \times 20(y_0 - 13) = 1200$$

which gives

$$y_0 = 13.18 \text{ in.}$$

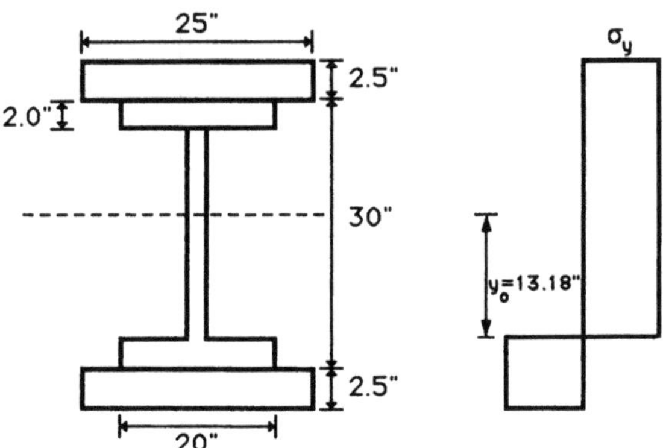

FIGURE 2.37. Computations of the effect of axial load on M_p of an I-section with a pair of cover plates.

Thus, we obtain

$$M_{pc} = 2\sigma_y A_p\left(15 + \frac{2.5}{2}\right) + 2\sigma_y(15 - 13.18)(20)\left(13.18 + \frac{15 - 13.18}{2}\right)$$

or

$$M_{pc} = (2)(36)(62.5)(16.25) + (2)(36)(1.82)(20)(14.09)$$

or

$$M_{pc} = 73{,}125 + 36{,}927 = 110{,}052 \text{ kip-in.}$$

Thus, the reduction in the plastic moment capacity due to axial load is

$$= 119{,}529 - 110{,}052 = 9{,}477 \text{ kip-in.}$$

Example 2.10.6. Show that for an I-section subjected to a shear force V, the full plastic moment about the strong axis is

$$M_{ps} = M_f + \sqrt{M_w\left(M_w - \frac{3}{4}\frac{V^2}{t_w\sigma_y}\right)}$$

where M_f and M_w are, respectively, the contributions of the flanges and the web to the full plastic moment in the absence of shear force, t_w is the thickness of the web, and σ_y is the yield stress in tension.

Solution: The full plastic moment capacity reduced for the presence of shear force can be expressed as

$$M_{ps} = M_f + M_w\frac{\sigma}{\sigma_y} = M_f + \sqrt{M_w^2\left(\frac{\sigma}{\sigma_y}\right)^2} \qquad (2.10.3)$$

in which σ is the normal stress that can be exerted on the web in the presence of a uniform shear stress τ and it can be written from the von Mises yield criterion (2.6.3) as

$$\left(\frac{\sigma}{\sigma_y}\right)^2 = 1 - \frac{3\tau^2}{\sigma_y^2} = 1 - \frac{3V^2}{d_w^2 t_w^2 \sigma_y^2} \qquad (2.10.4)$$

in which d_w is the depth of the web. Now substituting $(\sigma/\sigma_y)^2$ from Eq. (2.10.4) in Eq. (2.10.3), we have

$$M_{ps} = M_f + \sqrt{M_w^2\left(1 - \frac{3V^2}{d_w^2 t_w^2 \sigma_y^2}\right)}.$$

Using $M_w = \sigma_y t_w d_w^2/4$, we obtain

$$M_{ps} = M_f + \sqrt{M_w\left(M_w - \frac{3V^2}{4t_w\sigma_y}\right)}. \qquad (2.10.5)$$

Example 2.10.7. Determine the reduction in the plastic modulus about the strong axis of the I-section shown in Fig. 2.38 due to a shear force of 320 kips

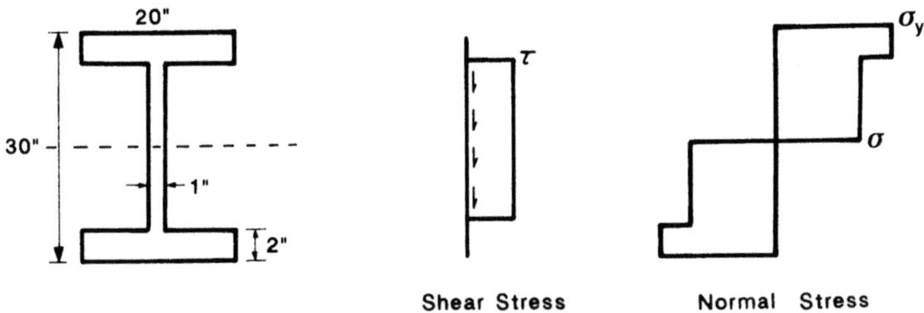

FIGURE 2.38. Plastic modulus of an I-section in the presence of shear force.

in the plane of web. Use the von Mises yield criterion. The section is made of steel with yield stress (a) 36 ksi and (b) 50 ksi.

Solution: By assuming that the shear stress is distributed uniformly over the web, the magnitude of shear stress can be determined as

$$\tau = \frac{V}{A_{\text{web}}} = \frac{320}{26 \times 1} = 12.31 \text{ ksi.}$$

(a) $\sigma_y = 36$ ksi: Using the von Mises yield criterion (3.6.3), the normal stress in the web can be determined from

$$\sigma^2 + 3\tau^2 = \sigma_y^2$$

which gives

$$\sigma = \sqrt{\sigma_y^2 - 3\tau^2} = \sqrt{(36)^2 - 3(12.31)^2} = 29.01 \text{ ksi.}$$

The reduced plastic moment or modulus of the section can be determined from Fig. 2.38 as

$$M_{ps} = M_p - M_{\text{web}}\left(1 - \frac{\sigma}{\sigma_y}\right)$$

or

$$Z_{ps} = Z - Z_{\text{web}}\left(1 - \frac{\sigma}{\sigma_y}\right).$$

From Example (2.10.1a), $Z = 1289$ in.^{3} and we have $Z_{\text{web}} = 2 \times 1 \times 13 \times 6.5 = 169$ in^3. Thus, we obtain

$$Z_{ps} = 1289 - 169\left(1 - \frac{29.01}{36}\right) = 1256 \text{ in.}^3$$

The percentage of reduction in the plastic modulus = $(1289 - 1256)/1289 = 2.56\%$

(b) $\sigma_y = $ **50 ksi:** The normal stress in the web is

$$\sigma = \sqrt{\sigma_y^2 - 3\tau^2} = \sqrt{(50)^2 - 3(12.31)^2} = 45.23 \text{ ksi.}$$

The reduced plastic modulus is

$$Z_{ps} = Z - Z_{\text{web}}\left(1 - \frac{\sigma}{\sigma_y}\right)$$

or

$$Z_{ps} = 1289 - 169\left(1 - \frac{45.23}{50}\right) = 1273 \text{ in.}^3$$

The percentage of reduction in the plastic modulus $= (1289 - 1273)/1289 = 1.24\%$.

Example 2.10.8. Use LRFD specifications to design a simply supported plate girder of A36 steel, with a web depth of 60 inches, to carry a uniformly distributed load of 34 kip/ft over a span of 50 feet. Assume that the girder has adequate lateral support.

Solution: The maximum moment in the beam is (Fig. 2.39)

$$M_{\max} = \frac{wL^2}{8} = \frac{(34)(50)^2}{8} = 10,625 \text{ kip-ft.}$$

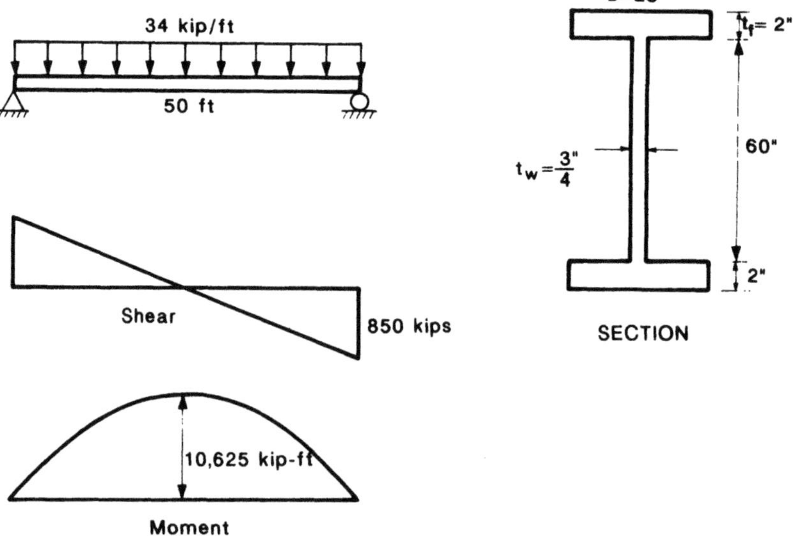

FIGURE 2.39. Design of a simply supported beam.

The maximum shear force in the beam is

$$V_{\max} = \frac{wL}{2} = \frac{(34)(50)}{2} = 850 \text{ kips.}$$

The required plastic modulus of the section is

$$Z = \frac{M_{\max}}{\phi_b \sigma_y} = \frac{10{,}625 \times 12}{0.9 \times 36} = 3935 \text{ in.}^3$$

and the required area of the web is

$$A_w = \frac{V_{\max}}{\phi_v \tau_y} = \frac{V_{\max}}{\phi_v 0.6 F_y} = \frac{850}{0.9 \times 0.6 \times 36} = 43.7 \text{ in.}^2$$

Required thickness of the web is $t_w = A_w/60 = 43.7/60 = 0.728$ in.

Try 3/4-inch-thick plate for web. To prevent local buckling of the web, we check

$$\frac{d_w}{t_w} = \frac{60}{3/4} = 80 < \frac{640}{\sqrt{F_y}} = \frac{640}{6} = 106.7 \quad \text{okay.}$$

Use $60 \times 3/4$-inch plate for the web.

The dimensions of flange plates should be such that the plastic modulus of a section is at least equal to the required plastic modulus, i.e.,

$$Z = bt_f(60 + t_f) + (0.75)\frac{(60)^2}{4} = 3935 \text{ in.}^3$$

Note that Z is not reduced due to the presence of shear stress because at the point of maximum moment, shear force is equal to zero. To prevent local buckling of flange plates, we must also have

$$\frac{b}{2t_f} \leq \frac{65}{\sqrt{F_y}} = 10.83$$

or

$$b \leq 21.66 t_f.$$

Try a conservative proportion of $b = 15t_f$. Substitution of this proportion in the preceding equation gives

$$900t_f^2 + 15t_f^3 + (0.75)\frac{(60)^2}{4} = 3935$$

or

$$t_f^3 + 60t_f^2 - 217 = 0.$$

Solving this equation by trial and error, we obtain

$$t_f = 1.87 \text{ in.}$$

Thus, we have

$$b = 15t_f = 28.05 \text{ in.}$$

Try 28- × 2-inch plates for flanges and check local buckling

$$\frac{b}{2t_f} = \frac{28}{2 \times 2} = 7 < \frac{65}{\sqrt{F_y}} \quad \text{okay.}$$

Use 28- × 2-inch plates for flanges. Dimensions of the recommended section are shown in Fig. 2.39.

Example 2.10.9. A straight steel beam is 5 feet long and is simply supported at its ends. It is made of rectangular section with 2-inch depth throughout. The width of the beam is tapering uniformly from 3 inches at the midspan to 1 inch at each end. If the yield stress of the steel used is 36 ksi, what uniformly distributed load will bring collapse of this beam? Assume that the effect of the shear force on the plastic moment capacity of the section is negligible.

Solution: For a rectangular cross section of varying width (Fig. 2.40), Z_x can be expressed as

$$Z_x = 2[(1 \times b_x)(0.5)] = b_x$$

where b_x between the left-hand end and midspan can be written as

$$b_x = 1 + \frac{2}{2.5}x = 1 + 0.8x.$$

Thus, Z_x is

$$Z_x = 1 + 0.8x$$

and the plastic moment capacity M_{px} of the beam is

$$M_{px} = \sigma_y Z_x = 36(1 + 0.8x) \tag{2.10.6}$$

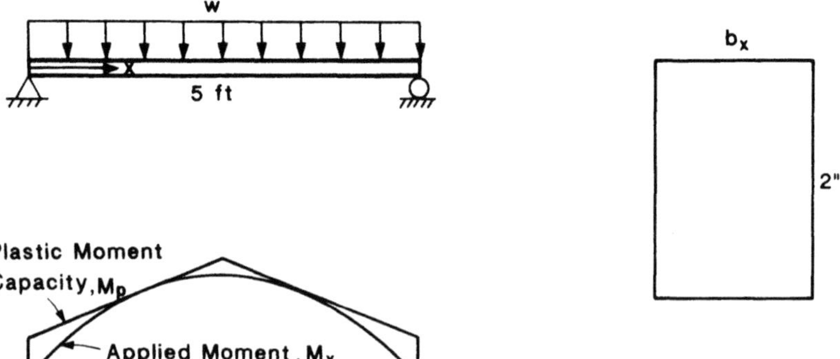

FIGURE 2.40. Load-carrying capacity of a tapered beam.

The applied moment at the section a distance x from one end due to the uniformly distributed load w is given by

$$M_x = \left[\frac{1}{2}(5w)x - \frac{1}{2}wx^2 \right]12$$

or

$$M_x = 30wx - 6wx^2. \tag{2.10.7}$$

The collapse value of w will now be controlled by the following two conditions:

(a) At collapse, the applied moment at the critical sections should be equal to the plastic moment capacity of the beam at these sections, i.e.,

$$M_x = M_{px}$$

or

$$30wx - 6wx^2 = 36(1 + 0.8x). \tag{2.10.8}$$

(b) The applied moment at other sections should not exceed the plastic moment capacity. This leads to the condition that the plastic moment capacity curve should enclose the applied moment diagram as shown in Fig. 2.40, and both diagrams should be tangent to each other at the critical sections, i.e.,

$$\frac{dM_x}{dx} = \frac{dM_p}{dx}$$

or

$$30w - 12wx = 36(0.8). \tag{2.10.9}$$

Solving Eqs. (2.10.8) and (2.10.9) for x and w, we obtain

$$w = 2.51 \text{ kip/ft}, \ x = 1.54 \text{ feet.}$$

Thus, at collapse, w is 2.51 kip/ft.

Example 2.10.10. A W14 × 53 beam made of A36 steel is built into a support at one end as shown in Fig. 2.41. Vertical loads are applied as shown. Determine the load factor against collapse (a) neglecting the effect of shear and (b) taking shear into account. Use the von Mises yield criterion and assume that the beam has adequate lateral support.

Solution: From the AISC manual, the following properties of W14 × 53 are noted: $Z = 87.1 \text{ in.}^3$,

$$M_p = \sigma_y Z = 36 \times 87.1 = 3136 \text{ kip-in.,}$$

$d_w = 12.6 \text{ in., and } t_w = 0.37 \text{ in.}$

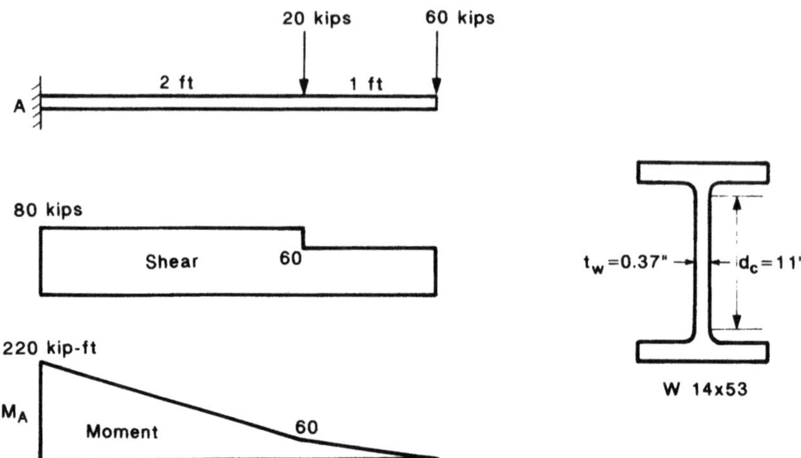

FIGURE 2.41. Load factor of a given cantilever beam.

From the shear force and bending moment diagrams (Fig. 2.41), we have

$$V_{\max} = V_A = 80 \text{ kips}$$

and

$$M_{\max} = M_A = 220 \text{ kip-ft} = 220 \times 12 = 2640 \text{ kip-in.}$$

(a) Neglecting the effects of shear: The load factor is

$$\lambda = \frac{M_p}{M_{\max}} = \frac{3136}{2640} = 1.188$$

(b) Taking shear into account: Since the maximum shear and moment both occur at the support, this point is critical. Assuming that the shear force is taken by the web only, shear stress τ can be computed as

$$\tau = \frac{V_{\max}}{A_w} = \frac{80}{12.6 \times 0.37} = 17.16 \text{ ksi.}$$

Using the von Mises yield criterion (2.6.3), the normal stress σ in the web is

$$\sigma = \sqrt{\sigma_y^2 - 3\tau^2} = \sqrt{(36)^2 - 3 \times (17.16)^2} = 20.31 \text{ ksi.}$$

Thus, we have

$$Z_{ps} = Z - Z_{web}\left(1 - \frac{\sigma}{\sigma_y}\right)$$

or

$$Z_{ps} = 87.1 - \left(\frac{1}{4}\right)(0.37)(12.36)^2\left(1 - \frac{20.31}{36}\right) = 80.94 \text{ in.}^3$$

and

$$M_{ps} = \sigma_y Z_{ps} = 36 \times 80.94 = 2914 \text{ kip-in.}$$

The load factor λ against bending is

$$\frac{M_{ps}}{M_{\max}} = \frac{2914}{2640} = 1.104.$$

The load factor λ against shear is

$$\frac{V_p}{V_{\max}} = \frac{12.6 \times 0.37 \times 36/\sqrt{3}}{80} = 1.211.$$

Thus, the overall load factor is

$$\lambda = 1.104.$$

Example 2.10.11. As originally designed, a 15-foot-wide balcony was to be supported by A36 W14 × 53 cantilever beams, so that each beam carried a uniform load of 1.2 kip/ft. It was then decided to add a balustrade, which would apply an additional load of 5 kips to the free end of each beam. Find (a) the load factor of the beams as originally designed, and (b) the dimensions of the symmetrical flange plates that must be added to provide a load factor of 2 when the balustrade is in position. Neglect the effects of shear force.

Solution: From Example 2.10.10, Z and M_p of W14 × 53 are

$$Z = 87.1 \text{ in.}^3, M_p = 36 \times 87.1 = 3136 \text{ kip-in.}$$

(a) Load Factor Before Balustrade is Added: From the bending moment diagram (Fig. 2.42), the maximum moment is

$$M_{\max} = 135 \text{ kip-ft} = 1620 \text{ kip-in.}$$

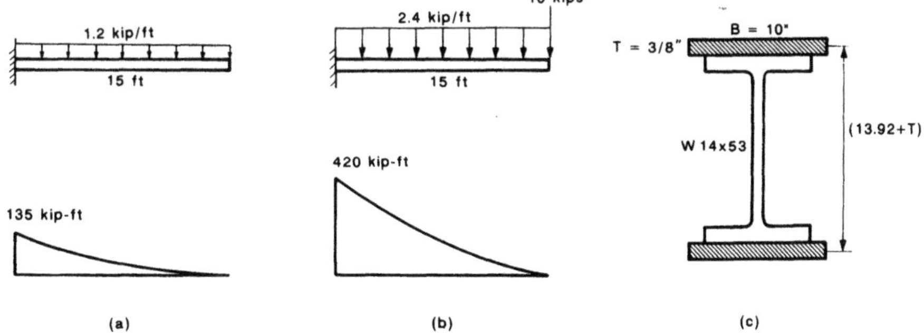

FIGURE 2.42. Design of flange cover plates to increase load-carrying capacity of a cantilever beam.

Thus the load factor is

$$\lambda = \frac{M_p}{M_{\max}} = \frac{3136}{1620} = 1.936$$

(b) Dimensions of Cover Plates Needed to Support Balustrade: The bending moment diagram for the beam with balustrade on and with a load factor $\lambda = 2$ is shown in Fig. 2.42(b). The maximum moment in this diagram is

$$M_{\max} = 420 \text{ kip-ft} = 5040 \text{ kip-in.}$$

$$\text{Required } Z = \frac{M_{\max}}{\sigma_y} = \frac{5040}{36} = 140 \text{ in.}^3$$

Z can be expressed in terms of the thickness and the width of the flange plates as [Fig. 2.42(c)]

$$Z = 87.1 + BT(13.92 + T).$$

Try a 10- $\times$ 3/8-inch plate. Then we have

$$Z = 87.1 + 10 \times \frac{3}{8}\left(13.92 + \frac{3}{8}\right) = 140.7 \text{ in.}^3 \quad \text{okay.}$$

Check local buckling of flange cover plates

$$\frac{B}{T} = \frac{10}{3/8} = 26.67 < \frac{190}{\sqrt{F_y}} = 31.67 \quad \text{okay.}$$

Use 10- $\times$ 3/8-inch flange cover plates as shown in Fig. 2.42(c).

2.11 Summary

The basic quantity required in any structural analysis of framed structures is the value of bending rigidity EI, which can be considered as the slope of the relationship between moment M and curvature Φ. In the *allowable stress design method*, which is based on elastic analysis, this quantity has a constant value and thus it presents no difficulties in the analysis and design process. But in the case of the *plastic design method*, which is based on plastic analysis, there are problems, because the actual moment-curvature response beyond the elastic range is nonlinear. It has been shown that the intensity of the axial load has a major influence on the shape of the moment-curvature curve and on the full plastic moment capacity. The stiffness in the elastic range does not vary with axial load but the variation of EI beyond the elastic limit is very significant. At the fully plastic limit state, the stiffness or the slope of the moment-curvature relation reduces to zero while the maximum moment remains at the full plastic moment M_{pc}.

As a practical approximation, the actual moment-curvature curve is replaced by two straight lines for which the stiffness is assumed not to be influenced at all by plastic yielding of the material until the full plastic moment is reached. At the full plastic state, the rotation of the section increases without limit at a constant moment along the horizontal portion of the curve. This is a very reasonable idealization for members of wide-flange shape, but appreciable errors in stiffness must be expected for cross-sectional shapes with a large value of shape factor. Nevertheless, the approximation is very good for estimating the *collapse load* of framed structures. Furthermore, this idealized cross-sectional behavior reduces the nonlinear structural analysis to a sequence of *elastic* and *"rusty"* hinge analysis. This rusty hinge is known as the *plastic hinge*, and its formation corresponds to fully plastic moment of the section of the structural member.

The plastic hinge idealization drastically simplifies the plastic analysis of framed structures and makes the full collapse load determination as a *quasi-static* process. It forms the basis of the *simple plastic theory*. All that is required in this simple theory is a knowledge of the value of the full plastic moment or the plastic hinge moment. This was described in detail in this chapter. As will be seen in the following chapters, if the full plastic moments of the various members of a frame are known, then the collapse load of that frame can be determined quickly in a direct manner, even if the frame is complex. Similarly, the design of a frame to carry given loads consists in the assignment of certain minimum values of full plastic moment to the members, which can also be achieved quickly in a direct manner.

However, several factors of secondary importance will prevent the member from reaching the full plastic moment. These factors include such things as axial load, shear, buckling, and connection details. These factors are not included in the *"simple"* plastic theory, but we must take them into account in practical design. In this chapter, the effects and characteristics of the following factors were discussed and appropriate design procedures provided for checking the suitability of the original simple plastic design:

- axial load and shear force that will reduce plastic moment.
- instability that may cause local buckling of thin-walled sections.
- connections that are properly proportioned to transmit plastic moment from one member to the other.

In addition, brittle fracture, repeated loading, and deflection limitation at working load must all be accounted for in an actual design to check the suitability of a design based on the simple plastic theory that neglects all these factors. Note that the necessity for considering these additional factors in the plastic design is not in any way different in principle from that in the conventional elastic design procedures.

References

2.1. Chen, W.F., and Han, D.J., *Plasticity for Structural Engineers*, Springer-Verlag, New York, 606 pp., 1988.

2.2. Hodge, P.G., Jr., "Interaction Curves for Shear and Bending of Plastic Beams," *Journal of Applied Mechanics*, 24, p. 453, 1957.

2.3. Drucker, D.C., "The Effect of Shear on the Plastic Bending of Beams," *Journal of Applied Mechanics*, 23, pp. 509–514, 1956.

2.4. Neal, B.G., "The Effect of Shear and Normal Forces on the Fully Plastic Moment of a Beam of Rectangular Cross-Section," *Journal of Applied Mechanics* 28, pp. 269–274, 1961.

2.5. Chen, W.F., and Lui, E.M., "Static Flange Moment Connections," *Journal of Constructional Steel Research*, 10, pp. 39–88, 1988.

2.6. Baker, L., and Heyman, J., *Plastic Design of Frames—1: Fundamentals*, Cambridge University Press, New York, 227 pp., 1980.

Problems

2.1. Show that the nondimensional M-Φ relationship for a beam of square cross section bent about a diagonal is

$$\frac{M}{M_y} = 2\left\{1 - \left(\frac{\Phi_y}{\Phi}\right)^2 + \frac{1}{2}\left(\frac{\Phi_y}{\Phi}\right)^3\right\}.$$

   (a) Plot the curve, indicating coordinates at $\Phi/\Phi_y = 1.0, 2.0, 4.0,$ and 10.0 $(1.0, 1.625, 1.891, 1.981)$.

   (b) On the same curve, plot the moment-curvature relationships for (i) a rectangle, (ii) W8 $\times$ 31, and (iii) the idealized curve (Fig. 2.5).

   (c) Find the shape factors for the three cross sections $(2.0, 1.5, 1.11)$.

   (d) Discuss the practical implications of the plastic hinge idealization of the M-Φ curve to consist of two straight lines.

2.2. Derive the expression for the plastic zone distribution of a uniformly loaded cantilever beam of rectangular cross section with $M_{\mathrm{max}} = M_p$. Sketch the result and find the plastic hinge length $(y/d = \sqrt{3[1 - (x/l)^2]}, l_p/l = 0.184)$.

2.3. Derive the expression for the plastic zone distribution shown in Fig. 2.6(c).

2.4. A simply supported beam of rectangular cross section is subjected to a concentrated load at one-third of the span. Plot the shape of the plastic zone at collapse of the beam. Determine plastic hinge length

$$\left(\text{left side } y/d = \sqrt{\frac{9x_1}{l}}; \text{ right side } y/d = \sqrt{\frac{9x_2}{2l}}, l_p/l = l/3\right).$$

2.5. Compute Z, S, and f for a box section with outside depth of 20 inches, wall thickness of 1/2 inch, and a width of 8 inches. Flexure is about the strong axis. Check local buckling requirements $(168.25 \text{ in.}^3, 133.23 \text{ in.}^3, 1.263)$.

2.6. Compute Z for a W14 $\times$ 132 shape bending about the strong axis, using the approximate expression (2.2.10). Compare the result with the exact value listed in the AISC manual $(232.3 \text{ in.}^3, 234 \text{ in.}^3)$

2.7. For a solid circular section with diameter d, we know $Z = d^3/6$ and $S = \pi d^3/32$, and for solid rectangular section with width b and depth d, we know $Z = bd^2/4$

and $S = bd^2/6$. Using these relationships, derive shape factor expressions for the following two shapes:

a. channel section bent about its strong axis, and

$$f = \frac{3d}{2}\left[\frac{bd^2 - (b - t_w)(d - 2t_f)^2}{bd^3 - (b - t_w)(d - 2t_f)^3}\right]$$

b. circular hollow tube

$$f = \frac{16D}{\pi}\frac{(D^3 - d^3)}{(D^4 - d^4)}.$$

2.8. A6- $\times$ 3-inch channel section has flange thickness 0.40 inch and web thickness 0.25 inch. Find M_{pc} bent about the minor axis under an axial thrust of 36 kips, assuming A36 steel. Show that the value of M_{pc} depends on whether the tips of the flanges are in tension or compression, and find both values (87.9 kip-in, 110.8 kip-in).

2.9. Select a wide-flange shape section of A36 steel that will transmit a sull plastic moment of 300 kip-in in the presence of an axial compression of 200 kips. Flexure is about the strong axis (W14 $\times$ 26).

2.10. For a W8 $\times$ 31 column bending about the strong axis, draw the theoretical $M_{pc} - P$ relationships. Compare the results with the approximate design equations at P/P_y equal to 0, 0.15, and 0.6.

(P/P_y)	Theoretical M_{pc}	Approximate M_{pc}
0	1	1
0.15	0.95	1
0.6	0.471	0.472

2.11. Select a member of wide-flange shape of A36 steel whose cross section will transmit a strong axis moment of 3000 kip-in., in the presence of a shear force of 150 kips (W24 $\times$ 55).

2.12. Two loads of 200 kips each are applied 1 foot from each end of a simply supported 10-foot-long beam. Select a member of wide-flange shape. Check its adequacy for shear and modify the design if necessary (PD/LRFD W24 $\times$ 84, PD/ASD W24 $\times$ 62).

2.13. A T-section, width 5 inches, depth 6 inches, composed of two equal rectangles 5 $\times$ 1 inch, is bent about the strong axis.
(a) Find Z, S, and f (15 in.^{3}, 8.33 in.^{3}, 1.8).
(b) Find M_{pc} due to an axial load of 100 kips. Assume A36 steel (498.3 kip-in.).

2.14. An I-section made of A36 steel with overall depth 12 inches, flanges 8 $\times$ 1 inch, web thickness 1/2 inch, is bent about (i) the strong axis and (ii) the weak axis.
(a) Find M_p assuming three rectangles (3618 kip-in., 1175 kip-in.).
(b) Calculate the percentage reductions in M_p due to an axial load of (i) 3.6 kips and (ii) 18 kips (2.2%, 0.28%, 40.7%, and 13.5%).

2.15. Calculate the percentage of reductions in M_p for the beam section in Problem 2.14 due to a shear force of 100 kips in the plane of the web when steel has a tensile yield stress of (i) 36 ksi and (ii) 50 ksi. Use the von Mises yield criterion (9.1%, 3.5%).

2.16. Cover plates of 10- $\times$ 1-inch are added to the flanges of the beam section in Problem 2.14. Find the increase in the full plastic moment M_p about the strong axis. Calculate the reduction in M_p due to an axial compression of 500 kips ($M_p = 8298$ kip-in., $M_{pc} = 6159$ kip-in.).

2.17. A fixed-ended beam of length L has a concentrated load Q at the left third-point. If the allowable stress is 24 ksi, the yield stress is 36 ksi, and the shape factor is 1.15, calculate:
(a) allowable working load Q_a (elastic design) ($3.91\ M_p/L$).
(b) yield load Q_y ($5.87\ M_p/L$).
(c) plastic limit load Q_p (plastic design) ($9M_p/L$).
(d) factor of safety against initial yielding (1.5).
(e) factor of safety against plastic collapse (load factor) (2.3).
(f) Give two reasons why Q_p is much greater than Q_y.
(g) Explain why erection forces will influence the calculation of Q_y but have no effect on the calculation of Q_p.
(h) If a maximum compressive residual stress $\sigma_{rc} = 10$ ksi is present in the beam section, what are the factors of safety now, against (i) initial yielding and (ii) ultimate collapse (1.08, 2.3).

2.18. A W14 $\times$ 34 fixed-ended beam is of 12-foot length and carries a concentrated load 60 kips at a distance 4 feet from one end. Find the load factor considering the effect of shear according to Equation (2.6.17). Assume a von Mises yield criterion and use A36 steel ($\lambda = 1.807$).

2.19. For an interior beam-to-column connection as shown in the inset of Fig. 2.24:
a. describe all possible failure modes.
b. list the corresponding LRFD rules to check against such failures.
c. sketch the shape of the plastic zones at collapse of the two beams. Determine plastic hinge length $L(1 - 1/f)$.

2.20. A W24 $\times$ 176 beam is connected to a column of same size at a right angle to transmit the full plastic moment at a corner. Detail the straight corner connection: (a) using a web doubler plate and (b) using a diagonal stiffener.

2.21. Two W18 $\times$ 76 beams are connected on opposite sides to a W14 $\times$ 82 column under a symmetric loading condition. Detail the stiffeners in the beam-to-column connection. Is the moment stiffening for the column adequate if the column must also transmit asymmetric moments from the two connecting beams?

2.22. Find the M_{pc} value for the hollow rectangular section shown in Fig. P2.22, where an eccentric axial compression load of 54 kips is applied 0.75 inch above the centroid of the section.

FIGURE P2.22

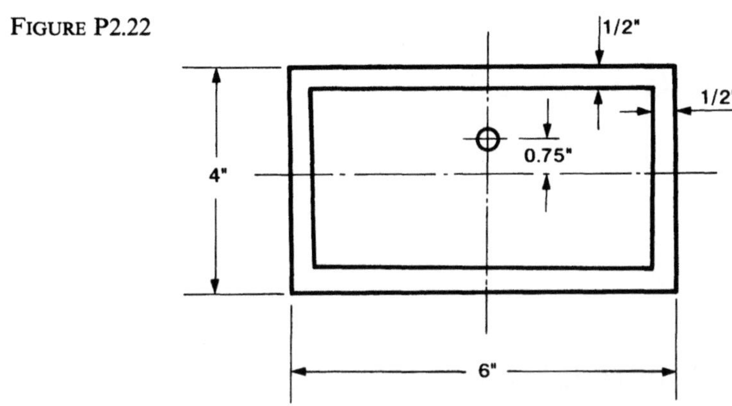

2.23. For a simply supported beam of rectangular section tapering uniformly from d at the midspan to $d/3$ at each end as shown in Fig. P2.23, find the region for which cover plates are needed.

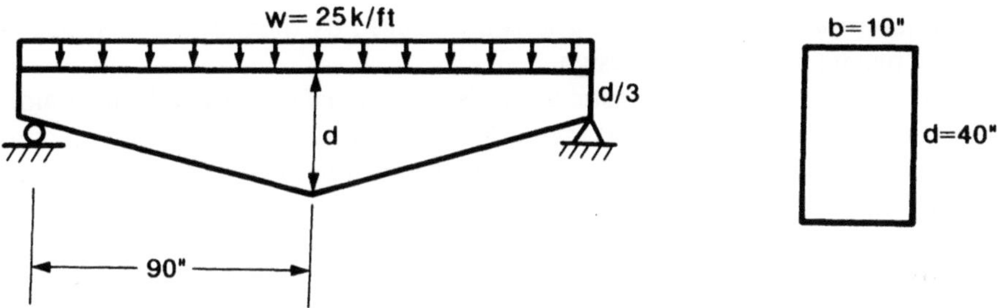

FIGURE P2.23

3
The Tools Used in Plastic Analysis and Design

3.1 Introduction

The methods of simple plastic analysis and design are based on two basic assumptions. The first assumes that the structure is made of a ductile material such as steel that is able to absorb large deformations beyond the elastic limit without the danger of fracture. The second is that the deflections of a structural system under loading are small such that the effect of this upon the overall geometry can be ignored. Herein, we shall discuss the practicality of these assumptions and the limitations introduced by them in the methods of simple plastic analysis and design.

The exact solution in a plastic analysis must satisfy the three basic conditions: *equilibrium, mechanism* (kinematics), and *plastic moment* (yielding) conditions. For simple structures such as the beams and portal frames discussed so far, they are simple enough to be solved by a "direct" approach or "visualized" readily. For more complex structures, it becomes more difficult to satisfy all these three conditions in order to obtain the exact solution immediately. In this situation it is natural to seek simple approximate methods of analysis for these structures, and some general principles and theorems with which the accuracy of these approximate solutions can be assessed. To this end, in this chapter, we shall use the *virtual work method* extensively to establish these fundamental theorems from which simple and approximate techniques of practical plastic methods are derived and developed.

3.2 The Assumption of Ductility of Steel

The fundamental property that makes possible the application of plastic analysis to structural steel design is that the structural material has sufficient ductility. The term *ductility* is defined here as the ability of a material to undergo a large deformation without a significant loss in strength. Structural steels, particularly the most commonly used A36 steel, have this property

120

in abundance. This ductility enables steel structures to reap the benefits of *plastification and moment redistribution* described in the preceding chapters and leads to a higher load-carrying capacity.

3.2.1 Stress-Strain Relationship of Various Types of Steels

The steel is almost entirely composed of iron. Other chemicals are added in small amounts to modify its physical properties such as strength and ductility. For example, the carbon is added to increase the yield strength but causes a reduction in the ductility of the steel.

Depending on the composition and manufacturing process, the structural steels for hot-rolled applications may be classified as carbon steels, high-strength and low-alloy steels, and quenched and tempered steels. Figure 3.1 shows typical stress-strain relationships for these three types of steels.

The carbon steels can further be subdivided into four categories: low carbon (less than 0.15%), mild carbon (0.15–0.29%), medium carbon (0.30–0.59%) and high carbon (0.60–1.7%). A36 steel has 0.25 to 0.29% carbon content. The stress-strain relationship of A36 steel is shown by curve (a) in the figure. High-strength low-alloy steels are obtained by adding small

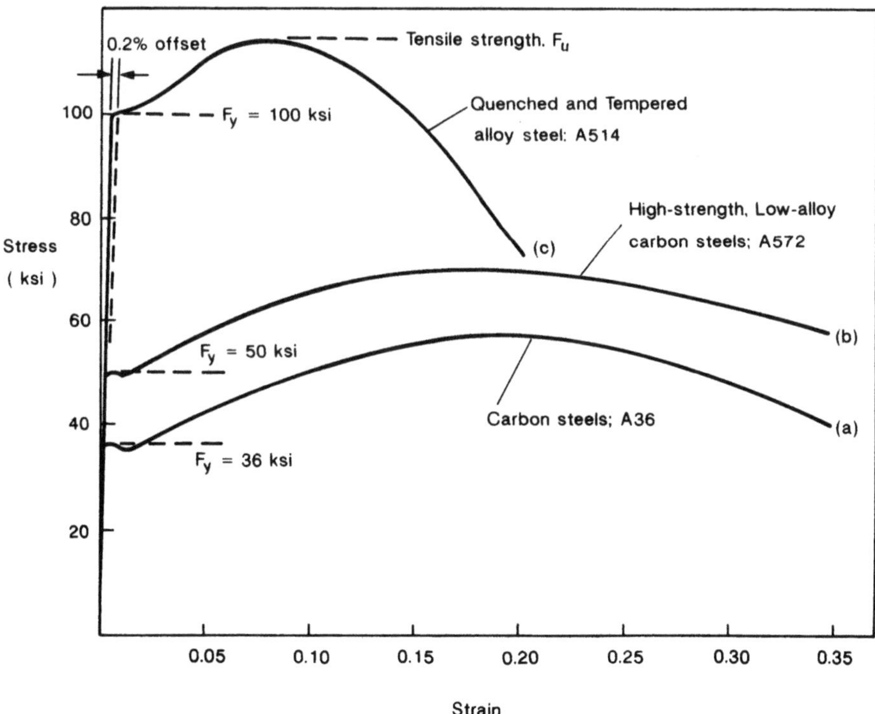

FIGURE 3.1. Typical stress-strain curves of steels.

amounts of alloys such as chromium, copper, manganese, nickel, and phosphorous to carbon steel. The addition of these noncarbon alloys increases strength without causing a significant loss in ductility as shown by curve (b) in Fig. 3.1. Both carbon-steel and high-strength low-alloy steel exhibit well-defined yield points as shown by curves (a) and (b). For both of these types, the strains of up to 8 to 15 times the elastic limit occur without any significant change in the stress. A later increase in strength is exhibited as the material strain hardens.

The strength and ductility of the steel can also be adjusted by a heat-treatment process consisting of *quenching* (rapid cooling with water or oil) and *tempering* (reheating and then allowing it to cool). The quenching results in a higher yield stress but reduces the ductility. The tempering, on the other hand, results in a lower strength but increases the ductility of the material. Curve (c) in Fig. 3.1 is the stress-strain curve of A514 steel obtained by a heat treatment of low-carbon steel. Note that the ductility of heat-treated high-strength steel is much lower than the low-strength A36 and A572 steels.

To ensure adequate ductility for plastic analysis and design, the AISC-LRFD specification (Chapter A) requires that the steel must exhibit a plastic plateau on the stress-strain curve, consequently, $F_y \leq 65$ ksi must be used.

3.2.2 Effects of Unloading and Strain Aging on the Stress-Strain Relationship

Elastic loading and unloading do not effect the stress-strain relationship of steel. However, when steel is unloaded after the yield strain is greatly exceeded, then reloading may give a stress-strain relationship different from that observed during an initial loading.

For example, if a specimen is loaded up to point C in Fig. 3.2 and unloaded to point D, then reloading will occur along path D, C, and E, thus causing a significant reduction in the available ductility. However, if reloading occurs after a certain period of time, the steel may exhibit a different stress-strain relationship due to a phenomenon known as *strain aging*. The strain aging, as shown in Fig. 3.2, restores the original shape of the stress-strain diagram, but the ductility is further reduced.

3.2.3 Idealized Stress-Strain Relationship

To simplify the analysis and design procedures, the actual stress-strain relationship of steels can be idealized as an elastic–perfectly plastic type with two straight lines as shown in Fig. 1.4. Up to the yield stress level, the material is elastic. After the yield stress is reached, the strain is assumed to increase infinitely without any change in stress. For both A36 and A572 steels, this idealization is conservative in the strain-hardening range. Note that both steels are perfectly ductile up to a strain of 0.35, compared with an infinite ductility assumed in the idealization. As will be shown in the forthcoming,

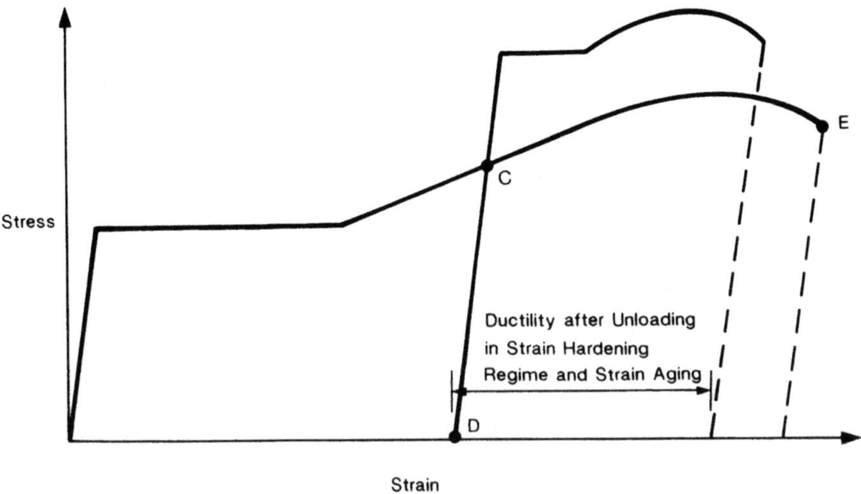

FIGURE 3.2. Effects of unloading in strain-hardening regime and strain aging on ductility of steel.

this ductility of 0.35 for both A36 and A572 steels is more than adequate to materialize the benefits of plastification and redistribution of moments in almost all practical steel structures.

3.2.4 Ductility Requirement for Plastification

The process of plastification enables a section of the members to realize its full plastic moment capacity by successive yielding of all the fibers in the section. Theoretically, the full plastic moment will be attained only if the extreme fibers are capable of sustaining the yield stress up to infinite strain. For all practical purposes, 99% of the full plastic moment capacity of a section can be attained at a curvature of about 2 to 8 times the initial yield curvature as shown in Fig. 2.3. In fact, the moment-carrying capacity of an actual section will exceed the theoretical full plastic moment M_p as soon as the extreme fibers reach the strain-hardening regime.

3.2.5 Ductility Requirements for Moment Redistribution

The process of moment redistribution enables a structural system to attain its plastic limit load by the successive development of plastic hinges to form a failure mechanism. To this end, the first plastic hinge developed in the process must be able to sustain a large rotation capacity with its full plastic moment capacity M_p, while plastic hinges are developed elsewhere in the structure. This required rotation capacity for a plastic hinge can be computed by the virtual work method or the hinge-by-hinge method. Details of these two procedures will be described in Chapters 6 and 7 respectively. LRFD guidelines of maximum width-to-thickness ratio λ_p for preventing local buckling of the cross section (Table 2.1 or LRFD Table B5.1) and of maximum spacing of laterally unbraced length L_{pd} for preventing lateral torsional buckling of members (Chapter F) are based on a rotation capacity of 3.0 times the elastic rotation capacity for nonseismic areas and 7 to 9 times the elastic rotation capacity for seismic areas.

3.3 The Assumption on Small Changes in Geometry of Structures

In simple plastic theory, as in elastic theory, the equilibrium equations are formulated on the basis of a structure's original undeformed geometry. However, the word "plastic" usually gives the impression that a large deflection would be involved at the plastic limit load. In the following, we will show through simple examples that defections at the plastic limit load are of the same order of magnitude as those at the elastic limit load.

Consider a fixed-ended beam with a concentrated load applied at one-third point. The deflections under concentrated load at elastic and plastic limits are determined by the hinge-by-hinge method in Example 1.8.2 as

$$\Delta_{el} = 0.0247 \frac{M_p L^2}{EI} \tag{3.3.1}$$

and

$$\Delta_{pl} = 0.0741 \frac{M_p L^2}{EI}. \tag{3.3.2}$$

Note that the deflection at the plastic limit load is about 3 times that at the elastic limit, while the plastic limit load is more than 33% higher than the elastic limit load. Similarly, the midspan deflection of a fixed-ended beam under a uniformly distributed lateral load at the elastic and plastic limit loads are found to be

$$\Delta_{el} = \frac{M_p L^2}{32 EI} \tag{3.3.3}$$

and

$$\Delta_{pl} = \frac{M_p L^2}{12 EI}. \tag{3.3.4}$$

For this case, the plastic limit deflection is only 2.7 times the elastic limit deflection while the plastic limit load is more than 33% higher than the elastic limit load.

3.4 The Equation of Virtual Work

3.4.1 The Equation

The *virtual work equation* relates a system of forces in equilibrium to a system of compatible displacements. Stated simply, if a body in equilibrium is given a set of small *compatible* displacement, then the work done W_E by the external loads on these external displacements is equal to the work done W_I by the internal forces on the internal deformations, i.e.,

$$W_E = W_I. \tag{3.4.1}$$

Note that the external displacements must be compatible with the internal deformations. However, these internal deformations need not be real, i.e., they need not correspond to any actual or possible state of equilibrium. The internal forces must be in equilibrium with the external forces, but they need not be the actual internal forces due to the external loads. They bear no relationships with the external or internal displacements. Any *equilibrium set* of forces may be used in the equation of virtual work. The structure can be arbitrarily distorted to produce a *displacement set* without reference to any loading system. Since the equilibrium set and the displacement set are not related in any way, the adjective *virtual* is used to describe their work equation.

In plastic methods, only mechanism-type deformations are considered in which internal deformations are assumed to be concentrated at plastic hinges, which are assumed to be connected by rigid members. As a result, the virtual work equation (3.4.1) for framed structures can be written in the explicit form as

$$
\begin{array}{c}
\text{Equilibrium Set} \\
\overbrace{\qquad\qquad} \\
\sum P_i \delta_i = \sum M_i \theta_i \\
\underbrace{\qquad\qquad} \\
\text{Displacement Set}
\end{array}
\tag{3.4.2}
$$

where P_i is an external load and M_i is the internal moment at a hinge location; both Ps and Ms together constitute an *equilibrium set* and therefore must be in equilibrium; δ_i is the displacement at the load point P_i and in direction of the load P_i; and θ_i is the rotation at a hinge location with the moment M_i, both δs and θs together constitute a *displacement set* and therefore must be compatible with each other.

3.4.2 Sign Convention

The use of the virtual work equation, particularly for the *moment check* used in the next chapter, requires proper signs with both moments and rotations. The following sign convention will be used when applying the virtual work equation (3.4.2): *The moments and rotations causing tension on the side of the dotted line are positive and vice-versa* (see Figs. 3.10 and 3.11).

In contrast, in calculating the actual plastic work done on a plastic collapse mechanism described in Chapter 2, the plastic moment M_p and the corresponding rotation θ are always positive for energy dissipation at a plastic hinge. Thus, no sign convention is needed for calculating the plastic work equation for a mechanism solution. This will be described in detail in Chapter 5.

3.4.3 Work Done by Distributed Loads

The external work done by concentrated loads can simply be determined as the product of the load and the corresponding displacement. However, when the load is distributed, Fig. 3.3(a), the external work W_E should be calculated by carrying out the following integration

$$W_E = \int_L (w\,dx)y \tag{3.4.3}$$

where w, dx, y, and L are shown in Fig. 3.3. If w is a uniformly distributed load, then w can be taken out of the integral. It follows that the remaining integral represents the area of the displacement diagram such as that shown in Fig. 3.3(b). In plastic methods, the rotations are concentrated at plastic hinges and the members between the hinges are straight. Thus, the work done

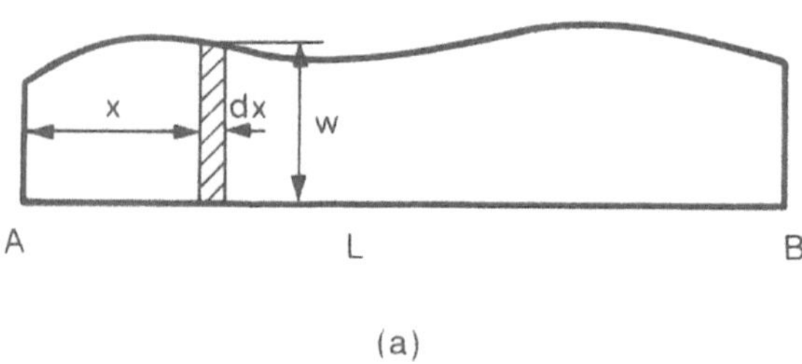

(a)

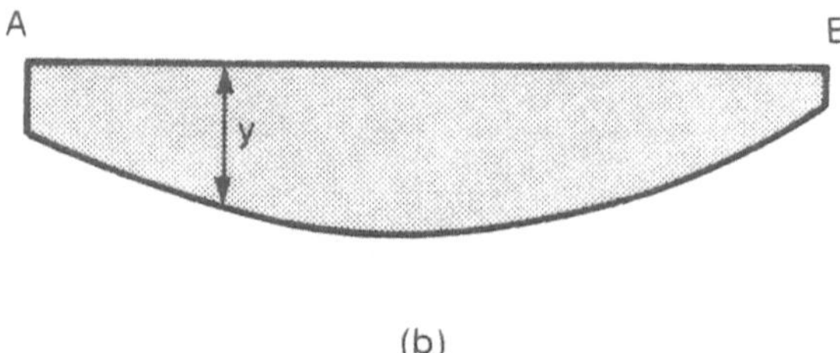

(b)

FIGURE 3.3. Work done by a distributed load: (a) distributed load and (b) deflection.

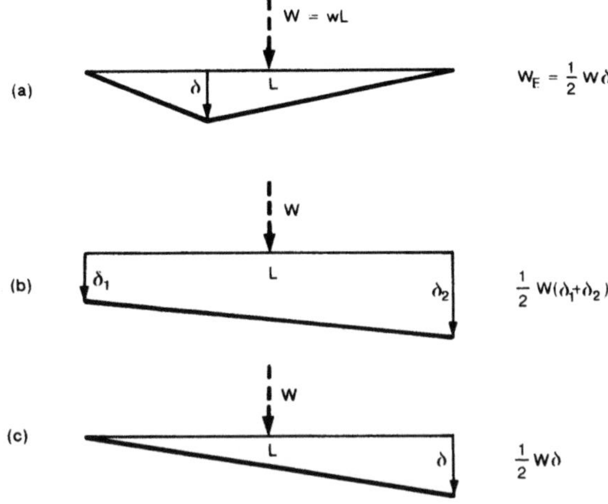

FIGURE 3.4. Work done by uniformly distributed loads on horizontal members.

can be found directly, without carrying out the integration for area, by simply calculating the maximum deflections of triangles and trapezoids as shown in Figs. 3.4(a), (b), and (c).

For inclined members under a vertical uniformly distributed load, such as those in gable frames (Fig. 3.5), W_E can be calculated as

$$W_E = \frac{1}{2}(W \cos \theta)\Delta \tag{3.4.4}$$

where $W = wL$ is the total uniformly distributed load on the member; and θ, Δ, and L are shown in Fig. 3.5(a). For practical applications, it is more convenient to use the formulas in terms of the vertical deflection δ than Δ as shown in of Fig. 3.5(a), (b), and (c) for the computation of external work W_E.

3.4.4 Applications of the Virtual Work Equation

In plastic methods, the virtual work equation has the following five major applications.

1. Obtain the geometrical relationships of mechanism motion by assuming appropriate equilibrium sets.
2. Make a moment check for a given mechanism by assuming appropriate displacement sets.
3. Prove the uniqueness, unsafe, and safe theorems.
4. Obtain bounding solutions: upper-bound load factors in analysis problems and lower-bound plastic moments in design problems.
5. Calculate deflections at collapse load.

The power and simplicity of the virtual work equation can be brought out best by simple examples. Herein, we shall present simple examples illustrat-

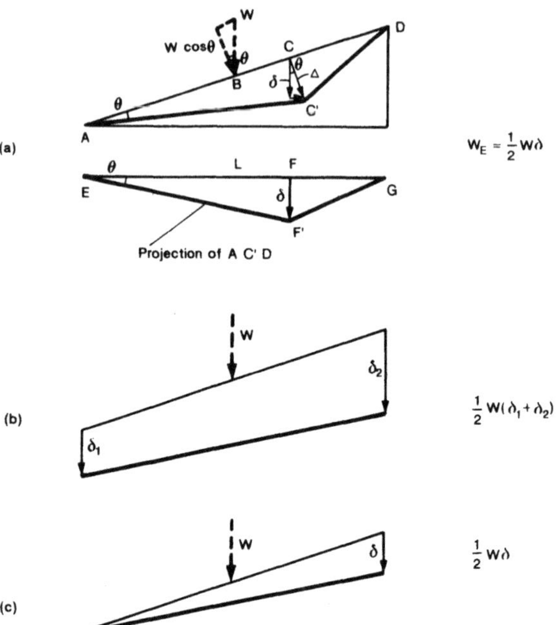

FIGURE 3.5. Work done by uniformly distributed loads on inclined members.

ing the first two applications. The plastic theory will be proved in the next section, followed by the calculations of upper and lower bounds on load factor and plastic moment of simple beams and frames in this chapter and of more complex structures in Chapters 4 and 5. Deflection calculations will be dealt with in Chapter 6.

Example 3.4.1. The mechanism shown in Fig. 3.6 is considered in a plastic frame analysis. Use the virtual work equation to determine the relationship between the angles θ_1, θ_2, and θ_3.

Solution: To relate θ_2 and θ_2, the equilibrium set shown in Fig. 3.7 is generated by simply applying a unit compressive internal force in member BC and then adding the necessary external moments to achieve moment equilibrium for the other two members. The virtual work done by this equilibrium set (Fig. 3.7) on the given displacement set (Fig. 3.6) gives the relationship as

$$20\theta_1 - 25\theta_2 = 0,$$

which leads to the angular relationship

$$\theta_1 = \frac{5}{4}\theta_2. \tag{3.4.5}$$

To relate θ_1 and θ_3, the equilibrium set shown in Fig. 3.8 is generated by first applying an internal compressive force in member CD with its vertical and horizontal components directly proportional to the member slope and

FIGURE 3.6. The frame mechanism to be analyzed.

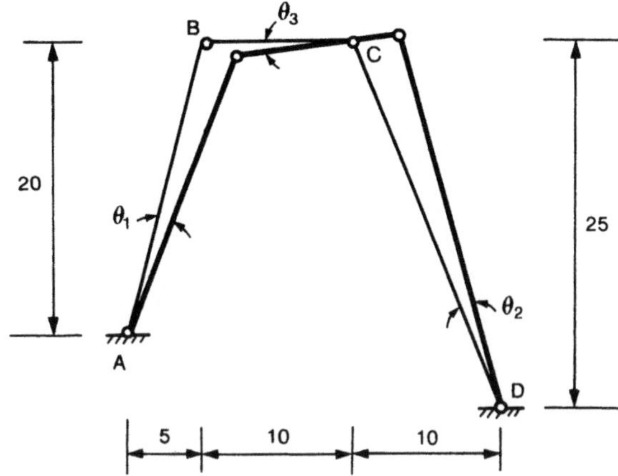

FIGURE 3.7. An equilibrium set for relating θ_1 and θ_2 of Fig. 3.6.

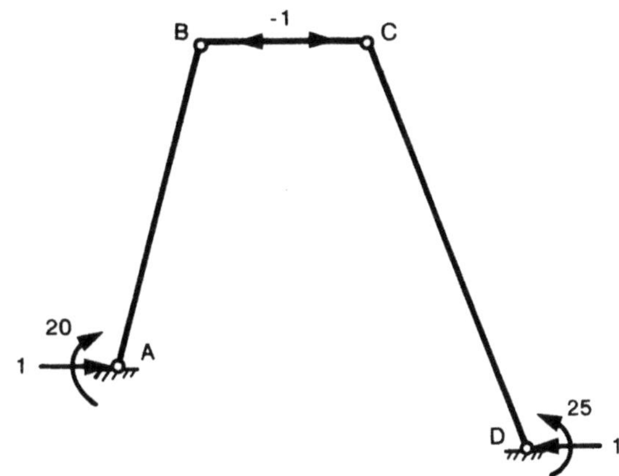

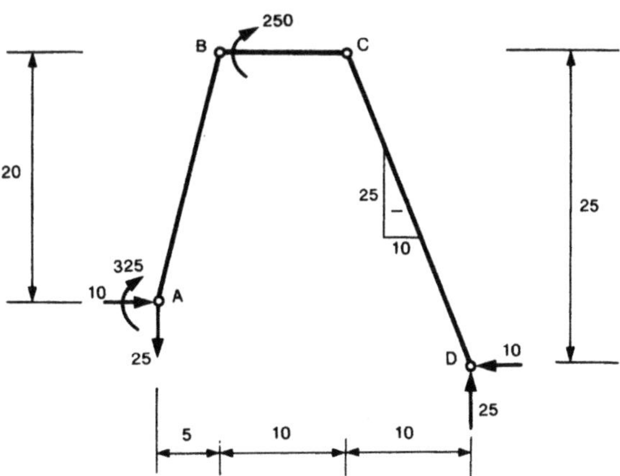

FIGURE 3.8. An equilibrium set for relating θ_1 and θ_3 of Fig. 3.6.

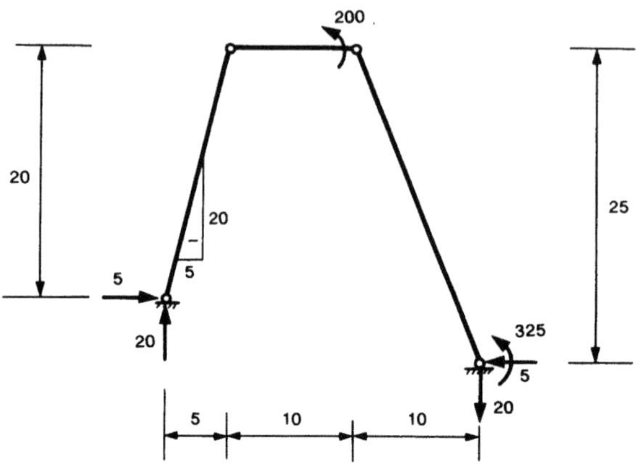

FIGURE 3.9. An equilibrium set for relating θ_2 and θ_3 of Fig. 3.6.

then adding external moments to meet equilibrium requirements for the other two members. The virtual work equation for these two sets now provides

$$325\theta_1 - 250\theta_3 = 0,$$

which leads to the desired angular relationship

$$\theta_3 = \frac{13}{10}\theta_1. \tag{3.4.6}$$

Alternatively, if we use the equilibrium set of Fig. 3.9, we obtain

$$200\theta_3 - 325\theta_2 = 0,$$

which gives an alternative relationship

$$\theta_3 = \frac{13}{8}\theta_2. \tag{3.4.7}$$

Equations (3.4.5) and (3.4.6) lead to the same relationship between θ_1 and θ_3 as in Eq. (3.4.7). Any equilibrium sets can be used for this purpose, but the procedure shown in Figs. 3.7 to 3.9 provides a simple and practical procedure of obtaining geometric relations easily.

Example 3.4.2. The mechanism of a rectangular frame shown in Fig. 3.10(a) provides an upper-bound solution of $P = 16M_p/3L$. This solution will be exact only if the moment condition ($M \leq M_p$) is satisfied everywhere in the frame. Use the virtual work equation to check the moments corresponding to the given plastic collapse mechanism.

Solution: The moment diagram corresponding to the given mechanism is shown in Fig. 3.10(b) with an unknown moment M_B. M_B can be determined by using either of the two displacement sets shown in Fig. 3.11 in conjunction with the equilibrium set of Fig. 3.10(b). Using the beam mechanism, the

FIGURE 3.10. The collapse mechanism and its corresponding moment diagram: (a) mechanism and (b) moment diagram.

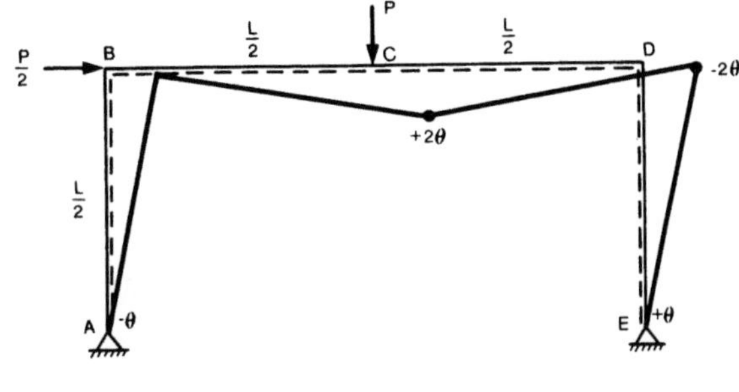

(a) Mechanism

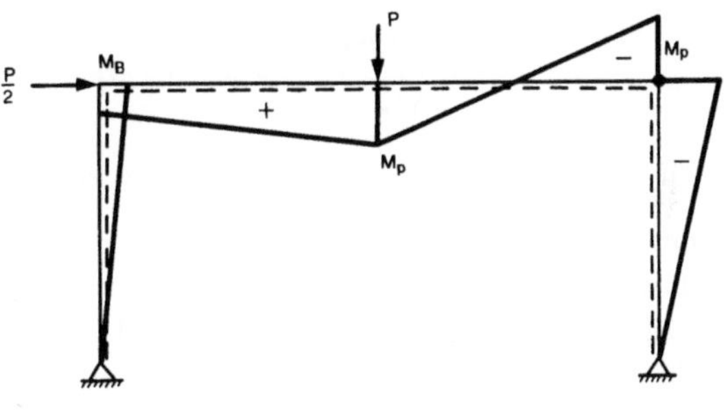

(b) Moment Diagram

FIGURE 3.11. Two possible geometry sets for finding M_B of Fig. 3.10(b): (a) beam mechanism and (b) sway mechanism.

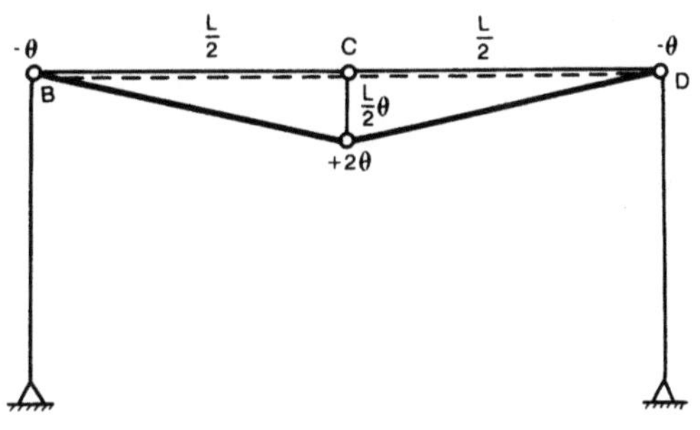

(a) Beam Mechanism

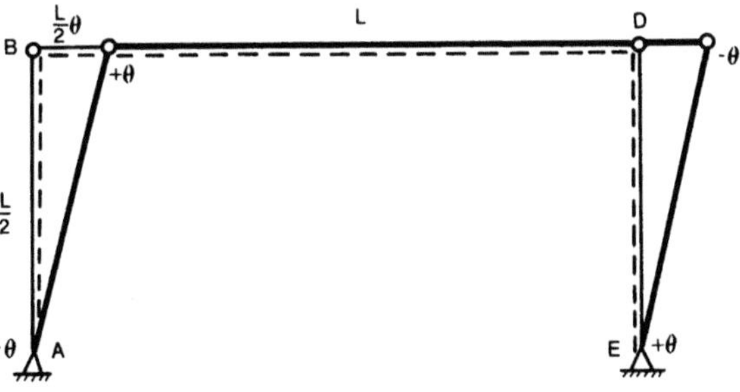

(b) Sway Mechanism

virtual work of Fig. 3.10(b) on Fig. 3.11(a) gives

$$P\left(\frac{L\theta}{2}\right) = (+M_B)(-\theta) + (+M_p)(+2\theta) + (-M_p)(-\theta).$$

Substituting $P = 16M_p/3L$, we have

$$M_B = \frac{M_p}{3}. \tag{3.4.8}$$

Since $M_B < M_p$, the moment condition is satisfied and the moment check is complete. Alternatively, using the sway mechanism, Fig. 3.11(b), and the equilibrium set of Fig. 3.10(b), we have the following virtual work equation:

$$\frac{P}{2}\left(\frac{L\theta}{2}\right) = (+M_B)(+\theta) + (-M_p)(-\theta),$$

which with $P = 16M_p/3L$ gives the same value of M_B as Eq. (3.4.6).

Example 3.4.3. The moment diagram corresponding to the plastic collapse mechanism of a gable frame is shown in Fig. 3.12 with unknown moments M_B, M_D, and M_E. Use the virtual work equation to perform the moment check. The collapse load P for this mechanism is $3M_p/5L$.

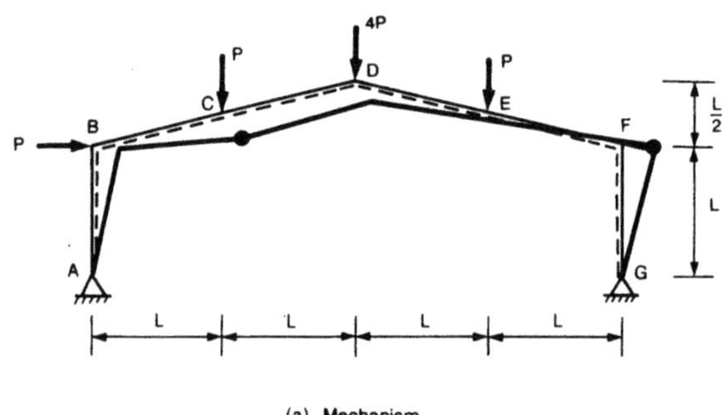

(a) Mechanism

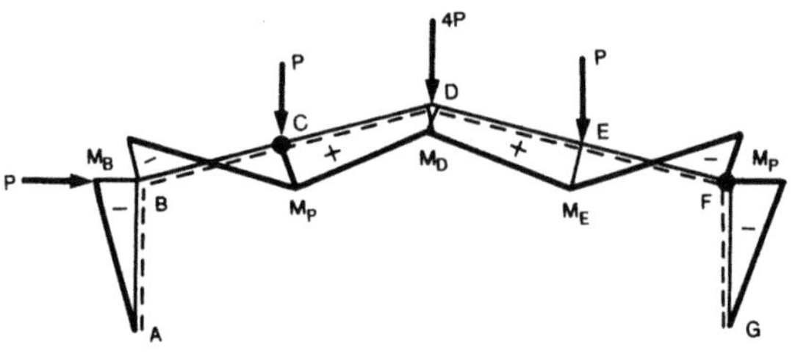

(b) Moment Diagram

FIGURE 3.12. A collapse mechanism and its corresponding moment diagram: (a) mechanism and (b) moment diagram.

Solution: The unknown moment M_B can be determined easily by assuming the sway mechanism of Fig. 3.13 in conjunction with the moment diagram of Fig. 3.12(b). The virtual work equation provides

$$P(L\theta) = (-M_B)(+\theta) + (-M_p)(-\theta).$$

With $P = 3M_p/5L$, M_B has the value

$$M_B = \frac{2}{5}M_p < M_p \quad \text{okay.}$$

M_D can be determined easily by using the beam mechanism of Fig. 3.14 with the moment diagram of Fig. 3.12(b). The virtual work equation gives

$$P(L\theta) = (-M_B)(-\theta) + (+M_p)(+2\theta) + (+M_D)(-\theta),$$

which with $P = 3M_p/5L$ and $M_B = (2/5)M_p$ provides

$$M_D = \frac{9}{5}M_p > M_p \quad \text{not okay.}$$

Since $M_D > M_p$, the moment condition is not satisfied and therefore $P = 3M_p/5L$ is an upper-bound solution, not the exact solution.

3.5 The Fundamental Theorems

Here, as in the elastic solution that must satisfy equilibrium, compatibility, and elastic moment-curvature conditions, the correct plastic solution must satisfy equilibrium, mechanism, and moment conditions. As the structure under examination becomes more complex, it becomes increasingly difficult to satisfy all three conditions simultaneously. In theory, it is always possible

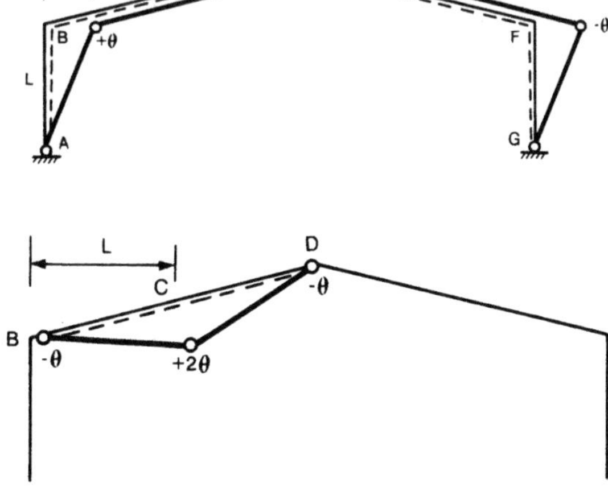

FIGURE 3.13. Sway mechanism for finding M_B of Fig. 3.12(b).

FIGURE 3.14. Beam mechanism for finding M_D of Fig. 3.12(b).

to write the relevant equilibrium equations compatible with a deformation system and satisfy the moment conditions. In practice, the resulting equations are virtually unsolvable for practicing engineers. The alternative to an exact solution is to evolve one in which only one or two of the three basic conditions are fully met so that an approximate solution can be obtained quickly. In order to judge the accuracy of these approximate solutions, several general principles and basic plastic theorems will be established in this section, against which the accuracy and meaning of the approximate methods can be measured.

All the basic theorems of plastic theory are based on the assumption that the loads acting on a structure will increase proportionally and will not be allowed to vary randomly and independently. These loads may be thought of as having their working values; as the loads increase, their working values are multiplied by a common factor λ, the *load factor*. The basic theorems to be established here are concerned with the load factor λ_c at the collapse of the structure. In establishing the theorems, the equation of virtual work is used extensively.

An important feature of plastic theory to be established in the forthcoming is the extent to which an engineer's intuition of structural behavior has been used in the development of practical solutions [2.3, 3.1–3.6]. It is the aim of this chapter to make full use of such intuitive ideas in conveying an understanding of the principles involved. These intuitive ideas are expressed first in formal statements, followed by proofs of the theorems of plasticity, after which methods of plastic analysis are illustrated by simple examples.

The more complex problems, which require a deeper appreciation of the methods of plastic analysis, are presented in Chapters 4, 5, and 6. Emphasis in these three chapters is placed on the methods of analysis that are suitable primarily to hand calculations, since a thorough understanding of plastic theory is best attained by the direct working of these examples. In Chapters 7 and 8, some methods suitable for computer application are introduced to provide the necessary transition from the current simple plastic methods to the more sophisticated analysis techniques that hold the promise of more realistic prediction of load effects and frame performance leading to a direct analysis of inelastic *strength* and *stability* for frame design. This is known as *advanced inelastic analysis*. It is a difficult subject in which plasticity and stability theories are taken together and where the structural engineering profession is going. Chapter 8 presents an introduction to this difficult subject and provides a balance view of the significance of stability theory in relation to plastic theory.

3.5.1 *Uniqueness Theorem*

Statement and Description. *The load factor λ_c at collapse has a definite value. In other words, as the loads on the structure are gradually increased, i.e., as the value of λ is increased, collapse occurs at one definite value of $\lambda = \lambda_c$.*

Proof: We shall show this by contradiction. First, assume the theorem is false. Thus, it is possible to have two collapse mechanisms of a structure under the given loads with two different load factors λ' and λ''. Both mechanisms will satisfy the equilibrium and moment conditions. Denote the displacements and rotations of the first mechanism by δ' and θ', and of the second one by δ'' and θ'', respectively. These displacement sets are compatible. The associated bending moment diagram corresponding to the first collapse mechanism, denoted by M', will be in equilibrium with the set of applied loads $\lambda'W$. This moment diagram satisfies the moment condition, $|M'| \leq M_p$. Similarly, the bending moment diagram corresponding to the second mechanism, denoted by M'', will be in equilibrium with the set of applied loads $\lambda''W$ and satisfy the moment condition, $|M''| \leq M_p$.

The actual plastic work for the first mechanism has the usual form

$$\sum (\lambda'W)\delta' = \sum M_p|\theta'|. \tag{3.5.1}$$

For the proportional loading case, λ' on the left-hand side of Eq. (3.5.1) can be taken out of the summation sign as

$$\lambda' \sum W\delta' = \sum M_p\theta'. \tag{3.5.2}$$

Now, applying the virtual work equation to the displacement set of the first mechanism and the equilibrium set of the second mechanism, we have

$$\lambda'' \sum W\delta' = \sum M''\theta'. \tag{3.5.3}$$

If the two mechanisms have certain common plastic hinges, then $M''\theta'$ terms on the right-hand side of the equation will be $M_p\theta'$ for the common plastic hinges and less than $M_p\theta'$ for other hinges in the first mechanism. Thus, we have

$$\lambda'' \sum W\delta' \leq \sum M_p|\theta'|. \tag{3.5.4}$$

Comparing Eq. (3.5.2) with expression (3.5.4), we have

$$\lambda'' \leq \lambda'. \tag{3.5.5}$$

Similarly, by comparing the virtual work done by the equilibrium set corresponding to mechanism 1 on mechanism 2, with the actual plastic work done for mechanism 2, we can write

$$\lambda' \leq \lambda''. \tag{3.5.6}$$

Inequalities (3.5.5) and (3.5.6) can be satisfied simultaneously only if the two load factors have the same value. This value is *unique* and denoted by λ_c, the *collapse load factor*.

We will now illustrate and prove the *uniqueness theorem* again using a simple structure, the fixed-ended beam, shown in Fig. 3.15(a). Here, the first mechanism and its associated moment diagram, and the second mechanism and its associated moment diagram are, respectively, shown in parts (b), (c), (d), and (e). First, we write the actual plastic work equation for the first

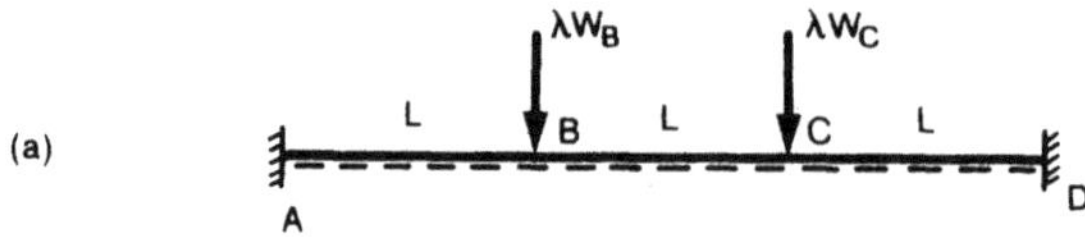

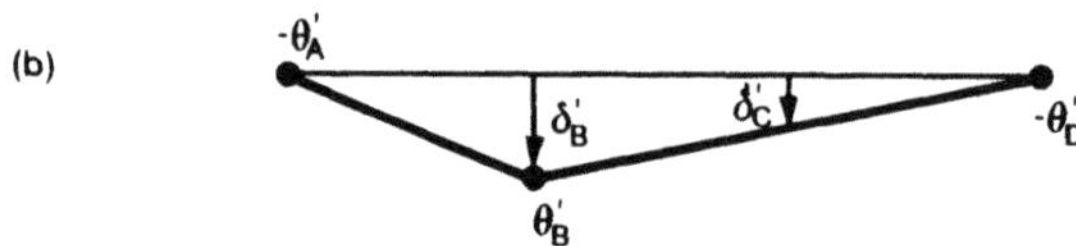

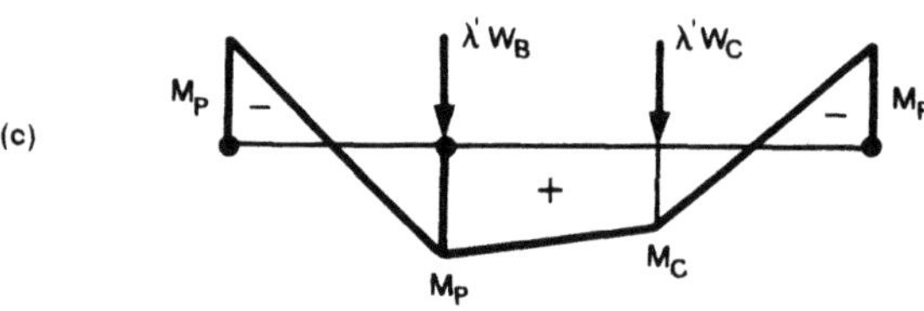

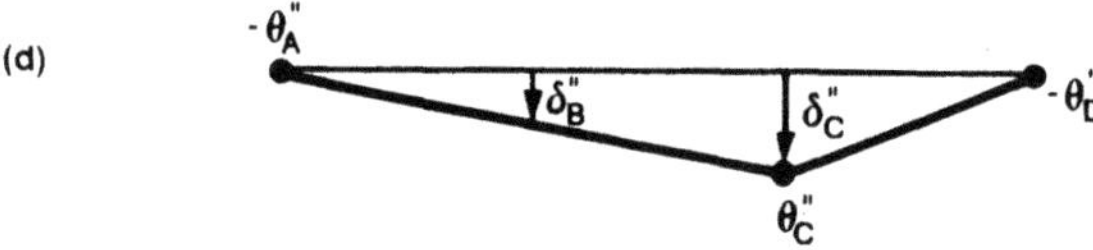

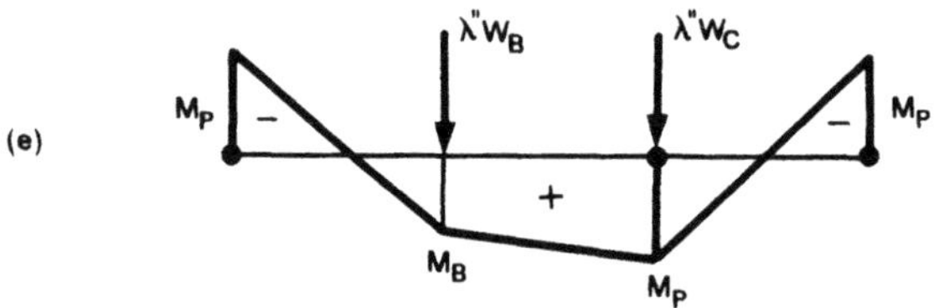

FIGURE 3.15. A fixed-ended beam, its mechanisms, and moment diagrams to illustrate the uniqueness theorem.

mechanism [parts (b) and (c)] and obtain

$$(\lambda' W_B)\delta'_B + (\lambda' W_C)\delta'_C = M_p|\theta'_A| + M_p|\theta'_B| + M_p|\theta'_D|. \qquad (3.5.7)$$

Now, we write the virtual work equation for the displacement set of the first mechanism in part (b) and the equilibrium set of the second mechanism in part (e) and get

$$(\lambda'' W_B)\delta'_B + (\lambda'' W_C)\delta'_C = (-M_p)(-\theta'_A) + (+M_B)(+\theta'_B) + (-M_p)(-\theta'_D). \qquad (3.5.8)$$

The subtraction of Eq. (3.5.8) from Eq. (3.5.7) leads to

$$(\lambda' - \lambda'')(W_B\delta'_B + W_C\delta'_C) = \theta'_B(M_p - M_B). \qquad (3.5.9)$$

Since M_B must be less than or at the most equal to M_p, Eq. (3.5.9) reduces to

$$\lambda' \geq \lambda''. \tag{3.5.10}$$

Similarly, write the plastic work equation for the second mechanism [parts (d) and (e)] and obtain

$$(\lambda'' W_B)\delta_B'' + (\lambda'' W_C)\delta_C'' = M_p|\theta_A''| + M_p|\theta_C''| + M_p|\theta_D''|. \tag{3.5.11}$$

Again, write the virtual work equation for the displacement set of the second mechanism [part (d)] with the equilibrium set of the first mechanism [part (c)] and obtain

$$(\lambda' W_B)\delta_B'' + (\lambda' W_C)\delta_C'' = (-M_p)(-\theta_A'') + (+M_C)(+\theta_C'') + (-M_p)(-\theta_D''). \tag{3.5.12}$$

The subtraction of Eq. (3.5.12) from Eq. (3.5.11) leads to

$$(\lambda'' - \lambda')(W_B\delta_B'' + W_C\delta_C'') = \theta_C''(M_p - M_c). \tag{3.5.13}$$

Since $M_C \leq M_p$, Eq. (3.5.13) reduces to

$$\lambda'' \geq \lambda'. \tag{3.5.14}$$

Inequalities (3.5.10) and (3.5.14) can be satisfied only when

$$\lambda'' = \lambda'. \tag{3.5.15}$$

Note that it has not been established that the collapse mechanism is unique or that the bending moment diagram at collapse is unique. In fact, it is possible to have two collapse mechanisms leading to the same collapse load factor λ_c. The uniqueness theorem simply states that the collapse load factor λ_c determined from the three basic conditions (mechanism, equilibrium, and moment) has a unique value.

3.5.2 Unsafe Theorem

Statement and Description. *If the collapse mechanism of a structure is guessed and its plastic collapse equation is written, then the load factor so computed will always be greater than, or at best equal to, the true value λ_c. It gives an unsafe solution. In other words, if all the loads are increased slowly in proportion to their working values, actual collapse would have already occurred before the formation of the guessed mechanism, unless it happens to be the correct mechanism.*

From the plastic work equation such as Eq. (3.5.7), we can determine λ for a given value of M_p (analysis problem); or we can determine M_p for a given value of λ (design problem). In the design sense, the unsafe theorem states that the value of M_p resulting from the analysis of an assumed mechanism will always be smaller than or at most equal to the actual required value.

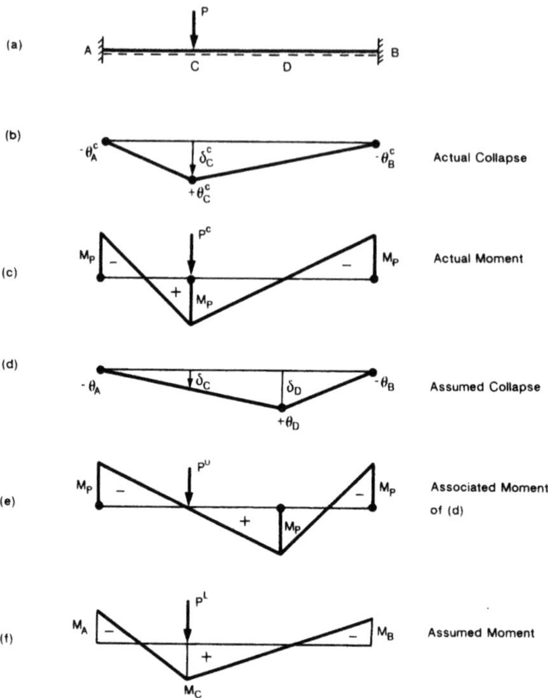

FIGURE 3.16. Mechanisms and moment diagrams for a fixed-ended beam for the illustration of unsafe and safe theorems.

Proof: Consider, for example, a fixed-ended beam with a concentrated lateral load P shown in Fig. 3.16(a). The actual collapse mechanism will form when plastic hinges are developed at two ends and at the location of the concentrated load P [Fig. 3.16(b)]. The actual moment diagram is shown in part (c). Now, assume an arbitrary collapse mechanism shown in part (d). The corresponding moment diagram is shown in part (e): First, we write the plastic work equation for the assumed mechanism in part (d) and we obtain

$$P^U\delta_C = M_p|\theta_A| + M_p|\theta_D| + M_p|\theta_B|. \tag{3.5.16}$$

Note that all the terms on the right-hand side are plastic work and are therefore positive. Now apply the virtual work equation to the assumed mechanism [part (d)] and the equilibrium set of the actual moment [part (c)] and obtain

$$P^c\delta_C = (-M_p)(-\theta_A) + (+M_D)(+\theta_D) + (-M_p)(-\theta_B). \tag{3.5.17}$$

Subtract Eq. (3.5.17) from Eq. (3.5.16) and get

$$(P^U - P^c)\delta_C = (M_p - M_D)\theta_D. \tag{3.5.18}$$

Since the moment diagrams in part (c) is the actual moment distribution, the moment condition must always be satisfied, i.e., $M_D \leq M_p$. Thus, Eq. (3.5.18) leads to the unsafe theorem

$$P^U \geq P^c \tag{3.5.19}$$

where P^c is the actual collapse load.

3.5.3 Safe Theorem

Statement and Description: *If a bending moment diagram in equilibrium with the applied external loads with load factor λ can be obtained such that the full plastic moment condition is not exceeded at any cross section of the structure, then the load factor λ computed from this moment diagram will be less than or at most equal to the true collapse load factor λ_c. In other words, if at a load factor λ it is possible to find a bending moment diagram that satisfies both the equilibrium and moment conditions but not necessarily the mechanism condition, then the structure will stand up and not collapse at that load factor, unless it happens to be the actual or correct solution.*

In the design problems if plastic moment M_p is determined from an assumed moment diagram described earlier, it will always be greater than or at least equal to (safe) the true required plastic moment. The engineers can intuitively visualize the distribution of moments and forces and calculate the corresponding plastic moment or plastic limit load. In fact, the basic concept of the safe theorem has been frequently used by practicing engineers in the design of structures without knowing the existence of such a theorem.

Proof: Again referring to the fixed-ended beam of Fig. 3.16(a), select an arbitrary equilibrium moment diagram shown in Fig. 3.16(f), with M_A, M_B, and M_C less than or equal to M_p. Apply the virtual work equation to the actual collapse mechanism of part (b) and the chosen equilibrium moment diagram of part (f) and obtain

$$P^L \delta_C^c = (-M_A)(-\theta_A^c) + (+M_C)(+\theta_C^c) + (-M_B)(-\theta_B^c). \tag{3.5.20}$$

Now, we write the actual plastic work equation for the collapse mechanism in part (b) and obtain

$$P^c \delta_C^c = M_p|\theta_A^c| + M_p|\theta_C^c| + M_p|\theta_B^c|. \tag{3.5.21}$$

Subtraction of Eq. (3.5.21) from Eq. (3.5.20) leads to

$$(P^L - P^c)\delta_C^c = (M_A - M_p)\theta_A^c + (M_C - M_p)\theta_C^c + (M_B - M_p)\theta_B^c. \tag{3.5.22}$$

Since M_A, M_B, and M_C are all less than or at most equal to M_p, it follows that

$$(P^L - P^c) \leq 0. \tag{3.5.23}$$

Thus, inequality (3.15.23) leads to the safe theorem

$$P^L \leq P^c. \tag{3.5.24}$$

3.5.4 Corollaries of the Safe and Unsafe Theorems

The following corollaries can be stated as a result of the safe and unsafe theorems.

1. *If a material of negligible self-weight is added to a structure or a restraint is imposed, the structure cannot thereby be weakened.*

    This follows from the safe theorem. Since the moment diagram for the collapse state of the unstrengthened structure must satisfy the moment condition, the same moment diagram will certainly satisfy the moment condition of the strengthened structure. Thus, the *strengthened structure* cannot collapse at a load less than that of the unstrengthened structure.

2. *The removal of a material or constraint from a structure will only weaken the structure.*

    This follows from the unsafe theorem. Since the mechanism condition must be satisfied at the collapse state of the unweakened structure, it will certainly be satisfied for the weakened structure under the same load. Thus, *weakened structure* cannot resist a load higher than the collapse load for the unweakened structure.

3. A useful corollary of the unsafe theorem is that *the true load factor at collapse is the smallest possible one that can be determined from a consideration of all possible mechanisms of collapse.* This fact is very useful in the method of "combination" of mechanisms, to be presented in Chapter 5.

In short, the three limit theorems can be summarized as follows:

$$\lambda = \lambda_c \begin{cases} \text{MECHANISM CONDITION } \lambda \geq \lambda_c \\ \text{EQUILIBRIUM CONDITION } \lambda \leq \lambda_c \\ \text{MOMENT CONDITION.} \end{cases}$$

3.6 Upper- and Lower-Bound Solutions Based on the Limit Theorems

The safe and unsafe limit theorems may be used together to obtain upper and lower bounds on the value of load factor λ_c in an analysis problem or on the value of the plastic moment M_p in a design problem. Herein, we shall illustrate the calculations of upper- and lower-bound techniques for both analysis and design problems with the use of a simple portal frame.

3.6.1 Analysis Example

Determine upper and lower bounds on the load factor λ_c for a rectangular frame shown in Fig. 3.17. Use the mechanism shown in Fig. 3.18.

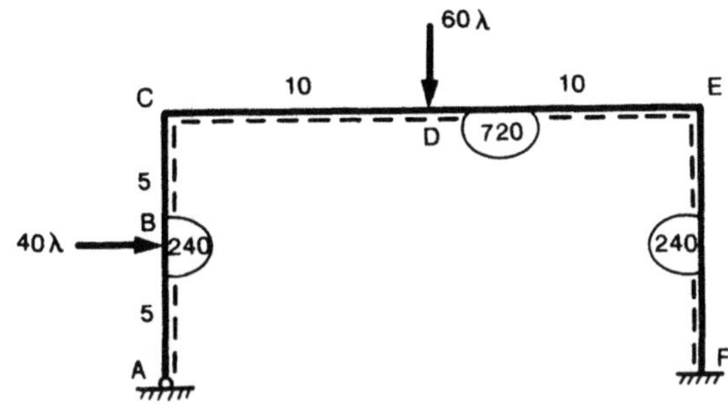

FIGURE 3.17. A simple rectangular frame with load factor λ.

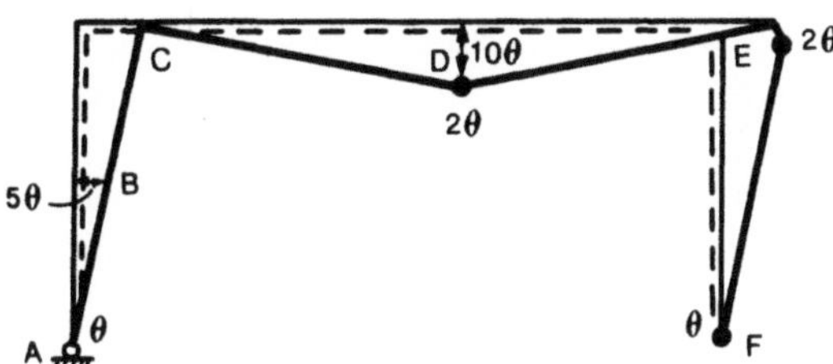

FIGURE 3.18. An assumed sway mechanism.

Solution: The assumed mechanism has three plastic hinges at D, E, and F. The plastic work equation for this mechanism can be written as

$$(40\lambda)(5\theta) + (60\lambda)(10\theta) = (720)(2\theta) + (240)(2\theta) + (240)(\theta). \qquad (3.6.1)$$

Note that plastic work is always positive, and there is no need to consider sign convention. Equation (3.6.1) provides an unsafe or upper-bound solution on the value of λ_c as

$$\lambda = 2.7. \qquad (3.6.2)$$

A lower bound can also be determined by checking the moments corresponding to this mechanism. This is described in the forthcoming. Moments at D, E, and F are known to be 720, 240, and 240, respectively. Unknown moments at B and C can be determined by the application of the virtual work equation as follows.

Moment at C: Apply the virtual work equation to the equilibrium and displacement sets shown in Fig. 3.19 and get

$$(+M_C)(-\theta) + (720)(+2\theta) + (-240)(-\theta) = (60)(2.7)(10\theta), \qquad (3.6.3)$$

which gives

$$M_C = 60. \qquad (3.6.4)$$

Moment at B: Apply the virtual work equation to the equilibrium and displacement sets of Fig. 3.19(b) and obtain

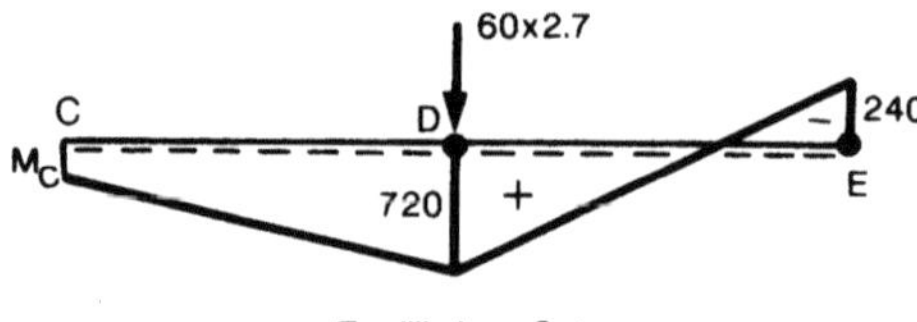

FIGURE 3.19. Equilibrium and geometry sets for (a) M_C and (b) M_B.

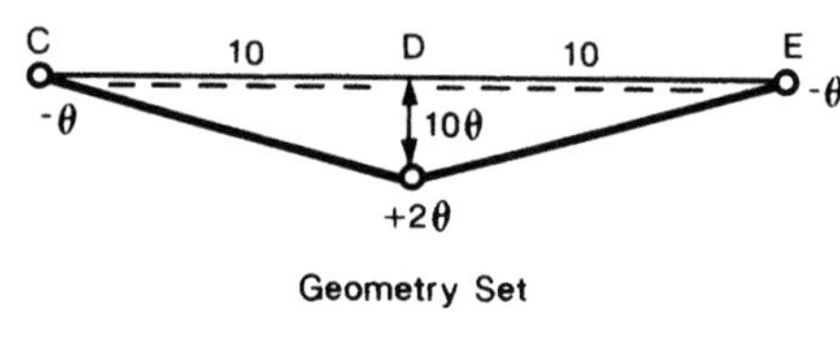

(a)

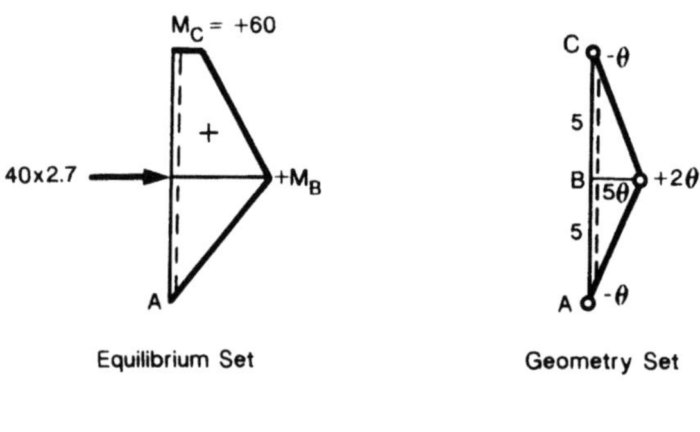

(b)

$$(60)(-\theta) + (+M_B)(+2\theta) = (40)(2.7)(5\theta), \tag{3.6.5}$$

which provides

$$M_B = 300. \tag{3.6.6}$$

Since $M_B > (M_p = 240)$ the moment condition was violated at B. However if we scale down the whole moment diagram of the frame by a factor $240/300 = 0.8$ by simply reducing the value of λ to $2.7 \times 0.8 = 2.16$, then the moment condition will be met throughout the frame. However, at this load level, the mechanism cannot develop. Since at $\lambda = 2.16$, both the equilibrium and moment conditions are satisfied without meeting the mechanism condition; it follows from the safe theorem that $\lambda = 2.16$ is a safe solution, i.e., 2.16 is a lower bound on the value of λ_c. Thus, for the assumed mechanism, lower and upper bounds are obtained as

$$2.16 \leq \lambda_c \leq 2.7. \tag{3.6.7}$$

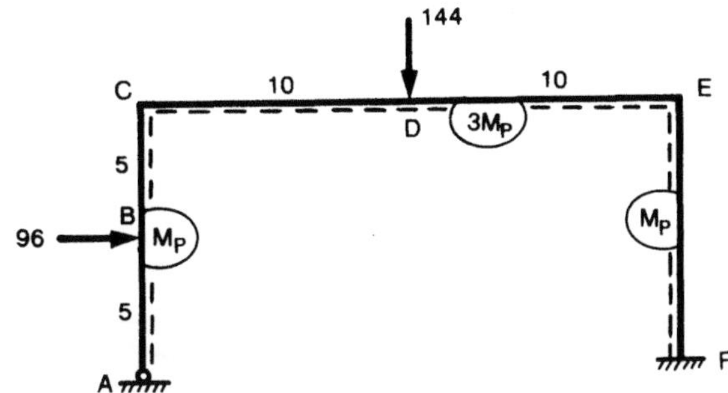

FIGURE 3.20. A simple rectangular frame with M_P to be determined.

The bounds can be further narrowed down by considering other mechanisms. Note that the actual collapse load factor for this frame is $\lambda = \lambda_c = 2.4$, corresponding to the formation of plastic hinges at B, E, and F.

3.6.2 Design Example

Determine upper and lower bounds on the value of M_p for the frame shown in Fig. 3.20. Again, use the mechanism of Fig. 3.18.

Solution: The plastic work equation for the assumed mechanism can be expressed as

$$(96)(5\theta) + (144)(10\theta) = (3M_p)(2\theta) + (M_p)(2\theta) + (M_p)(\theta), \qquad (3.6.8)$$

which gives

$$M_p = 213. \qquad (3.6.9)$$

This is an unsafe value of M_p, i.e., it is a lower bound on the correct value of M_p. The safe value, or upper bound, is determined by checking the moments in the frame. Moments at B and C are determined as follows.

Moment at C: Apply the virtual work equation to the equilibrium and displacement sets of Fig. 3.21(a), and obtain

$$(+M_C)(-\theta) + (640)(+2\theta) + (-213)(-\theta) = (144)(10\theta), \qquad (3.6.10)$$

which provides

$$M_C = 53. \qquad (3.6.11)$$

Moment at B: Apply the virtual work equation to the equilibrium and displacement sets of Fig. 3.21(b) and obtain

$$(53)(-\theta) + (+M_B)(+2\theta) = (96)(5\theta), \qquad (3.6.12)$$

which gives

$$M_B = 267. \qquad (3.6.13)$$

Since M_B is greater than $M_p = 213$, the moment condition was violated at B. However, if we assign $M_p = 267$ for the frame, then the moment condition

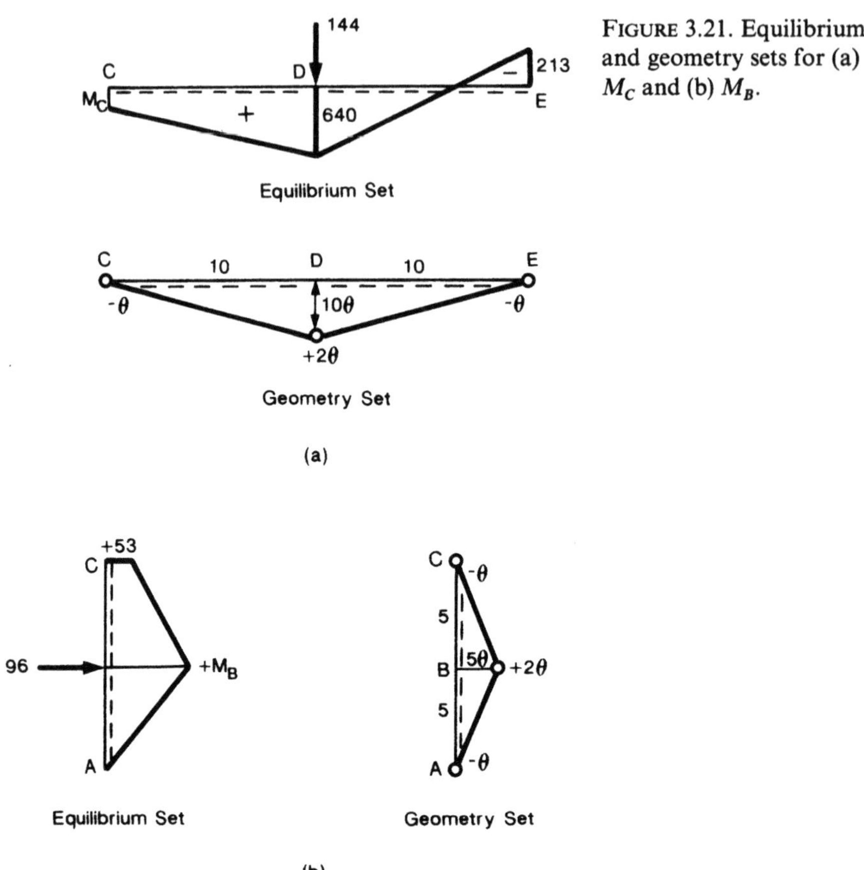

FIGURE 3.21. Equilibrium and geometry sets for (a) M_C and (b) M_B.

will be met throughout the frame. But with this larger section size, plastic hinges cannot develop at D, E, and F and thus no mechanism can form. Thus, $M_p = 267$ is an upper-bound solution on M_p. Thus, bounds on M_p are obtained as

$$213 \leq M_p \leq 267. \tag{3.6.14}$$

Recall from Example 3.6.1 that the exact value of M_p is 240.

3.7 Illustrative Examples

Herein, we shall present three different structures to further demonstrate the applications of the virtual work equation and/or determine their upper and lower bounds on the collapse loads.

Example 3.7.1. Due to settlement, end A of the frame shown in Fig. 3.22 is displaced vertically but $\Delta_v = 0.5$ inch and horizontally by $\Delta_h = 0.1$ inch.

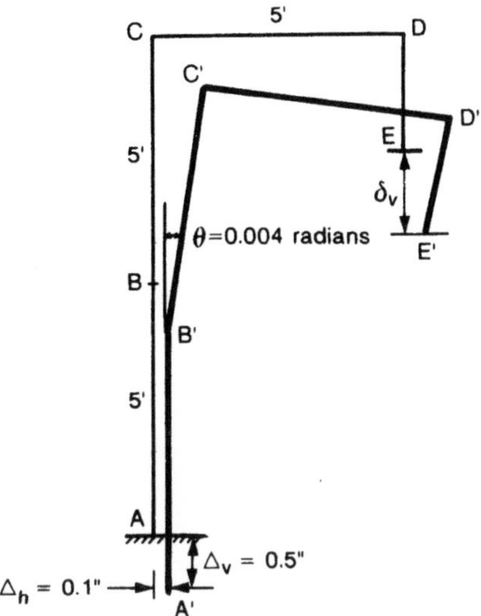

FIGURE 3.22. Displacements of a frame due to settlement and imperfection.

Also, due to initial imperfection in member AC, there is a clockwise rotation of $\theta = 0.004$ radian at point B. Determine the resulting vertical displacement δ_v of end E by the virtual work equation.

Solution: The first step to determine δ_v is to create an equilibrium set. To this end, apply a unit vertical load at point E as shown in Fig. 3.23. Now, apply the virtual work equation of the given displacement set to the created equilibrium set and obtain

$$(1)(\delta_v) + (0)(0.1) - (1)(0.5) = (-5 \times 12)(-0.004), \tag{3.7.1}$$

which gives

$$\delta_v = 0.5 + 0.24 = 0.74 \text{ inch.} \tag{3.7.2}$$

Example 3.7.2. All five members of the truss shown in Fig. 3.24 have the same cross section. Each bar can carry a yield load P_y in either tension or compression and can be extended or compressed indefinitely under this load. Derive an upper and lower bound on the collapse value of P.

Solution: Mechanism 1: Consider the mechanism with yielding of Bars AB, BC, AD, and CD, Fig. 3.25(a). The plastic work equation for this mechanism can be expressed as

$$4P_y\Delta_s = 2P^U\Delta. \tag{3.7.3}$$

From triangle $DD'D''$ in Fig. 3.25(a), Δ and Δ_s are related as

$$\Delta = \sqrt{2}\,\Delta_s. \tag{3.7.4}$$

Thus, Eq. (3.7.3) gives an upper bound

$$P^U = \sqrt{2}\,P_y. \tag{3.7.5}$$

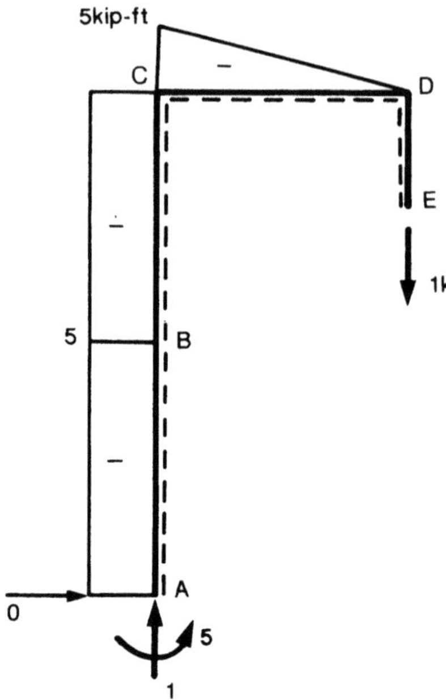

FIGURE 3.23. An equilibrium set for δ_v.

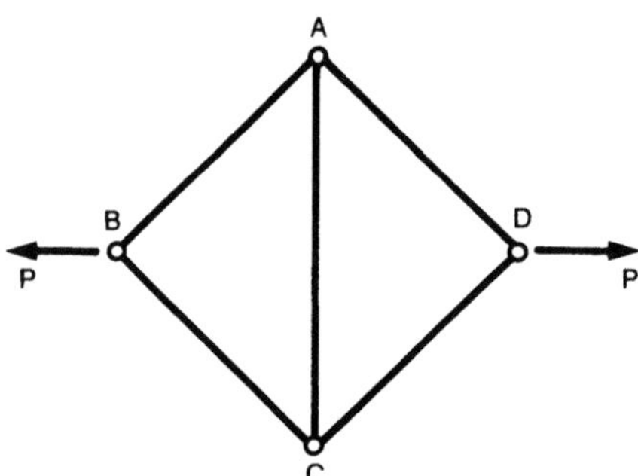

FIGURE 3.24. A square truss.

A lower bound for this mechanism can be found by carrying out an equilibrium check. The equilibrium of joint A [Fig. 3.25(b)] gives the force in vertical bar as

$$F_{AC} = \sqrt{2}\,P_y. \tag{3.7.6}$$

Since $F_{AC} > P_y$, the yield condition has violated in member AC. Thus, if we decrease the applied force by a factor $P_y/\sqrt{2}\,P_y$, then a lower-bound solution can be determined as

$$P^L = \frac{P^U}{\sqrt{2}} = P_y. \tag{3.7.7}$$

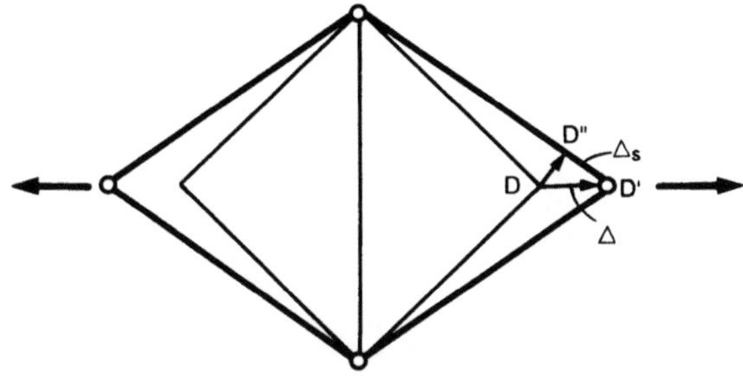

(a) Mechanism 1

FIGURE 3.25. Upper and lower bounds for mechanism 1: (a) mechanism and (b) equilibrium of joint A.

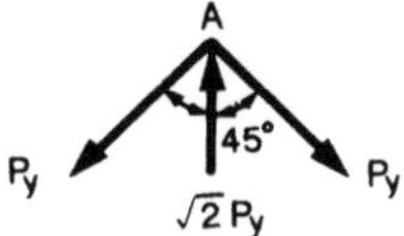

(b) Equilibrium of Joint A

Mechanism 2: The equilibrium check for mechanism 1 shows that yielding in the vertical bar AC is critical. So, we now try the mechanism in which only AC is yielded. The plastic work equation for this case is

$$2(P^U \Delta) = P_y(2\delta), \tag{3.7.8}$$

which gives

$$P^U = P_y \frac{\delta}{\Delta}. \tag{3.7.9}$$

To relate Δ to δ, an equilibrium set is created for member AD as shown in Fig. 3.26(b). Now, apply the virtual work equation of mechanism 2 to the created equilibrium set and obtain

$$(h)(\delta) + (b)(0) + (-b)(\Delta) + (h)(0) = 0. \tag{3.7.10}$$

Since $b = h$, we obtain $\Delta = \delta$ and Eq. (3.7.9) reduces to

$$P^U = P_y. \tag{3.7.11}$$

From the equilibrium of joint A of Fig. 3.26(c), the forces in all side bars are $P_y/\sqrt{2}$. Since all three conditions (equilibrium, moment, and mechanism) are now satisfied, $P = P_y$ is the exact collapse load. This can also be obtained by combining Eq. (3.7.7) with (3.7.11).

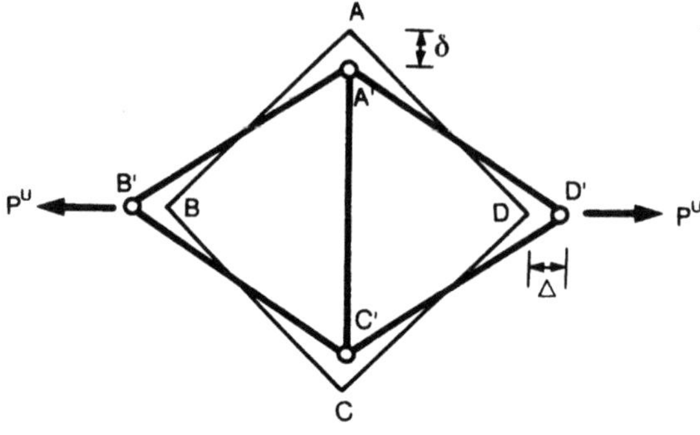

(a) Mechanism 2

FIGURE 3.26. Calculations of collapse load from mechanism 2: (a) mechanism 2; (b) an equilibrium set; and (c) equilibrium of join A.

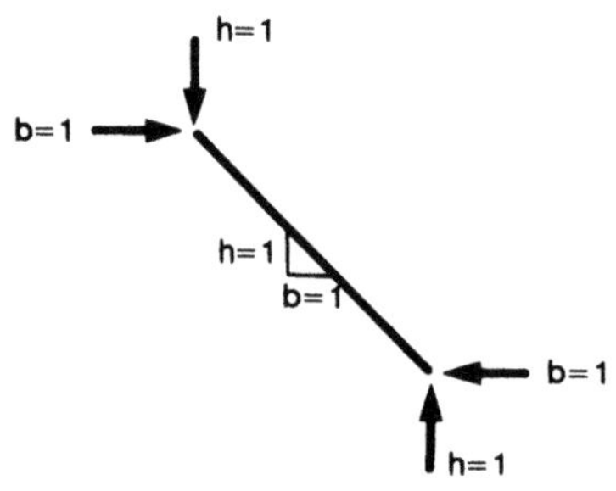

(b) An equilibrium Set

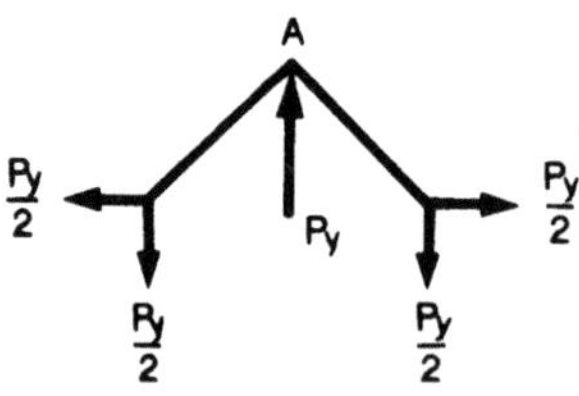

(c) Equilibrium of Joint A

Example 3.7.3. A rigid frame subjected to vertical distributed load and a concentrated horizontal load is shown in Fig. 3.27. Determine a lower and upper bound for the plastic moment M_p using the mechanism shown in Fig. 3.28.

Solution: We will first relate θ_1, θ_2, and θ_3, the angles defining the motion of the mechanism.

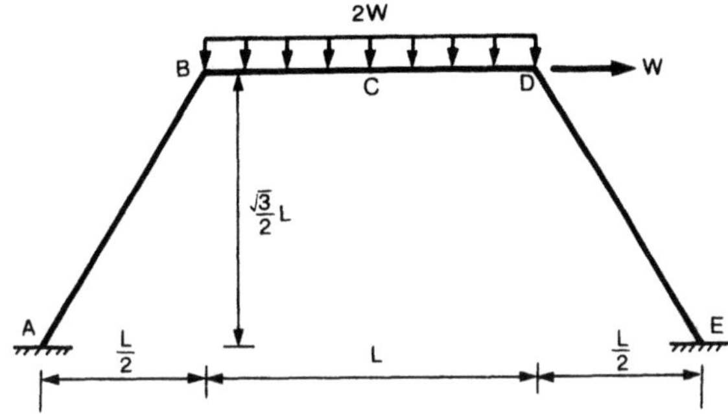

FIGURE 3.27. A fixed-ended frame.

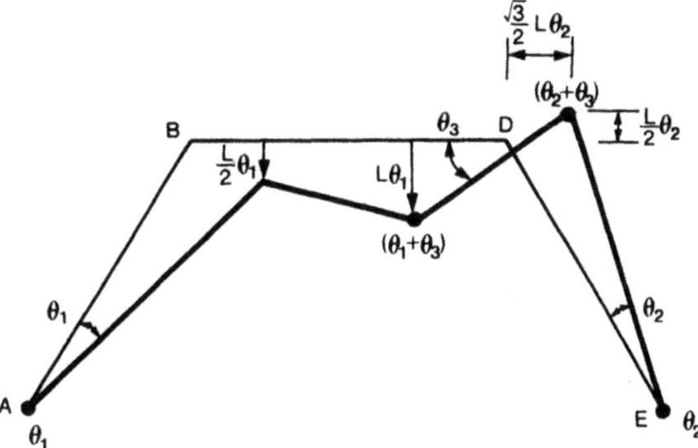

FIGURE 3.28. A sidesway mechanism.

θ_1 **and** θ_2**:** Apply the virtual work equation of the mechanism in Fig. 3.28 to the equilibrium set [Fig. 3.29(a)] created by applying a unit compressive axial force in segment CD and obtain

$$\left(\frac{\sqrt{3}}{2}L\right)\theta_1 - \left(\frac{\sqrt{3}}{2}L\right)\theta_2 = 0, \tag{3.7.12}$$

which gives

$$\theta_1 = \theta_2. \tag{3.7.13}$$

θ_1 **and** θ_3**:** Here we have created an equilibrium set by applying an axial force to member DE with horizontal and vertical components proportional to the horizontal and vertical projection of member DE, respectively [Fig. 3.29(b)]. Now the application of the virtual work equation to the mechanism of Fig. 3.28 and equilibrium set of Fig. 3.29(b) results in

$$\left(\frac{3\sqrt{3}L^2}{4}\right)(\theta_1) - \left(\frac{\sqrt{3}L^2}{4}\right)(\theta_3) = 0, \tag{3.7.14}$$

which gives

$$\theta_3 = 3\theta_1. \tag{3.7.15}$$

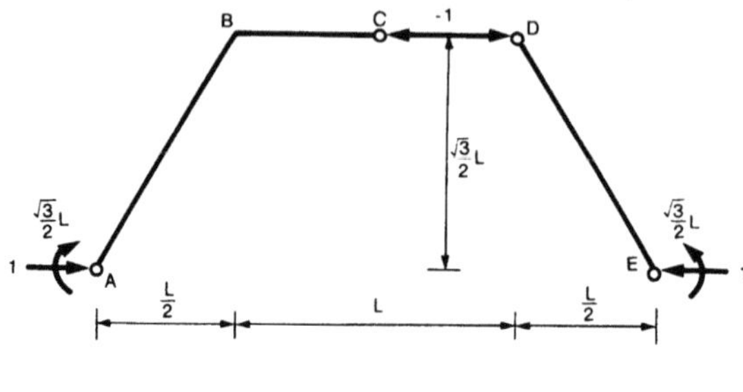

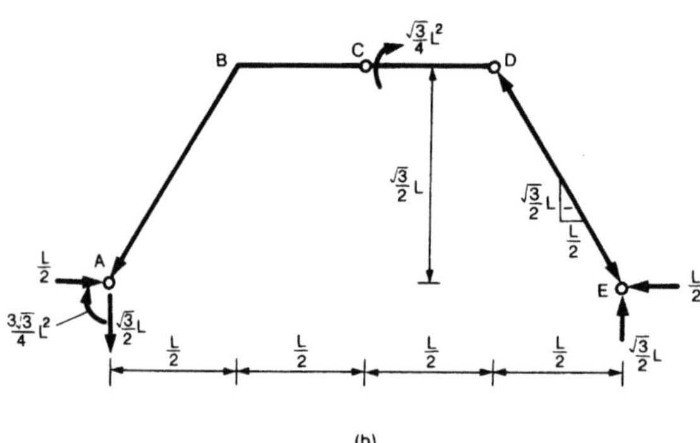

FIGURE 3.29. equilibrium sets for relating θ_1, θ_2, and θ_3.

Lower Bound: The plastic work equation for the mechanism Fig. 3.28 has the form

$$(W)\left(\frac{\sqrt{3}}{2}L\theta_2\right) + \frac{1}{2}W\left(\frac{L}{2}\theta_1 + L\theta_1\right) + \frac{1}{2}W\left(L\theta_1 - \frac{L}{2}\theta_2\right)$$
$$= M_p\theta_1 + M_p(\theta_1 + \theta_3) + M_p(\theta_2 + \theta_3) + M_p\theta_2. \tag{3.7.16}$$

Substituting θ_2 and θ_3 in terms of θ_1 and simplifying, we obtain

$$M_p = 0.187WL. \tag{3.7.17}$$

This is a lower-bound value of M_p.

Upper Bound: An upper bound can be determined by checking the moment at B and the maximum moment in member BD.

Moment at B: Apply the virtual work equation to the equilibrium and displacement sets of Fig. 3.30(a) and obtain

$$(+M_B)(-\theta) + (+M_p)(+2\theta) + (-M_p)(-\theta) = \frac{1}{2}(2W)\left(\frac{L}{2}\theta\right). \tag{3.7.18}$$

Substituting $M_p = 0.187WL$ and simplifying, we have

$$M_B = 0.061WL < (M_p = 0.187WL) \quad \text{okay.} \tag{3.7.19}$$

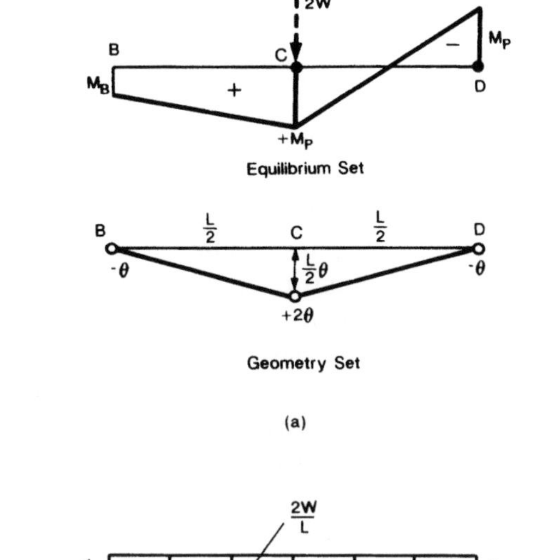

FIGURE 3.30. Moment check for member *BD*.

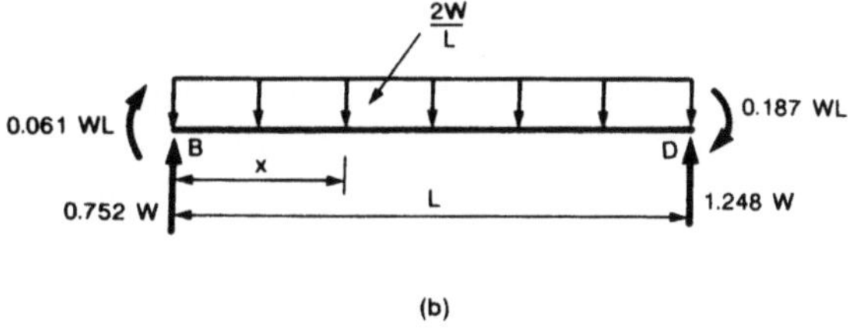

Maximum Moment in Member BD: The moment in member *BD* can be expressed as [Fig. 3.30(b)]

$$M = 0.061WL + 0.752Wx - \left(\frac{2W}{L}\right)\frac{x^2}{2}. \qquad (3.7.20)$$

By equating dM/dx to zero, the location of the maximum moment from B is obtained as

$$x_0 = 0.376L. \qquad (3.7.21)$$

Substituting this value of x_0 in Eq. (3.7.20), we have

$$M_{\max} = WL[0.061 + 0.283 - 0.141] = 0.203WL. \qquad (3.7.22)$$

Thus, if we assign $M_p = 0.203WL$, then both equilibrium and moment conditions are met, but the mechanism cannot form, since moments at points A, C, D, and E are all less than M_p. So, $M_p = 0.203WL$ is an upper-bound value of M_p. Thus, we have

$$0.187WL \le M_p \le 0.203WL. \qquad (3.7.23)$$

3.8 Summary

The methods of simple plastic analysis and design are based on two basic assumptions. The first assumes that the material is ductile. The stress-strain behavior of structural steel has a large amount of ductility, which can be fully realized through the process of plastification and redistribution of forces and moments in a structural analysis. The second assumption is that the displacements in a structure under loading are small so that the equilibrium equations can be formulated on the basis of the original geometry. For most practical structures, the displacements at the plastic limit load are not large and are of the same order of magnitude as those at the elastic limit load.

Here, as in the elastic solution that must satisfy equilibrium, compatibility, and moment-curvature conditions, the correct plastic solution must satisfy equilibrium, mechanism, and moment conditions. However, as the structure becomes more complex, it becomes more difficult to obtain the exact solution that satisfies all three basic conditions. It is more convenient to obtain a close approximate solution rapidly while satisfying only some of the conditions. To judge the nature of these types of solutions, some general principles and theorems are needed. For this purpose, we have described and proved three fundamental theorems of plastic methods, namely, *uniqueness, safe, and unsafe theorems*. According to the uniqueness theorem, if the loads are increased proportionately, then the collapse will occur at one define and *unique* value of the load factor when all three basic conditions are satisfied. The *unsafe* theorem states that a guessed mechanism, if not correct, will overestimate the load factor and underestimate the required plastic moment capacity. In contrast, the *safe* theorem states that a moment diagram satisfying the moment condition and in equilibrium with the applied loads, if not corresponding to a mechanism, will underestimate the load factor and overestimate the required plastic moment capacity. The safe and unsafe theorems can be used to establish quickly and easily close upper and lower bounds of the load factor or the required plastic moment capacity of a structure.

The conditions required to establish an upper- or lower-bound solution in the simple plastic theory are summarized as follows:

a. Upper-bound solution, which gives an unsafe, or correct, value of the collapse load.
   1. A valid mechanism of collapse must exist such that it satisfies the mechanical boundary conditions (mechanism condition).
   2. The internal dissipation of energy at plastic hinges must equal the expenditure of energy due to the external loads (work equation).
   3. All deformations take place at the plastic hinge locations and the material stays rigid between plastic hinges. The amount of plastic hinge rotation at each location is defined by the mechanism. The direction of the plastic rotation at each location in turn defines the direction of the plastic moment required to calculate the dissipation of energy at the plastic hinge location. This shows that the dissipation of energy at each plastic hinge location is always positive. (This is known as the yield conditions and its associate flow rule in the theory of plasticity.)
b. Lower-bound solution, which gives an oversafe, or correct, value of the collapse load.
   1. A complete moment distribution must be found everywhere in the

structure satisfying the equations of equilibrium (equilibrium condition).

2. The forces and moments at the ends must satisfy the boundary conditions (static boundary conditions).
3. At no place in the structure the moment condition is violated (moment condition).

From these rules, it can be seen that the upper-bound technique is based on the mechanism or work approach, while the lower-bound technique is based on the equilibrium approach. Both approaches are based on an engineers' intuitive approach and are very powerful, since they are backed by repeated experiments and years of experience. These *two* alternative approaches to an exact solution within the simple plastic theory will be described in detail in the following two chapters as the "equilibrium method" and the "work method," respectively, in Chapters 4 and 5.

The *equation of virtual work* provides a powerful tool in the plastic methods. Its usefulness has been demonstrated here by its use in

a. proving the uniqueness, safe, and unsafe theorems.
b. obtaining the geometrical relationships of motions of mechanism of structures by creating proper equilibrium sets.
c. making moment checks corresponding to assumed collapse mechanisms by selecting appropriate displacement sets.
d. obtaining upper and lower bounds for the load factor or the required plastic moment capacity of structures based on an assumed mechanism.

In writing the virtual work equation, especially for its use in the *moment check*, the moments and rotations must be accompanied with a proper sign convention. It is not necessary, however, to consider sign convention in a plastic work equation for an assumed mechanism since all terms contributing to the plastic energy dissipation are always positive.

References

3.1. Drucker, D.C., Greenberg, H.J., and Prager, W., "The Safety Factor of an Elastic-Plastic Body in Plane-Strain," *Transactions*, ASME, 73, pp. 371–378, 1951.
3.2. Drucker, D.C., Greenberg, H.J., and Prager, W., "Extended Limit Design Theorems for Continuous Media," *Quarterly of Applied Mathematics*, 9, pp. 381–389, Jan. 1952.
3.3. Feinberg, S.M., "The Principle of Limiting Stress," *Prikladnaya Matematika i Mekhanika*, 12, p. 63, 1948.
3.4. Gvozdev, A.A., "The Determination of the Value of the Collapse Load for Statically Indeterminate Systems Undergoing Plastic Deformations," *Proceedings of the Conference on Plastic Deformations*, Akademiia Nauk SSSR, Moscow Leningrad, USSR, p. 19, 1938; English translation by R.M. Haythornthwaite, *International Journal of Mechanical Sciences*, 1(4), p. 322, July 1960.
3.5. Horne, M.R., "Fundamental Propositions in the Plastic Theory of Structures," *Journal of the Institution of Civil Engineers*, 34, p. 174, April 1960.

3.6. Horne, M.R., and Morris, L.J., *Plastic Design of Low-Rise Frames*, MIT Press, Cambridge, Mass. p. 238, 1982.

Problems

3.1. Why is the correct mechanism the one that corresponds to the lowest load?

3.2. Demonstrate the upper- and lower-bound theorems with the aid of a beam, loaded at the third points, fixed at one end but simply supported at the other.

3.3. The propped cantilever beam AB as shown in Fig. P3.3 is of constant cross section with plastic moment M_p and total length $4a$. Determine the upper and lower limits on P_0 by considering three mechanisms with plastic hinges at (a) B and D, (b) B and E, and (c) C and E.

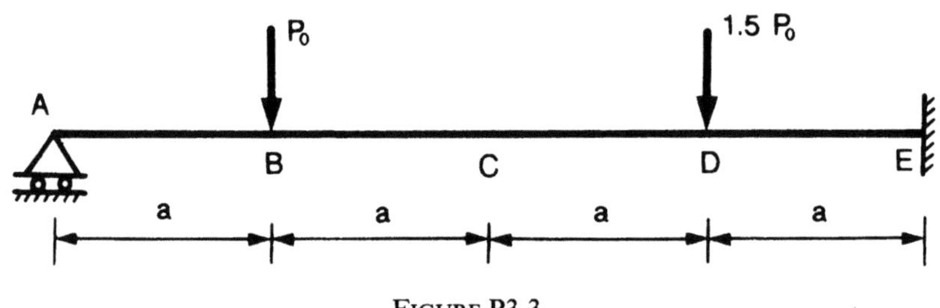

FIGURE P3.3

3.4. If end A of the beam in Problem 3.3 is also fixed, determine the three sets of upper and lower bounds on P_0.

3.5. The beam of Problem 3.3 is loaded by a uniform load q per unit length. Determine the upper and lower bounds on q using a mechanism with plastic hinges at C and D.

3.6. What is the effect on q in Problem 3.5 of settlement of support A?

3.7. If end A of the beam in Problem 3.5 is also fixed find q.

3.8. The fixed-ended beam as shown in Fig. P3.8 is subjected to a uniformly distributed load q. Find the upper and lower bounds on q by considering mechanisms with plastic hinges at (a) A, B, and C; (b) A, C, and D; and (c) A, B, and D.

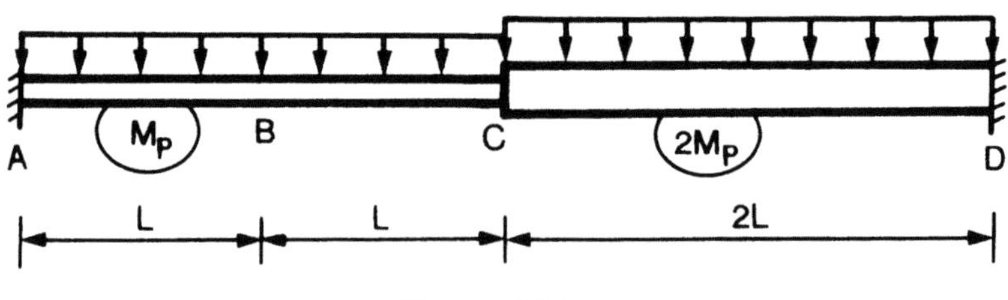

FIGURE P3.8

3.9. If the frame of Fig. 3.27 is subjected to a vertical concentrated load $2W$ at C and a horizontal concentrated load W at B, determine the upper and lower bounds on W by considering a mechanism with plastic hinges at A, C, D, and E.

3.10. A rectangular frame is subjected to a vertical concentrated load $1.5W$ at C and a horizontal concentrated load W at D as shown in Fig. P3.10. Find the upper and lower bounds on W by considering a mechanism with plastic hinges at A, C, D, and E.

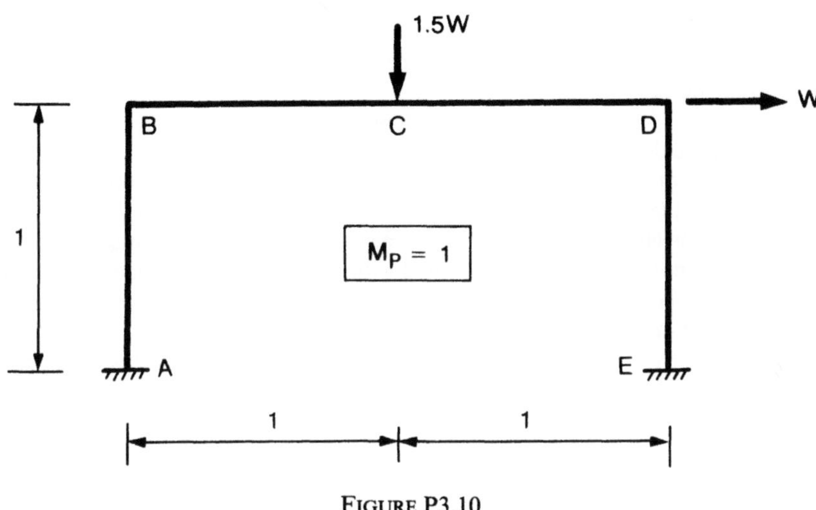

FIGURE P3.10

3.11. If the vertical concentrated load in Problem 3.10 is replaced by a uniformly distributed load P, determine the upper and lower bounds on W by considering mechanism with plastic hinges at A, C, D, and E, when (a) $P = W/4$ and (b) $P = W$.

3.12. Member BD is added to the square truss of Fig. 3.24. Determine the upper and lower bounds on P for mechanisms with yielding of (a) members AB, BC, CD, DA, and BD and (b) members AC and BD.

3.13. The loads and member plastic moment capacities of a two-bay frame are as

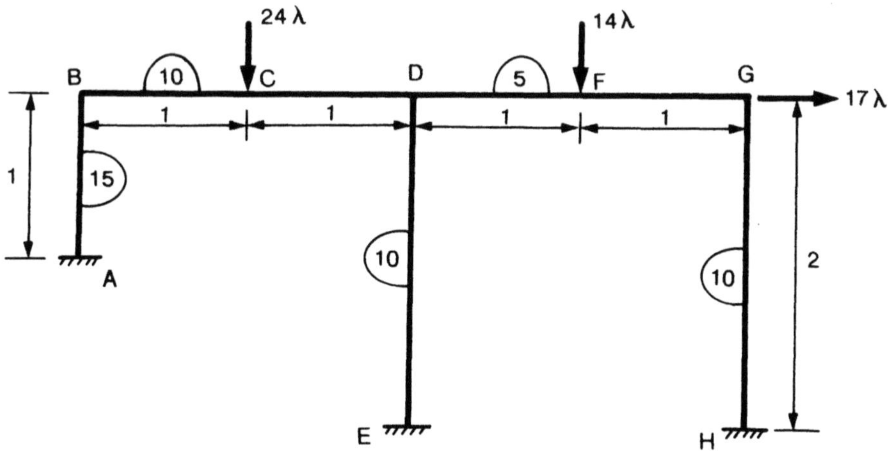

FIGURE P3.13

shown in Fig. P3.13. Determine the upper and lower limit of λ by considering a mechanism with plastic hinges at A, C, D, E, F, G, and H.

3.14. Define "rotation capacity" and "ductility." For a fixed-ended beam subjected to a uniformly distributed lateral load, what is the required rotation capacity and ductility in order that the computed plastic limit load will be reached?

3.15. Why is no elastic energy at a plastic hinge location included in the work equation such as Eq. (3.6.1)?

3.16. Under what circumstances does the conventional elastic design become a "lower-bound" solution.

3.17. Explain why erection forces and support settlements influence the elastic but not plastic design of a structure.

3.18. Demonstrate the uniqueness theorem with the aid of a simply supported beam, loaded at the third points that: (a) the collapse load of the beam is unique; (b) the collapse mechanism leading to the same collapse load in not unique; and (c) the bending moment diagram associated with the same collapse load is not unique.

3.19. Can the plastic design method be applied to reinforced concrete frames? Explain. Can it be applied to steel bridges? Explain.

3.20. Explain why plastic theorems are not applicable to structural problems involving instability.

4
Equilibrium Method

4.1 Introduction

The preceding chapter was concerned with basic assumptions and theorems used in the plastic analysis and design of steel structures. In the following two chapters we shall present two basic methods of plastic analysis and design. In this chapter, we shall present the plastic analysis and design technique known as the "equilibrium method," based on the lower-bound theorem. In the next chapter, the plastic analysis and design technique known as the "work method," based on the upper-bound theorem, will be presented.

As the name implies, in the equilibrium method, the relationship between the strength of a steel structure and the applied loads is found by adjusting the unknown redundants in an indeterminate structure such that the *equilibrium* condition is always satisfied and the *moment* condition is not violated, and the *mechanism* condition may or may not be satisfied. The equation formed in this way is called the statically admissible "equilibrium equation" and gives the relationship between the structure strength and the applied loads for a particular set of *assumed* redundant moments.

It is, however, the task of the analyst/designer to seek the best set of redundant moments that gives the largest applied load-carrying capacity (or the smallest required plastic moment) of the structure. In fact, the best set of assumed redundant moments corresponds to the formation of a plastic failure mechanism. Although any set of assumed redundants will give a safe or lower-bound solution, the critical set is not generally apparent and requires physical intuition combined with the use of differential calculus and algebraic techniques. This is described in this chapter.

Herein, we will begin by reexamining the lower-bound theorem that forms the basis of the equilibrium method. Then we will discuss the steps involved in obtaining statically admissible equilibrium equations and making the mechanism checks. Next, we will demonstrate the use of the equilibrium method for plastic design and analysis of simple beams, rectangular portal frames, and gable frames. Finally, for plastic analysis and design of large structures, we will present a practical procedure of the equilibrium approach.

In this and subsequent chapters, for moments and rotations we will follow the sign convention used in Chapter 3, which is repeated here: the moments and rotations causing tension on the side of the member marked with dotted lines will be positive, and vice-versa.

4.2 Basis of the Method

For a statically determinate structure, the equilibrium conditions provide a complete moment diagram. For these structures, the elastic or plastic limit load can be determined directly by equating the critical moment in the equilibrium moment diagram to the yield or plastic moment of the section. However, for indeterminate structures, equilibrium equations always lead to the moment diagram in terms of unknown redundants. In the elastic analysis, these unknowns are determined from compatibility conditions, which often make the analysis complicated. In the plastic analysis, however, the analyst has the freedom to choose the values of unknown redundants in the moment equilibrium equations, which often lead to a quick safe solution to the problem. The solution will of course be exact only if the chosen values of the redundant moments result in a plastic collapse mechanism.

The fact that any values of redundants yield a safe solution is based on the lower-bound theorem of plastic analysis described in Chapter 3. The theorem is restated here: a load computed on the basis of an equilibrium moment distribution in which moments are nowhere greater than M_p is less than or equal to the true plastic limit load.

The theorem gives lower bounds on, or safe values of, the limit or collapse load; the maximum lower bound is the limit load itself. For example, the loads determined by the hinge-by-hinge analysis in the examples in Chapter 1 are lower bounds to the limit loads. In the example of fixed-ended beam in Section 1.4, the stage 1 moment diagram satisfies the equilibrium and plastic moment conditions, thus, the corresponding load is a lower-bound solution and is of course lower than the plastic limit load. The stage 2 moment diagram satisfies the equilibrium and moment conditions as well as the mechanism condition, thus the corresponding load is equal to the plastic limit load. Similarly, in Example 1.8.2, moment diagrams at the points of one and two plastic hinges satisfy the equilibrium and plastic moment conditions, thus the corresponding loads are lower bounds on the plastic limit load. When a third hinge is formed, the moment diagram satisfies the equilibrium and moment conditions as well as the mechanism condition, thus, the corresponding load is equal to the plastic limit load.

Before the most critical solution of a particular equilibrium can be found, the equilibrium equations must be formed first, either graphically or algebraically, followed by a proper selection of redundant moments. As a consequence, we will present the equilibrium method in two stages: (a) drawing the equilibrium moment diagram and writing the moment equations in terms of redundants and (b) adjusting the redundants in such a way that both plastic moment and mechanism conditions are astisfied. These two stages of analysis procedure are elaborated on in the next two sections.

4.3 Moment Equilibrium Equations

The moment equilibrium equations of a statically indeterminate structure are obtained as follows:

(a) Select redundant(s). Enough freedom must be introduced at support points or in the structure to produce a simple determinate structure.
(b) Draw a moment diagram for the determinate structure under applied loads. Moment diagrams are drawn along all members comprising the structure, following the usual sign convention for the moment.
(c) Draw a moment diagram for the determinate structure under redundant(s). This is the restraining moment diagram induced by the actual continuity at the redundant points.
(d) Superimpose moment diagrams of steps (b) and (c). The true moment at any point is given by the difference in the ordinates of the two diagrams.
(e) Write moment equations at critical sections of the structure using the moment diagram of step (d).

The following examples demonstrate these steps of obtaining the moment equilibrium equations.

Example 4.3.1. For a fixed-ended beam shown in Fig. 4.1(a), obtain the moment equilibrium equation in terms of unknown redundants.

Solution: The degree of indeterminacy for the beam shown in Fig. 4.1 is two. The redundants may be selected as the two end moments M_1. The resulting simply supported beams under the uniform load w and under the redundants M_1 are, respectively, shown in parts (b) and (c) of Figure 4.1. The moment diagram for the determinate structure under the uniformly distributed load W is shown in part (d) with the moment at the center equal to

$$M = \frac{WL}{8}.$$
(4.3.1)

The moment diagram under the redundants is shown in part (e). The moment diagrams of parts (d) and (e) are combined in part (f).

Note that the moment diagrams are plotted on the tension side of the member. We will use this sign convention throughout this book. From Fig. 4.1(f), the critical moment in the beam is

$$M_{cr} = \frac{WL}{8} - M_1.$$
(4.3.2)

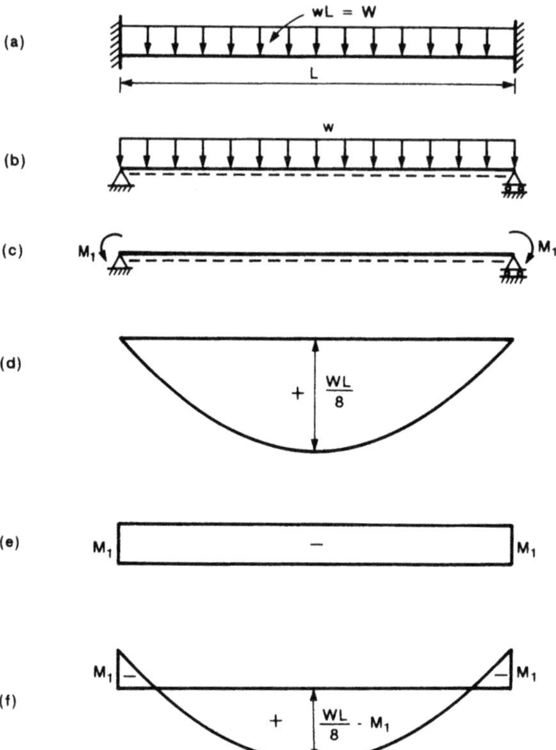

FIGURE 4.1. Bending moment diagrams and critical moment in terms of redundant moments M_1 for a fixed-ended beam.

Example 4.3.2. For the two-span continuous beam shown in Fig. 4.2(a), obtain the moment equilibrium equation in terms of unknown redundant moment.

Solution: The degree of indeterminancy of the beam in Fig. 4.2(a) is one. The redundant may be conveniently selected as the moment at midsupport. Let this moment be M_1. The resulting determinate beam under the two concentrated loads and the redundant moment are shown in parts (b) and (c), and their moment diagrams are shown, respectively, in parts (d) and (e). These diagrams are then superimposed in part (f) and redrawn on a single straight base line in part (g). The diagrams in (f) and (g) are identical, from which we obtain the critical moment in the beam as

$$M_{cr} = \frac{Pa(L - a)}{L} - \frac{M_1 a}{L}. \tag{4.3.3}$$

4.4 Mechanism Check

Once the moment equilibrium equations in terms of the unknown redundants are set up, the plastic limit load (or plastic moment) of a structure can be determined by adjusting the redundants (or the restraining moment dia-

FIGURE 4.2. Bending moment diagrams and critical moments in terms of redundant moment M_1 for a two-span continuous beam.

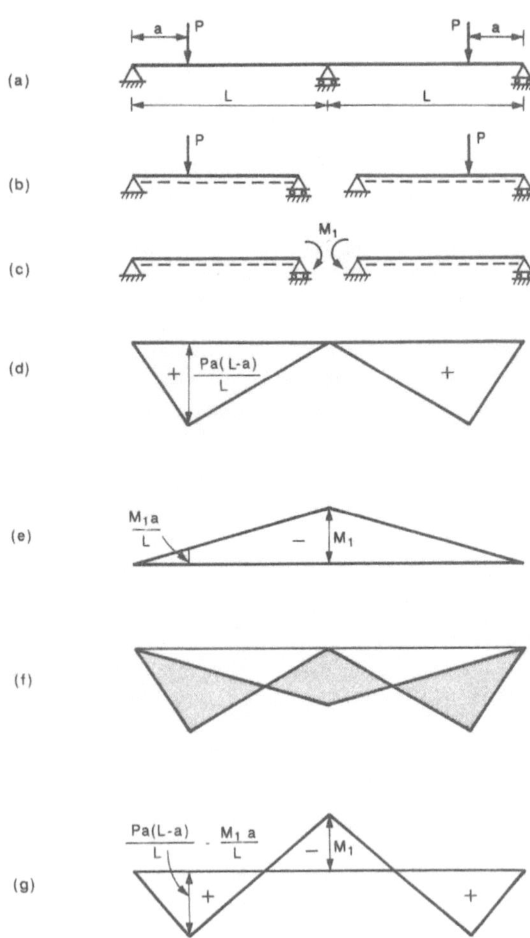

gram) until the resulting true moment value is maximum and equal to the plastic bending strength to be furnished, at enough points in the structure to reduce it to a mechanism. The following steps may be used to achieve this goal.

(a) Select value(s) of redundant(s) such that the plastic moment condition at maximum moment locations is not violated at any point in the structure.
(b) Determine the load or plastic moment corresponding to the selected redundant(s).
(c) Check for the formation of a mechanism. If a plastic collapse mechanism condition is met, then the computed load (or plastic moment) is the exact plastic limit load. Otherwise, it is a lower-bound or safe solution to the exact limit load.
(d) Adjust the redundant(s) and repeat steps (a) to (c) until the exact plastic limit load (or plastic moment) is obtained.

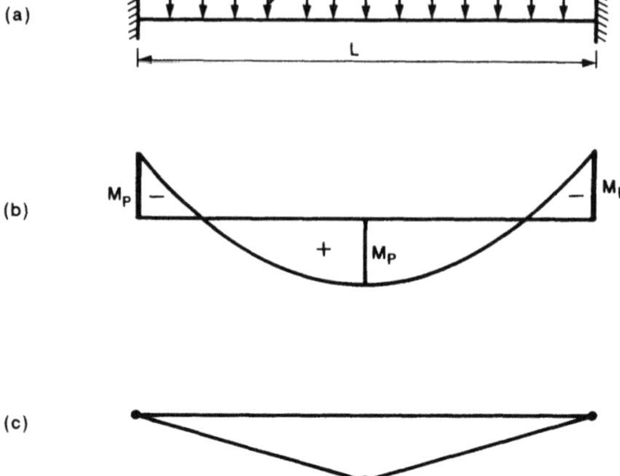

FIGURE 4.3. Plastic limit state of the fixed-ended beam of Fig. 4.1.

Assuming the value of redundant(s) in the structure is equivalent to visualizing how the structure will carry the applied loads or distribute the applied loads among its components or parts. The most efficient (moment) distribution corresponds to the formation of a failure mechanism. Any other moment distribution will be less efficient and will therefore result in a lower load (or safe plastic moment).

The complete procedure for determining the plastic limit load from the moment equilibrium equation(s) in terms of unknown redundant(s) will be demonstrated by the following two examples.

Example 4.4.1. Determine the plastic limit load of the fixed-ended beam shown in Fig. 4.3(a). Use the moment diagram and equilibrium equation obtained in Example 4.3.1.

Solution: In this case, it is obvious by inspection that the required bending strength is equal to one-half of the maximum determinate (simple span) moment that would be produced by the limit load W_c, which is attained when the available plastic bending capacity M_p is reached at all three hinge points. For more complex structures and loading, the appropriate choice of redundant (or restraining) moment is not obvious. An arbitrary choice will lead to a safe solution. This is illustrated here.

A safe solution of the fixed-ended beam can be determined from Eq. (4.3.2) by selecting a value for the redundant moment M_1. For example, if we select $M_1 = 0$, the moment diagram is reduced to the one shown in Fig. 4.1(d). By equating the critical moment in this diagram to the plastic moment capacity of the beam M_p, we have

$$M_{cr} = \frac{WL}{8} = M_p,$$ (4.4.1)

which gives $W = W_1$ as

$$W_1 = \frac{8M_p}{L}. \qquad (4.4.2)$$

The moment diagram in Fig. 4.1(d) has only one plastic hinge developed at midspan, but three hinges are required to form a failure mechanism for the fixed-ended beam. It follows that W_1 is a lower-bound solution to the exact plastic limit load. The assumed moment distribution is obviously not an efficient way to carry the applied uniform load. The actual beam will adjust itself better to carry the higher load.

To show this point, we now assume the unknown redundant moment M_1 equal to the largest possible moment capacity as in the midspan section. Thus, from Fig. 4.1(f), we obtain

$$M_{cr} = \frac{WL}{8} - M_1 = M_p, \qquad (4.4.3)$$

which gives $W = W_2$ as

$$W_2 = \frac{16M_p}{L}. \qquad (4.4.4)$$

The improved moment diagram [Fig. 4.3(b)] results in the formation of three plastic hinges and thus leads to the formation of a failure mechanism as shown in Fig. 4.3(c). Therefore, W_2 is the exact plastic limit load.

Note that if a value of M_1 less than M_p such as $M_p/2$ is assumed, no failure mechanism will be developed and the corresponding load ($W_3 = 12M_p/L$) will be a better lower bound than that of the first lower-bound value W_1.

Example 4.4.2. Determine the plastic limit load of a two-span continuous beam shown in Fig. 4.4(a). Use the moment diagram and equilibrium equation obtained in Example 4.3.2.

(a)

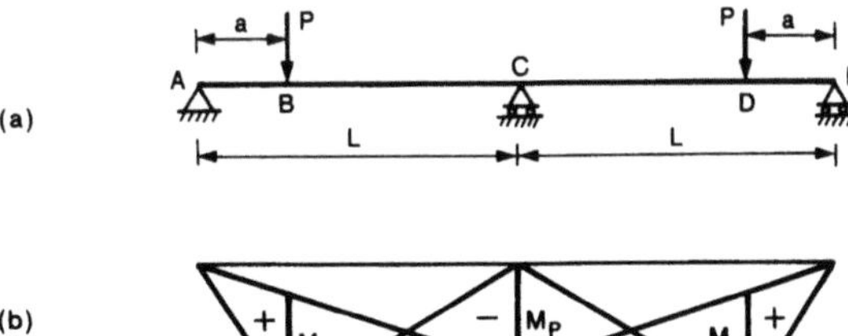

(b)

(c)

FIGURE 4.4. Plastic limit state of the two-span continuous beam of Fig. 4.2.

Solution: In constructing the determinate moment diagrams for the continuous beam, the freedom at the midsupport is assumed to be provided by means of a real hinge.

A lower bound on the plastic limit load can easily be found from Eq. (4.3.3) by selecting a value of the midsupport moment M_1. If we select $M_1 = 0$, for example, the moment diagram is shown in Fig. 4.2(d). By equating maximum moment in this diagram to the plastic moment capacity of the beam, we have

$$M_{cr} = \frac{Pa(L - a)}{L} = M_p, \tag{4.4.5}$$

which gives $P = P_1$ as

$$P_1 = \frac{M_p L}{a(L - a)}. \tag{4.4.6}$$

The moment diagram in Fig. 4.2(d) will develop one plastic hinge in each span. Since two plastic hinges in each span are required to develop a plastic failure mechanism, the load P_1 is a lower-bound solution. On the other hand, by setting redundant moment M_1 equal to the plastic moment M_p, and by equating the maximum moment in Fig. 4.2(g) to the plastic moment, we have

$$M_{cr} = \frac{Pa(L - a)}{L} - M_p \frac{a}{L} = M_p, \tag{4.4.7}$$

which gives $P = P_2$ as

$$P_2 = \frac{M_p(L + a)}{a(L - a)}.$$

Since a sufficient number of plastic hinges has formed in the beam [Fig. 4.4(b)], resulting in the formation of a failure mechanism shown in Fig. 4.4(c), P_2 is the exact plastic limit load.

4.5 Design of Simple Beams

The basic steps involved in the equilibrium method have been described and illustrated in Sections 4.3 and 4.4. Herein, we shall use the equilibrium method to carry out a complete plastic analysis and design of steel beams. The load-carrying capacity of beams may be significantly affected by the presence of shear force in the beam and the possible lateral torsional buckling of the beam if not properly designed. These two effects on beam strength and their considerations in design are described in the forthcoming.

4.5.1 Shear Force

In many cases, beams are subjected to high shear force. For example, a short beam under central load or a long beam under a concentrated load applied near its support may develop a high shear-to-moment ratio. In such cases, shear force will decrease the plastic moment capacity of the beam section and cause larger deflections than might otherwise be expected. Design and analysis of such beams must therefore consider the effect of shear force on the yielding of the section. This may be achieved simply by replacing the plastic moment capacity M_p with the reduced plastic moment capacity M_{ps} in the procedure given in Sections 4.3 and 4.4. The calculation of M_{ps} was described in Section 2.6. Examples 4.5.2 and 4.5.3 in the forthcoming illustrate how the shear force can be included in the plastic analysis and design of beams.

Since high shear and moment values frequently occur in regions of localized yielding, the beneficial effects of strain-hardening usually enable beams of wide flange and I shapes to reach the full plastic moment M_p. Experimental evidence shows that design of beams based on M_{ps} is too conservative because of the effect of strain-hardening in an actual beam. Therefore a design rule to account for the influence of shear force may be obtained by a consideration only of the maximum shear force V_p to prevent "failure" due to excessive shear deformations.

Therefore, as far as shear is concerned, the full plastic bending strength M_p may be used in design, provided the total transverse shear on the wide-flange section at plastic limit load, in kips, is no more than

$$V_p = \frac{F_y}{\sqrt{3}} t_w (d - 2t_f) = \frac{F_y}{\sqrt{3}} (t_w d) \frac{(d - 2t_f)}{d}$$

or

$$V_p = \frac{F_y}{\sqrt{3}} (t_w d) \frac{1}{1.07} = 0.54 F_y t_w d \approx 0.55 F_y t_w d \tag{4.5.1}$$

where t_w = web thickness in inches, d = section depth, t_f = flange thickness, and $d/(d - 2t_f) = 1.07$. The webs of columns, beams, and girders shall be reinforced by stiffeners or a doubler plate if the shear force V at plastic limit load exceeds the shear strength V_p (Section N5, Chapter N, ASD, 1989).

4.5.2 Lateral Torsional Buckling

Any member of I-shape sections bent about its strong axis may be susceptible to lateral torsional buckling if the distance between points of lateral support is excessive [4.4–4.6, 4.9]. The effect of lateral-torsional buckling (Fig. 4.5) of a beam is similar to the effect of local buckling of a cross section (Section 2.7). Both effects may prevent the beam from attaining its full plastic moment capacity, and if the plastic moment capacity is attained, they may prevent the beam from sustaining its moment capacity up to the desired rotation capacity. The local buckling occurs due to a high width-to-thickness ratio of the component elements of a cross section, while the lateral-torsional buckling occurs due to a high unbraced slenderness ratio of a narrow beam.

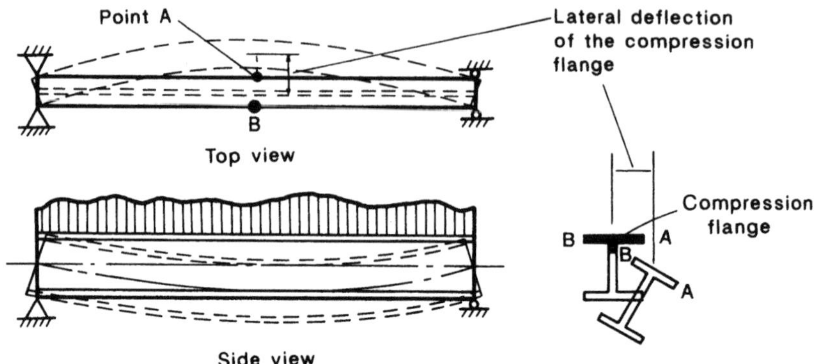

FIGURE 4.5. Lateral torsional buckling of a wide-flange beam loaded in the plane of web.

The schematic relationship between the unbraced length and the moment-carrying capacity and rotation capacity of a beam are illustrated in, respectively, the upper and lower part of Fig. 4.6. When the unbraced length is long, the member will fail by elastic lateral-torsional buckling and the rotation capacity will be small. When the unbraced length is short, the full plastic moment will be attained or exceeded and the section will be able to deliver much larger rotation capacity. For intermediate unbraced length, the inelastic buckling occurs and the rotation capacity is higher than that for elastic buckling. For satisfactory performance, the rotation capacity of a beam (or a member of a structure) should be equal to or greater than the hinge angle required to form a mechanism.

To ensure the proper moment and rotation capacities, the LRFD specification has defined three sets of limiting unbraced lengths: L_{pd}, L_p, and L_r. When unbraced length $L_b \leq L_{pd}$, the member is suitable for plastic design, i.e., it will be able to sustain plastic moment up to the rotation capacity necessary to form a failure mechanism (taken as three times the yielding rotation by LRFD). When $L_{pd} < L_b \leq L_p$, the member will attain the plastic moment but it may not sustain it for the desired rotation capacity. When $L_p < L_b \leq L_r$, the member will fail by inelastic lateral torsional buckling without attaining plastic moment. When $L_b > L_r$, the member will fail by elastic lateral torsional buckling.

The elastic limiting length L_r is obtained from the elastic lateral torsional buckling moment M_{cr} for beams of various cross-sectional shapes. For example, for I-shaped members under uniform moment, the elastic solution of the governing differential equation gives [4.1]

$$M_{cr} = \frac{\pi}{L_b}\sqrt{EI_y GJ + \left(\frac{\pi E}{L_b}\right)^2 I_y C_w} \tag{4.5.2}$$

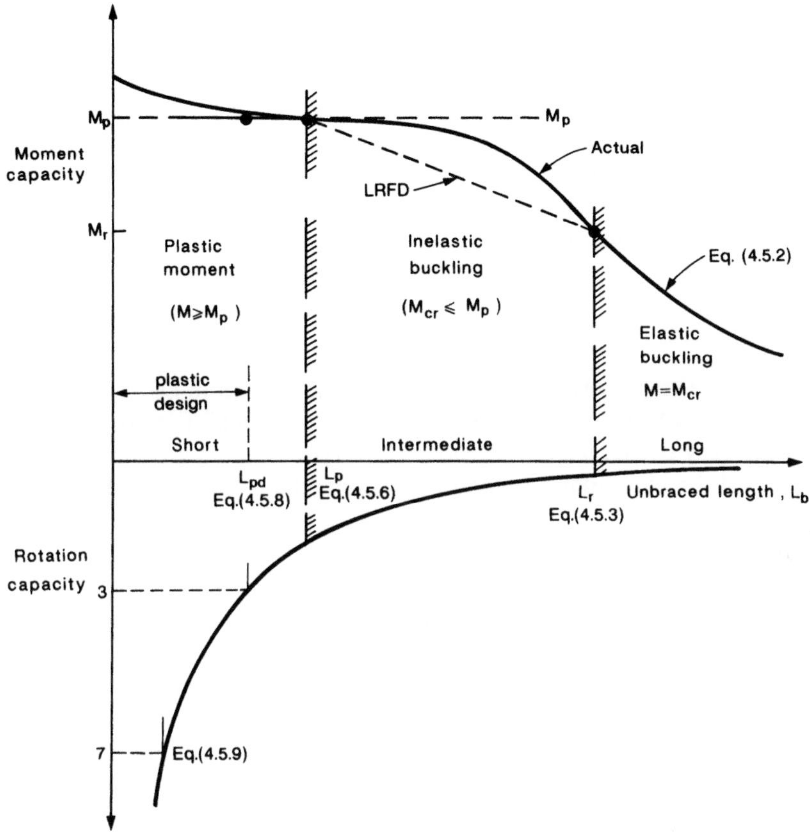

FIGURE 4.6. Moment and rotation capacity of a beam with its unbraced length.

where L_b is unbraced length; E and G are, respectively, Young's modulus in ksi and shear modulus in ksi of steel; I_y, J, and C_w are, respectively, the moment of inertia about the weak axis in in.[4], torsional constant in in.[4], and warping constant in in.[6] for the section. By equating $M_{cr} = (F_y - F_r)S_x$ in this equation and then rearranging the resulting equation, $L_r = L_b$ is obtained as

$$L_r = \frac{r_y X_1}{(F_y - F_r)} \sqrt{1 + \sqrt{1 + X_2(F_y - F_r)^2}} \tag{4.5.3}$$

where F_y is yield stress in ksi, F_r is the maximum compressive residual stress in ksi, r_y is the radius of gyration of the section about the minor axis in in., and

$$X_1 = \frac{\pi}{S_x} \sqrt{\frac{EAGJ}{2}} \tag{4.5.4}$$

$$X_2 = 4 \frac{C_w}{I_y} \left(\frac{S_x}{GJ} \right)^2 \tag{4.5.5}$$

in which S_x is the elastic section modulus about the strong axis in in.[3] and A is the cross-sectional area of the beam in in[2].

The limiting length L_p is obtained from the inelastic critical moment for beams. The inelastic critical moment is obtained by replacing EI_y, GJ, and EC_w in Eq. (4.5.2) by effective bending rigidity $(EI_y)_e$, torsional rigidity $(GJ)_e$, and warping rigidity $(EC_w)_e$. These effective values are estimated by using the tangent modulus concept. The recommended value of L_p for members of I-shaped and channel sections is:

$$\frac{L_p}{r_y} = \frac{300}{\sqrt{F_{yf}}} \qquad (4.5.6)$$

where F_{yf} is the yield stress of the flange of the section in ksi.

Equation (4.5.6) is valid for the usual case of "open" I-shape sections but becomes very conservative for "close" box-type or solid sections. For example, for solid rectangular bars and box-beams the recommended value is:

$$\frac{L_p}{r_y} = \frac{3,750}{M_p} \sqrt{JA} \qquad (4.5.7)$$

in which M_p is the plastic moment in kip-in., A is the cross-sectional area in in.[2], and J is the torsional constant in in[4].

Tests of I-shape sections have shown that the maximum effective spacing between points of lateral support adjacent to a plastic hinge is affected by several factors, the most important of which is the moment gradient, i.e., the change in moment over this distance. The limiting unbraced length L_{pd} for plastic analysis can be obtained from a beam model in which the beam is partly elastic and partly strain hardened. The solution of the resulting differential equation with some simplifications can be expressed in terms of the moment ratio M_1/M_p at the first adjacent bracing hinge moment as

$$\frac{L_{pd}}{r_y} = \frac{3,600 + 2,200M_1/M_p}{F_y} \qquad (4.5.8)$$

where M_1 is smaller moment at the end of unbraced length of beam in kip-in. and M_1/M_p is positive when the moments cause reverse curvature. In Eq. (4.5.8), L_{pd} is measured from a plastic hinge location to the next adjacent bracing point, and r_y is the radius of gyration of the member with respect to its weak axis. Equation (4.5.8) assumes that the moment diagram within the unbraced length next to the plastic hinge locations is reasonably linear. For nonlinear diagrams between braces, judgement should be used in choosing a representative ratio. All plastic hinge locations associated with the failure mechanism shall be braced to resist lateral and torsional displacements. Equation (4.5.8) is developed to provide a rotation capacity at least three times that of the initial yield curvature, which is sufficient for most applications. However, in the areas of high seismicity, rotations of seven to nine times the yield curvature may be required. For such cases, LRFD

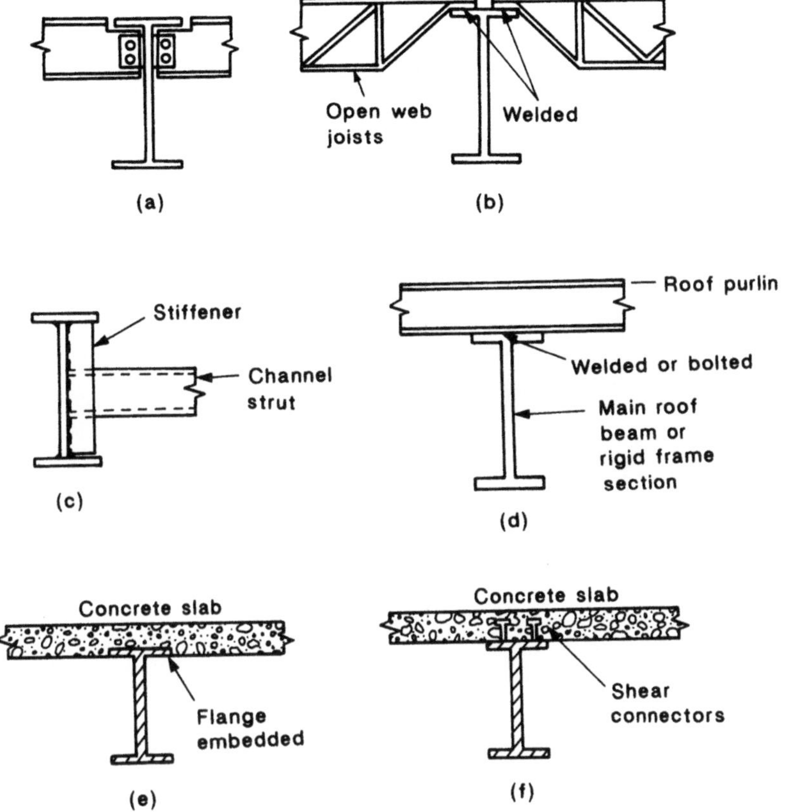

FIGURE 4.7. Types of lateral support for beams.

recommends

$$\frac{L_{pd}}{r_y} = \frac{146}{\sqrt{F_y}}.$$
(4.5.9)

Adequate lateral support to beams or structural members may be provided at intervals by cross beams, cross frames, struts, and roof purlins [Fig. 4.7(a–d)]. It may also be provided by embedment of compression flange in a concrete floor slab [Fig. 4.7(e and f)].

Note that at the region of the last plastic hinge, a large rotation capacity is not required. Therefore, the lateral support in this region can be provided at a spacing more than L_{pd}. The lateral support requirements for the region of the last hinge to form, and for regions not adjacent to a plastic hinge, are no different from those of elastic design. Its flexural design strength shall be determined in accordance with its respective length: L_{pd}, L_p, or L_r (Section F1, LRFD).

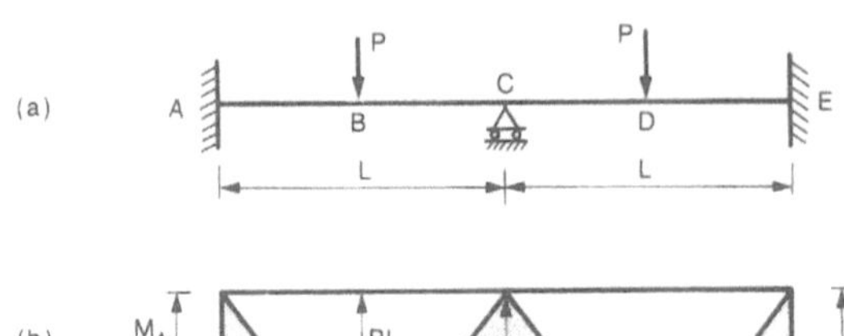

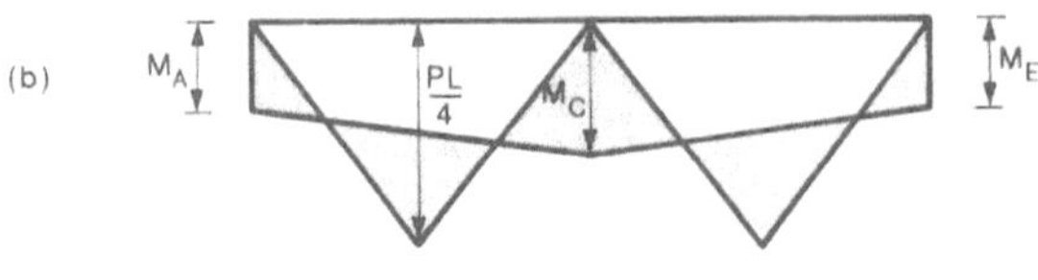

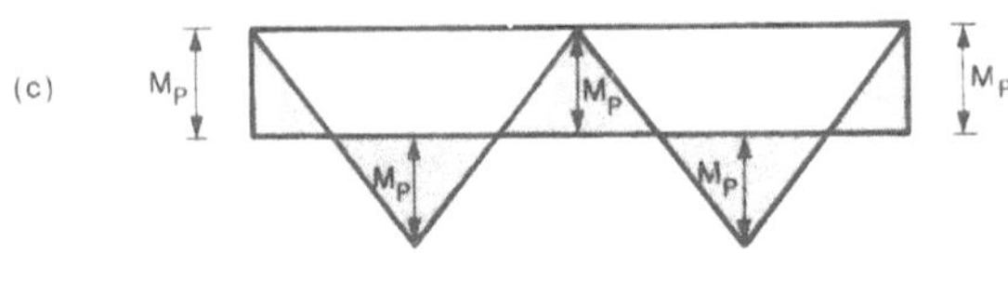

FIGURE 4.8. Plastic limit load of a fixed-ended beam with a central prop.

Example 4.5.1. A beam with plastic moment capacity M_p and length $2L$ has fixed supports at its ends and rests on a central support. Equal concentrated loads are applied at the center of each span as shown in Fig. 4.8(a). Determine the value of the applied loads at the plastic collapse state. Assume that the effect of shear force on plastic moment capacity is negligible.

Solution: The three redundants for the beam may be selected as two end moments M_A and M_E, and the central support moment M_C. Due to symmetry, the end moment M_A is equal to M_E. The moment diagram in terms of the two unknowns M_A and M_C is shown in Fig. 4.8(b).

In this moment diagram, if we select $M_A = M_E = M_C$ equal to plastic moment M_p of the beam, then the moment diagram becomes Fig. 4.8(c). By equating the critical moment at B to M_p, we have

$$M_B = \frac{PL}{4} - M_p = M_p, \tag{4.5.10}$$

which gives

$$P = \frac{8M_p}{L}. \tag{4.5.11}$$

At this value of P, nowhere does the moment exceed the plastic moment M_p [Fig. 4.8(c)], and the moments at A, B, C, D, and E are all equal to the plastic moment M_p, thus forming plastic hinges at these points and leading

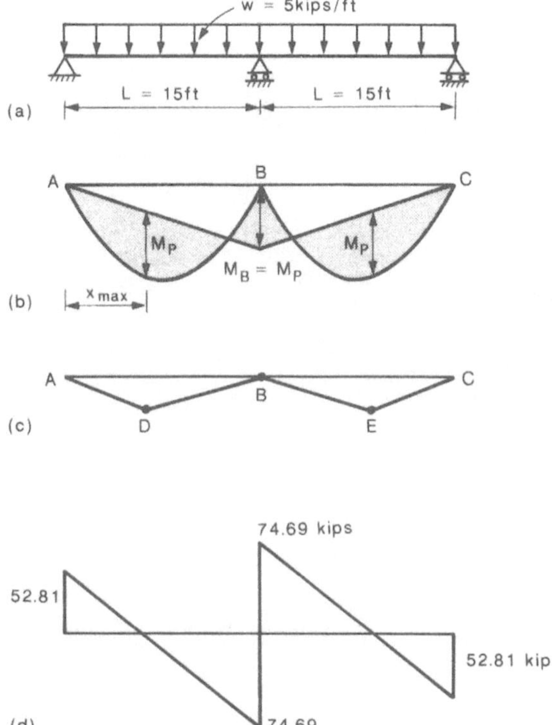

FIGURE 4.9. Design of a two-span continuous beam with distributed load.

to the formation of a failure mechanism as shown in Fig. 4.8(d). Equation (4.5.11) therefore gives the exact value of the limit load P.

Example 4.5.2. A beam of uniform section, 30 feet long, is to be part of a floor system. It is simply supported at its ends and rests on a central support as shown in Fig. 4.9(a). The dead and live loads together amount to a uniformly distributed load of 5 kips/ft. Design a suitable section of A36 steel for this beam. Provide a load factor of 1.70. Assume that the beam has adequate lateral support.

Solution: To select a suitable section for the beam, it is necessary to determine the required plastic moment (or section modulus) and shear capacity of the section. The required plastic moment capacity will be determined by the equilibrium method.

The degree of redundancy for the structure is one. The redundant may be selected as moment M_B at the central support. The moment diagram in terms of M_B is shown in Fig. 4.9(b). The conditions of plastic moment and mechanism of Fig. 4.9(c) will be satisfied if M_B and the in-span maximum moments in the beam simultaneously become equal to plastic moment M_p. The equation for the moment in the first span is obtained by superimposing the moment due to the applied loads and the moment due to the redundant M_B $(=M_p)$.

$$M = \frac{wL}{2}x - \frac{wx^2}{2} - \frac{M_p}{L}x \tag{4.5.12}$$

where x is the distance from end A. By equating the derivative of Eq. (4.5.12) to zero, the location of maximum moment can be found as

$$x_{\max} = \frac{L}{2} - \frac{M_p}{wL}. \tag{4.5.13}$$

Substituting $x = x_{\max}$ and $M = M_p$ in Eq. (4.5.12) and then solving for M_p, we have

$$M_p = \frac{wL^2}{2}[3 - \sqrt{8}] \tag{4.5.14}$$

or

$$M_p = \frac{(5)(1.7)(15)^2}{2}[3 - \sqrt{8}] = 164.07 \text{ kip-ft.}$$

The required plastic modulus $Z = M_p/F_y = [(164.07)(12)]/36 = 54.69$ in.^{3}.

The shear force diagram of the beam is shown in Fig. 4.9(d). The maximum shear force is at the central support and has the value

$$V_B = 74.69 \text{ kips}$$

$$\text{required area of web} = \frac{V_B}{\tau_y} = \frac{74.69}{36/\sqrt{3}} = 3.59 \text{ in.}^2$$

Try W16 × 36

$$Z = 64.0 \text{ in.}^3$$

$$A_w = d_w t_w = 15 \times 0.295 = 4.42 \text{ in.}^2$$

$$\text{shear stress in web} = \frac{74.69}{4.42} = 16.89 \text{ ksi.}$$

The plastic modulus reduced due to the presence of the shear force is

$$Z_{ps} = Z - Z_w\left[1 - \frac{\sqrt{\sigma_y^2 - 3\tau^2}}{\sigma_y}\right]$$

or

$$Z_{ps} = 64.0 - \frac{(0.295)(15)^2}{4}\left[1 - \frac{\sqrt{(36)^2 - 3(16.89)^2}}{36}\right]$$

$$= 64.0 - 6.92 = 57.07 \text{ in.}^3 > 54.69 \text{ in.}^3$$

Use W16 × 36.

Example 4.5.3. A three-span continuous beam is to support loads of 100 kips at the middle of each span as shown in Fig. 4.10(a). For architectural reasons,

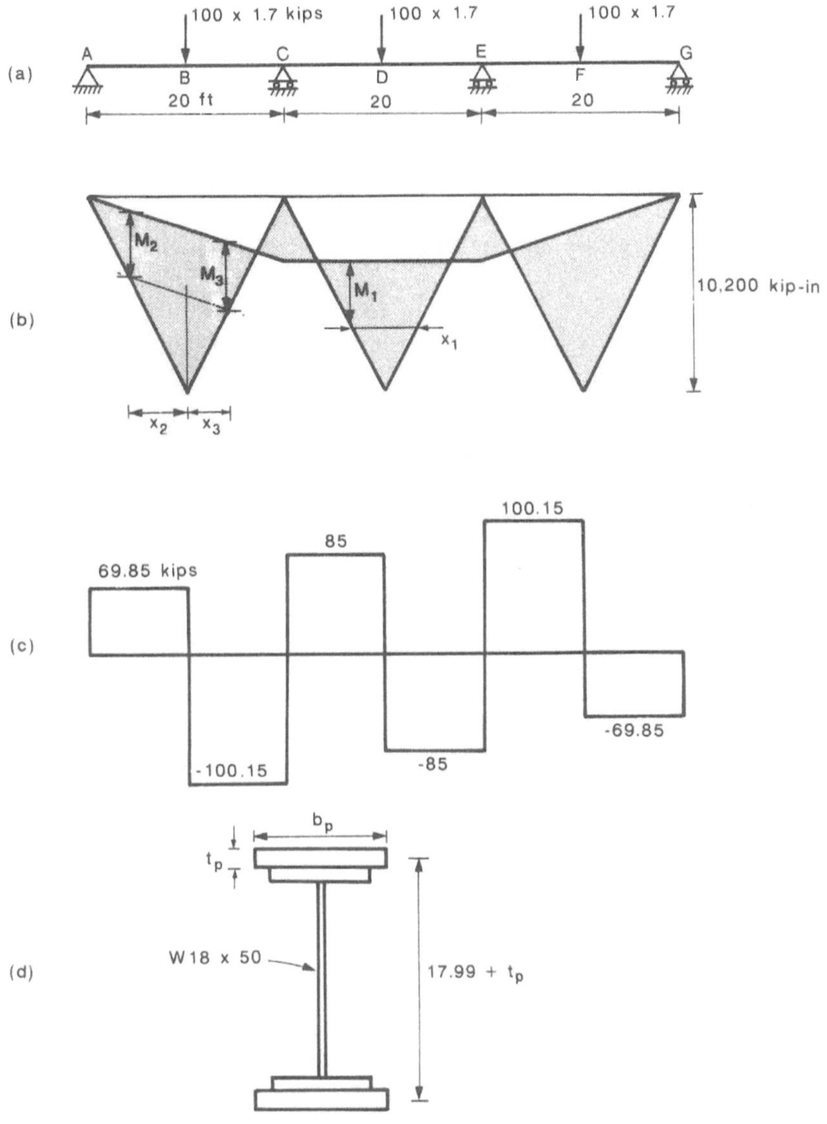

FIGURE 4.10. Design of cover plates.

the overall depth of the beam is limited to 20 inches. Show that A36 W18 × 50 with cover plates in the regions under concentrated loads is adequate. Determine the sizes and lengths of the cover plates. Use a load factor of 1.70. Assume that the beam has adequate lateral support.

Solution: To design the cover plates, first we will draw bending moment and shear force diagrams of the beam. The degree of indeterminacy for the beam is two. The redundants can be chosen as M_C and M_E. The moment diagram in terms of these two unknowns is shown in Fig. 4.10(b). First, assume that the reduction in the plastic moment capacity due to shear force is negligible, so that the unknown moments at C and E can be selected as the plastic moment of W18 × 50:

$$M_C = M_E = M_p = \sigma_y Z = (36)(101) = 3636 \text{ kip-in.}$$

Corresponding to these values of M_C and M_E, the shear force diagram is shown in Fig. 4.10(c). The shear forces at B and F are

$$V_B = V_F = 100.15 \text{ kips.}$$

Shear stress τ at these points is calculated by assuming that τ is uniformly distributed over the web of W18 × 50

$$\tau_B = \tau_F = \frac{V_F}{d_w t_w} = \frac{100.15}{(16.85)(0.355)} = 16.74 \text{ ksi} < \frac{36}{\sqrt{3}} \text{ ksi.}$$

Thus, at these points, the plastic moment capacity of W18 × 50 is reduced to

$$M_{ps} = M_p - \sigma_y Z_w \left[1 - \frac{\sqrt{\sigma_y^2 - 3\tau^2}}{\sigma_y} \right]$$

or

$$M_{ps} = 3636 - 36 \left[\frac{0.355(16.85)^2}{4} \right] \left[1 - \frac{\sqrt{(36)^2 - 3(16.74)^2}}{36} \right] = 3267 \text{ kip-in.}$$

Now the cover plates in the middle and end spans can be designed as follows.

Cover Plates for Middle Span: Moment at D is

$$M_D = \frac{(170)(20)(12)}{4} - 3267 = 6934 \text{ kip-in.}$$

Shear stress in the web at D is

$$\tau = \frac{85}{(16.85)(0.355)} = 14.21 \text{ ksi} < \frac{36}{\sqrt{3}} = 21.6 \text{ ksi,} \quad \text{okay.}$$

Thus, in the middle span, M_{ps} of W18 × 50 is taken as

$$M_{ps} = 3636 - 36 \left[\frac{0.355(16.85)^2}{4} \right] \left[1 - \frac{\sqrt{(36)^2 - (3)(14.21)^2}}{36} \right] = 3391 \text{ kip-in.}$$

The plastic modulus to be provided by the cover plates is therefore

$$Z_{\text{plates}} = \frac{6961 - 3391}{36} = 99.17 \text{ in.}^3$$

From Fig. 4.10(d), Z_{plates} can be written as

$$Z_{\text{plate}} = b_p t_p [17.99 + t_p] = 99.17 \text{ in.}^3$$

Trying $t_p = 0.5$ in., b_p is 10.75 in. Therefore, use 10.75- × 1/2-inch plates.

The length of cover plates x_1 [Fig. 4.10(b)] is determined simply by equating moment M_1 in the middle span to M_{ps} of W18 $\times$ 50 in the middle span

$$M_1 = 10{,}200\frac{(10 - x_1/2)}{10} - 3267 = 3391 \text{ kip-in.,}$$

which gives

$$x_1 = 6.95 \text{ ft.}$$

Therefore provide 7-foot-long 11-inch $\times$ 1/2-inch cover plates under the load in the middle span.

Cover Plates for End Spans: The moments at B and F are

$$M_B = M_F = \frac{170(20)(12)}{4} - \frac{3267}{2} = 8567 \text{ kip-in.}$$

In the end spans, M_{ps} is

$$M_{ps} = 3{,}267 \text{ kip-in.}$$

Plastic modulus to be provided by cover plates is therefore

$$Z_{\text{plates}} = \frac{8567 - 3267}{36} = 147.23 \text{ in.}^3$$

From Fig. 4.10(d), Z_{plates} can be expressed as

$$Z_{\text{plates}} = b_b t_p (17.99 + t_p) = 147.23 \text{ in.}^3$$

Trying $t_p = 0.75$-inches, b_p is 10.54 inches. Use 10.5- $\times$ 0.75-inch cover plates.
The length of cover plates $x = x_2 + x_3$ [Fig. 4.10(b)], is determined by equating bending moments M_2 and M_3 in the end spans to the respective M_{ps} of W18 $\times$ 50. To do so, start by calculating M_{ps} at A. The shear force at A is:

$$V_A = 69.85 \text{ kips.}$$

Then the shear stress τ at these points is calculated by assuming τ is uniformly distributed over the web of W18 $\times$ 50

$$\tau_A = \frac{V_A}{d_w t_w} = \frac{69.85}{16.85 \times 0.355} = 11.68 < \frac{36}{\sqrt{3}} \text{ksi.}$$

Thus, the moment capacity of W18 $\times$ 50 is reduced to

$$M_{ps} = M_p - \sigma_y Z_w \left[1 - \frac{\sqrt{\sigma_y^2 - 3\tau^2}}{\sigma_y} \right]$$

$$= 3636 - 36 \left[\frac{0.355(16.85)^2}{4} \right]\left[1 - \frac{\sqrt{36^2 - 3(11.68)^2}}{36} \right] = 3479 \text{ kip-in.}$$

Now, M_2 has the value

$$M_2 = \frac{(10 - x_2)}{10}\left(10{,}200 - \frac{3{,}267}{2}\right) = 3479,$$

Which gives

$$x_2 = 5.94 \text{ ft},$$

and M_3 has the value

$$M_3 = \frac{10{,}200(10 - x_3)}{10} - \frac{3267(10 + x_3)}{20} = 3267,$$

which gives

$$x_3 = 4.48 \text{ ft}.$$

Thus, the length of the cover plates in end spans is

$$x = 5.94 + 4.48 = 10.42 \text{ ft}.$$

Therefore, provide 10-foot 5-inch-long, 10.5-inch $\times$ 3/4-inch cover plates under loads in the two end spans.

Example 4.5.4. The beam in Example 4.5.2 is designed by the simple plastic theory. The lateral support is provided at vertical supports and the location of the plastic hinges. Do we need to provide additional lateral supports for adequate rotation capacity to form a plastic failure mechanism? If yes, at what location would you provide the lateral supports?

Solution: Check Segments *AD* and *EC* [Fig. 4.9(c)]: From Eq. (4.5.8), maximum spacing of lateral supports is

$$L_{pd} = \frac{3600 + 2200M_1/M_p}{F_y}r_y = \frac{3600}{36}(r_y) = 100r_y.$$

For W16 $\times$ 36, $r_y = 1.52$ in. Therefore, the maximum permissible spacing is

$$L_{pd} = \frac{(100)(1.52)}{12} = 12.67 \text{ ft}.$$

From Fig. 4.9(d), the zero shear location is

$$AD = EC = x_{\max} = \frac{(52.81)(15)}{52.81 + 74.67} = 6.21 \text{ ft} < 12.67 \text{ ft}, \quad \text{okay}.$$

Check Segments *DB* and *BE* [Fig. 4.9(c)]:

$$L_{pd} = \frac{3600 + 2200\left(\dfrac{164}{164}\right)}{36}(1.52) = 244.9 \text{ in.} = 20.41 \text{ ft}$$

$$DB = BE = AB - AD = 8.79 \text{ ft} < 20.41 \text{ ft}, \quad \text{okay}.$$

Therefore, there is no need to provide additional lateral supports. Since LRFD requires that all plastic hinge locations associated with the failure mechanism be braced to resist lateral and torsional displacements, it follows that lateral supports must be provided for the plastic hinge locations D and E in Fig. 4.9(c).

4.6 Design of Portal Frames

In the preceding section, the equilibrium method was applied for a complete design of steel beams. Herein, we shall use the same procedure for a complete design of simple portal frames including the possible yielding effect for members under tensile axial forces and yielding as well as instability effects for members under compressive axial forces. For one- or two-story frames, the effect of frame instability can generally be ignored in a routine plastic analysis procedure. For multistory frames, the frame instability effect must be included in the calculations of maximum strength and in the design of the bracing system and frame members. A direct second-order elastic-plastic hinge analysis procedure will be described in Chapter 8.

4.6.1 Tensile Axial Force

The presence of an axial force in general reduces the moment-carrying capacity of a member. If the axial force is tensile, the equilibrium method can be modified simply by replacing M_p (full plastic moment capacity) with M_{pc} (the plastic moment capacity reduced for the presence of axial load, Section 2.5). AISC-LRFD recommends that the following interaction equations be satisfied to consider the effect of tensile axial force in the design of doubly and singly symmetric members.

For $P/\phi P_n \geq 0.2$

$$\frac{P}{\phi P_n} + \frac{8}{9}\frac{M_x}{\phi_b M_{nx}} \leq 1.0. \tag{4.6.1}$$

For $P/\phi P_n < 0.2$

$$\frac{P}{2\phi P_n} + \frac{M_x}{\phi_b M_{nx}} \leq 1.0 \tag{4.6.2}$$

where P = applied axial force; $P_n = P_y$ is the available axial strength; M is the applied end moment; M_n is the available beam-bending capacity including the effect of lateral torsional buckling, and is equal to M_p when the beam is adequately braced against lateral-torsional instability; $\phi = \phi_t$ = the resistance factor for tension = 0.9; and ϕ_b = the resistance factor for bending = 0.9.

4.6.2 Compressive Axial Force

The plastic theory assumes that failure of the entire frame in the formation of a mechanism is not preceded by failure of compression members due to instability. Consequently, after the frame has been designed and the members selected, the compression members must be checked to assure that applied axial force and end moments can be carried by the member in a stable manner. Furthermore, any plastic hinges forming at the ends must have adequate rotation capacity.

The compressive axial force affects both yielding and instability of a member. The effect on the yielding is the same as for tensile axial force. Instability may be caused by two factors. The first is known as the P-δ effect, and the second is known as the P-Δ effect. The P-δ effect, also known as the *individual member instability effect*, is due to the lateral deflections within the member. For braced multistory frames, provisions must be made to include the P-δ effect in the design of bracing system and frame members. The P-Δ effect, also known as the *frame instability effect*, is due to the lateral translation of an end of the member. For unbraced multistory frames, the P-Δ effect must be included directly in the calculations of maximum strength. Although the strength of an isolated compression member subjected to axial force and bending moment can be predicted with relative ease, the instability problem becomes exceedingly complex when the compression member is a part of the framework. A complete solution of this latter problem requires a *second-order inelastic analysis*. In structures designed on the basis of plastic or elastic first-order analysis, simplified procedures are available in AISC-LRFD for checking the suitability of framed compression members considering the instability effects using the interaction Eqs. (4.6.1) and (4.6.2) for flexure and compression in symmetric shapes. This is given in the forthcoming.

The P-δ effect is shown in Fig. 4.11. Figure 4.11(a) shows a member with joint translation prevented and subjected to end moments and lateral loads. The lateral deflection δ_I for this member may be calculated on the basis of the original straight configuration and is known as the *first-order deflection*. Figure 4.11(b) shows the same member subjected to end moments, lateral

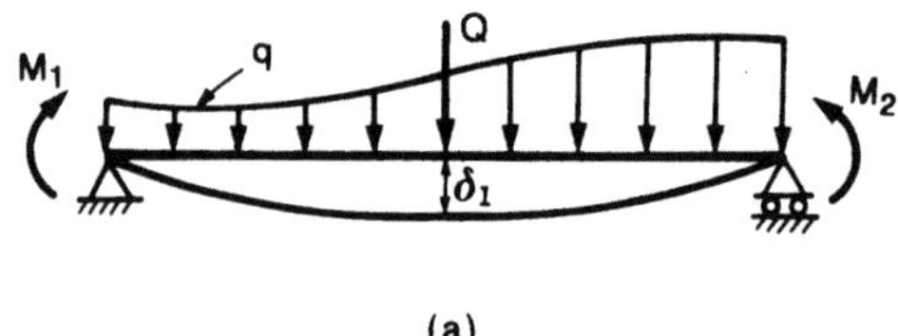

(a)

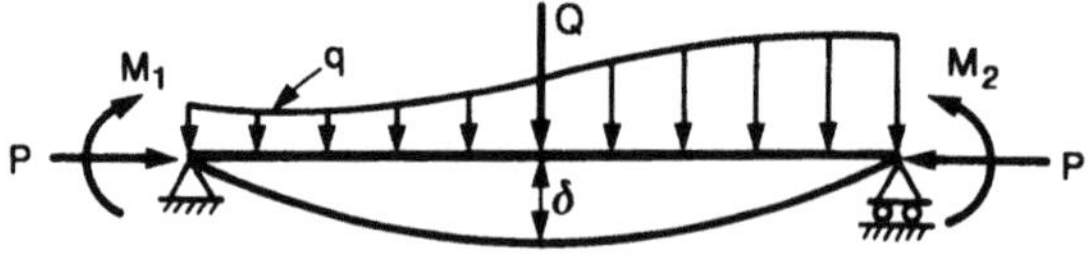

(b)

FIGURE 4.11. P-δ effect.

FIGURE 4.12. *P-Δ* effect.

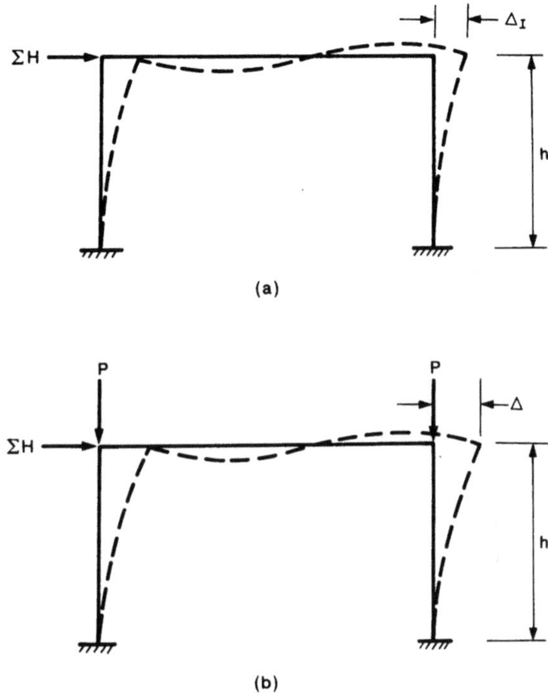

loads, and the axial force. In this case, the axial force P will interact with the first-order lateral deflection δ_I caused by the end moments and lateral loads and will amplify the first-order lateral deflection and first-order moments. The $P\text{-}\delta$ effect will be considered in Chapter 8, where a direct second-order elastic-plastic hinge analysis procedure will be described.

The *P-Δ effect* is illustrated in Fig. 4.12 [4.1–4.6]. Figure 4.12(a) shows an unbraced frame subjected to lateral forces $\sum H$. The frame deflects laterally until an equilibrium position is reached. This lateral deflection Δ_I can be calculated on the basis of the original undeformed configuration of the frame and is known as the *first-order deflection*. Figure 4.12(b) shows the same frame subjected to the combined lateral forces $\sum H$ and gravity loads $\sum P$. In this case, the gravity loads will interact with the lateral deflection caused by $\sum H$ and will introduce the additional $P\text{-}\Delta$ moment to the ends of the columns. This in turn amplifies the initial lateral deflection Δ_I and the first-order moments. Here, as the $P\text{-}\delta$ effect, the $P\text{-}\Delta$ effect will be considered in Chapter 8 with the use of a direct second-order analysis.

The effects of instability are included in AISC-LRFD interaction Eqs. (4.6.1) and (4.6.2) by applying B_1 and B_2 factors to $(M = B_1 M_{nt} + B_2 M_{lt})$ in which M_{nt} is the required flexural strength in the member, assuming there is no lateral translation of the frame, and M_{lt} is the required strength in the member as a result of the lateral translation of the frame only, by substituting

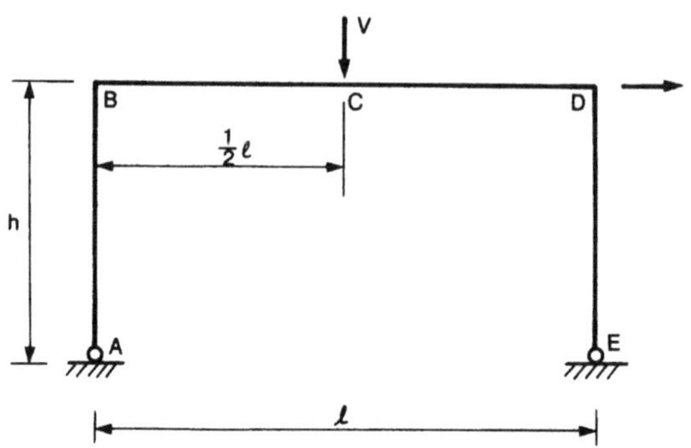

FIGURE 4.13. Pinned-ended portal frame.

$\phi = \phi_c = 0.85$ and replacing P_n with the compressive axial strength of the member. LRFD recommends the following equations for the compressive axial strength of a member:

For $\lambda_c \leq 1.5$

$$P_n = 0.658^{\lambda_c^2} P_y. \tag{4.6.3}$$

For $\lambda_c > 1.5$

$$P_n = \left[\frac{0.877}{\lambda_c^2}\right] P_y \tag{4.6.4}$$

where

$$\lambda_c = \frac{1}{\pi} \frac{KL}{r} \sqrt{\frac{F_y}{E}} \tag{4.6.5}$$

in which KL/r is the governing effective slenderness ratio about the plane of buckling, F_y is the yield stress, and E is the Young modulus of steel.

Note that for plastic design, λ_c should not be greater than $1.5K$ and the axial force in the member due to factored loads should not exceed $0.75\,A_g F_y$, where A_g is the gross cross-sectional area of the member.

Example 4.6.1. A pinned-ended rectangular portal frame *ABCDE* is subjected to factored external loads V and H as shown in Fig. 4.13. All the members of the frame *AB*, *BD*, and *DE* are made of the same section. Determine the limit values of H in terms of plastic moment M_p when

(i) $l/h = 1$ and $V/H = 1/3$.
(ii) $l/h = 1$ and $V/H = 3$.
(iii) $l/h = 3$ and $V/H = 1/3$.
(iv) $l/h = 3$ and $V/H = 3$.

Neglect the effects of shear force, lateral torsional buckling, and axial load on member strength.

Solution: The frame has one degree of redundancy. The redundancy for this structure can be chosen as the horizontal reaction at E. Figure 4.14 shows the

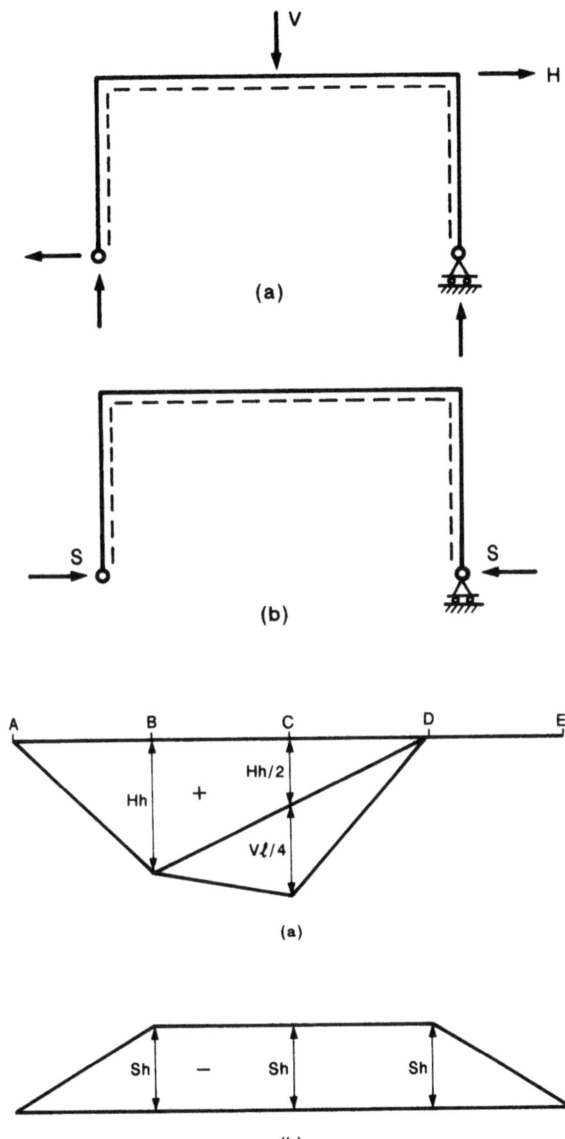

FIGURE 4.14. Determinate portal frame loaded by (a) applied forces and (b) redundant forces.

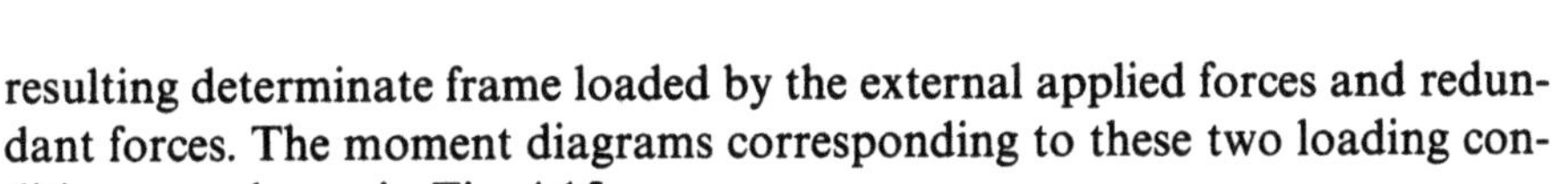

FIGURE 4.15. Moment diagram corresponding to (a) applied forces and (b) redundant forces.

resulting determinate frame loaded by the external applied forces and redundant forces. The moment diagrams corresponding to these two loading conditions are shown in Fig. 4.15.

Now the horizontal reaction S should be selected in such a manner that all three conditions of equilibrium, plastic moment, and mechanism are satisfied. Formation of two plastic hinges is necessary to form a failure mechanism. The hinges can possibly be formed at B, C, and D. Let us select

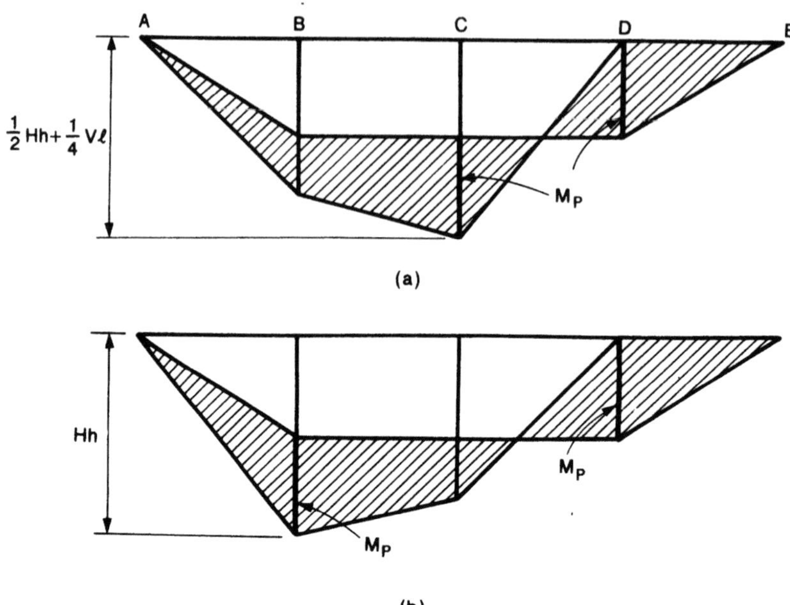

FIGURE 4.16. Moment diagrams for combined loading: (a) plastic hinges at C and D and (b) plastic hinges at B and D.

$$S = \frac{M_p}{h}$$

so that one hinge forms at D. Corresponding to this value of S, the moment at B and C can be expressed as

$$M_B = Hh - M_p$$

$$M_C = \frac{Hh}{2} + \frac{Vl}{4} - M_p.$$

The location of the second hinge will depend on the relative magnitude of V, H, l, and h. Figure 4.16 shows the bending moment diagram corresponding to the combined loading. Figure 4.16(a) corresponds to the second hinge at C and Fig. 4.16(b) corresponds to the second hinge at B. For various combinations of V, H, l, and h, H is calculated in the following.

Case I: $l/h = 1$ and $V/H = 1/3$—Corresponding to $l = h$ and $V = H/3$, we have

$$M_B = Hh - M_p,$$

$$M_C = \frac{Hh}{2} + \frac{Hh}{12} - M_p = \frac{7}{12}Hh - M_p.$$

Since $|M_B| > |M_C|$, the second hinge will form at B [Fig. 4.16(b)] and the corresponding value of H is

$$H = \frac{2M_p}{h}.$$

Case II: When $l/h = 1$ and $V/H = 3$—Corresponding to $l = h$ and $V = 3H$, we have

$$M_B = Hh - M_p$$

$$M_C = \frac{Hh}{2} + \frac{3}{4}Hh - M_p = \frac{5}{4}Hh - M_p.$$

Since $|M_C| > |M_B|$, the second hinge will form at C [Fig. 4.16(a)] and the corresponding value of H is

$$H = \frac{1.6M_p}{h}.$$

Case III: $l/h = 3$ and $V/H = 1/3$—Corresponding to $l = 3h$ and $V = H/3$, we have

$$M_B = Hh - M_p$$

$$M_C = \frac{Hh}{2} + \frac{Hh}{4} - M_p = \frac{3}{4}Hh - M_p.$$

Since $|M_B| > |M_C|$, the second hinge will form at B [Fig. 4.16(b)] and the corresponding value of H will be

$$H = \frac{2M_p}{h}.$$

Case IV: $l = 3h$ and $V = 3H$—Corresponding to $l = 3h$ and $V = 3H$, we have

$$M_B = Hh - M_p$$

$$M_C = \frac{Hh}{2} + \frac{3H}{4}(3h) - M_p = 2.75Hh - M_p.$$

Since $|M_C| > |M_B|$, the second hinge will form at C [Fig. 4.16(a)] and the corresponding H is

$$H = 0.727\frac{M_p}{h}.$$

To summarize, we have (i) $l/h = 1$ and $V/H = 1/3$, $H = 2\,M_p/h$; (ii) $l/h = 1$ and $V/H = 3$, $H = 1.6\,M_p/h$; (iii) $l/h = 3$ and $V/H = 1/3$, $H = 2\,M_p/h$; and (iv) $l/h = 3$ and $V/H = 3$, $H = 0.727\,M_p/h$.

Example 4.6.2. The frame in Example 4.6.1 has $l = h = 20$ feet and $V/H = 3$. All members are made of W16 × 45. Determine the limit values of V and H

(a) without considering the effect of axial force.
(b) considering the effect of axial force assume $B_1 = B_2 = 1.0$, $K_x = 1$, $K_y = 1$.

Column DE is braced in the middle against buckling about weak-axis.

Solution: (a) Without the effect of axial force: From Example 4.6.1, for $l/h = 1$ and $V/H = 3$, we have

$$H = 1.6 \frac{M_p}{h}.$$

From the AISC-LRFD manual, for W16 × 45, we have

$$M_{px} = Z_x F_y = 82.3 \times 36 = 2{,}963 \text{ kip-in.},$$

resulting in

$$H = 1.6 \frac{2{,}963}{(20)(12)} = 19.75 \text{ kips}$$

$$V = 3H = (3)(19.75) = 59.25 \text{ kips.}$$

(b) With the effect of axial force: Member *BD* (Fig. 4.13): The axial force in member BD is

$$T = H - S = 1.6 \frac{M_p}{h} - \frac{M_p}{h} = 0.6 \frac{M_p}{h} = 7.41 \text{ kips.}$$

For W16 × 45, the yield axial force is

$$P_y = (13.3)(36) = 478.8 \text{ kips.}$$

Now, since

$$\frac{P}{\phi_t P_n} = \frac{7.41}{(0.9)(478.8)} = 0.018 < 0.2.$$

The reduced plastic moment capacity M_{pc} can be determined from the following interaction equation

$$\frac{P}{2\phi_t P_n} + \frac{M_{pc}}{\phi_b M_p} \leq 1.0,$$

which gives

$$M_{pc} = \left(1 - \frac{0.018}{2}\right)\phi_b M_p = 0.991 \phi_b M_p.$$

The axial force effect is negligible.

Member _DE_ (Fig. 4.13): From Fig. 4.13(a), we have

$$H = 19.75 \text{ kips}$$

$$V = 59.25 \text{ kips.}$$

Axial force in member _DE_ is

$$P = \frac{V}{2} + \frac{Hh}{l} = \frac{V}{2} + H = 49.38 \text{ kips.}$$

Since column _DE_ is supported in the middle against buckling about the weak axis, $(KL)_y = 10$ feet. Thus, we have

$$\lambda_{cy} = \frac{1}{\pi} \frac{(KL)_y}{r_y} \sqrt{\frac{F_y}{E}} = \frac{1}{\pi} \frac{(10)(12)}{1.57} \sqrt{\frac{36}{30,000}}$$

or

$$\lambda_{cy} = 0.843.$$

For buckling about the strong axis, we have

$$\lambda_{cx} = \frac{1}{\pi} \frac{(KL)_x}{r_x} \sqrt{\frac{F_y}{E}} = \frac{1}{\pi} \frac{(1)(20)(12)}{6.65} \sqrt{\frac{36}{30,000}} = 0.398.$$

So, buckling about the weak axis controls and we have $\lambda_c = \lambda_{cy}$. As $\lambda_c \leq 1.5$, P_n is

$$P_n = 0.658^{\lambda_c^2} P_y = 0.658^{(0.843)^2}(478.8) = 355.6 \text{ kips.}$$

The ratio $P/\phi_c P_n$ is

$$\frac{P}{\phi_c P_n} = \frac{49.38}{(0.85)(355.6)} = 0.163.$$

Now the moment M_{pc} (reduced for the effect of the axial load) can be determined from the following interaction equation

$$\frac{P}{2\phi_c P_n} + \frac{M_{pc}}{\phi_b M_{nx}} \leq 1$$

or

$$\frac{0.163}{2} + \frac{M_{pc}}{0.9 M_p} = 1.0,$$

which gives

$$M_{pc} = 0.827 M_p.$$

Now, H and V become

$$H = 1.6 \frac{M_{pc}}{h} = \frac{(1.6)(0.827)(2963)}{(20)(12)} = 16.34 \text{ kips}$$

$$V = 3H = 49 \text{ kips.}$$

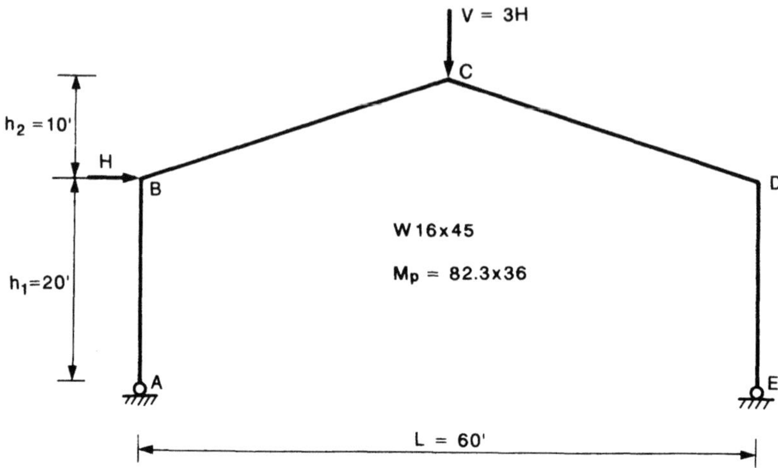

FIGURE 4.17. Pinned-ended gabled frame.

Note that 17% reduction in H and V is partly due to axial load and partly due to inclusion of ϕ factors in the computation for the latter case.

Example 4.6.3. A pinned-ended gable portal frame *ABCDE* is subjected to factored external loads V and H as shown in Fig. 4.17. If all the members are made of W16 × 45, determine the limit values of V and H. Neglect the effects of shear force, lateral torsional buckling, and axial load on member strength.

Solution: The gable frame has one degree of redundancy. The horizontal reaction at E is chosen as the redundant. Figures 4.18(a) and (b) show, respectively, the determinate frame loaded by external applied forces V and H and the redundant forces. The moment diagrams corresponding to these two loading conditions are shown in Fig. 4.19. Now the horizontal reaction S should be selected in such a manner that all three conditions of equilibrium, plastic moment, and mechanism are satisfied. Formation of two plastic hinges is necessary to form a failure mechanism of the structure. The hinges can possibly be formed at B, C, and D. Let us select

$$S = \frac{M_p}{h_1} = \frac{(82.3)(36)}{(20)(12)} = 12.35 \text{ kips}$$

so that one hinge forms at D. Corresponding to this value of S, the moments at B and C can be expressed as (Fig. 4.19)

$$M_B = Hh_1 - Sh_1 = 20H - (12.35)(20) = 20H - 247$$

$$M_C = \frac{Hh_1}{2} + \frac{VL}{4} - S(h_1 + h_2) = 10H + 45H - (12.35)(30) = 55H - 370.5.$$

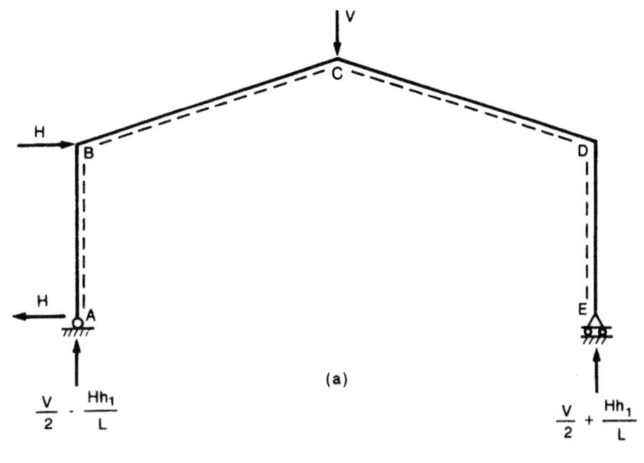

FIGURE 4.18. Determinate gabled frame loaded by (a) applied forces and (b) redundant force.

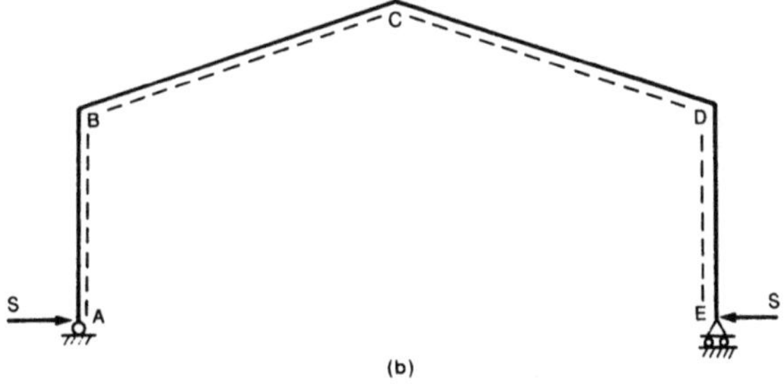

Assuming the second hinge at B, we have

$$M_B = M_p = \frac{2962.8}{12} = 20H - 247,$$

which gives

$$H = 24.7 \text{ kips}$$

$$M_C = (55)(24.7) - 370.5 = 988 \text{ kip-ft} = 11{,}856 \text{ kip-in}, \ |M_C| > M_p.$$

Therefore, the second hinge will form at C and the corresponding H can be determined by equating M_C and M_p:

$$M_C = M_p = \frac{2962.8}{12} = 55H - 370.5,$$

which gives

$$H = 11.23 \text{ kips}$$

and

$$V = 3H = 33.69 \text{ kips}.$$

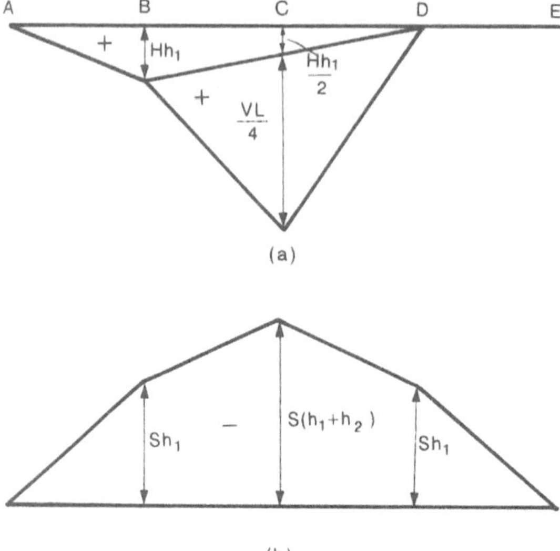

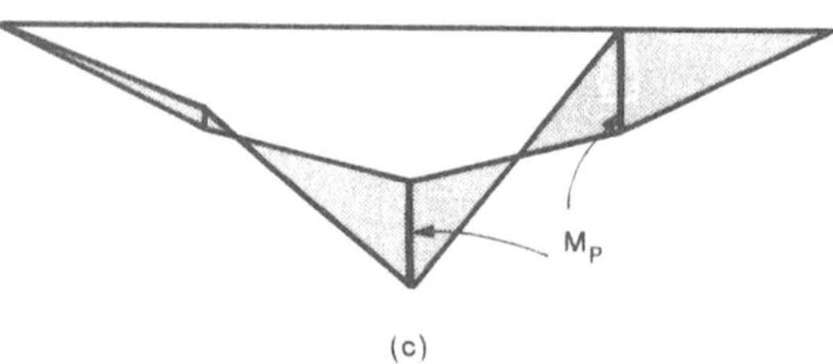

FIGURE 4.19. Moment diagrams corresponding to (a) applied forces; (b) redundant force; and (c) combined loading.

Note that with proper bracing of the columns against buckling, the reduction in V and H due to axial load is less than 5%.

4.7 Practical Procedure for Large Structures

The equilibrium method is convenient for plastic analysis and design when the number of redundants is small. With an increase in the redundants, the method in itself becomes too involved. Selection of the right combination of redundants demands increasingly high intuition. For plastic analysis and design of large structures that will have a much larger number of redundants, we will present a practical procedure of the equilibrium method. The procedure consists of the following steps:

(a) Select the redundants.

(b) Obtain the moment diagram of statically determinate structure under the applied loads.

(c) Obtain the moment diagram of a statically determinate structure under the redundant forces.

(d) Assume a failure mechanism for the structure.

(e) Combine moment diagrams of steps (b) and (c) to obtain moments at the plastic hinge locations and equate these moments to the plastic moment of the section.

(f) Solve the equations resulting from step (e) to determine the selected redundants and the plastic limit load or the required plastic moment for the structure.

(g) Check the plastic moment condition in the entire structure.

(h) If the plastic moment condition is satisfied for the structure, then the solution is exact. If not, proceed further to determine the upper and lower limits on the solution.

(i) If it is an analysis problem and the plastic limit load is to be determined, then the load determined in step (f) will be an upper bound because the solution satisfies only the mechanism condition. A lower bound to the exact limit load is determined by

$$P_{lo} = P_{up} \frac{M_p}{M_{\max}} \tag{4.7.1}$$

where P_{up} is an upper bound to the exact solution, $M_{\max}$ is the maximum moment in the structure, and M_p is the plastic moment capacity of section at the point of maximum moment.

Equation (4.7.1) provides a lower bound to the exact solution, because the moment diagram corresponding to this load has been scaled down to meet the moment condition. Thus, the solution so obtained satisfies both equilibrium and plastic moment conditions but not the mechanism condition and is therefore a lower bound to the limit load.

(j) If it is a design problem and plastic moment capacity has to be determined, then the plastic moment capacity determined in step (f) at the hinge locations will be unsafe to carry the applied loads, while the maximum moment that exceeds the plastic moment capacity in other parts of the structure will be a safe solution. This follows from the logic that if the plastic moment for all sections in the entire structure is increased to the maximum moment obtained from the equilibrium diagram, then both equilibrium and plastic moment conditions are satisfied but not the mechanism condition. Therefore, the plastic moment capacity so obtained provides a safe design when compared with the exact plastic moment required for the structure.

The following examples demonstrate this procedure. The sign convention used in these examples is that a moment causing tension inside the frame is a

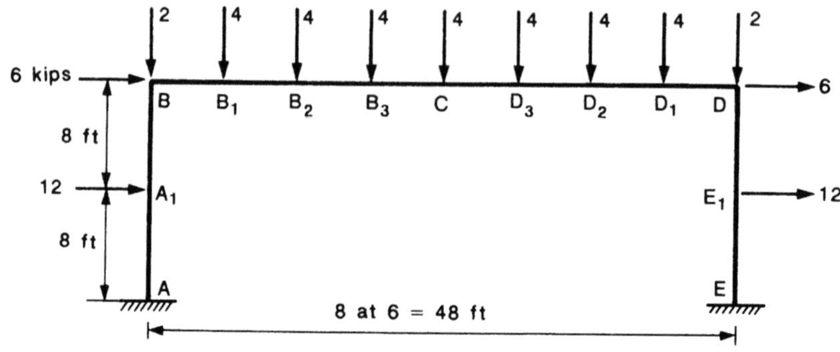

FIGURE 4.20. Portal frame with fixed supports.

positive moment and a moment causing opening of a joint is a positive moment, and vice versa.

Example 4.7.1. The frame shown in Fig. 4.20 has same cross section throughout. All joints are rigid and the two base supports are fixed. Find the required plastic moment M_p. Neglect the effects of axial force, shear force, and lateral torsional buckling on the strength of members.

Solution: The frame has three degrees of redundancy. The three internal forces at midspan C are chosen as the redundants. Figures 4.21 (a) and (b) show the determinate structures under the external forces and the redundant forces, respectively. In order to preserve symmetry, the vertical load at C has been cut into two equal halves. The moments due to external applied loads (free moments) and the moment due to redundants are tabulated in Table 4.1. The next step is to assume a failure mechanism and equate the moments at the plastic hinge locations to plastic moment.

Mechanism 1: To begin, try the side-sway mechanism with plastic hinges at the two joints B and D and the two supporting bases A and E. By equating moments at these points to the plastic moment, we have

$$M_A = -384 + 16T + 24S + M = -M_p$$

$$M_B = -192 + 24S + M = M_p$$

$$M_D = -192 - 24S + M = -M_p$$

$$M_E = M - 24S + 16T = M_p.$$

(Note that moment at a joint is positive when it causes opening of the joint.)
These four equations are solved for the four unknowns:

$$M_p = 96 \text{ kip-ft}$$

$$M = 192 \text{ kip-ft}$$

$$S = 4 \text{ kips}$$

$$T = 0.$$

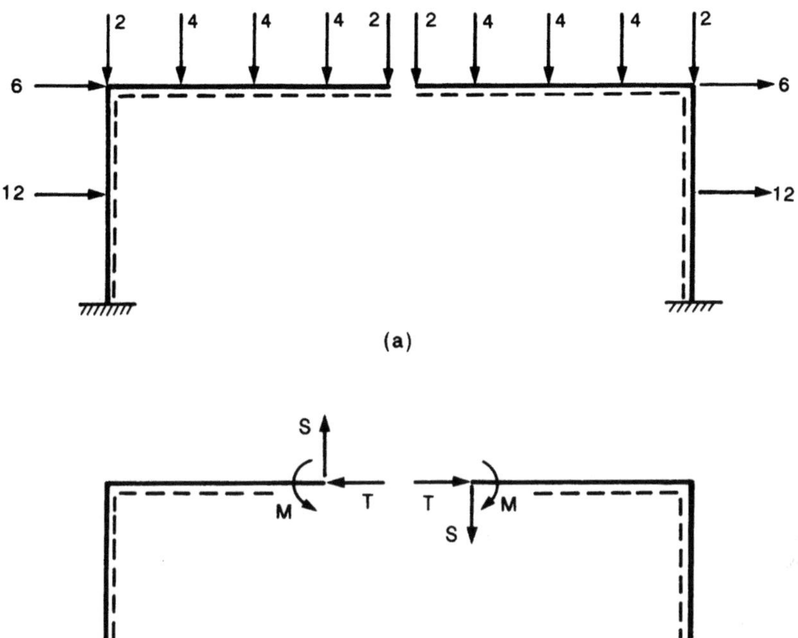

(a)

(b)

FIGURE 4.21. Determinate frame subjected to (a) applied forces and (b) redundant forces.

The moments in the frame corresponding to these values of unknowns are also listed in Table 4.1. The plastic moment condition is violated at many points. The absolute maximum moment in the frame is 204 kip-ft. Thus this mechanism provides both upper and lower bounds

$$96 \le M_p \le 204.$$

Mechanism 2: Try the combined mechanism with plastic hinges at the two supporting bases A and E, beam midspan C, and right top joint D. By equating moments at these points to the plastic moment, we have

$$M_A = -384 + M + 24S + 16T = = -M_p$$

$$M_c = M = M_p$$

$$M_D = -192 + M - 24S = -M_p$$

$$M_E = M - 24S + 16T = M_p.$$

TABLE 4.1. Example 4.7.1. portal frame calculations

Joint	A	A_1	B	B_1	B_2	B_3	C	D_3	D_2	D_1	D	E_1	E
Free	-384	-240	-192	-108	-48	-12	0	-12	-48	-108	-192	-144	0
T	$16T$	$8T$	0	0	0	0	0	0	0	0	0	$8T$	$16T$
S	$24S$	$24S$	$24S$	$18S$	$12S$	$6S$	0	$-6S$	$-12S$	$-18S$	$-24S$	$-24S$	$-24S$
M	M	M	M	M	M	M	M	M	M	M	M	M	M
Total (mechanism 1)	-96	48	96	156	192	204	192	156	96	12	-96	-48	96
Total (mechanism 2)	-128	-16	0	68	112	132	128	100	48	-28	-128	-48	128
Total (mechanism 3)	-129.2	-18.4	-3.6	64.8	108.8	129.3	125.6	98	46	-29.6	-129.2	-48	129.6

These four equations are solved for the four unknowns as

$$M_p = M = 128 \text{ kip-ft}$$

$$S = 2.67 \text{ kips}$$

$$T = 4 \text{ kips}.$$

The moments in the frame corresponding to these unknown values are also listed in Table 4.1. The plastic moment condition is violated at B_3 just to the left of the center of the beam. The upper and lower bounds corresponding to this mechanism are

$$128 \le M_p \le 132.$$

Mechanism 3: Try the combined mechanism with plastic hinges C switched to B_3 [Fig. 4.22(a)].

The new set of collapse equations is

$$M_A = -384 + M + 24S + 16T = -M_p$$

$$M_{B_3} = -12 + 6S + M = M_p$$

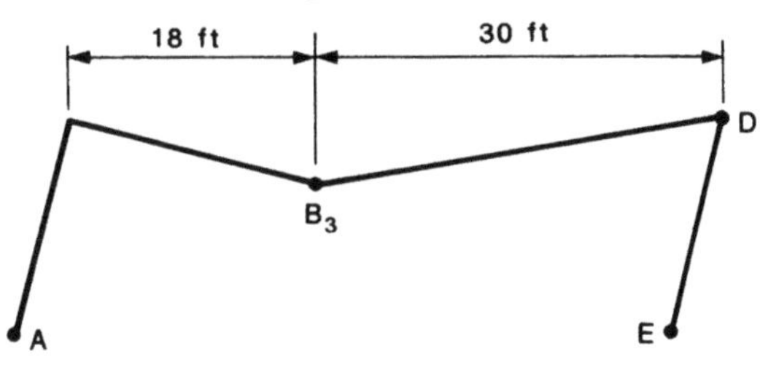

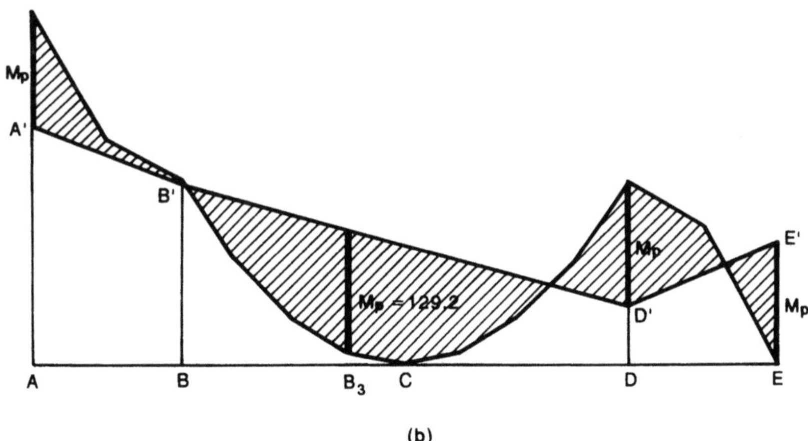

FIGURE 4.22. Moment diagram corresponding to the failure mechanism: (a) failure mechanism and (b) moment diagram.

$$M_D = -192 - 24S + M = -M_p$$

$$M_E = M - 24S + 16T = M_p.$$

These four equations give the following values of the four unknowns:

$$M_p = 129.2 \text{ kip-ft}$$

$$M = 125.6 \text{ kip-ft}$$

$$S = 2.615 \text{ kip-ft}$$

$$T = 4.15 \text{ kip-ft.}$$

Corresponding to these values of unknowns, the moments in the frame are tabulated in Table 4.1 and plotted in Fig. 4.22. The plastic moment condition is satisfied everywhere. Since only concentrated loads are acting on the frame, the moment diagram between various joints and concentrated loads is linear. For such cases, if the plastic moment condition is satisfied at the joints and at the concentrated load locations $(A, A_1, B, B_1, B_2, B_3, C, D_3, D_2, D_1, E_1, E)$, it will automatically be satisfied between these locations.

Since all three conditions of equilibrium, mechanism, and plastic moment are satisfied, $M_p = 129.2$ kip-fit $= 1550.4$ kip-in. is the exact required plastic moment for the frame.

Example 4.7.2. The gable frame shown in Fig. 4.23 has same cross section throughout. All joints are rigid and the two base supports are fixed. Loads having the magnitudes shown are actually uniformly distributed over the members. Find the required plastic moment M_p so that the load factor is 1.4. Neglect the effects of axial force, shear force, and lateral torsional buckling on member strength.

Soultion: The frame has three redundants. These redundants are chosen as the three internal forces at joint C. Figure 4.24 (a) and (b) show the determi-

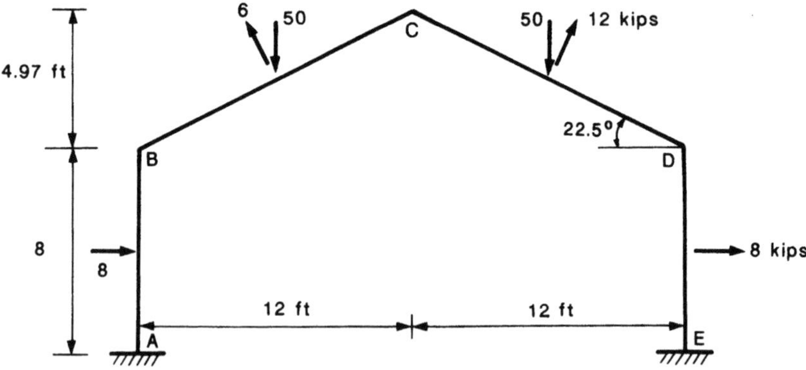

FIGURE 4.23. Gable frame with fixed supports subjected to distributed loads.

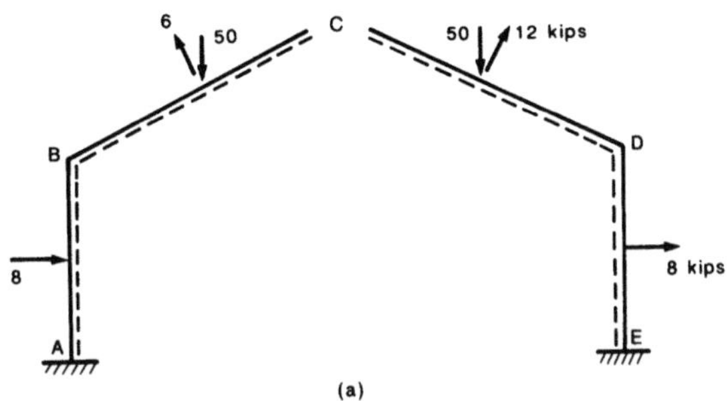

FIGURE 4.24. Determinate gable frame subjected to (a) applied forces (distributed) and (b) redundant forces.

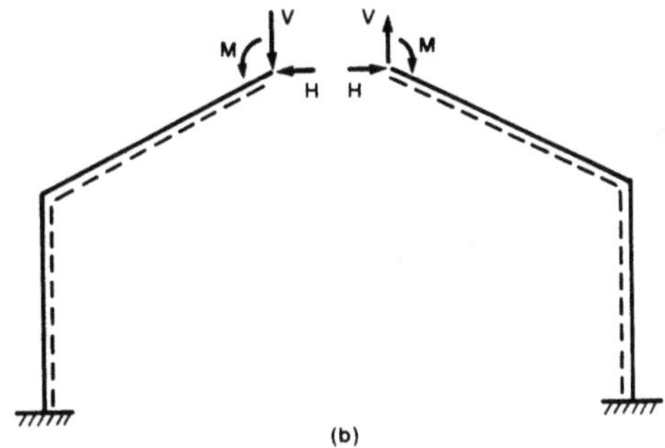

TABLE 4.2. Example 4.7.2. gable frame calculations

Joint	A	B	C	D	E
Free moments	−274.66	−261.03	0	−222.07	−153.33
M_c	M	M	M	M	M
V	$-12V$	$-12V$	0	$12V$	$12V$
H	$12.97H$	$4.97H$	0	$4.97H$	$12.97H$

nate structures under the external forces and the redundant forces, respectively. The bending moment due to external applied loads (free moments) and redundants at various points are tabulated in Table 4.2.

First, try the failure mechanism with column AB remaining vertical while other three members collapsed to the right with hinges at B, C, D, and E. By equating the moments at these points to the plastic moment, we have

$$M_B = -261.03 + M - 12V + 4.97H = -M_p$$

$$M_C = M = M_p$$

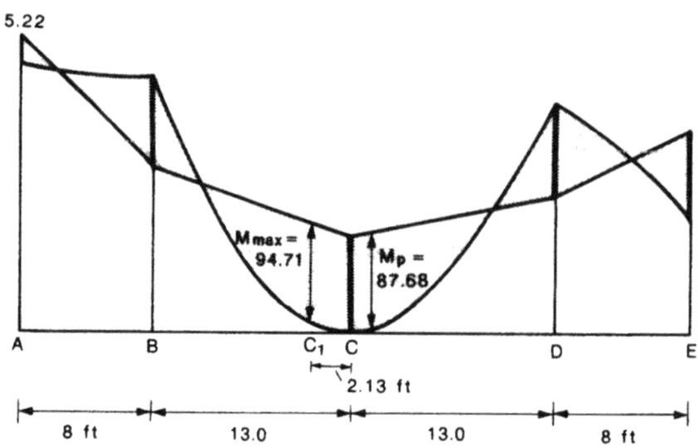

Figure 4.25. Moment diagram corresponding to a mechanism with plastic hinges at B, C, D, and E.

$$M_D = -222.07 + M + 12V + 4.97H = -M_p$$

$$M_E = -153.33 + M + 12V + 12.97H = M_p.$$

(The moment at a joint is positive when it causes opening of the joint.)

These four equations are solved to determine the four unknown values as

$$M_C = M = M_p = 87.68 \text{ kip-ft}$$

$$V = -1.62 \text{ kips}$$

$$H = 13.32 \text{ kips.}$$

We now check the moment at A to see if the plastic moment condition is satisfied:

$$M_A = -274.66 + 87.68 - 12(-1.62) + (12.97)(13.32) = 5.22 \text{ kip-ft}$$

Since $|M_A| < M_p = 87.68$ kip-ft, this mechanism is okay. However, since the actual loads are distributed, we must also check the maximum moment within the member. From the bending moment diagram (Fig. 4.25), it appears that the maximum moment may occur in the member BC. The moment M_x in the right half of BC can be expressed as

$$M_x = 87.68 + V'x - \frac{wx^2}{2}$$

where V' is the resultant shear force at C acting perpendicular to BC and is

$$V' = 13.32 \sin 22.5° + 1.62 \cos 22.5° = 6.6 \text{ kips,}$$

w is the distributed load acting perpendicular to BC and is

$$w = \frac{50 \cos 22.5° - 6}{12.98} = 3.1 \text{ kips/ft,}$$

and x is the distance along the member from point C to C_1 (the point of maximum moment). Now M_x along the member BC can be written as

$$M_x = 87.68 + 6.6x - 3.1\frac{x^2}{2}.$$

To determine the maximum moment location, we set $dM_x/dx = 0$. Thus

$$\frac{dM_x}{dx} = 6.6 - 3.1x = 0,$$

which gives

$$x = 2.13 \text{ ft.}$$

Thus, the maximum moment at C has the value

$$M_{c_1} = 87.68 + 6.6 \times 2.13 - 3.1(2.13)^2/2$$

$$= 87.68 + 14.06 - 7.03 = 94.71 \text{ kips-ft}$$

and this mechanism provides the following upper and lower bounds

$$87.68 \le M_p \le 94.71.$$

Now try the mechanism with plastic hinges at B, C_1, D, and E. The collapse equations become

$$M_B = -261.03 + M - 12V + 4.97H = -M_p$$

$$M_{C_1} = -7.03 + M - 1.97V + 0.82H = M_p$$

$$M_D = -222.07 + M + 12V + 4.97H = -M_p$$

$$M_E = -153.33 + M + 12V + 12.97H = M_p.$$

The solution of these four equations gives

$$M = 82.48 \text{ kip-ft}$$

$$H = 13.9 \text{ kips}$$

$$V = -1.624 \text{ kips}$$

$$M_p = 89.96 \text{ kip-ft} = 1,080 \text{ kip-in.}$$

In addition to the equilibrium and mechanism conditions, the plastic moment condition is also satisfied. Thus, $M_p = 1080$ kip-in. is exact. Multiplying the plastic moment M_p by the load factor 1.4, we have

$$M_p = 1,080 \times 1.4 = 1,512 \text{ kip-in.}$$

Example 4.7.3. The frame shown in Fig. 4.26 has rigid joints and is rigidly fixed to foundations at A and E. The frame has a uniform plastic moment capacity of M_p. Investigate the value of W at which collapse will just occur for the load system shown. Present the solution in such a way that the mode of collapse and the value of W at collapse are indicated for any value of α. Neglect the instability effects and the effects of axial load and shear force on the moment capacity.

Solution: The redundants for the frame are chosen as the three internal forces at joint C. Figure 4.27 shows the determinate gable frame under the applied forces and redundant forces, respectively. The free moments and moments

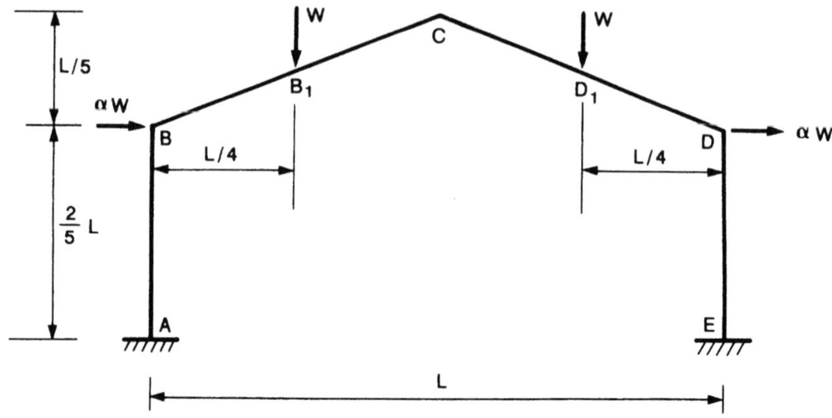

FIGURE 4.26. Gable frame with fixed supports and subjected to concentrated loads.

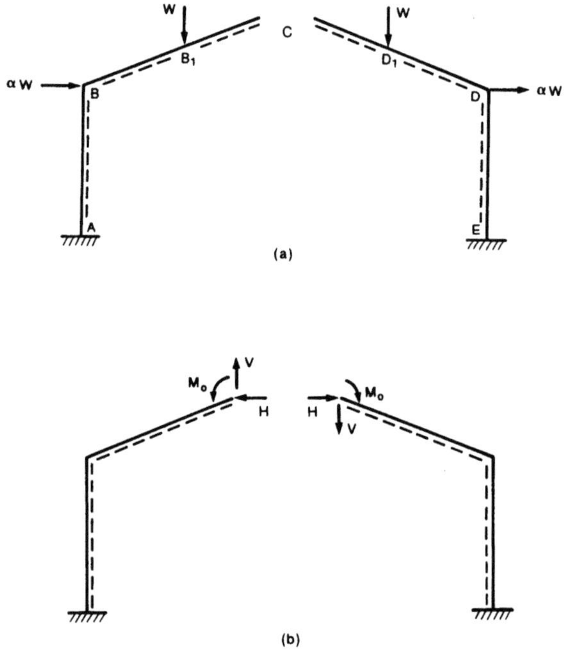

FIGURE 4.27. Determinate gable frame subjected to (a) applied forces (concentrated) and (b) redundant forces.

due to redundant forces are tabulated in Table 4.3. To determine W for various ranges of α, we will try all possible mechanisms.

Mechanism 1: Try the failure mechanism with column AB remaining vertical while other three members collapsed to the right with plastic hinges at B, C, D, and E. By equating moments at these points to the plastic moment, we have

TABLE 4.3. Example 4.7.3. gable frame calculations

Joint	A	B	C	D	E
Free moments	$-WL\left(\dfrac{1}{4}+\dfrac{2\alpha}{5}\right)$	$-WL/4$	0	$-WL/4$	$-WL\left(\dfrac{1}{4}-\dfrac{2\alpha}{5}\right)$
M_0	M_0	M_0	M_0	M_0	M_0
V	$\dfrac{1}{2}VL$	$\dfrac{1}{2}VL$	0	$-\dfrac{1}{2}VL$	$-\dfrac{1}{2}VL$
H	$\dfrac{3}{5}HL$	$\dfrac{1}{5}HL$	0	$\dfrac{1}{5}HL$	$\dfrac{3}{5}HL$

$$M_B = -\frac{WL}{4} + M_0 + \frac{VL}{2} + \frac{HL}{5} = -M_p$$

$$M_C = 0 + M_0 + 0 + 0 = M_p$$

$$M_D = -\frac{WL}{4} + M_0 - \frac{VL}{2} + \frac{HL}{5} = -M_p$$

$$M_E = -WL\left(\frac{1}{4} - \frac{2\alpha}{5}\right) + M_0 - \frac{VL}{2} + \frac{3HL}{5} = M_p.$$

Solve these four equations and determine the four unknowns:

$$M_0 = M_p = \frac{WL}{12} + \frac{\alpha WL}{15}$$

$$V = 0$$

$$H = \frac{5W}{4} - 10\frac{M_p}{L}$$

or

$$H = \frac{5W}{4} - \frac{10}{L}\left(\frac{WL}{12} + \frac{\alpha WL}{15}\right) = \frac{5W}{12} - \frac{2\alpha W}{3}.$$

Now we must ascertain that $|M| \leq M_p$ at points B_1, D_1, and A. Since $V = 0$, the moments at B_1 and D_1 are the same. At B_1 and D_1, we have

$$M_{B1} = 0 + M_0 + \frac{1}{4}VL + \frac{1}{10}HL \leq M_p$$

$$M_{D1} = 0 + M_0 - \frac{1}{4}VL + \frac{1}{10}HL \leq M_p.$$

Substituting values of M_0, M_p, and H and solving for α, we have

$$\alpha \geq 0.625.$$

At A, we have

$$M_A = -WL\left(\frac{1}{4} + \frac{2\alpha}{5}\right) + M_0 + \frac{1}{2}VL + \frac{3}{5}HL \geq -M_p,$$

which gives

$$\alpha \leq 0.25;$$

$\alpha \geq 0.625$ and $\alpha \leq 0.25$ cannot be satisfied together, so this mechanism is not possible.

Mechanism 2: Try the side-sway mechanism with plastic hinges at A, C, D, and E. The corresponding collapse equations are

$$M_A = -WL\left(\frac{1}{4} + \frac{2\alpha}{5}\right) + M_0 + \frac{1}{2}VL + \frac{3}{5}HL = -M_p$$

$$M_C = 0 + M_0 + 0 + 0 = M_p$$

$$M_D = -\frac{WL}{4} + M_0 - \frac{VL}{2} + \frac{1}{5}HL = -M_p$$

$$M_E = -WL\left(\frac{1}{4} - \frac{2\alpha}{5}\right) + M_0 - \frac{1}{2}VL + \frac{3}{5}HL = M_p.$$

Solving these four equations for M_0, V, H, and M_p, we have

$$M_0 = M_p = \frac{1}{16}WL + WL\left(\frac{3\alpha}{20}\right)$$

$$V = -\frac{W}{8} + \frac{\alpha W}{2}$$

$$H = \frac{5}{16}W - \frac{\alpha W}{4}.$$

Moments at B_1, D_1, and B must be less than M_p. Thus

$$M_{B1} = M_0 + \frac{VL}{4} + \frac{1}{10}HL \leq M_p.$$

Substituting values of M_0, V, M_p, and H and then solving for α, we have

$$\alpha \leq 0.$$

Similarly, for moment at D_1 and B, we have

$$M_{D1} = M_0 - \frac{VL}{4} + \frac{HL}{10} \leq M_p$$

$$M_B = -\frac{wL}{4} + M_0 + \frac{VL}{2} + \frac{HL}{5} \leq M_p,$$

which give

$$\alpha > 0.417, \alpha < 5/4.$$

Since the conditions $\alpha < 0$, $\alpha > 0.417$, and $\alpha < 5/4$ cannot be satisfied simultaneously, this again is not a possible mechanism.

Mechanism 3 (Fig. 4.28): Try the mechanism with column AB remaining vertical while other members collapsed to the right with plastic hinges at B, B_1, D, and E. The resulting collapse equations are

$$M_B = -\frac{WL}{4} + M_0 + \frac{1}{2}VL + \frac{1}{5}HL = -M_p$$

$$M_{B1} = 0 + M_0 + \frac{1}{4}VL + \frac{1}{10}HL = M_p$$

$$M_D = -\frac{WL}{4} + M_0 - \frac{1}{2}VL + \frac{1}{5}HL = -M_p$$

$$M_E = -WL\left(\frac{1}{4} - \frac{2\alpha}{5}\right) + M_0 - \frac{1}{2}VL + \frac{3}{5}HL = M_p.$$

Solution of these four simultaneous equations gives us

$$V = 0$$

$$H = \frac{1}{2}W - \frac{4}{5}\alpha W$$

$$M_p = \frac{1}{10}WL + \frac{1}{25}\alpha WL$$

$$M_0 = \frac{WL}{20} + 0.12\alpha WL.$$

Now we use the conditions: $|M_A| \le M_p$, $|M_C| \le M_p$, and $|M_{D1}| \le M_p$.

Since $V = 0$, $M_{D1} = M_{B1} = M_p$, there is no need to check moment at D_1. At A, we have either $M_A \le M_p$

$$M_A = -WL\left(\frac{1}{4} + \frac{2\alpha}{5}\right) + M_0 + \frac{1}{2}VL + \frac{3}{5}HL \le M_p.$$

Substituting values of M_0, V, M_p, and H and solving for α, we have

$$\alpha \ge 0$$

or $M_A \ge -M_p$

$$M_A = -WL\left(\frac{1}{4} + \frac{2\alpha}{5}\right) + M_0 + \frac{1}{2}VL + \frac{3}{5}HL \ge -M_p,$$

which gives

$$\alpha \le 0.278.$$

At C, we have either $M_C \le M_p$

$$M_C = M_0 = \frac{1}{20}WL + 0.12\alpha WL \le M_p,$$

which gives

$$\alpha \le 0.625$$

or $M_C \ge -M_p$

$$M_C = M_0 = \frac{1}{20}WL + 0.12\alpha WL \ge -M_p,$$

which gives

$$\alpha \ge -0.938.$$

This condition cannot control, since α cannot be negative. Thus, this mechanism is valid for the range

$$0 \le \alpha \le 0.278$$

with

$$W = \frac{M_p}{L(0.1 + 0.04\alpha)}.$$

Mechanism 4 (Fig. 4.28): Try the side-sway mechanism with plastic hinges at A, B, D, and E. The corresponding collapse equations are

$$M_A = -WL\left(\frac{1}{4} + \frac{2\alpha}{5}\right) + M_0 + \frac{1}{2}VL + \frac{3}{5}HL = -M_p$$

$$M_B = -\frac{1}{4}WL + M_0 + \frac{1}{2}VL + \frac{1}{5}HL = M_p$$

$$M_D = -\frac{1}{4}WL + M_0 - \frac{1}{2}VL + \frac{1}{5}HL = -M_p$$

$$M_E = -WL\left(\frac{1}{4} - \frac{2\alpha}{5}\right) + M_0 - \frac{1}{2}VL + \frac{3}{5}HL = M_p.$$

Solution of these four equations gives

$$M_0 = \frac{WL}{4}$$

$$M_p = \frac{1}{5}\alpha WL$$

$$V = \frac{2}{5}\alpha W$$

$$H = 0.$$

Now imposing the conditions $|M_C| \le M_p$, $|M_{B1}| \le M_p$, and $|M_{D1}| \le M_p$. At C, we have either $M_C \le M_p$

$$M_C = \frac{1}{4}WL \le M_p = \frac{1}{5}\alpha WL,$$

which gives

$$\alpha \ge 1.25,$$

or $M_C \ge -M_p$

$$M_C = \frac{1}{4}WL \ge -M_p = -\frac{1}{5}\alpha WL,$$

which gives

$$\alpha \ge -1.25.$$

At B_1, we have either $M_{B1} \le M_p$

$$M_{B1} = M_0 + \frac{1}{4}VL + \frac{1}{10}HL \le M_p,$$

which gives

$$\alpha \ge 2.5,$$

or $M_{B1} \ge -M_p$

$$M_{B1} = M_0 + \frac{1}{4}VL + \frac{1}{10}HL \ge -M_p,$$

which gives

$$\alpha \ge -0.833.$$

At D_1, we have either $M_{D1} \le M_p$

$$M_{D1} = M_0 - \frac{1}{4}VL + \frac{1}{10}HL \le M_p,$$

which gives

$$\alpha \ge 0.833,$$

or $M_{D1} \ge -M_p$

$$M_{D1} = M_0 - \frac{VL}{4} + \frac{HL}{10} \ge -M_p,$$

which gives

$$\alpha \geq -2.5.$$

So this mechanism is valid for

$$\alpha \geq 2.5$$

with

$$W = \frac{5M_p}{\alpha L}.$$

Mechanism 5 (Fig. 4.28): Try the side-sway mechanism with plastic hinges at A, B_1, D, and E. The corresponding collapse conditions are

$$M_A = -WL\left(\frac{1}{4} + \frac{2\alpha}{5}\right) + M_0 + \frac{1}{2}VL + \frac{3}{5}HL = -M_p$$

$$M_{B1} = M_0 + \frac{1}{4}VL + \frac{1}{10}HL = M_p$$

$$M_D = -\frac{1}{4}WL + M_0 - \frac{1}{2}VL + \frac{1}{5}HL = -M_p$$

$$M_E = -WL\left(\frac{1}{4} - \frac{2\alpha}{5}\right) + M_0 - \frac{1}{2}VL + \frac{3}{5}HL = M_p.$$

Solution of these four simultaneous equations is

$$M_p = \frac{1}{16}WL + \frac{7}{40}\alpha WL$$

$$V = -\frac{1}{8}W + \frac{9}{20}\alpha W$$

$$H = 0.3125W - 0.125\alpha W$$

$$M_0 = 0.0625WL + 0.075\alpha WL.$$

Now impose the conditions: $M_B \leq M_p$, $M_B \geq -M_p$, $M_C \leq M_p$, and $M_C \geq -M_p$. At B, we have

$$M_B = -\frac{WL}{4} + M_0 + \frac{1}{2}VL + \frac{1}{5}HL \leq M_p.$$

Substituting values of M_0, V, H, and M_p and then solving for α, we have

$$\alpha \leq 2.5$$

or

$$M_B = -\frac{WL}{4} + M_0 + \frac{1}{2}VL + \frac{1}{5}HL \geq -M_p,$$

which gives

$$\alpha \geq 0.278.$$

At C, we have

$$M_C = M_0 = 0.0625WL + 0.075\alpha WL \leq M_p,$$

which gives

$$\alpha \geq 0,$$

or

$$M_C = M_0 \geq -M_p,$$

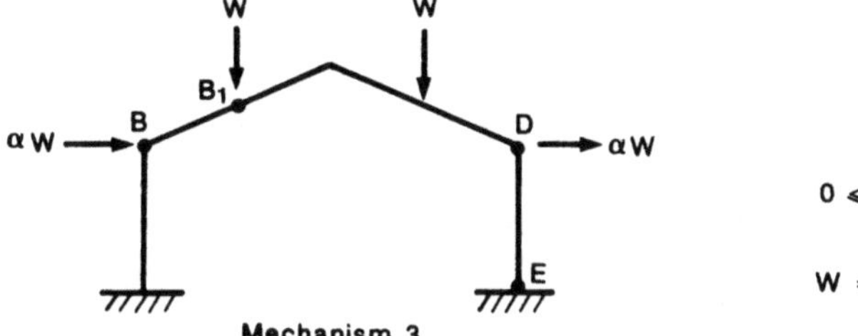

$$W = \frac{M_P}{L(0.1 + 0.04\alpha)}$$

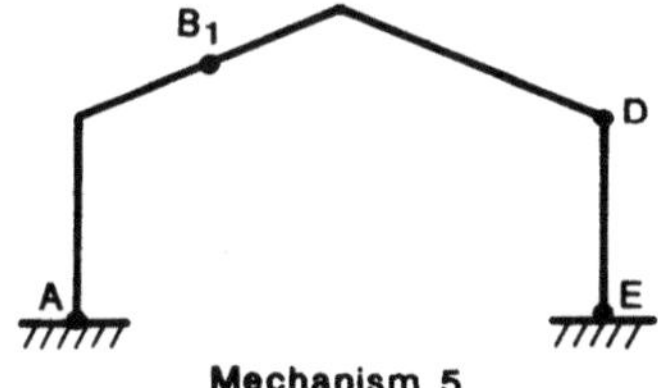

$$W = \frac{M_P}{L\left(\dfrac{1}{16} + \dfrac{7}{40}\alpha\right)}$$

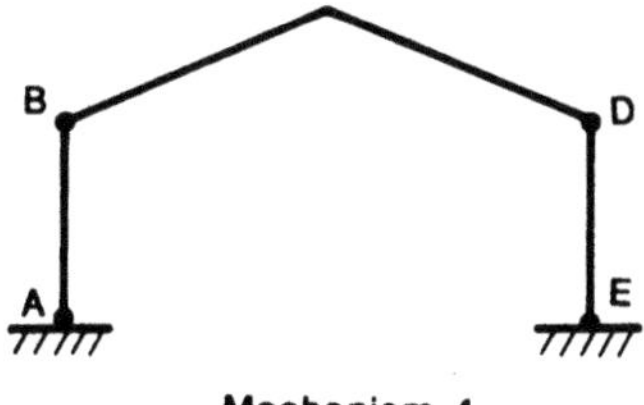

$$W = \frac{5\,M_P}{\alpha L}$$

FIGURE 4.28. Failure mechanisms and limit loads for various values of α.

which gives

$$\alpha \geq -0.5.$$

So this mechanism is valid for the range

$$0.278 \leq \alpha \leq 2.5$$

with

$$W = \frac{M_p}{L(1/16 + 7\alpha/40)}.$$

The failure mechanisms and the limit loads for the full range of α are summarized in Fig. 4.28.

4.8 Examples of Portal and Gable Frames

In this section, we will present two examples. One is a portal frame, the other is a gable frame. The effects of axial load, lateral torsional buckling, and shear force on member strength, which were neglected in Section 4.7, are considered here.

Example 4.8.1. Portal Frame: The rigid rectangular frame is subjected to applied loads shown in Fig. 4.29. All the members are made of W16 × 45. First, determine the limit value of the load factor λ without considering the effects of axial load, lateral torsional buckling, and shear force on member strength. Then evaluate these effects and provide the necessary reinforcements to minimize these effects. Assume $B_1 = B_2 = 1.0$, and $K_x = K_y = 1$.

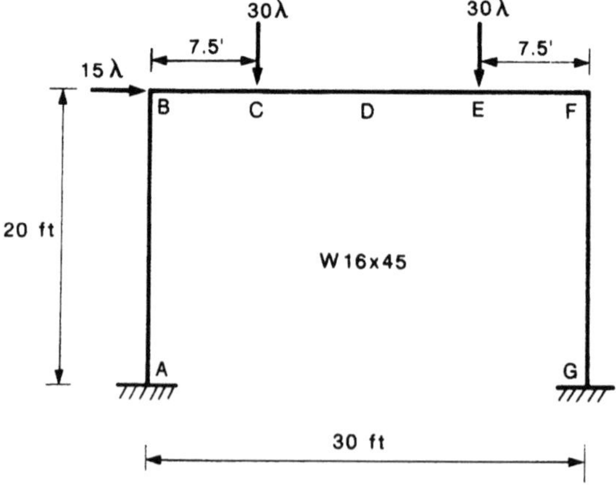

FIGURE 4.29. Portal frame.

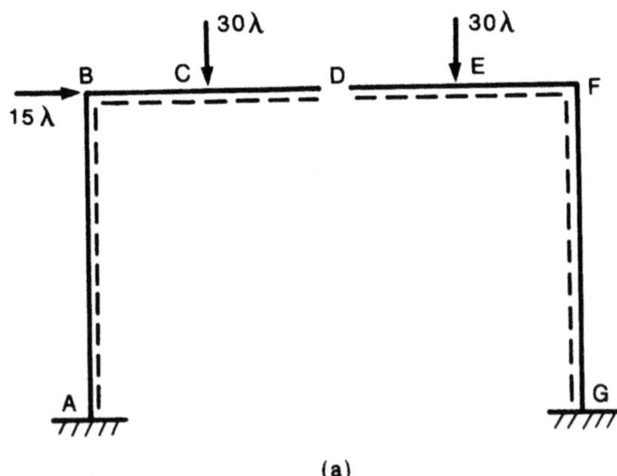

FIGURE 4.30. Determinate frame subjected to (a) applied forces and (b) redundant forces.

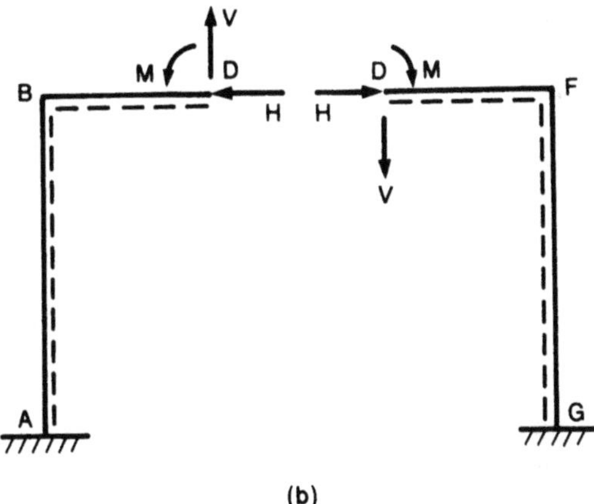

Solution: The frame has three redundants. We shall make a cut at D and consider the three internal forces at this cut as redundants. The bending moments due to external applied forces and redundant forces (Fig. 4.30) are tabulated in Table 4.4.

First, try the side-sway mechanism shown in Fig. 4.31 with plastic hinges at A, C, F, and G. From Table 4.4, the collapse equations can be written as

$$M_A = -525\lambda + M + 15V + 20H = -M_p$$

$$M_C = M + 7.5V = M_p$$

$$M_F = -225\lambda + M - 15V = -M_p$$

$$M_G = -225\lambda + M - 15V + 20H = M_p.$$

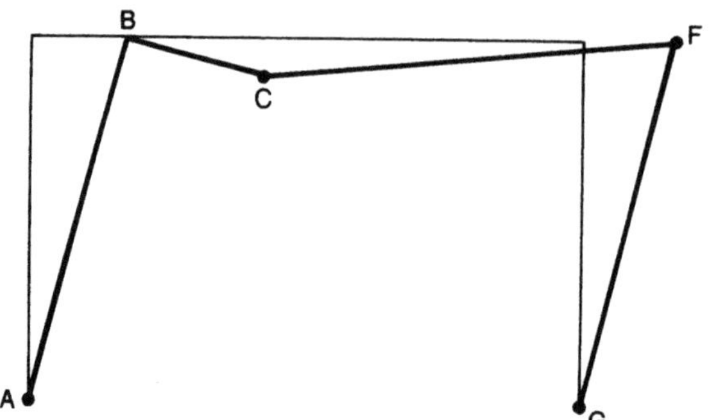

FIGURE 4.31. Plastic mechanism for the portal frame of Fig. 4.29.

TABLE 4.4. Example 4.8.1. portal frame calculations

Point	A	B	C	D	E	F	G
Free moment	-525λ	-225λ	0	0	0	-225λ	-225λ
M	M	M	M	M	M	M	M
V	$15V$	$15V$	$7.5V$	0	$-7.5V$	$-15V$	$-15V$
H	$20H$	0	0	0	0	0	$20H$

The solution of these four equations is

$$H = \frac{M_p}{10}$$

$$\lambda = \frac{7}{900} M_p$$

$$V = \frac{M_p}{90}$$

$$M = \frac{11}{12} M_p.$$

To ensure the satisfaction of the plastic moment condition, we must check the moments at B, D, and E:

$$M_B = -225\lambda + M + 15V = -\frac{2M_p}{3}, |M_B| < M_p, \quad \text{okay}$$

$$M_D = M = \frac{11}{12} M_p, |M_D| < M_p \quad \text{okay}$$

$$M_E = M - 7.5V = \frac{10}{12} M_p, |M_E| < M_p \quad \text{okay.}$$

FIGURE 4.32. Internal forces in the frame of Fig. 4.31: (a) axial force, (b) shear force, and (c) bending moment.

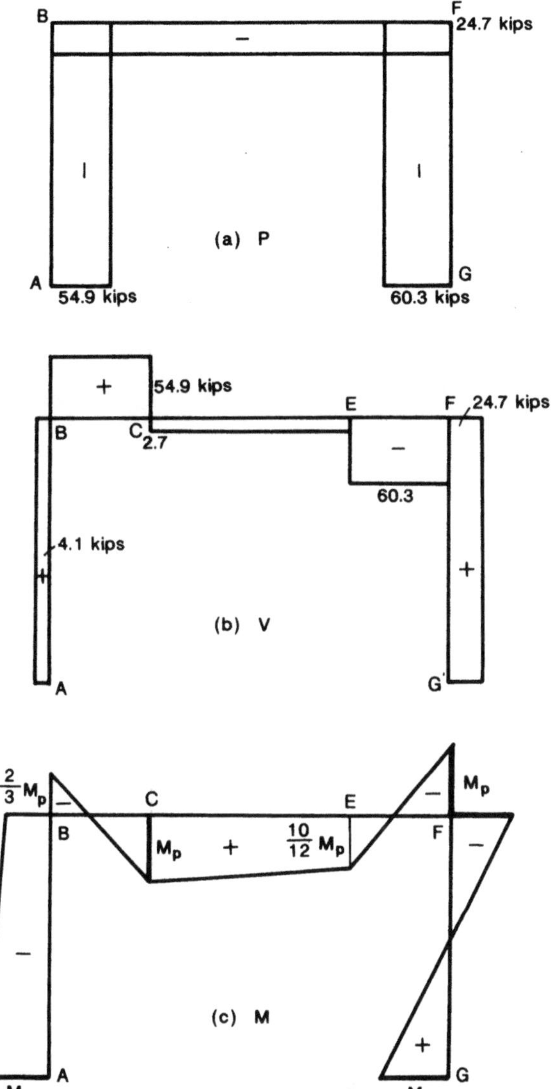

The axial load, shear force, and bending moment diagrams of the frame are shown in Fig. 4.32.

From the AISC manual, for W16 × 45, we have

$$M_p = M_{px} = F_y Z_x = (36)(82.3) = 2963 \text{ kip-in.} = 247 \text{ kip-ft}$$

$$A = 13.3 \text{ in.}^2, r_y = 1.57 \text{ in.}, r_x = 6.65 \text{ in.}, I_x = 586 \text{ in.}^4$$

Thus, without considering the effects of axial load, we have

$$\lambda = \frac{7}{900} M_p = \frac{7}{900}(247) = 1.92.$$

Since all three conditions of equilibrium, mechanism, and plastic moment are satisfied, $\lambda = 1.92$ is the exact solution if secondary effects are not considered.

Now we shall evaluate the secondary effects and provide the necessary reinforcements to minimize these effects.

Axial Force: Member FG is the critical member [see Fig. 4.32(a)] and the axial force in this member is

$$P_{FG} = 30\lambda + V = 30\lambda + \frac{M_p}{90} = (30)(1.92) + \frac{247}{90} = 60.3 \text{ kips.}$$

The yield axial load P_y is

$$P_y = (13.3)(36) = 478.8 \text{ kips.}$$

For strong-axis buckling and weak-axis buckling, we have

$$\lambda_{cx} = \frac{1}{\pi} \frac{KL}{r_x} \sqrt{\frac{F_y}{E}} = \frac{1}{\pi} \frac{(20)(12)}{6.65} \sqrt{\frac{36}{30,000}} = 0.398$$

and

$$\lambda_{cy} = \frac{1}{\pi} \frac{(20)(12)}{1.57} \sqrt{\frac{36}{30,000}} = 1.686,$$

the weak-axis buckling controls, thus giving

$$P_n = \left[\frac{0.877}{\lambda_c^2}\right] P_y = 147.7 \text{ kips.}$$

For the plastic design, $\lambda_c < 1.5K$. Also with such a high value of λ_c, the reduction in the axial capacity of member FG is very large. To increase the axial capacity, we need to provide bracings against weak-axis buckling at a spacing of 5 feet such that

$$\lambda_c = \frac{1.686}{4} = 0.422 < 1.5, \quad \text{okay.}$$

The weak-axis buckling still controls, so the axial capacity of the member is

$$P_n = 0.658^{\lambda_c^2} P_y = 444.4 \text{ kips.}$$

Since

$$\frac{P}{\phi_c P_n} = \frac{60.3}{(0.85)(444.4)} = 0.160 < 0.2.$$

So the moment capacity of the member can be determined from

$$\frac{P}{2\phi_c P_n} + \frac{M_x}{\phi_b M_{nx}} \leq 1.0$$

where $\phi_b = 0.9$, $M_{nx} = M_{px} = 2963$ kip-in., and $M_x = M_{pc}$. The reduced plastic moment M_{pc} is determined from LRFD interaction equation as

$$\frac{M_x}{\phi_b M_{nx}} = \frac{M_{pc}}{0.9 M_p} \leq \left(1 - \frac{0.160}{2}\right)$$

$$M_{pc} = (0.92)(0.9) M_p = 0.828 M_p.$$

which will in turn reduce the load factor from $\lambda = 1.92$ to $\lambda = 7 M_{pc}/900 = 1.59$.

Lateral Torsional Buckling: For plastic design, the maximum unbraced length is

$$L_{pd} = \frac{(3{,}600 + 2{,}200 M_1/M_p)}{F_y} r_y.$$

We will now check L_{pd} for all portions of the frame and provide the lateral support if necessary.

Portion AB [Fig. 4.32(c)]

$$L_{pd} = \left[\frac{3600 - 2200\left(\frac{2}{3}\right)}{36}\right] 1.57 = 93 \text{ in.} > 60 \text{ in.} \quad \text{okay.}$$

Lateral supports at a spacing of 5 feet are okay.

Portion BC

$$L_{pd} = \left[\frac{3600 + 2200\left(\frac{2}{3}\right)}{36}\right] 1.57 = 221 \text{ in.} > 7.5 \times 12 = 90 \text{ in.} \quad \text{okay.}$$

There is no need to provide any lateral support, but the lateral support at the plastic hinge location C is required by the LRFD.

Portion CE

$$L_{pd} = \left[\frac{3600 - 2200\left(\frac{10}{12}\right)}{36}\right] 1.57 = 77 \text{ in.} < 15 \times 12 = 180 \text{ in.,} \quad \text{not okay.}$$

So we provide lateral supports at a spacing of 5 feet.

Portion EF

$$L_{pd} = \left[\frac{3600 + 2200\left(\frac{10}{12}\right)}{36}\right] 1.57 = 237 \text{ in.} > 7.5 \times 12 = 90 \text{ in.} \quad \text{okay.}$$

There is no need to provide any lateral support.

Portion FG

$$L_{pd} = \left[\frac{3600 + 2200(1)}{36}\right] 1.57 = 253 \text{ in.} > 20 \times 12 = 240 \text{ in.} \quad \text{okay.}$$

Lateral supports at a spacing of 5 feet are okay.

Shear Force: As far as shear is concerned, the full plastic bending strength of the section may be used in design if the total transverse shear on the section at the plastic limit load is no more than V_p (Eq. 4.5.1):

$$V_p = 0.55 F_y t_w d = 0.55 \times 36 \times 0.345 \times 16.13 = 110 \text{ kips.}$$

Since the shear forces in the frame are all smaller than the shear strength, no web stiffeners or doubler plates are needed for frame members.

Example 4.8.2. Gable Frame: A gable frame is subjected to uniformly distributed loads as shown in Fig. 4.33. Select an appropriate section for the members. Consider the effects of axial force, lateral torsional buckling, and shear force on member strength. Assume $B_1 = B_2 = 1.0$ and $K_x = K_y = 1$ to check column buckling.

Solution: The degree of redundancy of the frame is one. Take the horizontal reaction at E as the redundant. The resulting determinate frame subjected to applied and redundant forces is shown in Fig. 4.34. The bending moments in the determinate frame due to applied forces and redundant forces are tabulated in Table 4.5.

First, try the mechanism with plastic hinges at D and C. From Table 4.5, the collapse equations can be written as

$$-20S = -M_p$$

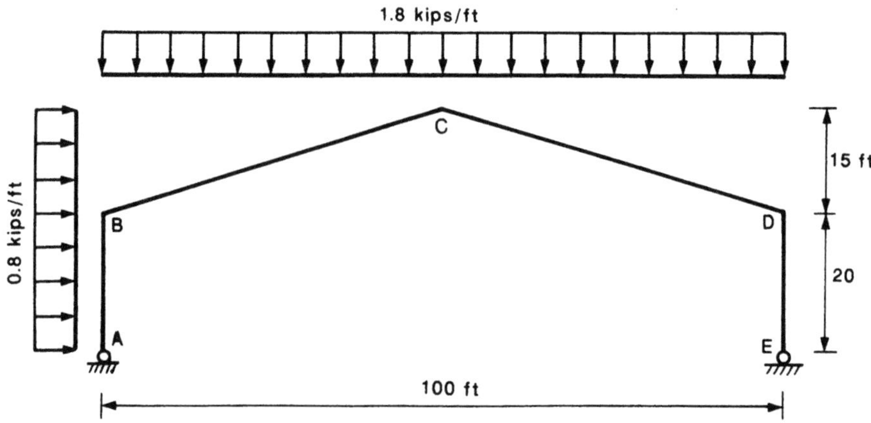

FIGURE 4.33. Gable portal frame subjected to uniformly distributed load.

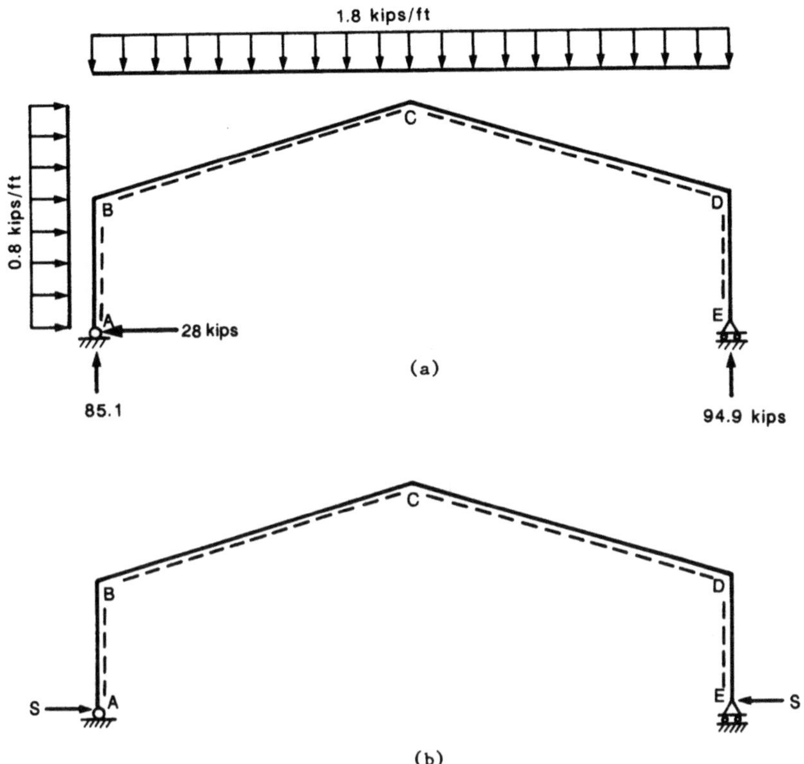

FIGURE 4.34. Determinate structure subjected to (a) applied forces and (b) redundant forces.

TABLE 4.5. Example 4.8.2. gable frame calculations

Joint	A	B	C	D	E
Free moment	0	400	2495	0	0
S	0	$-20S$	$-35S$	$-20S$	0

and

$$-35S + 2495 = M_p.$$

Solving these two equations for S and M_p, we have

$$M_p = 907 \text{ kip-ft}$$

$$S = 45.4 \text{ kips.}$$

The moment diagram corresponding to these values of M_p and S is shown in Fig. 4.35. We must now check moments at B and in between B and C

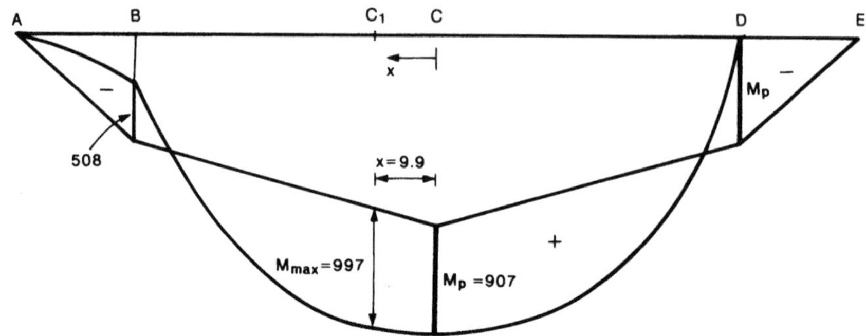

FIGURE 4.35. Moment diagram corresponding to a mechanism with plastic hinges at C and D.

(because the load is distributed):

$$M_B = 400 - 20S = -508 \text{ kip-ft}, \quad |M_B| \le M_p = 907 \text{ kip-ft}, \quad \text{okay.}$$

Moment M_x at a horizontal distance x from C toward B (Fig. 4.35) can be expressed by taking the free body of the right portion of the frame C_1CDE as

$$M_x = 94.9(50 + x) - \frac{(1.8)(50 + x)^2}{2} - \frac{(0.8)(0.3x)^2}{2} - (35 - 0.3x)(45.4)$$

or

$$M_x = -0.94x^2 + 18.52x + 906.$$

To determine the maximum moment location, we set

$$\frac{dM_x}{dx} = 0 = -1.88x + 18.52$$

and obtain

$$x = 9.9 \text{ ft}$$

and

$$M_{\max} = -0.94(9.9)^2 + 18.52(9.9) + 906 = 997 \text{ kip-ft.}$$

Since the plastic moment condition is violated at C_1, $M_p = 907$ kip-ft is not an exact solution. However, this mechanism does provide the following upper and lower bounds on M_p:

$$907 \le M_p \le 997.$$

Now, try the mechanism with hinges at C_1 and D. Corresponding to this mechanism, the collapse equations are

$$M_D = -20S = -M_p$$

$$M_{C1} = (94.9)(59.9) - \frac{(1.8)(59.9)^2}{2} - (0.8)(2.97)^2 - 32S = M_p$$

or

$$M_{C1} = 2448 - 32S = M_p.$$

The solution of these two equations gives

$$M_p = 942 \text{ kip-ft}$$

$$S = 47.1 \text{ kips.}$$

The required plastic modulus $Z_x = M_p/F_y = (942/36)(12) = 314$ in.^{3} Try W30 $\times$ 116 throughout the frame.

From the AISC manual, for W30 $\times$ 116, we have

$$Z_x = 378 \text{ in.}^3, r_x = 12 \text{ in.}, r_y = 2.19 \text{ in.}, A = 34.2 \text{ in.}^2, I_x = 4930 \text{ in.}^4$$

To avoid the effects of weak-axis column buckling and lateral torsional buckling, we shall provide bracings at a spacing of 5 feet in all the members. We now evaluate the effects of axial force, lateral torsional buckling, and shear force on member strength.

Axial Force: Member DE is the critical member and the axial force in this member is

$$P_{DE} = V_E = 94.9 \text{ kips.}$$

The yield axial yield load for W30 $\times$ 116 is

$$P_y = (34.2)(36) = 1231 \text{ kips.}$$

For strong-axis and weak-axis buckling, the values of λ_c are

$$\lambda_{cx} = \frac{1}{\pi} \frac{(KL)_x}{r_x} \sqrt{\frac{F_y}{E}} = \frac{1}{\pi} \frac{1 \times 20 \times 2}{12} \sqrt{\frac{36}{30,000}} = 0.221$$

$$\lambda_{cy} = \frac{1}{\pi} \frac{(KL)_y}{r_y} \sqrt{\frac{F_y}{E}} = \frac{1}{\pi} \frac{(1)(5)(12)}{2.19} \sqrt{\frac{36}{30,000}} = 0.302.$$

The weak-axis buckling controls. Thus $\lambda_c = \lambda_{cy} = 0.302$. Now since $\lambda_c < 1.5$, the axial capacity of the member is

$$P_n = 0.658^{\lambda_c^2} P_y = 0.658^{(0.302)^2}(1231) = 1185 \text{ kips.}$$

Since

$$\frac{P}{\phi_c P_n} = \frac{94.9}{(0.85)(1185)} = 0.094 < 0.2.$$

The reduced moment capacity of the member can be determined from

$$\frac{P}{2\phi_c P_n} + \frac{M_x}{\phi_b M_{nx}} = 1.0$$

where the first-order moment obtained from plastic analysis, i.e., $M_x =$

$M_{pc} = 942$ kip-ft:

$$\frac{P}{2\phi_c P_n} + \frac{M_x}{\phi_b M_{nx}} = \frac{0.094}{2} + \frac{942}{0.9M_p} = 1.0,$$

which gives the required plastic moment

$$M_p = 1113.7 \text{ kip-ft} = 13{,}364 \text{ kip-in.}$$

The adjusted plastic modulus

$$Z = \frac{13{,}364}{36} = 371.2 < 378 \text{ in.}^3, \quad \text{okay.}$$

Lateral Torsional Buckling:

$$L_{pd} = \frac{(3600 + 2000M_1/M_p)}{F_y} r_y.$$

Take a conservative value of M_1/M_p as -1

$$L_{pd} = \frac{(3600 - 2200)}{36}(2.19) = 85 \text{ in.} > 60 \text{ in.}, \quad \text{okay.}$$

Shear Force: Member CD is the critical member and the maximum transverse shear force in member CD is

$$V_{CD} = (94.9)\frac{50}{\sqrt{(50)^2 + (15)^2}} = 90.9 \text{ kips.}$$

The maximum shear strength allowed according to Eq. (4.5.1) is

$$V_p = 0.55F_y t_w d = 0.55 \times 36 \times 0.565 \times 30.01 = 335.7 \text{ kips,}$$

which is sufficient to support the shear load without general yielding of the web, the modification of plastic moment capacity is therefore not required by specification. However, to see the influence of shear force on plastic bending strength, we shall follow the procedure described in Section 2.6, Chapter 2, and compute first the shear stress as

$$\text{shear stress in web} = \frac{V_{CD}}{d_w t_w} = \frac{90.9}{(28.31)(0.565)} = 5.68 \text{ ksi.}$$

The plastic modulus reduced due to the presence of shear force is

$$Z_{ps} = Z - Z_w\left[1 - \frac{\sqrt{\sigma_y^2 - 3\tau^2}}{36}\right]$$

$$= 378 - \frac{(0.565)(28.31)^2}{4}\left[1 - \frac{\sqrt{(36)^2 - (3)(5.68)^2}}{36}\right]$$

$$Z_{ps} = 378 - 4.3 = 373.7 \text{ in.}^3 > 371.2 \text{ in.}^3, \quad \text{okay.}$$

Use W30 $\times$ 116.

4.9 Summary

In the theory of plastic analysis and design, three fundamental conditions must be satisfied when a structure is at its plastic collapse state.

1. The structure is in equilibrium with respect to the given loads and their reactions.
2. At no point in the structure does the moment produced by the loading exceed the available plastic bending strength.
3. The structure is on the verge of becoming a mechanism.

Based on these three conditions, two basic methods of plastic analysis and design are available. The equilibrium method presented in this chapter is based on the lower-bound theorem of limit analysis. The work method to be presented in the following chapter is based on the upper-bound theorem. The equilibrium method involves two stages of operation. In the first stage, the moment equilibrium equations are formed in terms of applied loads and unknown redundants. In the second stage, the redundants are selected such that the plastic moment condition will not be violated, while the mechanism condition may not be satisfied. A safe or lower-bound solution can be obtained quickly by assuming any set of values of redundants. The best solution or the highest limit load corresponds to the formation of a failure mechanism. A successful application of the second stage of operation requires physical intuition combined with the use of differential calculus and algebraic technique.

The equilibrium method is applied to the analysis and design of simple beams, rectangular frames, and gable frames. The consideration of the effect of shear force on the plastic moment capacity in the design of beams is illustrated. Cover plates are designed for a continuous beam having a limited depth for architectural reasons. The LRFD specifications for the limiting unbraced lengths for preventing lateral torsional buckling are presented, and their use in actual frame design is illustrated.

The effects of axial load on the strength and stability of frames are considered. To this end, the LRFD interaction equations for beam-columns are presented, and their use in designing rectangular and gable frames is illustrated.

For the practical design of large-framed structures involving a larger number of redundants, a practical equilibrium procedure is presented. In this procedure, we first assume a failure mechanism and then equate the moments at the plastic hinge locations to the plastic moment. The resulting simultaneous equations are solved for the unknown redundants. This leads to the solution of the required plastic moment capacity or the plastic limit load. The results so obtained are used to check moments at other critical sections. This check provides upper and lower bounds on the exact solution. The procedure is repeated by assuming a new mechanism until the coincidental upper and lower bounds.

References

4.1. Chen, W.F., and Lui, E.M., *Structural Stability: Theory and Implementation*, Elsevier, New York, 486 pp., 1987.

4.2. Chen, W.F., and Lui, E.M., *Stability Design of Steel Frames*, CRC Press, Boca Raton, FL, 380 pp., 1991.

4.3. Cheong-Siat-Moy, F., "Consideration of Secondary Effects in Frame Design," *Journal of Structural Division*, ASCE, 103, ST10, pp. 2005–2019, 1972.

4.4. Rosenblueth, E., "Slenderness Effects in Buildings," *Journal of Structural Division*, ASCE, 91, ST1, pp. 229–252, 1965.

4.5. Stevens, L.K., "Elastic Stability of Practical Multistory Frames," *Proceedings of Civil Engineering*, 36, London, England, 1967.

4.6. Yura, J.A., *Elements for Teaching Load and Resistance Factor Design*, AISC, Chicago, IL, 1987.

4.7. Galambos, T.V., "Inelastic Lateral Buckling of Beams," *Journal of Structural Division*, ASCE, 89, ST5, pp. 217–242, 1963.

4.8. Galambos, T.V., and Fukumoto, Y., "Inelastic Lateral-Torsional Buckling of Beam-Columns," *Bulletin No. 115*, Welding Research Council, July 1966.

4.9. Nethercot, D.A., and Trahair, N.S., "Inelastic Lateral Buckling of Determinate Beams," *Journal of Structural Division*, ASCE, 102, ST4, pp. 701–717, 1976.

4.10. Salmon, C.G., and Johnson, J.E., *Steel Structures: Design and Behavior, Emphasizing Load and Resistance Factor Design*, third edition, Harper and Row, New York, p. 1086, 1990.

Problems

4.1. A beam having uniform cross section with full plastic moment M_p and length $2L$ rests on simple supports at its ends and on a central prop. It carries four equal loads W, symmetrically arranged about the center of length, one at the center of each span and one a distance $L/8$ from each end of the beam. What would be the value of W at collapse? $[24M_p/5L]$.

4.2. If the beam of Problem 4.1 is carrying four equal loads W, symmetrically arranged about the center of the beam, one at the center of each span and one at a distance $3L/8$ from each end of the beam, determine the value of W at collapse $[88M_p/27L]$.

4.3. A fixed-ended beam with a prop at the center has a uniform cross section with full plastic moment M_p and length $2L$. Equal concentrated loads are applied at a distance aL from each fixed end. What would be their value a collapse? $[2/[a(1-a)]M_p/L]$.

4.4. A fixed-ended beam has uniform section with full plastic moment M_p and length L. It is subjected to a uniformly distributed load W and to a concentrated load $0.5W$ at a distance of $L/3$ from one end. Find the value of W at collapse $[9M_p/l]$.

4.5. A uniform beam of length L and full plastic moment M_p is simply supported at one end and fixed at the other end. A concentrated load W may be applied anywhere in the span. Find the values of M_p corresponding to the most unfavorable position of the load W $[M_p = 0.172WL]$.

4.6. A beam has uniform cross section with full plastic moment M_p and length L and is built in at one end and simply supported at the other. It carries a concentrated load W a distance a from the built-in end. Show that at collapse, W has the value

$$\frac{2L - a}{a(L - a)} M_p.$$

Show that if both ends had been built in, the load at collapse would have increased in the ratio

$$\frac{2L}{2L - a}.$$

4.7. A beam $ABCD$ of uniform section throughout has full plastic moment M_p and is pinned to four supports, thus forming a continuous beam of three equal spans each with length L. A load W is applied at the center of each span. Find the value of W that causes collapse $[6M_p/L]$.

4.8. The continuous beam shown in Fig. P4.8 has three equal spans carrying central point loads. There is no change of beam section between supports but the plastic moment resistance of the outer spans is only two-thirds that of the central span, which is M_p. At what value of W does the collapse occur? $[8M_p/L]$.

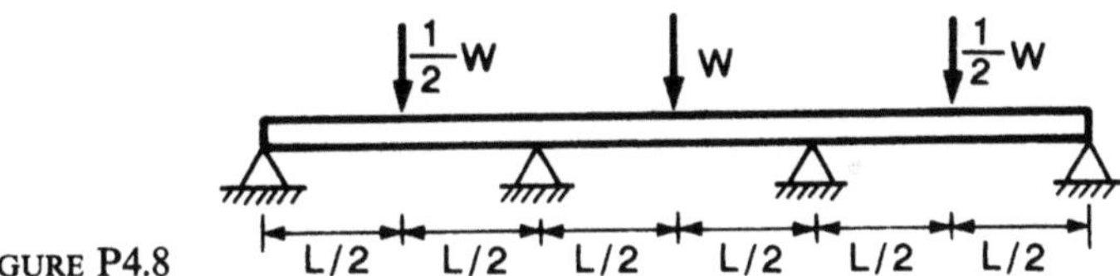

FIGURE P4.8

4.9. A uniform continuous beam with full plastic moment M_p rests on five simple supports A, B, C, D, and E such that $AB = 6L$, $BC = CD = 8L$, and $DE = 10L$. Each span carries a concentrated load at its midpoint, these loads being W on AB, W on BC, $1.4W$ on CD, and $0.5W$ on DE. Find the value of W that will just cause collapse $[5M_p/7L]$.

4.10. A fixed-ended beam of span L is to be designed for collapse under a single central point load λW. For a distance $(1/2)kL$ from each end of the span, the fully plastic moment is to be r times the value for the remainder of the span. The weight per unit length of the beam is

$$\beta\left[1 + \frac{32M}{\lambda WL}\right]$$

where M is the fully plastic moment at the section in question, and β is a consant. Determine the ratios r and k for minimum weight of the beam, and find the saving in material as compared with a design using a uniform beam. Neglect the weight of the beam $[r = 5/3, k = 1/4, 10\%]$.

4.11. A load of 150 kips is applied 4 feet from the end of a fixed-ended beam 16 feet long. Select a wide-flange section. Include the effect of shear force on plastic moment capacity in the design of the beam.

4.12. A fixed-ended W16 × 40 beam 20 feet long is reinforced at its ends with cover plates welded to top and bottom flanges. These plates are 5-in. wide and 3/8 in. thick. It carries a concentrated load W at midspan. Find the value of W taking into account the effect of shear force on the plastic moment capacity.

4.13. How long must the plates be in Problem 4.12 so that this ultimate load can be reached. Does the beam require lateral supports for adequate rotation capacity to form a plastic failure mechanism? If yes at what locations.

4.14. Find a suitable section of A36 steel for the beam of Example 4.5.2 if the ends of the continuous beam are fixed instead of simply supported [W14 × 30].

4.15. A load of 5 kip/ft is to be carried over the three spans shown in Fig. P4.15 with a load factor of 1.7. It is decided to use a wide-flange section with $F_y = 36$ ksi, running continuously over all spans with added cover plates running continuously over support C into the spans BC and CD. Find the size of the basic section and length and area of the cover plates. Include the effect of shear force on the plastic moment capacity of the beam.

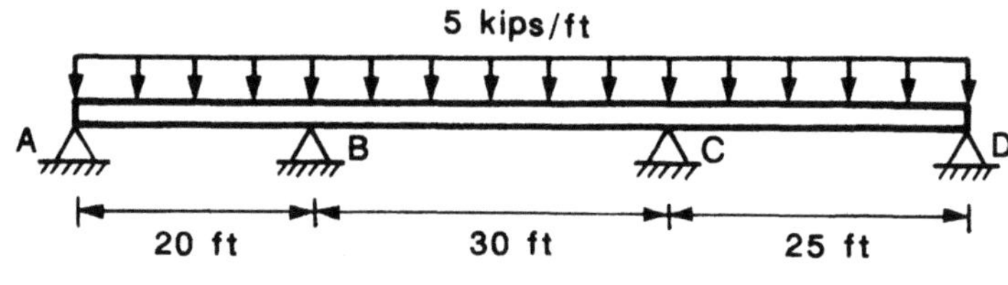

FIGURE P4.15

4.16. The beam designed in Problem 4.15 has lateral support at the vertical supports. Do we need additional lateral supports? If yes, at what locations?

4.17. The beam ABC shown in Fig. P4.17 is to be designed to carry the given loads with a load factor 1.7. Using A36 steel, design the beam continuous from A to C by taking into account effect of shear force on the plastic moment capacity. Determine the location of lateral supports if necessary.

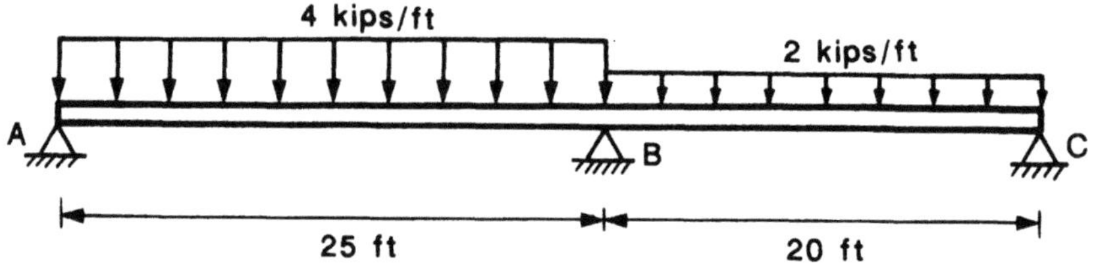

FIGURE P4.17

4.18. The beam in Problem 4.17 can be designed by using a smaller section continuous from A to C and by adding symmetrical flange reinforcement where necessary in span AB. Select a wide-flange section and find the length and size of the cover plates needed.

4.19. In a fixed-base rectangular portal frame $ABCD$, the column AB is of height 16 feet and column DC is of height 24 feet and beam BC is 16 feet and horizontal. All the members of the frame have the same full plastic moment M_p. The beam BC carries a central concentrated vertical load of 70 kips and a concentrated horizontal load of 28 kips is applied at C. Select an appropriate wide-flange section. Consider the effects of (a) axial load including stability of the columns AB and CD and (b) shear force on moment capacity of beam BC. Provide the necessary lateral support to prevent lateral torsional buckling of the beam.

4.20. Repeat Problem 4.19 when the horizontal load of 28 kips is reversed in direction.

4.21. The frame of Fig. 4.26 has $L = 40$ ft, $\alpha = 0.5$, and $W = 50$ kips. Select an appropriate wide-flange section by taking into account effects of axial load and shear force. Provide the lateral supports if necessary to prevent lateral torsional buckling. Use LRFD rules.

4.22. For the gable frame shown in Fig. 4.26 and assuming the following conditions $\alpha = 1.0$, $L = 20$ ft, A36 steel, W18 × 46 section:

(a) Estimate the change in the value of W caused by positioning the plastic hinge at B_1 correctly if the vertical load W on span BC is actually uniformly distributed [$x = 4$ ft, $M_p = 4.45W$, 6.7%].
(b) Estimate the change in the value of W caused by considering the effect of axial load on the value M_p, using the LRFD beam-column equation with $B_1 = B_2 = 1.0$ [$M_{pc} = 2703$ kip-in., 5.2%].
(c) Determine the location of lateral supports if necessary to prevent lateral torsional buckling [A, B, B_1, C, D_1, D, E].

4.23. The rigid frame shown in Fig. P4.23 is subjected to a vertical distributed load and a concentrated horizontal load. Determine a lower and upper bound for the plastic moment M_p using the mechanism $ACDE$ [$65.79 \leq M_p \leq 71.71$].

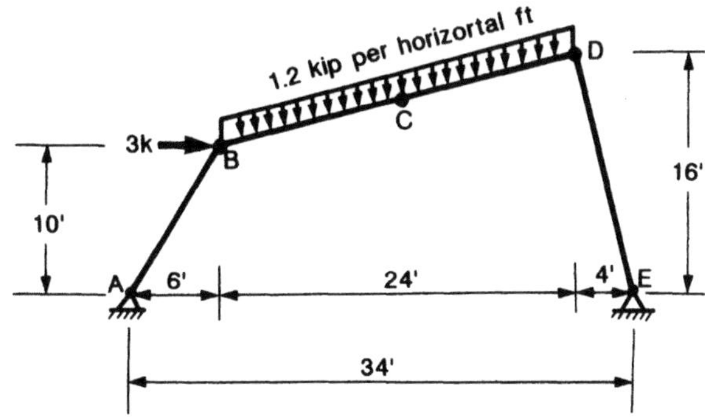

FIGURE P4.23

4.24. For the thick ring shown in Fig. P4.24, where $r \gg t$, find the collapse load P in terms of M_p (neglect axial force effect) $[P = 4M_p/(r + t/2)]$.

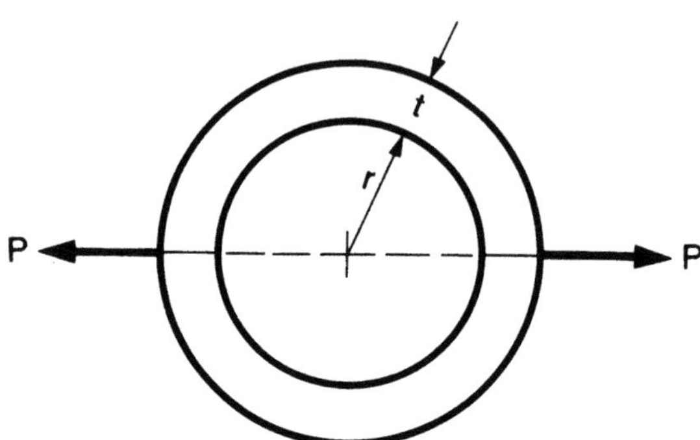

FIGURE P4.24

4.25. Using the mechanism AB_1CD shown in Fig. P4.25, we find that the load factor is $\lambda = 2.4$.

(a) Carry out your moment check.
(b) Design members AB and BC assuming that the frame is braced at B_1, B, C_1, and C [(a) $M_B = 0$, $M_{C1} = 600$ kft, $P_{AB} = 60$ kips; (b) member AB, W21 × 50, member BC, W30 × 99].

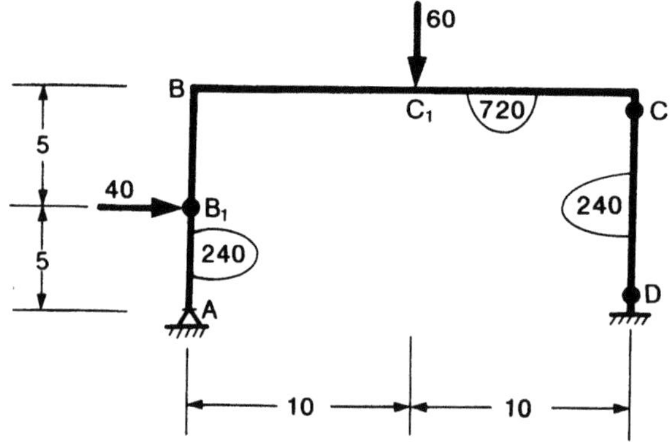

FIGURE P4.25

5
Work Method

5.1 Introduction

The equilibrium method described in the preceding chapter can easily be used in the plastic analysis and design of simple frames where the number of redundants is small. As the number of redundants increases, it becomes more difficult to draw the bending moment diagram of the structure, and thus more difficult to use the equilibrium method. For such structures, the work method of plastic analysis is more appropriate and affords a simpler solution. As the name implies, the relation between the strength of a frame and the applied loads in the work method is found by assuming that there is no overall loss of energy as the frame under failure loads undergoes a small change in displacement. Thus, by postulating a *valid* failure mechanism, an equation can be formed by equating the external work done by the applied loads through the displacements to the internal dissipation of energy at the plastic hinge locations. The interal dissipation of energy is the sum of the products of the plastic moment at each hinge and the corresponding angular change required to effect a small movement of the failure mechanism. The external work is the sum of the products of the component of the small displacement of the failure mechanism in line with the applied load and the corresponding applied load. The equation formed in this way is called the *work equation* and the corresponding collapse load or the required plastic moment capacity can be determined by solving the work equation [1.8, 5.1–5.5]. The computed load for the particular assumed failure mechanism is exact if a moment check is performed and shows that the plastic moment condition is not violated anywhere in the frame.

In this chapter, we will first describe the basis of the work method and then present the formulation of the work equation and the procedure for a moment check, the two major steps in the work method. Then, we will demonstrate the use of the work method for the analysis and design of simple frames. Next, we will describe simple methods for calculating the geometrical relations of failure mechanisms. A practical method of combining independent mechanisms is then presented, which facilitates the determination of plastic limit load or required plastic moment capacity of multistory and multbay frames. Finally, we will show how simple modifications can be made to the present procedure for the presence of distributed loads.

5.2 Basis of the Method

The load computed on the basis of an assumed failure mechanism is never less than the exact plastic limit load of the structure. This fact is based on the upper-bound theorem of limit analysis described in Chapter 3 and is now restated here: A load computed from the work equation on the basis of an assumed failure mechanism will always be greater than or at best equal to the plastic limit load.

The upper-bound theorem states that if a mode of failure exists, the structure will not stand up. The computed loads are upper bounds on, or unsafe values of, the limit or collapse loading. The minimum upper bound is the limit load itself. The work method has the following two major steps:

(a) Assume a failure mechanism and form the corresponding work equation from which an upper-bound value of the plastic limit load or an unsafe value of the plastic moment can be found.
(b) Write the equilibrium equations for the assumed failure mechanism and check the moments to see whether the plastic moment condition is met everywhere in the structure.

These two major steps will be elaborated in the following two sections.

5.3 Work Equation

The work equation can be regarded as an energy balance statement in mathematical form for the structure under collapse loads undergoing a small change in displacement and hinge rotation. A work equation can be formed for an assumed mechanism by equating the summation of expenditure of energy due to the movement of each applied load W_i through a distance δ_i or $\sum W_i \delta_i$ to the summation of internal dissipation of energy in rotating each plastic hinge through an angle θ_i, at the constant plastic moment M_{pi}, or $\sum M_{pi} \theta_i$,

$$\sum W_i \delta_i = \sum M_{pi} \theta_i \tag{5.3.1}$$

where the left-hand summation extends over all the loads and the right-hand summation extends over all the plastic hinges. The internal dissipation of energy is always positive, regardless of the direction of hinge rotation. Thus, there is no need to establish the signs for moments and plastic hinge rotations for calculating the internal energy dissipation. This is in contrast to the equilibrium method and moment check procedure, which require the proper establishment of signs for moments and rotations in the application of the virtual work equation. Herein, for the moment check, we shall use the following sign convention: Moment-causing tension on the dotted-line side of the member is positive and vice-versa, and rotation causing opening on the dotted-line side of the members is positive and vice-versa.

The following three examples have been chosen to show the techniques of calculating each of the two work quantities, to form the work equation, and to obtain an upper bound of the plastic limit load corresponding to an assumed failure mechanism. Later in this chapter, we will show that for a given frame and loading all the possible mechanisms can be obtained as different combinations of a comparatively small number of independent mechanisms, which are readily identified for a given frame and loading. The determination of plastic collapse loads by the method of combining mechanisms is described in Section 5.8.

5.3.1 A Simply Supported Continuous Beam

Example 5.3.1. Obtain the plastic limit load of the two-span continuous beam shown in Fig. 5.1(a).

Solution: Plastic hinges can possibly be formed at sections B, C, and D. Because of symmetry there is only one possible mechanism, as shown in Fig. 5.1(b). If the plastic hinge at point A is rotated through a small angle θ, then by geometry, the plastic hinges at B, C, and D are rotated through an angle equal to 2θ. The external work W_E is done by the two loads moving vertically downward. The small vertical distances are computed in terms of angle θ as

$$\Delta_B = \Delta_D = \frac{L}{2}\theta. \tag{5.3.2}$$

Thus, the total external work done is

$$W_E = P\left(\frac{L}{2}\theta\right) + P\left(\frac{L}{2}\theta\right). \tag{5.3.3}$$

The internal energy W_I is dissipated at each of the plastic hinges. The energy dissipation at each plastic hinge is equal to the plastic moment at that hinge times the angle through which it rotates. Thus, the total internal energy dissipation W_I can be written as

$$W_I = M_p(2\theta) + M_p(2\theta) + M_p(2\theta). \tag{5.3.4}$$

Note that angular and linear displacements are assumed merely as differential values; hence the dimension of the undeformed beam can be used in the computation, as would be done in elastic analysis. By equating W_E to W_I, we have formed the work equation

$$2\left(\frac{PL}{2}\theta\right) = 3M_p(2\theta)$$

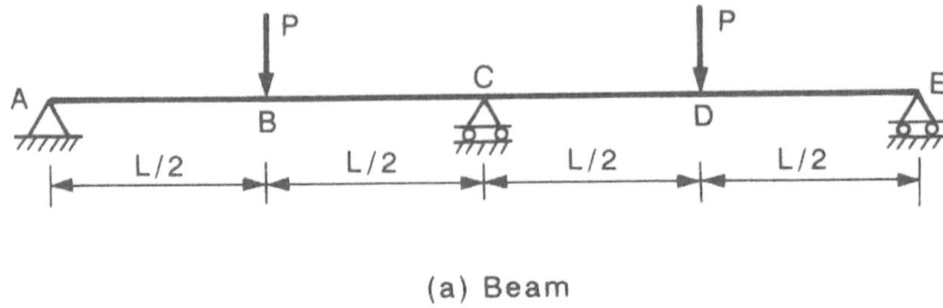

(a) Beam

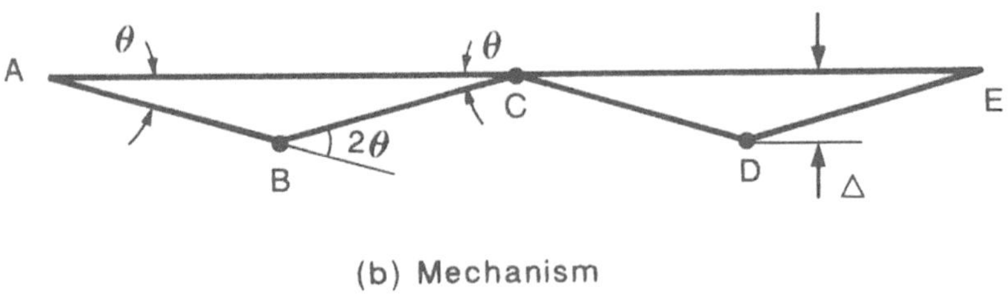

(b) Mechanism

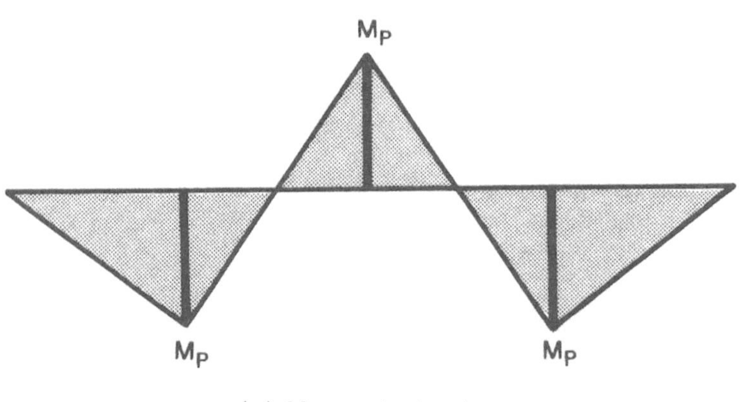

(c) Moment check

FIGURE 5.1. Mechanism analysis of two-span continuous beam (Example 5.3.1): (a) beam, (b) mechanism, and (c) moment check.

from which we obtain an upper-bound solution to the plastic limit load

$$P = \frac{6M_p}{L}. \tag{5.3.5}$$

Since moments at A, B, C, D, and E are known, the moment diagram for the beam can be constructed as shown in Fig. 5.1(c). Since the plastic moment condition ($M \le M_p$) is met everywhere in the beam, it follows that the solution $P = 6M_p/L$ is exact.

5.3.2 A Pinned-Fixed Continuous Beam

Example 5.3.2. Obtain the plastic limit load of the unsymmetrical two-span continuous beam shown in Fig. 5.2(a).

Solution: Plastic hinges can possibly be formed at Sections B, C, D, and E of Fig. 5.2(a).

Two possible mechanisms are shown in Fig. 5.2(b). One involves the failure of beam $A - C$ and the other of beam $C - E$. For mechanism 1, if θ is the angle of rotation at A, then the rotation at C is also equal to θ. The angular discontinuity at B is 2θ. The vertical displacement at B is equal to the rotation at A times the distance between A and B, i.e., $\Delta = \theta(L/2)$. For mechanism 2, if θ is the angle of rotation at C, then, the rotation at E is $\theta/2$. The discontinuity at D is the sum of angles at C and E. The vertical displacement of D is $\theta\,(L/3)$.

The work equation for the left-hand beam mechanism is obtained by equating the external energy work to the internal energy dissipation as

$$(0.75P)\left(\frac{L}{2}\theta\right) = M_p[2\theta + \theta], \tag{5.3.6}$$

which gives an upper-bound solution corresponding to the left-hand beam mechanism as

$$P_1 = P\frac{8M_p}{L}. \tag{5.3.7}$$

Similarly, the work equation for the right-hand beam mechanism is

$$(2P)\left(\frac{L}{3}\theta\right) = M_p(\theta) + 2M_p(1.5\theta) + 2M_p\left(\frac{\theta}{2}\right), \tag{5.3.8}$$

which gives a lower and thus better upper-bound solution for the continuous beam as

$$P_2 = P = \frac{7.5M_p}{L}. \tag{5.3.9}$$

In the next section, we shall perform a moment check on mechanism 2 to show that Eq. (5.3.8) gives the exact plastic limit load.

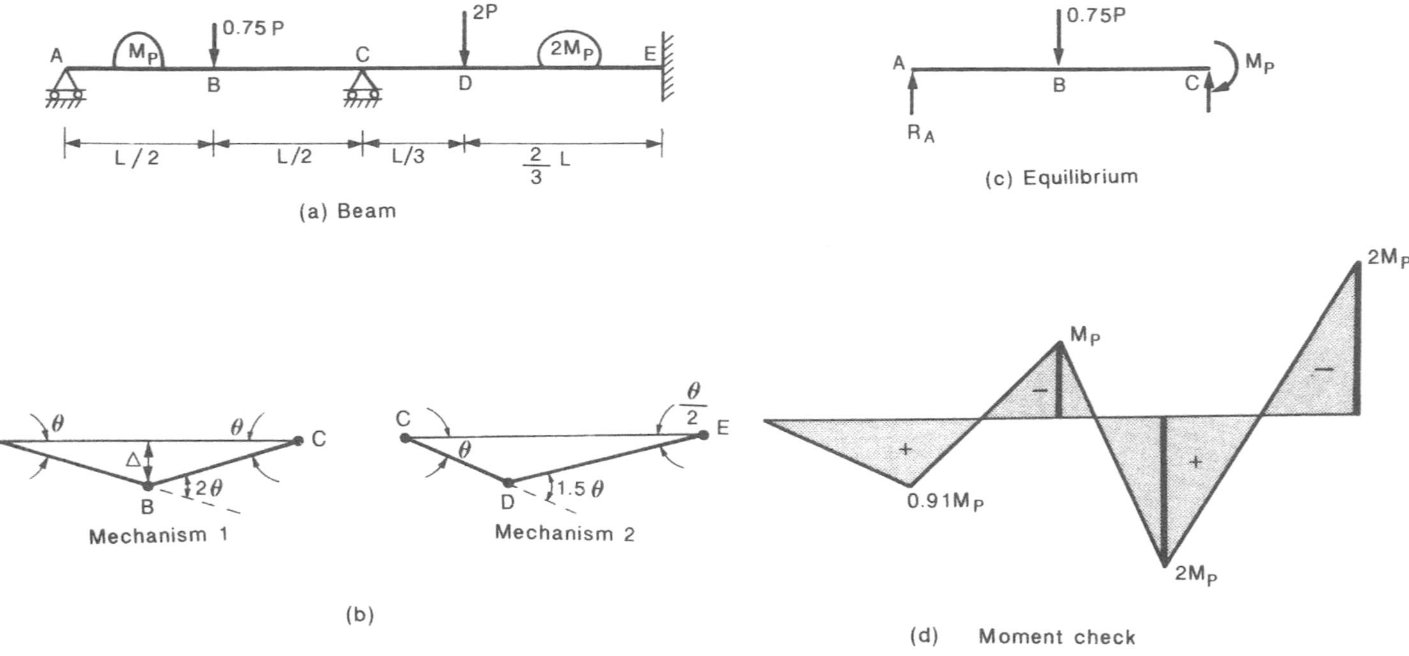

FIGURE 5.2. Mechanism analysis of an unsymmetrical two-span continuous beam (Example 5.3.2): (a) beam, (b) mechanism 1 (left) and mechanism 2, (c) equilibrium, and (d) moment check.

5.3.3 A Pinned-Based Portal Frame

Example 5.3.3. Obtain the plastic limit load for the portal frame shown in Fig. 5.3(a).

Solution: Plastic hinges can possibly be formed at Sections B, C, and D. Three possible mechanisms are shown in Figs. 5.3(b), (c), and (d). The work equation corresponding to each of these three mechanisms is given later.

Referring to the beam mechanism (mechanism 1) shown in Fig. 5.3(b), we have

$$P\left(\frac{L}{2}\theta\right) = M_p\theta + M_p(2\theta) + M_p(\theta), \tag{5.3.10}$$

which gives an upper-bound solution as

$$P_1 = \frac{8M_p}{L}. \tag{5.3.11}$$

For the side-sway mechanism (mechanism 2) shown in Fig. 5.3(c), since both plastic hinges rotate the same amount θ, we have

$$\frac{P}{2}\left(\frac{L}{2}\theta\right) = M_p\theta + M_p\theta, \tag{5.3.12}$$

which gives another upper-bound solution as

$$P_2 = \frac{8M_p}{L}. \tag{5.3.13}$$

For the combined mechanism (mechanism 3) shown in Fig. 5.3(d), it is composed of three links—segment AC, segment CD, and column DE. Using the geometrical relationships shown, we obtain (see Section 5.6.1)

$$P\left(\frac{L}{2}\theta\right) + \frac{P}{2}\left(\frac{L}{2}\theta\right) = M_p(2\theta) + M_p(2\theta), \tag{5.3.14}$$

which gives the lowest upper-bound solution of the three assumed collapse mechanisms

$$P_3 = \frac{16}{3}\frac{M_p}{L}. \tag{5.3.15}$$

The lowest value P_3 is the plastic limit load P_p of the frame. To be sure that no other possible mechanisms are overlooked, it is necessary to check that the plastic moment condition ($M \leq M_p$) is not violated anywhere in the frame. This will be done in the forthcoming.

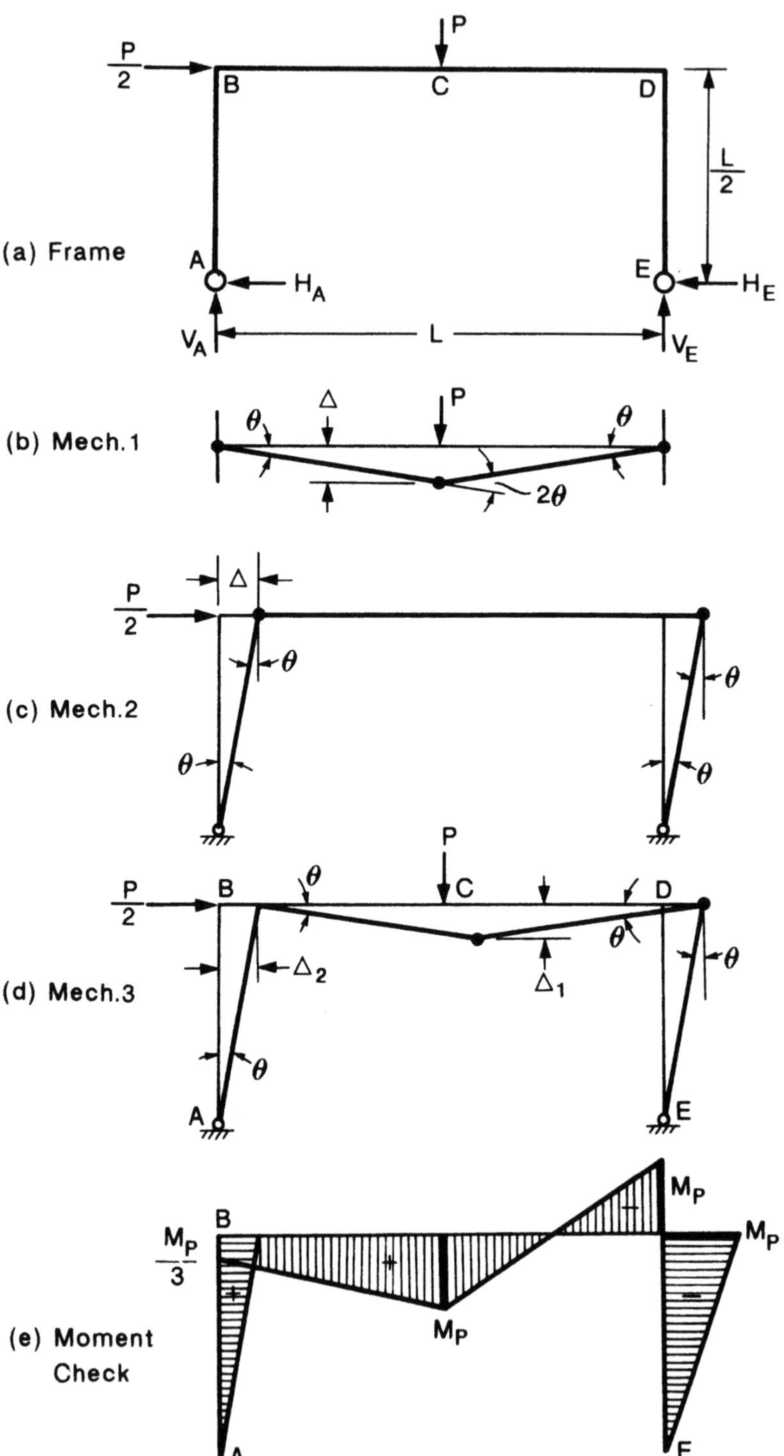

FIGURE 5.3. Mechanism analysis of pinned-based portal frame (Example 5.3.3): (a) frame, (b) mechanism 1, (c) mechanism 2, (d) mechanism 3, and (e) moment check.

5.4 Moment Check

The work equation gives an upper bound to the exact plastic limit load. It is therefore necessary to check and see whether the moment condition $M \leq M_p$ is met throughout the structure for the assumed mechanism. Otherwise, we may overlook a more favorable mechanism, which may give a lower load. Thus, if the moment condition cannot be met for the assumed mechanism, a fresh guess as to the collapse mechanism is made again, but now it is guided by the results of the moment check, and the process is repeated.

For an assumed mechanism, the structure can be determinate or indeterminate. The number of indeterminacy I of the structure at collapse load can be determined from the following rule

$$I = X - (M - 1) \tag{5.4.1}$$

where X is the number of redundancies in the original structure and M is the number of plastic hinges necessary to develop the mechanism.

The design that leads to an indeterminate structure at collapse is probably not the most efficient design, since in theory the material can be saved somewhere in the structure to bring the moments in the inderminate parts of the structure up to their fully plastic values.

However, partial collapse mechanisms often occur in practice, and their moment check procedures are more tedious. The procedures of making a moment check for a detertminate or indeterminate structure at collapse are briefly described in the following.

5.4.1 Determinate Structures

If the structure at collapse is determinate, simple statics or the virtual work equation can be used to determine the moments in all parts of the structure. The moment checks by simple statics and the virtual work equation are illustrated in Examples 5.4.1 and 5.4.2.

5.4.2 Indeterminate Structures

If the structure at collapse is indeterminate, both the simple statics and the virtual work equation can be used for the moment check. The virtual work equation can be used to express all unknown moments in terms of redundant moments in the redundant portions of the collapsing structure. The resulting moment diagram, which is in equilibrium with the applied loads, permits us to check the plastic moment condition.

5.4.3 Illustrative Examples

Example 5.4.1. Make a moment check for the right-hand beam mechanism [Fig. 5.2(b)] of the unsymmetrical two-span continuous beam of Eexample 5.3.2 using (a) simple statics and (b) virtual work equation.

Solution:

(a) *Simple statics:* For the right-hand beam mechanism, plastic hinges form at C, D, and E. The moments at these three locations are equal to the plastic moment capacity of the sections at these locations [Fig. 5.2(d)]. The moment at B is unknown and can be determined by considering the free body diagram of portion AC as shown in Fig. 5.2(c) from which the reaction R_A is

$$R_A = \frac{0.75P}{2} - \frac{M_p}{L}. \tag{5.4.2}$$

Substituting $P = P_2 = 7.5\, M_p/L$ from Eq. (5.3.8), we have

$$R_A = \left(\frac{0.75}{2}\right)\left(\frac{7.5M_p}{L}\right) - \frac{M_p}{L} = \frac{1.81M_p}{L}. \tag{5.4.3}$$

Thus, the central unknown moment M_B has the value

$$M_B = \frac{1.81M_p}{L}\frac{L}{2} = 0.91M_p < M_p. \tag{5.4.4}$$

Since the moment is not greater than the plastic moment capacity M_p anywhere in the beam, $P = 7.5M_p/L$ is the exact plastic collapse load.

(b) *Virtual work equation:* The unknown moment M_B is determined by equating the virtual work done by the applied load to the virtual internal work done by moments on the left-hand beam mechanism as the virtual displacements and rotations, using the usual sign convention for moment ($M_B = +M_B$, $M_C = -M_p$) and rotations ($\theta_B = +2\theta$, $\theta_C = -\theta$) in the virtual work equation

$$(0.75P)(\Delta) = M_B\theta_B + M_C\theta_C \tag{5.4.5}$$

or

$$(0.75P)\left(\frac{L}{2}\theta\right) = (+M_B)(+2\theta) + (-M_p)(-\theta), \tag{5.4.6}$$

which gives the unknown central moment at B as

$$M_B = \frac{1}{2}\left[\frac{0.75}{2}PL - M_p\right]. \tag{5.4.7}$$

Substituting $P = P_2 = 7.5\, M_p/L$, we have the central moment

$$M_B = \frac{1}{2}\left[\frac{0.75}{2}\left(\frac{7.5M_p}{L}\right)L - M_p\right] = 0.91M_p \tag{5.4.8}$$

$M_B < M_p$, okay.

Example 5.4.2. Make a moment check for the combined mechanism of the portal frame of Example 5.3.3 [Fig. 5.3(d)] using (a) simple statics and (b) the virtual work equation.

Solution:

(a) *Simple statics:* For the combined mechanism, plastic hinges form at C and D. Therefore moments at C and D are equal to the plastic moment capacity at these locations [Fig. 5.3(e)]. The moment at B is unknown and can be determined by first determining the horizontal reactions at A and E. The horizontal reaction at E is obtained from the free body diagram of column DE as

$$H_E = \frac{2M_p}{L}. \tag{5.4.9}$$

It follows that the horizontal reaction at A is

$$H_A = \frac{P}{2} - \frac{2M_p}{L}. \tag{5.4.10}$$

Substituting $P = P_3 = (16/3)(M_p/L)$ from Eq. (5.3.14), we have

$$H_A = \frac{16M_p}{3L}\frac{1}{2} - \frac{2M_p}{L} = \frac{2}{3}\frac{M_p}{L}. \tag{5.4.11}$$

Thus, the unknown beam end moment at B has the value

$$M_B = H_A\frac{L}{2} = \frac{2M_p}{3L}\frac{L}{2} = \frac{M_p}{3}.$$

The resulting moment diagram for the frame is shown in Fig. 5.3(e). Since the moment is less than M_p everywhere in the frame, $P = 16M_p/3L$ is the exact collapse load of the frame.

(b) *Virtual work equation:* The unknown moment M_B can be determined directly by equating the virtual work done by the applied loads on the beam mechanism and the internal virtual work done by the moments at the collapse load $P = P_3$ (mechanism 3) on the beam mechanism, in the usual sign convention for moments ($M_B = +M_B$, $M_C = +M_p$, $M_D = -M_p$) and rotations ($\theta_B = -\theta$, $\theta_C = +2\theta$, $\theta_D = -\theta$):

$$P(\Delta) = M_B\theta_B + M_C\theta_C + M_D\theta_D \tag{5.4.12}$$

or

$$P\left(\theta\frac{L}{2}\right) = (+M_B)(-\theta) + (+M_p)(+2\theta) + (-M_p)(-\theta), \tag{5.4.13}$$

which gives

$$M_B = -\frac{PL}{2} + 3M_p. \tag{5.4.14}$$

Substituing $P = P_3 = 16M_p/3L$, we have the unknown beam end moment at B as

$$M_B = -\frac{1}{2}\left(\frac{16}{3}M_p\right) + 3M_p = \frac{M_p}{3} < M_p, \quad \text{okay.} \tag{5.4.15}$$

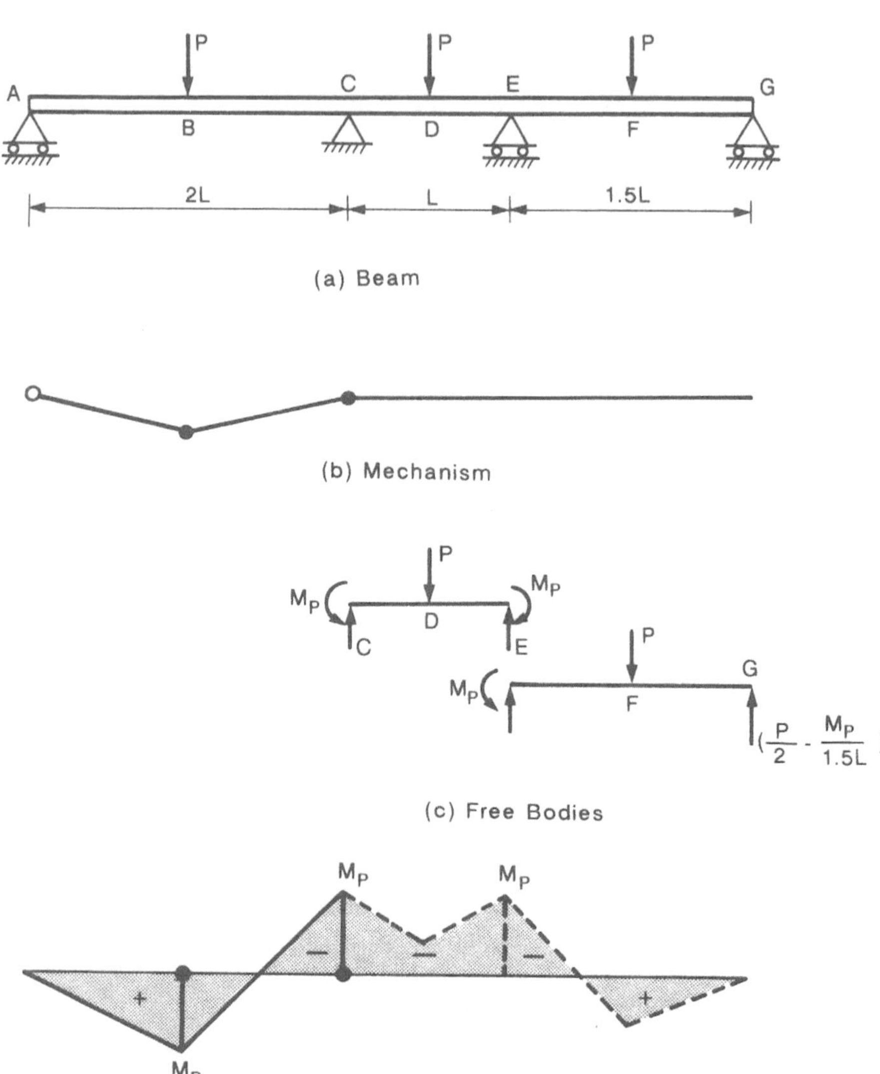

(a) Beam

(b) Mechanism

(c) Free Bodies

(d) Moment check

FIGURE 5.4. Moment check by the trial-and-error method (Example 5.4.3): (a) beam, (b) mechanism, (c) free bodies, and (d) moment check.

Example 5.4.3. A three-span continuous beam has a uniform section with plastic moment capacity M_p and is subjected to concentrated loads in each span as shown in Fig. 5.4(a). The lowest upper bound is found to be $P = 3M_p/L$ from the mechanism shown in Fig. 5.4(b). Make a moment check.

Solution:

(a) *Simple statics:* The degree of indeterminacy at the plastic collapse load is

$$I = X - (M - 1) = 2 - (2 - 1) = 1. \tag{5.4.16}$$

The moments at D, E, and F are unknown. Try the redundant moment at E to be $-M_p$. Then the values of M_D and M_F can be determined by considering the free body diagrams of segments CE and EG in Fig. 5.4(c). From the free body CE, the unknown central moment M_D can be found as

$$M_D = \frac{PL}{4} - M_p = \frac{L}{4}\left(\frac{3M_p}{L}\right) - M_p = -\frac{M_p}{4}. \tag{5.4.17}$$

Similarly, from the free body EG, the unknown central moment M_F can be found as

$$M_F = \left(\frac{P}{2} - \frac{M_p}{1.5L}\right)(0.75L) = \left(\frac{3M_p}{L}\right)\frac{(0.75L)}{2} - \frac{M_p}{1.5L}(0.75L) = \frac{5}{8}M_p. \tag{5.4.18}$$

The resulting moment diagram is shown by the dashed line in Fig. 5.4(d). Since $M \leq M_p$ throughout the beam, the limit load $P = 3M_p/L$ is the exact solution. Note that the *plastic moment check* does not require the redundant moment M_E be determined by an elastic analysis. It only requires to show that there exists a value of M_E such that it is in equilibrium with the applied loads and the resulting moment diagram does not violate the moment condition. Also, note that a more efficient use of material will result if the design is revised to supply only the required plastic moment for other noncollapsing spans.

(b) *Virtual work equation:* Assume the moment at E is the redundant (Fig. 5.5). Now, by the virtual work equation, the unknown moments at D and F can be expressed in terms of the plastic moment M_p and the redundant moment M_E.

The moment at D is determined by applying the virtual work equation to the equilibrium and geometry sets shown in Figs. 5.5(b) and (c):

$$(-M_p)(-\theta) + (+M_D)(+2\theta) + (+M_E)(-\theta) = \left(\frac{3M_p}{L}\right)\left(\frac{L}{2}\theta\right), \tag{5.4.19}$$

which gives the central moment for the left-hand beam as

$$M_D = \frac{1}{4}M_p + \frac{1}{2}M_E. \tag{5.4.20}$$

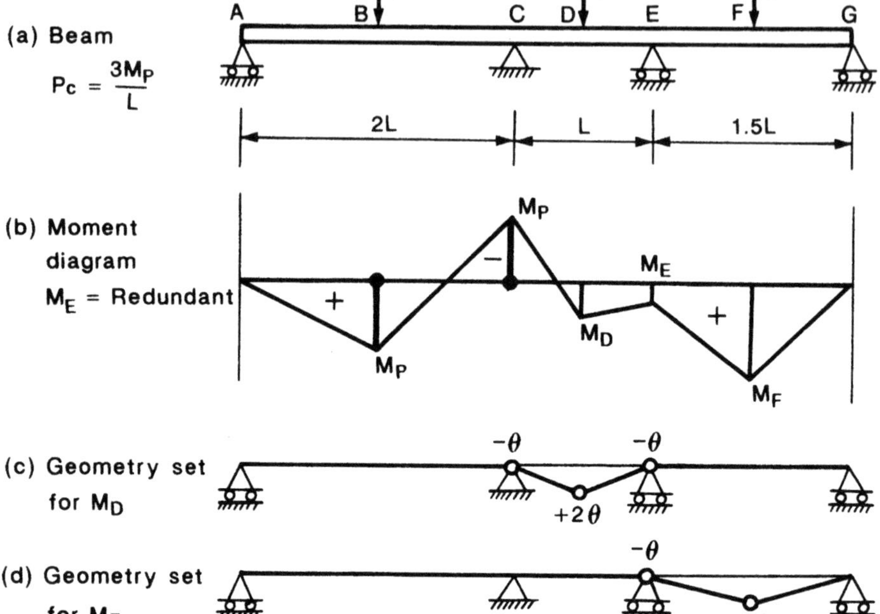

FIGURE 5.5. Moment check by the virtual work equation for an indeterminate structure after a partial collapse (Example 5.4.3): (a) beam, $P_c = 3M_P/L$; (b) moment diagram, M_E = redundant; (c) geometry set for M_D; and (d) geometry set for M_F.

Similarly, the moment at F is determined by applying virtual work equation to the equilibrium and geometry sets shown in Figs. 5.5(b) and (d):

$$(+M_E)(-\theta) + (+M_F)(+2\theta) = \left(\frac{3M_p}{L}\right)\left(\frac{1.5L}{2}\theta\right), \tag{5.4.21}$$

which gives the central moment for the right-hand beam as

$$M_F = \frac{9M_p}{8} + \frac{1}{2}M_E. \tag{5.4.22}$$

The condition that the absolute values of the moments at D, E, and F be less than M_p leads to

$$-M_p \leq \frac{1}{4}M_p + \frac{1}{2}M_E \leq M_p \tag{5.4.23}$$

$$-M_p \leq M_E \leq M_p \tag{5.4.24}$$

$$-M_p \leq \frac{9}{8}M_p + \frac{1}{2}M_E \leq M_p. \tag{5.4.25}$$

These inequalities can be expressed as

$$-\frac{5}{2}M_p \le M_E \le \frac{3}{2}M_p \tag{5.4.26}$$

$$-M_p \le M_E \le M_p \tag{5.4.27}$$

$$-\frac{17}{4}M_p \le M_E \le -\frac{1}{4}M_p. \tag{5.4.28}$$

Inequalities (5.4.26) to (5.4.28) are equivalent to

$$-M_p \le M_E \le -\frac{1}{4}M_p, \tag{5.4.29}$$

which indicates that there exists an M_E value in the range given by Eq. (5.4.29) corresponding to which the absolute values of the moments at D and F are less than M_p so that the moment check is complete.

5.5 Design of Rectangular Portal Frame

The axial force in columns will reduce the moment-carrying capacity of the columns as described in Section 4.6. Herein, we shall apply the work method for the design of rectangular frames and consider the column buckling effects of axial load in the plastic design of these frames using the LRFD interaction equations presented in Section 4.6.

Example 5.5.1. A rectangular portal frame is subjected to factored vertical loads as shown in Fig. 5.6. Find the required M_p, show the collapse mechanism, and perform a moment check. Select an A36 W section for the columns

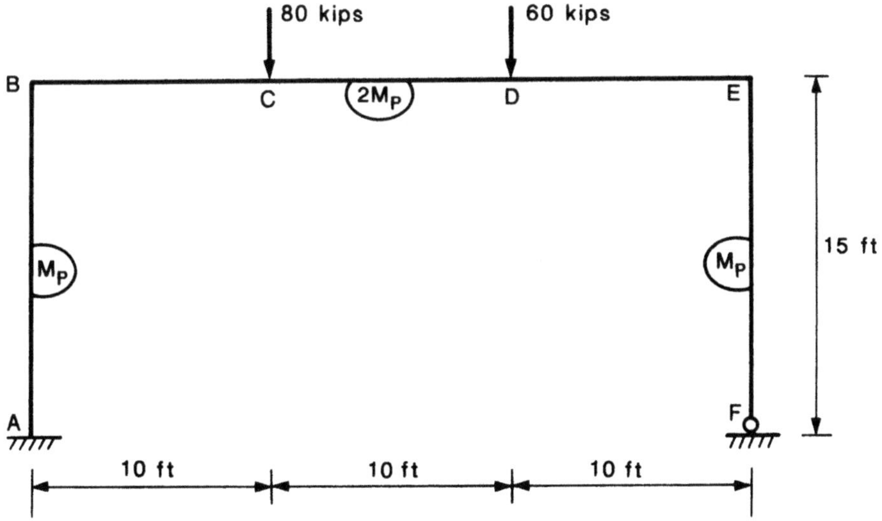

FIGURE 5.6. Design of rectangular portal frame with two concentrated loads (Example 5.5.1).

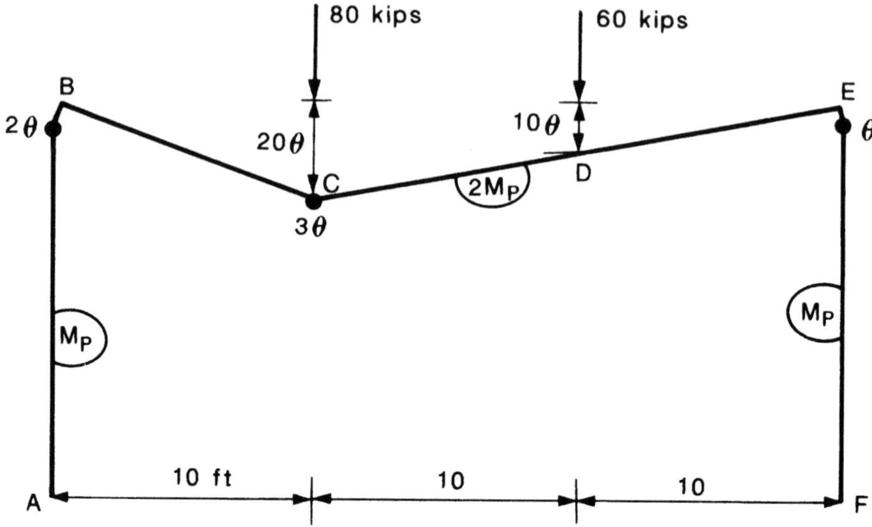

FIGURE 5.7. Collapse mechanism for the frame in Figure 5.6.

and check the effects of axial load using the LRFD interaction equations for columns AB and EF. Assume $B_1 = B_2 = 1$, and $K_x = K_y = 1$ for both columns.

Solution: Consider the beam mechanism shown in Fig. 5.7. By equating the external work due to applied loads to the internal energy dissipation, we have the work equation for the beam mechanism as

$$80(20\theta) + 60(10\theta) = M_p(2\theta) + 2M_p(3\theta) + M_p(\theta),$$

which gives

$$M_p = 244 \text{ kip-ft.}$$

Next, we perform a moment check. For the mechanism under consideration, the hinges are formed at B, C, and E. So the moments at B, C, and E are known to be $-M_p$, $+2M_p$, and $-M_p$, respectively. The moment at A will be determined by applying the virtual work equation to equilibrium and geometric sets chosen as shown in Fig. 5.8(a) as

$$80(0) + 60(0) = (+M_A)(-\theta) + (-M_p)(+\theta) + (-M_p)(-\theta) + (0)(\theta),$$

which gives the unknown fixed-end moment at A as

$$M_A = 0 < M_p, \quad \text{okay.}$$

The equation for the moment at D is obtained by applying the virtual work equation to equilibrium and geometry sets selected as shown in Fig. 5.8(b) as

$$(80)(0) + (60)(10\theta) = (+2M_p)(-\theta) + M_D(+2\theta) + (-M_p)(-\theta),$$

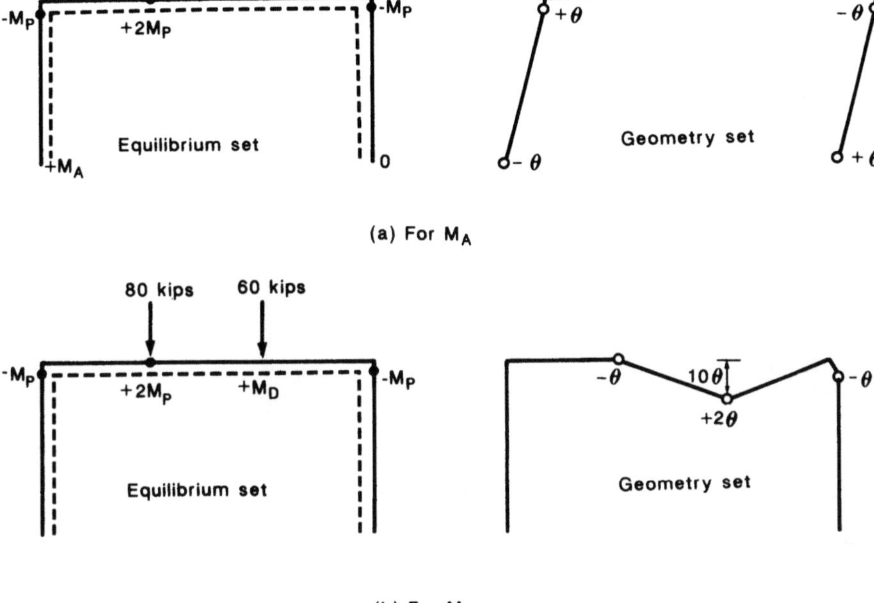

FIGURE 5.8. Moment check for the collapse mechanism in Figure 5.7: (a) for M_A and (b) for M_D.

which gives the unknown beam moment at D as

$$M_D = \frac{1}{2}(600 + M_p) = \frac{1}{2}(600 + 244) = 422 \text{ kip-ft} < 2M_p = 488, \quad \text{okay.}$$

Design of Columns: Since columns AB and EF are subjected to combined bending and axial compression as shown in Fig. 5.9, we shall design the columns for an equivalent moment

$$M_{eq} = M_x + P\frac{d}{2}.$$

Assume depth of section $d = 18$ in. and obtain

$$M_{eq} = 244 + (73.3)\frac{18}{(12)(2)} = 299 \text{ kip-ft}$$

$$\text{required } Z_x = \frac{M_{eq}}{\phi_b F_y} = \frac{(299)(12)}{(0.9)(36)} = 111 \text{ in.}^3$$

Try W18 × 55, which has the values

$$A = 16.2 \text{ in.}^2$$

$$Z_x = 112 \text{ in.}^3$$

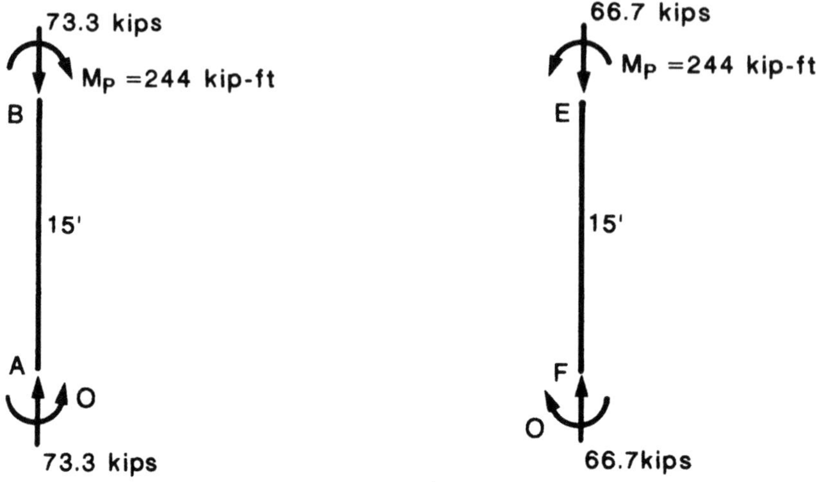

FIGURE 5.9. Forces in columns AB and EF in Figure 5.6.

$$r_x = 7.41 \text{ in.}, \ r_y = 1.67 \text{ in.}$$

$$\phi_b M_p = \frac{(0.9)(112)(36)}{12} = 302 \text{ kip-ft.}$$

Check Strength of Column AB: The values of slenderness parameters λ_{cx} and λ_{cy} are calculated as

$$\lambda_{cx} = \frac{KL}{r_x \pi} \sqrt{\frac{F_y}{E}} = \frac{(1)(15)(12)}{(7.41)\pi} \sqrt{\frac{36}{29,000}} = 0.272$$

$$\lambda_{cy} = \frac{KL}{r_y} \sqrt{\frac{F_y}{E}} = \frac{(1)(15)(12)}{1.67\pi} \sqrt{\frac{36}{29,000}} = 1.209.$$

The buckling about the weak axis controls. So $\lambda_c = \lambda_{cy} = 1.209 < 1.5$; thus the axial capacity of the column is calculated from

$$P_n = 0.658^{\lambda_c^2} P_y = 0.658^{(1.209)^2}(36)(16.2) = 316.3 \text{ kips.}$$

Now, we have an axial load ratio equal to

$$\frac{P}{\phi_c P_n} = \frac{73.3}{(0.85)(316.3)} = 0.273 > 0.2.$$

So we shall use the following interaction equation to check the capacity of the member

$$\frac{P}{\phi_c P_n} + \frac{8}{9}\left(\frac{M}{\phi_b M_p}\right) < 1$$

or

$$0.273 + \left(\frac{8}{9}\right)\left(\frac{244}{302}\right) = 0.991 < 1.0, \quad \text{okay.}$$

Since the axial load in column EF is smaller than that in column AB, W18 × 55 is also sufficient for column EF. So use W18 × 55.

5.6 Calculation of Geometrical Relations

The calculations of displacements in the direction of applied loads and rotations at the plastic hinge locations are simple for beams, but they may become somewhat tedious for complex frames. In such cases, the rigid-body motion of a failure mechanism may be found easily by the methods known at the *instantaneous center* and the *virtual work equation* [1.8, 5.1]. These two methods will be described and illustrated in the forthcoming.

5.6.1 *Instantaneous Center*

Consider first the application of the instantaneous center method to the combined mechanism (mechanism 3) of Fig. 5.3(d) or Fig. 5.10. As the frame moves sideways, segment A-B-C rotates as a rigid body around the base at A. Member D-E rotates about point E. The center of rotation of C-D is yet unknown and is to be determined by considering the movement of the ends of the segment.

Point D is constrained to move perpendicular to line DE. Thus, its center of rotation as part of segment CD must be somewhere along line E-D extended. Point C rotates about A, since it is a part of segment A-B-C. Therefore, it must move normal to line A-C and its center of rotation as part of C-D must lie along A-C extended. The intersection point I satisfies both conditions and therefore segment C-D must rotate about point I which is called the *instantaneous center* of rotation of segment C-D. Once the instantaneous centers for all rigid parts are found, we can determine the relevant displacements and angles of rotation between the connected rigid parts in the following manner.

The mechanism angles at the plastic hinges of Fig. 5.10 are determined as follows. Since the mechanism movement is infinitely small, lines $AC'I$ and $ED'I$ are tantamount to the straight lines ACI and EDI drawn through the hinge points of the undeformed frame. The infinitely small displacement CC' can be assumed as linear and perpendicular to ACI, and hence a common tangent to arcs of circles having points A and I for centers. Likewise, the displacement DD' had both points E and I as centers of rotation. If the rotation angle at column base E is θ, then the horizontal motion of point D is equal to $\theta(L/2)$. Since ABC and CDI are similar triangles and $DE = DI$, it follows that the rotation of C-D about I is $\theta(L/2)/(L/2) = \theta$. Since the lengths A-C and C-I are equal, the rotation at A is equal to rotation at I, which is equal to θ. The total rotation at D is equal to 2θ, the sum of rotations for parts ED and CD. Similarly, the total rotation at C is also equal to 2θ, the sum of rotations for parts A-B-C and C-D.

The vertical movement at point C is determined by considering the rotation of segment A-B-C about A. Since A-B-C is a rigid body, the angle at B

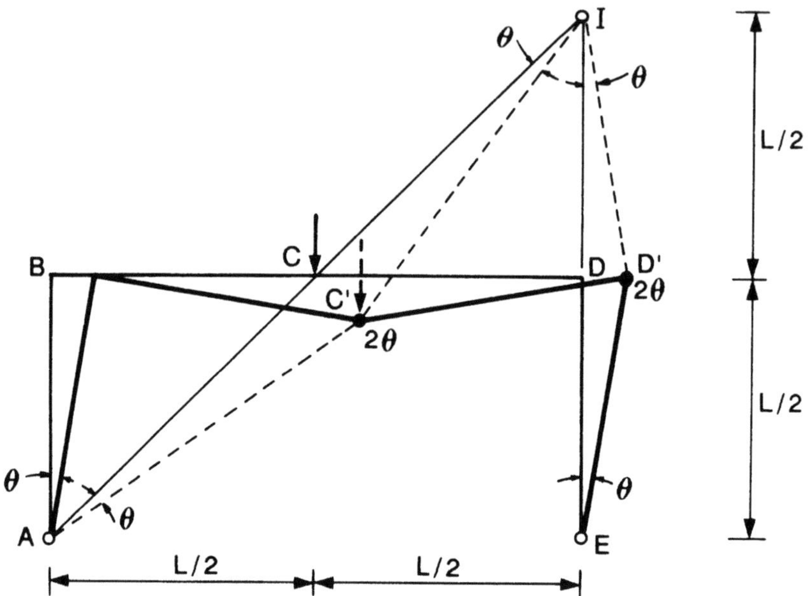

FIGURE 5.10. Instantaneous center for a sway mechanism.

remains a right angle and the rotation of *B-C* with respect to the horizontal is equal to θ. The vertical motion of point *C* is therefore $\theta L/2$.

Further applications of the instantaneous center method to gable frames are given in Example 5.6.1.

5.6.2 Virtual Work Equation

The caculations of geometrical relations of a mechanism can also be made easily by the use of the virtual work equation. This is illustrated in the following. Consider the gable frame shown in Fig. 5.11(a). The deflected shape is shown by the dashed line in the figure. It is clear that the mechanism has two degrees of freedom and hence two independent angles of rotation, ϕ_{AB} and ϕ_{DE}. All other angles (ϕ_{BC} and ϕ_{CD}) can be expressed in terms of ϕ_{AB} and ϕ_{DE}.

ϕ_{BC} will be computed first. Figure 5.11(b) shows an equilibrium system of external forces and moments applied to the mechanism. The equilibrium system is obtained in the following way. Bar *CD* is assumed to be under axial load with vertical component equal to r_2 and horizontal component equal to n (i.e., proportional to the slope of bar *CD*). It produces vertical reactions r_2 and horizontal reactions n at the two supports. The moment equilibrium is then established at the remaining joints of the mechanism. They are equal to nh_1, $(nr_1 + mr_2)$ and nh_2 at joints *A*, *B*, and *E*, respectively.

If the equilibrium system shown in Fig. 5.11(b) undergoes the displacements shown in Fig. 5.11(a), then the virtual work equation gives

$$nh_1\phi_{AB} + (nr_1 + mr_2)\phi_{BC} - nh_2\phi_{DE} = 0. \tag{5.6.1}$$

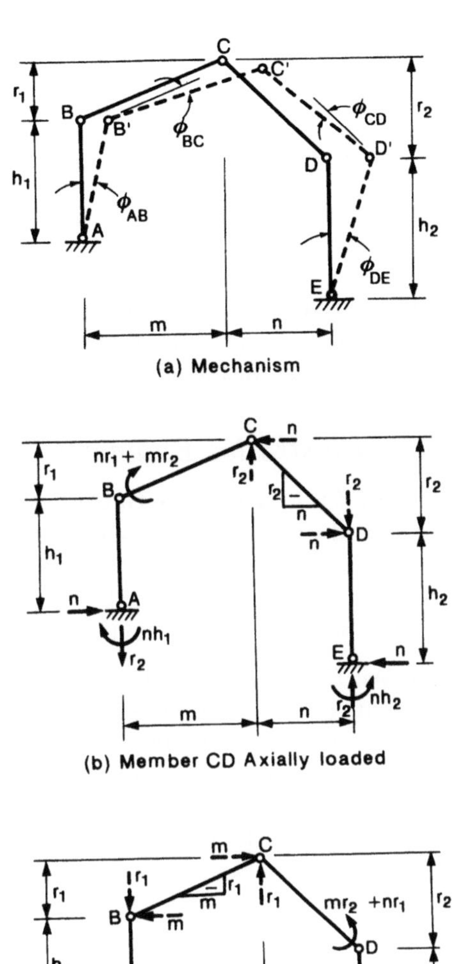

FIGURE 5.11. Calculation of geometric relations by the virtual work equation: (a) mechanism, (b) member *CD* axially loaded, and (c) member *BC* axially loaded.

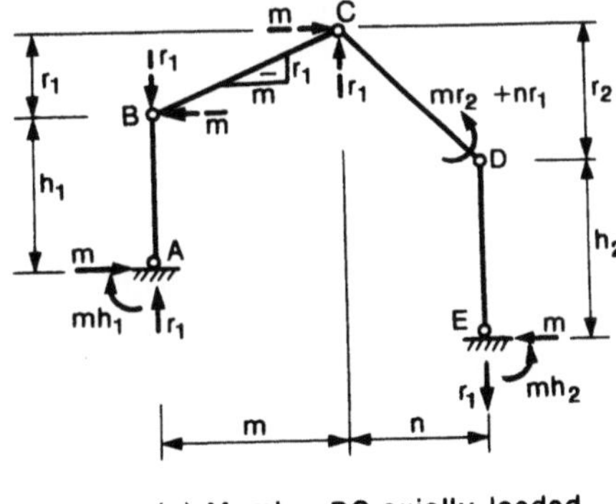

The rotation of bar *BC* can then be expressed as

$$\phi_{BC} = \frac{n(h_2\phi_{DE} - h_1\phi_{AB})}{mr_2 + nr_1}.$$

(5.6.2)

Similarly, bar *BC* can be assumed to act as an axially loaded member. Figure 5.11(c) summarizes the resulting equilibrium system. The virtual work equation for the equilibrium system in (c) doing work on the mechanism in

(a) then yields

$$mh_1\phi_{AB} + (mr_2 + nr_1)\phi_{CD} - mh_2\phi_{DE} = 0. \tag{5.6.3}$$

The rotation of bar CD can then be expressed as

$$\phi_{CD} = \frac{m(h_2\phi_{DE} - h_1\phi_{AB})}{mr_2 + nr_1}. \tag{5.6.4}$$

The virtual work equation will be applied later to a shed-like gable frame in Example 5.6.2 to show its power and simplicity in obtaining the needed relationship for plastic analysis.

5.6.3 *Illustrative Examples*

Example 5.6.1. **A Regular Gable Frame:** A gable frame is subjected to horizontal and vertical loads as shown in Fig. 5.12. Determine the displacements in the direction of loads and rotations at the plastic hinge locations corresponding to the mechanism shown. Also determine the upper bound load P corresponding to this mechanism.

Solution: Denote the small rotation of member $F - G$ about point G by θ. Segment $A - B - C$ rotates as a rigid body about point A by an unknown angle. To find the instantaneous center I of segment $C - D - F$, we need to find the common point about which both ends rotate. Point F is constrained to move normal to line $G - F$ and will have its center along line $G - F$ extended. Similarly, the center of C will move along $A - C$ extended. Thus point I is the intersection of $G - F$ extended and $A - C$ extended.

By geometry, the length $I - G$ is equal to $5L$. $F - I$ is therefore equal to $4L$. Since the horizontal displacement of point F is θL, the rotation at I is $\theta(G - F/F - I) = \theta/4$. By similar triangles, the ratio $C - I$ to $A - C$ is $3 : 1$. Thus, the rotation at A is $(\theta/4)(3/1) = (3/4)\theta$.

Mechanism angles and displacements in the direction of load may now be computed. The rotation at F is the sum of rotations of FG and FDC, i.e., $(\theta + \theta/4) = (5/4)\theta$. The rotation at C is the sum of rotations of CDF and ABC, i.e., $(\theta/4 + 3/4\theta) = \theta$. The displacements of horizontal load, left vertical load, and right vertical load are, respectively, $\Delta_1 = (3\theta/4)(L)$, $\Delta_2 = (\theta/4)(3L)$, $\Delta_3 = (\theta/4)(L)$.

The correctness of Δ_2 and Δ_3 may be checked by working out the geometry of similar triangles ICF in Fig. 5.12(a) and the one shown in Fig. 5.12(b). The vertical component of the mechanism motion of point C is equal to the rotation about instantaneous center I times the distance to that center measured normal to the line of action.

The collapse load for this mechanism is determined from the work equation [Fig. 5.12(a)]

$$P\Delta_1 + (2P)\Delta_2 + (2P)\Delta_3 = M_p\theta_C + M_p\theta_F$$

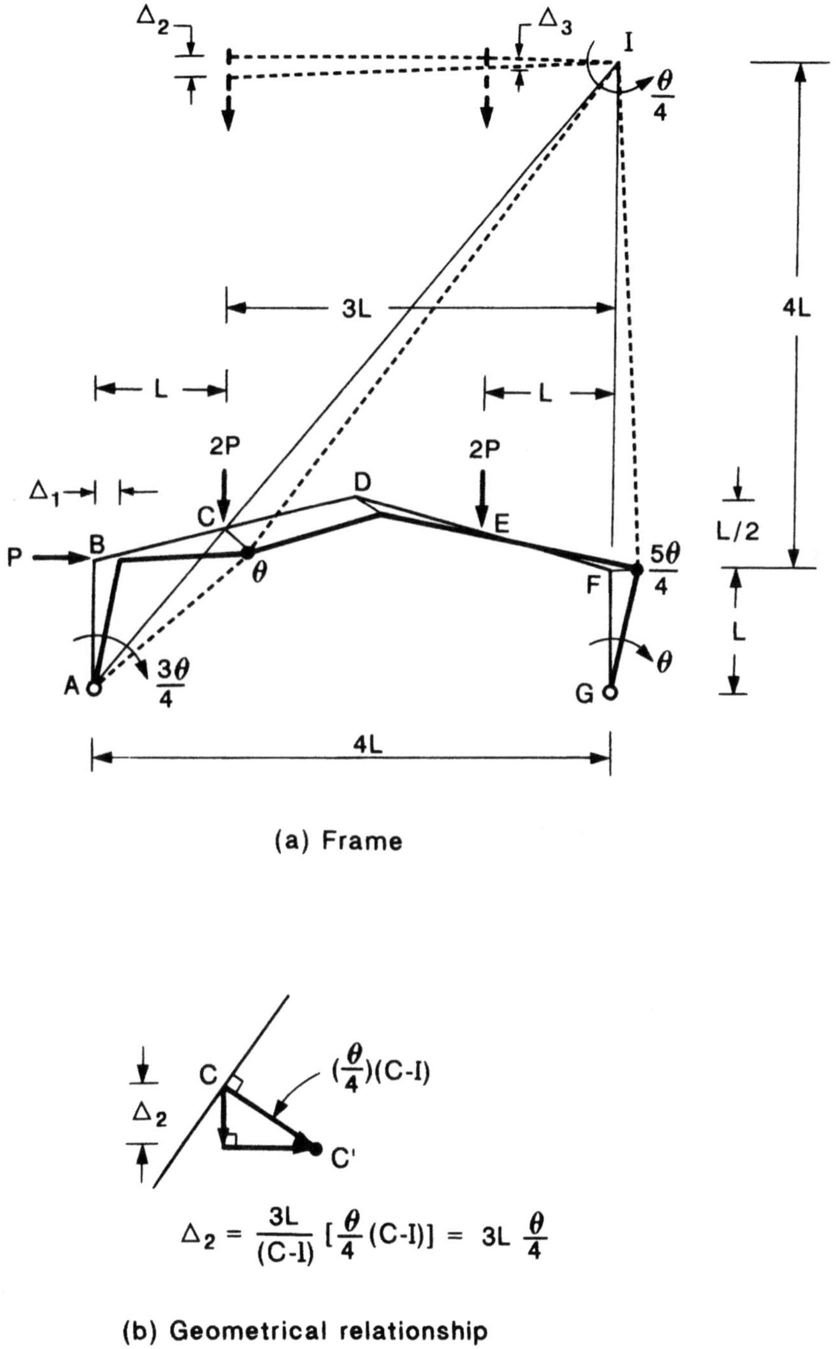

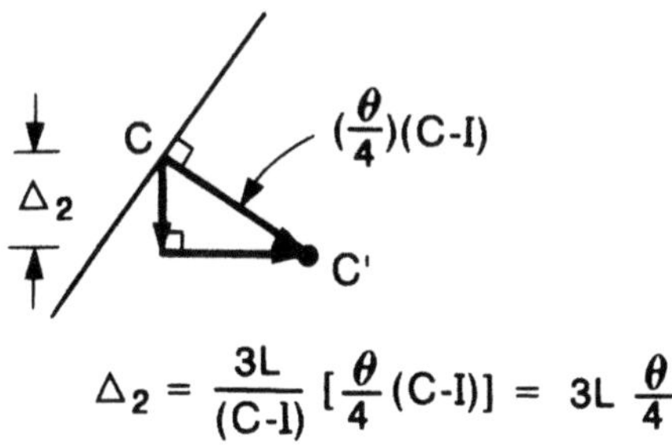

(a) Frame

$$\Delta_2 = \frac{3L}{(C\text{-}I)} \left[\frac{\theta}{4}(C\text{-}I) \right] = 3L \frac{\theta}{4}$$

(b) Geometrical relationship

FIGURE 5.12. Instantaneous center for a gable frame mechanism (Example 5.6.1): (a) frame and (b) geometrical relationship.

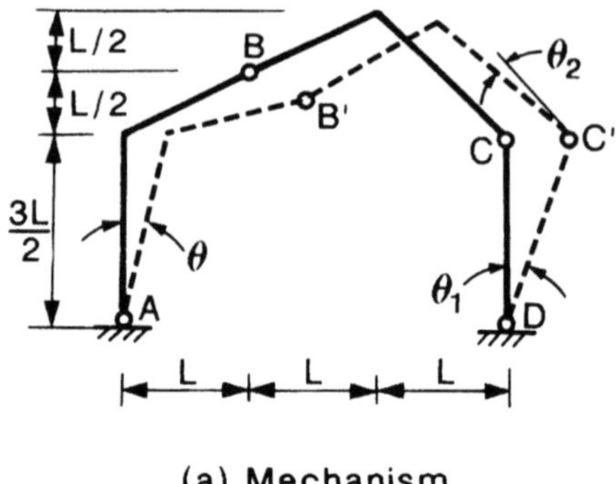

(a) Mechanism

FIGURE 5.13. Calculations of geometric relations by the virtual work equation (Example 5.6.2): (a) mechanism, (b) member *BC* axially loaded, and (c) member *CD* axially loaded.

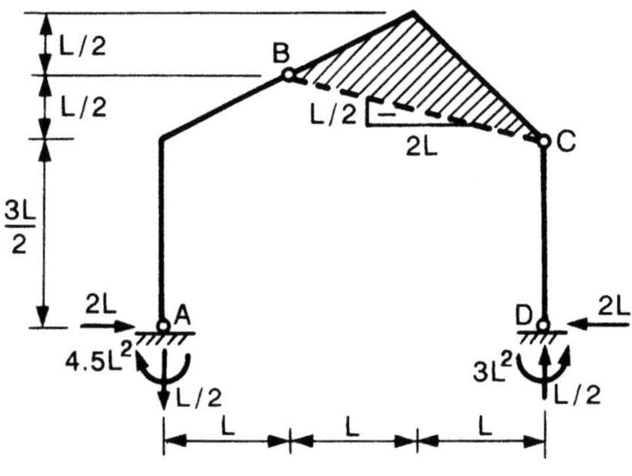

(b) Member BC axially loaded

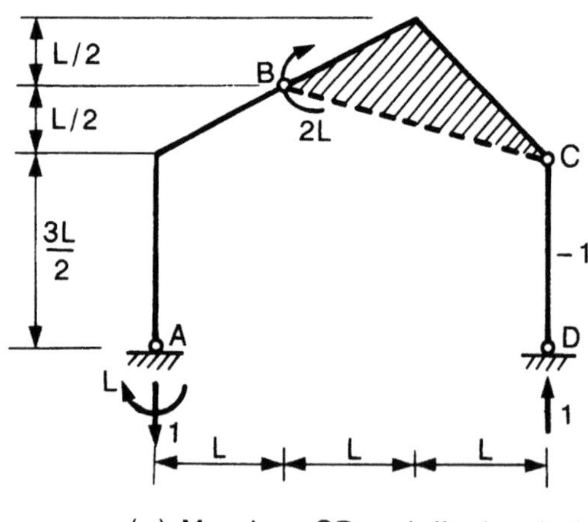

(c) Member CD axially loaded

or

$$P\left(\frac{3\theta L}{4}\right) + 2P\left(\frac{3\theta L}{4}\right) + 2P\left(\frac{\theta L}{4}\right) = M_p(\theta) + M_p\left(\frac{5}{4}\theta\right), \qquad (5.6.5)$$

which gives an upper-bound solution corresponding to the side-sway mechanism assumed in Fig. 5.12:

$$P^u = \frac{9}{11}\frac{M_p}{L}. \qquad (5.6.6)$$

Example 5.6.2. **A Shed Gable Frame:** A shed-type gable frame with four hinges is subjected to displacement as shown in Fig. 5.13(a). Using the virtual work equation, determine the total rotations at the hinge locations in the frame.

Solution: First, we shall determine the hinge rotation at D. Assume that BC (shaded triangle) is an axially loaded compression member with vertical component $L/2$ and horizontal component $2L$ [Fig. 5.13(b)]. The resulting equilibrium system is shown in Fig. 5.13(b). If this equilibrium system undergoes the displacement shown in Fig. 5.13(a), the virtual work equation gives the relationship between the two rotational angles θ and θ_1 as

$$(4.5L^2)\theta - (3L^2)\theta_1 = 0, \qquad (5.6.7)$$

which furnishes

$$\theta_1 = \frac{3}{2}\theta. \qquad (5.6.8)$$

For determining the rotation of member BC, assume that CD is an axially loaded compression member with vertical component -1 and horizontal component zero. The resulting equilibrium system is shown in Fig. 5.13(c). When this equilibrium system is subjected to the displacements shown in Fig. 5.13(a), the virtual work equation leads to

$$L\theta - 2L\theta_2 = 0, \qquad (5.6.9)$$

which expresses the rotational angle θ_2 of member BC in terms of the rotational angle of member AB as

$$\theta_2 = (1/2)\theta. \qquad (5.6.10)$$

The total hinge rotation at B is equal to the sum of rotations of AB and BC, i.e., $[\theta + (\theta/2)] = (3/2)\theta$. Similarly, the total rotation at C is $(\theta_1 + \theta_2) = 2\theta$.

5.7 Gable Frames

The gable frames involve more complicated geometry than that of rectangular frames. Herein, we shall solve two gable frames by the work method.

Example 5.7.1. **Analysis of a Gable Frame Subject to Concentrated Loads:** The frame shown in Fig. 5.14 is composed of members with a full plastic

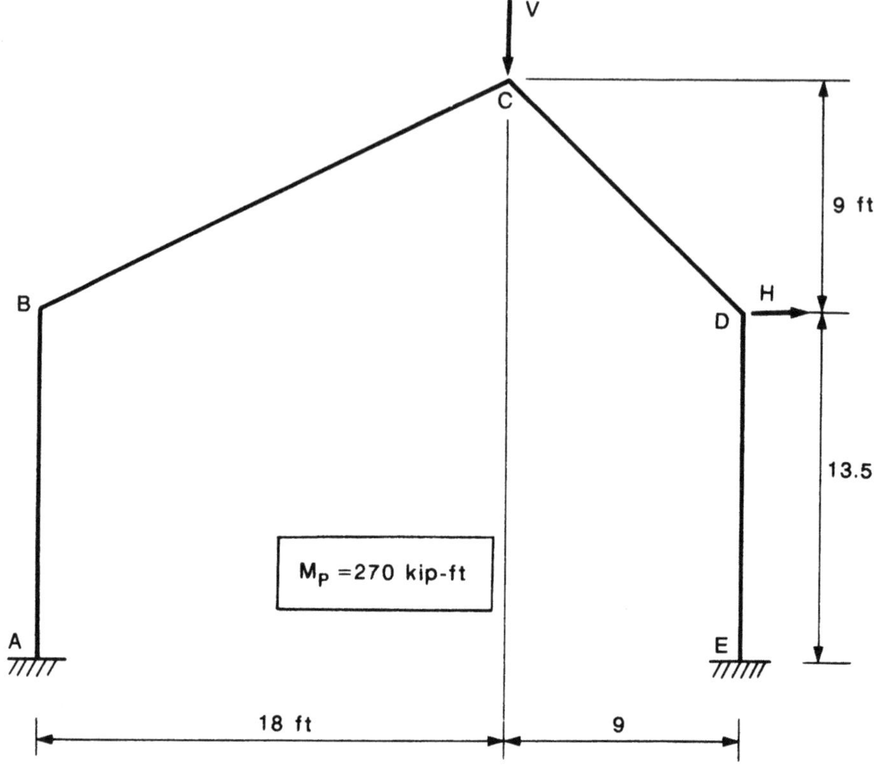

FIGURE 5.14. Gable frame subjected to concentrated loads (Example 5.7.1).

moment capacity of 270 kip-ft and has fixed feet and full-strength joints. The concentrated loads are as shown. Plot a graph (interaction diagram) from which positive values of V and H just causing collapse can be read.

Solution: To plot the interaction diagram, we will evaluate the strength of the frame against three basic modes of collapse.

1. Mechanism with Hinges at A, C, D, and E: The motion of this side-sway mechanism is shown in Fig. 5.15. The instantaneous center O for Member CD is located at the intersection of AC and ED extended. It is convenient to express the motion of the mechanism in terms of rotation θ of member CD about the instantaneous center O.

From similar triangles ACC_1 and OCC_2, we have

$$\frac{OC_2}{CC_2} = \frac{C_1 A}{C_1 C},$$

which gives

$$OC_2 = \frac{C_1 A}{C_1 C} CC_2 = \frac{(22.5)(9)}{18} = 11.25 \text{ ft.}$$

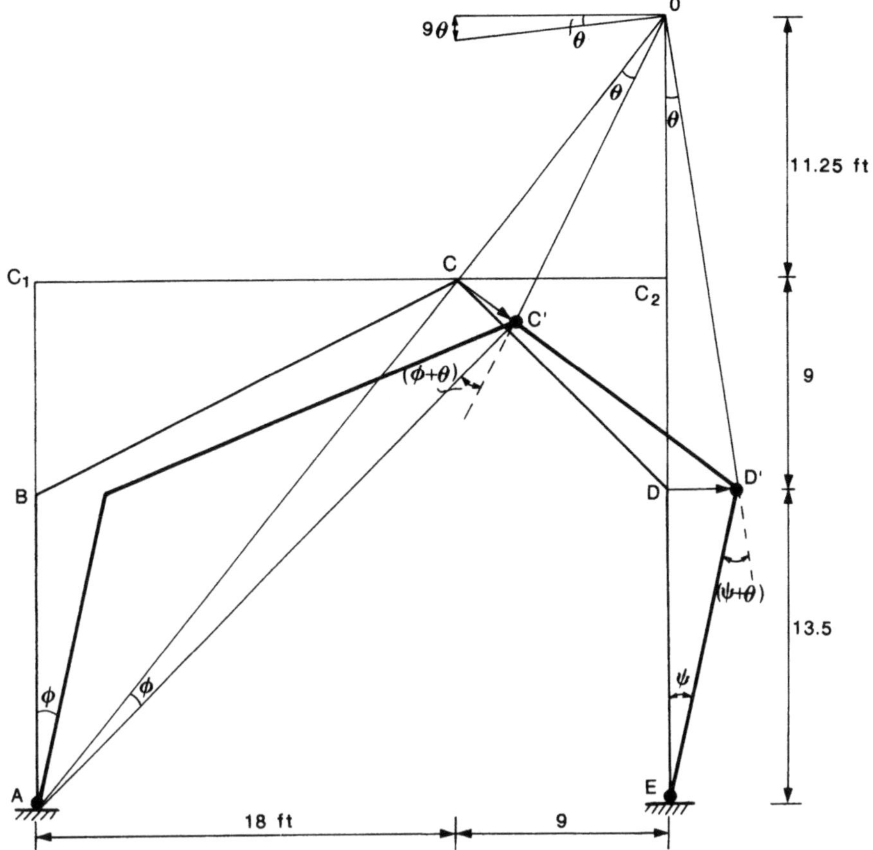

FIGURE 5.15. A mechanism with hinges at joints A, C, D, and E.

From triangles ACC' and $CC'O$, we have

$$(AC)(\phi) = (OC)(\theta),$$

which leads to the relationship between the angles ϕ and θ as

$$\phi = \frac{OC}{AC}\theta = \frac{CC_2}{C_1 C}\theta = \frac{9}{18}\theta = \frac{\theta}{2}.$$

Similarly, from triangles ODD' and EDD', the rotation at E is given by

$$(DE)(\psi) = \theta(OD),$$

which expresses the angle ψ in terms of the angle θ as

$$\psi = \theta\frac{OD}{DE} = 1.5\theta.$$

From the known hinge rotations and displacements of loads, the work equa-

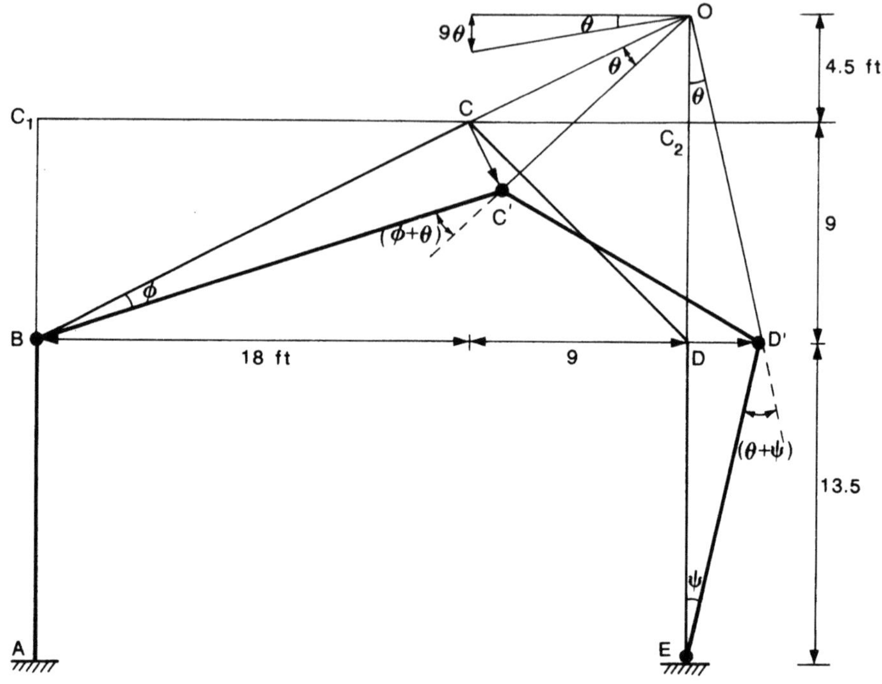

FIGURE 5.16. A mechanism with hinges at joints B, C, D, and E.

tion for this side-sway mechanism can be written as

$$V(9\theta) + H(13.5\psi) = M_p[\phi + (\phi + \theta) + (\theta + \psi) + \psi].$$

Substituting values of ψ and ϕ and simplifying, we obtain

$$V + 2.25H = 180. \tag{5.7.1}$$

2. Mechanism with Hinges at B, C, D, and E: Figure 5.16 shows the motion corresponding to this side-sway mechanism with no sway in the left-hand column AB. Again, we express all plastic hinge rotations and displacements at load points in terms of the rotation θ of member CD about the instantaneous center O.

From similar triangles BCC_1 and CC_2O, we have

$$\frac{OC_2}{CC_2} = \frac{BC_1}{C_1C},$$

which gives

$$OC_2 = \frac{BC_1}{C_1C}(CC_2) = \frac{9}{18}(9) = 4.5 \text{ ft.}$$

From triangles BCC' and $CC'O$, we have

$$(BC)(\phi) = (OC)(\theta),$$

Which expresses ϕ in terms of θ as

$$\phi = \frac{OC}{BC}(\theta) = \frac{OC_2}{BC_1}\theta = \frac{4.5}{9}\theta = \frac{\theta}{2}.$$

Similarly, from triangles ODD' and EDD', we have

$$(DE)(\psi) = (OD)(\theta),$$

which shows that

$$\psi = \frac{OD}{DE}\theta = \frac{13.5}{13.5}\theta = \theta.$$

Now the work equation for this mechanism can be written as (Fig. 5.16)

$$V(9\theta) + H(13.5\psi) = M_p[\phi + (\phi + \theta) + (\theta + \psi) + \psi].$$

Substituting values of ψ and ϕ and simplifying, we obtain

$$V + 1.5H = 150. \tag{5.7.2}$$

3. Mechanism with Hinges at A, B, D, and E: The hinge rotations and displacements corresponding to this simple side-sway mechanism are shown in Fig. 5.17. The rotation of all hinges is θ. The horizontal load moves by 13.5θ

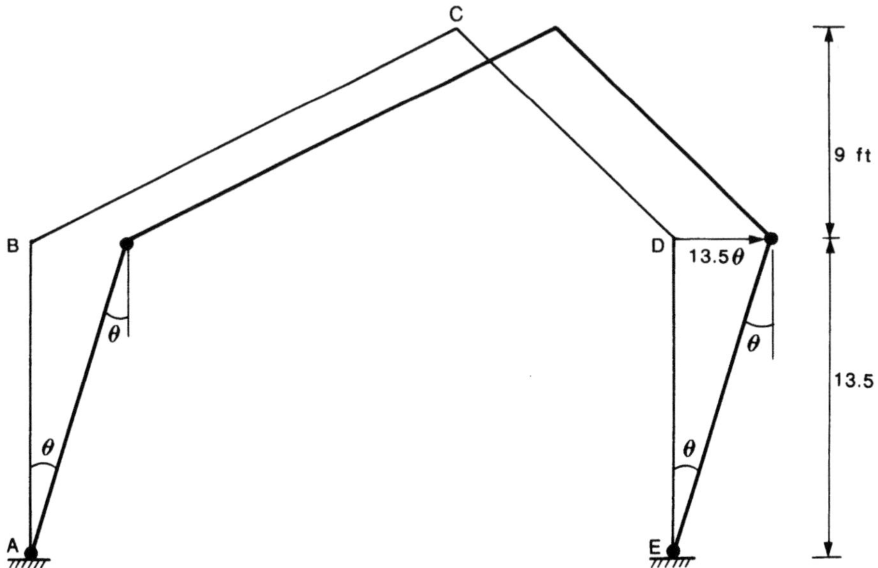

FIGURE 5.17. A mechanism with hinges at joints A, B, D, and E.

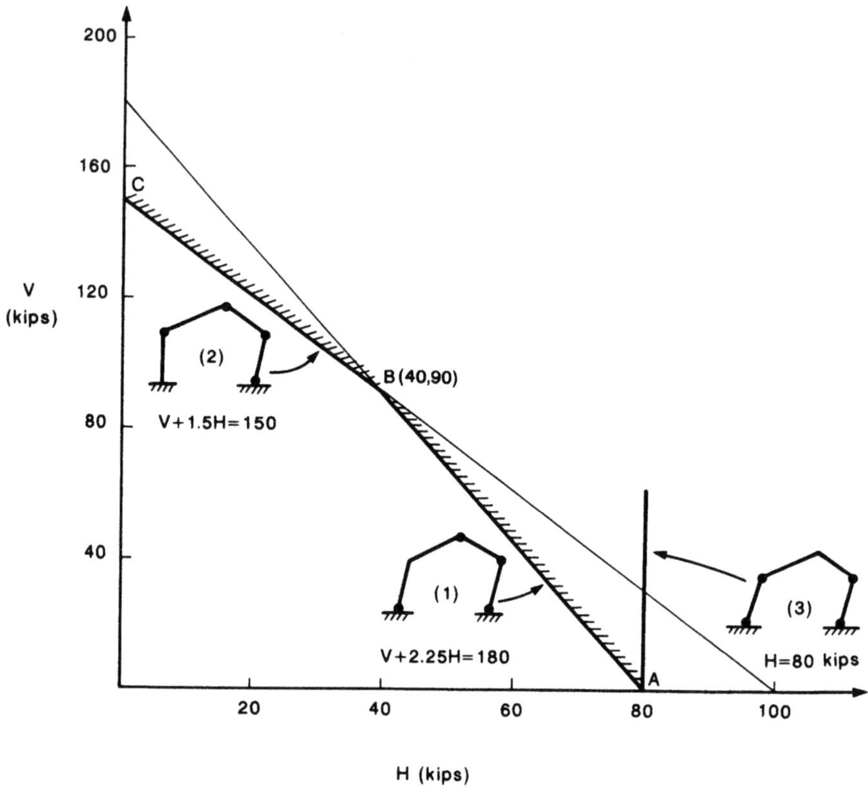

FIGURE 5.18. Interaction diagram for load-carrying capacity of the gable frame in Figure 5.14.

but the vertical load has no vertical movement. The work equation thus has the simple form

$$H(13.5\theta) = M_p(\theta + \theta + \theta + \theta)$$

or

$$H = 80 \text{ kips.} \qquad (5.7.3)$$

The interaction equations corresponding to these three mechanisms are plotted in Fig. 5.18. The positive values of V and H, which just cause collapse of the frame, may be read from the solid shaded line.

By carrying out a moment check for these three side-sway mechanisms, it can be shown that mechanism 1 is valid for the portion AB of the interaction curve ($H \geq 40, V \leq 90$), mechanism 2 is valid for the portion BC ($H \leq 40$, $V \geq 90$), and mechanism 3 is valid only when $V = 0$.

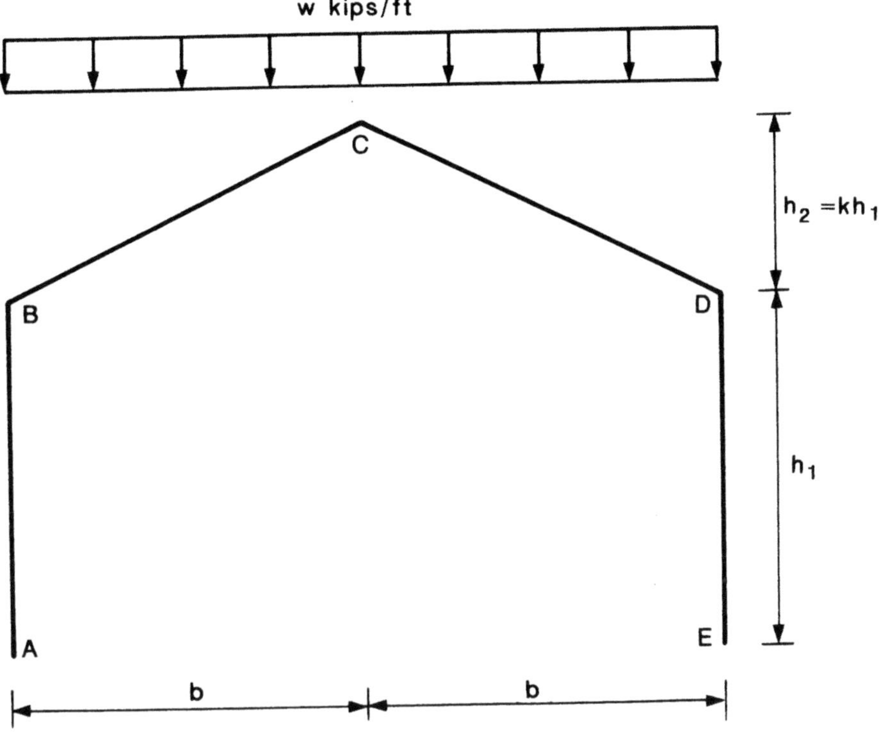

FIGURE 5.19. A gable frame subjected to distributed load with (1) fixed supports and (2) pinned supports (Example 5.7.2).

Example 5.7.2. **Design of a Gable Frame Subject to Uniformly Distributed Load:** A gable frame shown in Fig. 5.19 has a uniform section and is to be designed by simple plastic theory to carry the uniformly distributed load *w*. Two designs are made, one for a frame with pinned feet and the other for fixed feet. Show that the ratio of full plastic moments for the two designs is

$$\left[\frac{1 + \sqrt{(1 + 2k)}}{1 + \sqrt{(1 + k)}}\right]^2$$

Solution: The problem will be solved in three stages. We will first find the required plastic moment for the frame with fixed ends and then solve for the pinned-ended case. The ratio of the two plastic moments gives the desired form.

(a) *Plastic Moment for the Fixed-Ended Case:* Consider the symmetric mechanism shown in Fig. 5.20(a). The plastic hinges are formed at A, B, B_1, D_1, D, and E. Due to symmetry, points B_1 and D_1 move vertically downward. The

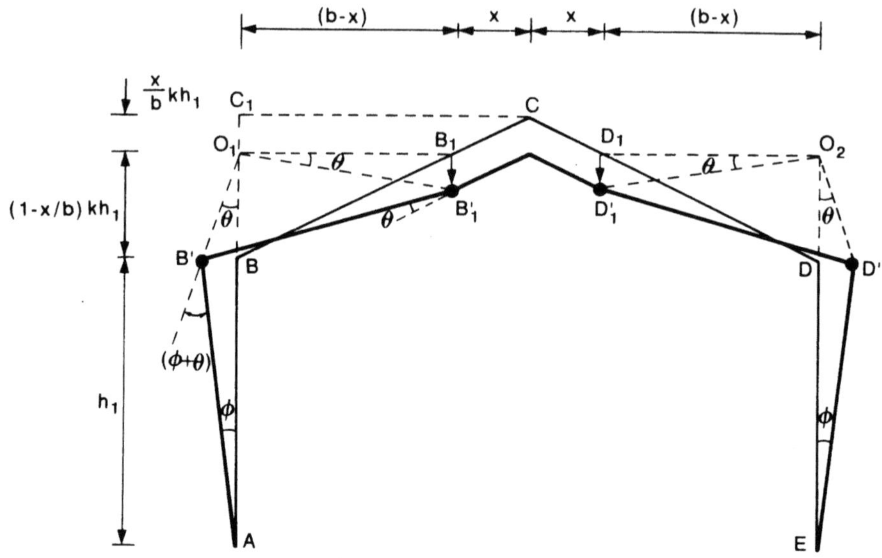

(a) Rotation of hinges

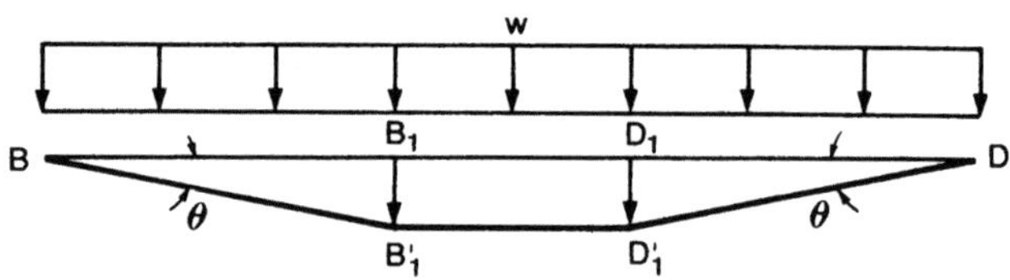

(b) Motion of the load

FIGURE 5.20. A symmetric failure mechanism for the uniformly loaded gable frame in Figure 5.19: (a) rotation of hinges and (b) motion of the load.

instantaneous center O_1 of the segment BB_1 is at the intersection of AB extended and a horizontal line through B_1. Similarly, the instantaneous center for the segment $D_1 D$ is at O_2.

From the similar triangles $CC_1 B$ and $O_1 B_1 B$, the instantaneous center O_1 can be located as

$$\frac{BO_1}{B_1 O_1} = \frac{BC_1}{C_1 C},$$

which gives

$$BO_1 = \frac{BC_1}{C_1 C}(B_1 O_1) = (kh_1)\frac{(b-x)}{b} = (kh_1)\left(1 - \frac{x}{b}\right).$$

Then the length $C_1 O_1$ is obtained as

$$C_1 O_1 = BC_1 - BO_1 = kh_1 - (kh_1)\left(1 - \frac{x}{b}\right) = \frac{x}{b}kh_1.$$

The angles ϕ and θ can be related by considering the triangles ABB' and $O_1 BB'$, which gives

$$\phi = \frac{O_1 B}{AB}\theta = \frac{\left(1 - \frac{x}{b}\right)(kh_1)}{h_1}(\theta) = \left(1 - \frac{x}{b}\right)k\theta.$$

In Fig. 5.20(b), the vertical displacement $B_1 B_1'$ is

$$B_1 B_1' = D_1 D_1' = (b - x)\theta.$$

The total internal work done at all hinge locations is

$$W_I = M_p[\phi + (\phi + \theta) + \theta](2).$$

By expressing ϕ in terms of θ, we have

$$W_I = 4M_p\theta\left[\left(1 - \frac{x}{b}\right)k + 1\right]. \tag{5.7.4}$$

The total external work done by the distributed load is [Fig. 5.20(b)]

$$W_E = \frac{1}{2}w(b - x)B_1 B_1' + w(2x)B_1 B_1' + \frac{1}{2}w(b - x)D_1 D_1'.$$

Substituting $B_1 B_1'$ and $D_1 D_1'$, we have the total external work as

$$W_E = (b - x)[w(b - x) + w(2x)]\theta$$

or

$$W_E = w(b^2 - x^2)\theta. \tag{5.7.5}$$

Equating the total external work (5.7.5) to the total internal energy dissipation (5.7.4), we find the desired plastic moment capacity as

$$M_p = \frac{w(b^2 - x^2)\theta}{4\left[\dfrac{(b - x)k + b}{b}\right]\theta} = \frac{w(b^2 - x^2)b}{4[k(b - x) + b]}. \tag{5.7.6}$$

The value of M_p can be maximized by equating its derivative to zero, i.e.,

$$\frac{dM_p}{dx} = \frac{4[k(b - x) + b](-2wbx) - wb(b^2 - x^2)(-4k)}{\{4[k(b - x) + b]\}^2} = 0,$$

which results in the condition for x:

$$kx^2 - 2b(k + 1)x + kb^2 = 0.$$

Solving for x, we find

$$x_{cr} = \frac{2b(k + 1) \pm \sqrt{4b^2(k + 1)^2 - 4k^2b^2}}{2k}$$

or

$$x_{cr} = b\left[1 + \frac{1}{k} \pm \sqrt{\frac{2}{k} + \frac{1}{k^2}}\right].$$

Since $x_{cr} \leq b$, x_{cr} has the value

$$x_{cr} = b\left(1 + \frac{1}{k} - \sqrt{\frac{2}{k} + \frac{1}{k^2}}\right) = \frac{b}{k}[(k + 1) - \sqrt{2k + 1}].$$

Substituting $x = x_{cr}$ in Eq. (5.7.6), we obtain the maximum plastic moment as

$$M_p = \frac{wb}{4} \frac{b^2 - \dfrac{b^2}{k^2}[(k + 1) - \sqrt{2k + 1}]^2}{\left\{k\left(b - \dfrac{b}{k}[(k + 1) - \sqrt{2k + 1}]\right) + b\right\}} \tag{5.7.7}$$

and a proper simplification leads to the required plastic moment capacity for the frame with fixed ends as

$$M_p = \frac{wb^2}{2k^2}[(k + 1) - \sqrt{2k + 1}]. \tag{5.7.8}$$

(b) *Plastic Moment for the Pinned-Ended Case:* Consider the same mechanism as that of the fixed-ended case. The total internal work W_I for the pinned-ended case is

$$W_I = M_p[(\phi + \theta) + \theta](2).$$

Substituting the value of ϕ in terms of θ, we have

$$W_I = 2M_p\left[\left(1 - \frac{x}{b}\right)k + 2\right]\theta. \tag{5.7.9}$$

The total external work is still the same as that for the fixed-ended case [Eq. (5.7.5)]. By equating the internal and external work, M_p can be expressed in the simple form as

$$M_p = \frac{wb(b^2 - x^2)\theta}{2[k(b - x) + 2b]\theta}. \tag{5.7.10}$$

Again, setting the derivative of M_p to zero, we obtain the condition for the critical locations for the plastic hinges B_1 and D_1, as shown in Fig. 5.20, as

$$\frac{dM_p}{dx} = 0 = \frac{2[kb - kx + 2b]wb(-2x) - wb(b^2 - x^2)(-2k)}{4[k(b - x) + 2b]^2}$$

or

$$kx^2 - 2b(k + 2)x + kb^2 = 0.$$

Solving for x, we find

$$x_{cr} = \frac{2b(k + 2) - \sqrt{4b^2(k + 2)^2 - 4k^2b^2}}{2k}$$

or

$$x_{cr} = \frac{b}{k}[(k + 2) - 2\sqrt{k + 1}]. \tag{5.7.11}$$

Substituting x_{cr} so obtained in Eq. (5.7.10), we have the desired plastic moment capacity as

$$M_p = \left[\frac{wb}{2}\right]\left[\frac{b^2 - \dfrac{b^2}{k^2}[(k + 2) - 2\sqrt{k + 1}]^2}{k\left(b - \dfrac{b}{k}[(k + 2) - 2\sqrt{(k + 1)}]\right) + 2b}\right].$$

Simplifying, the required plastic moment capacity M_p for the frame with pinned ends reduces to

$$M_p = \frac{wb^2}{2k^2}\left[\frac{-4k - 4 + 4(k + 2)\sqrt{k + 1} - 4k - 4}{2\sqrt{k + 1}}\right]$$

or

$$M_p = \frac{wb^2}{k^2}[(k + 2) - 2\sqrt{k + 1}]. \tag{5.7.12}$$

(c) *Ratio of the Two Plastic Moments* (α):

$$\alpha = \frac{M_{p,\text{pinned}}}{M_{p,\text{fixed}}} = \frac{\dfrac{wb^2}{k^2}[(k + 2) - 2\sqrt{k + 1}]}{\dfrac{wb^2}{2k^2}[(k + 1) - \sqrt{2k + 1}]} = \frac{2[(k + 2) - 2\sqrt{k + 1}]}{[(k + 1) - \sqrt{2k + 1}]}$$

or the ratio of the two plastic moments can be written in the simple form:

$$\alpha = \left[\frac{1 + \sqrt{1 + 2k}}{1 + \sqrt{1 + k}}\right]^2. \tag{5.7.13}$$

5.8 The Combination of Mechanisms

The basic concept underlying the method of combining mechanisms is that for a given frame and loading, every possible collapse mechanism can be obtained as some combination of a certain number of independent mechanisms. Once these independent mechanisms have been identified, a work equation is written for each combination and the corresponding collapse load is determined. The lowest load among those obtained by considering all the possible combinations of the independent mechanisms is the correct plastic limit or collapse load. The final confirmation of the validity of the best combination is made by performing a moment check, which may also indicate further adjustments that need to be made.

5.8.1 Number of Independent Mechanisms

If the number of independent mechanisms is known in advance, then the combination could be made in a systematic manner and there would be less likelihood of overlooking a possible combination. The number of possible independent mechanisms n for a frame can be determined from

$$n = N - R \tag{5.8.1}$$

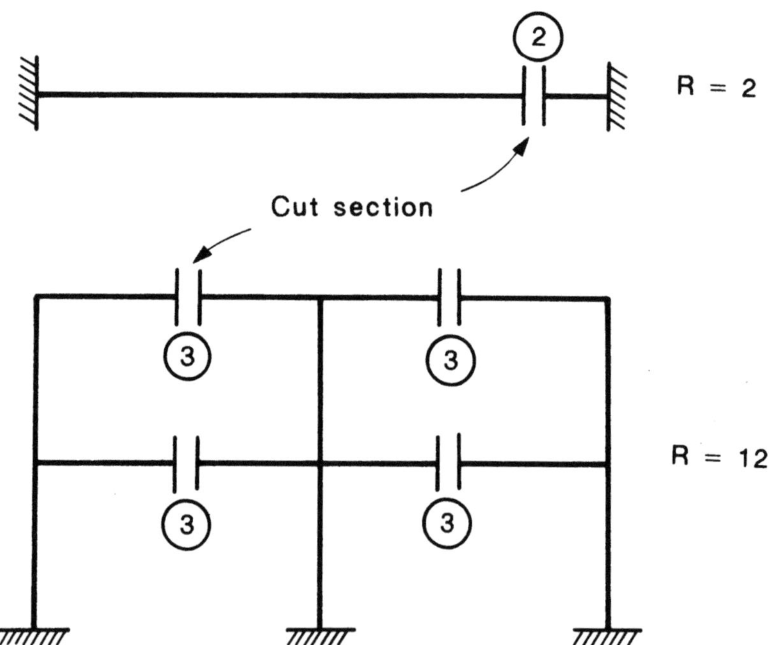

FIGURE 5.21. Determination of the number of redundants.

where N is the number of critical sections at which plastic hinges might form under the particular loading system and R is the degree of redundancy of the structure.

For a frame under concentrated loads, the critical sections will occur at loading points and joints. In order to determine the number of redundants R for a frame, it is necessary to cut sufficient supports and structural members such that all loads are carried out by simple beam or cantilever action. The number of redundants is then equal to the number of forces and moments required to restore continuity. Figure 5.21 shows two examples. The cuts made in each of the structures reduce them to either cantilevers or simply supported elements. The fixed-ended beam requires a shear force and a moment to restore continuity at the cut section, and thus $R = 2$. In the two-story structure, an axial force, a shear force, and a moment are required for continuity at each cut section, and thus $R = 12$.

5.8.2 Types of Mechanism

For convenience of reference to different modes of failure, the following types of mechanisms are shown in Fig. 5.22, using the structure shown in part (a) of this figure.

(a) *Beam mechanism:* For possible beam mechanisms for the structure in Fig. 5.22(a) are shown in Fig. 5.22(b).
(b) *Panel mechanism:* The motion of the mechanism is due to side-sway as shown in Fig. 5.22(c).
(c) *Gable mechanism:* This mechanism involves spreading of column tops with respect to bases as shown in Fig. 5.22(d).
(d) *Joint mechanism:* This mechanism, shown in Fig. 5.22(e), forms at the junction of three or more members and represents motion under the action of an applied moment.
(e) *Combined mechanism:* The combined mechanism may be a partial collapse mechanism as shown in Fig. 5.22(f) or it may be a complete collapse mechanism as shown in Fig. 5.22(g). In the partical collapse mechanism, the frame at failure is still indeterminate in the noncollapsing portion, while in the collapsing portion, it becomes determinate.

5.8.3 Method of Combination

The basic principle of combination is to see whether the independent mechanisms can be combined to form a mechanism that gives an even lower value of collapse load. To this end, the combinations are selected in such a way that the external work becomes a maximum and the internal work becomes a minimum [1.8, 2.3]. In this way, the lowest possible load can be obtained. Thus, the procedure in the combination is generally to involve mechanism motion by as many applied loads as possible and in the meantime to elimi-

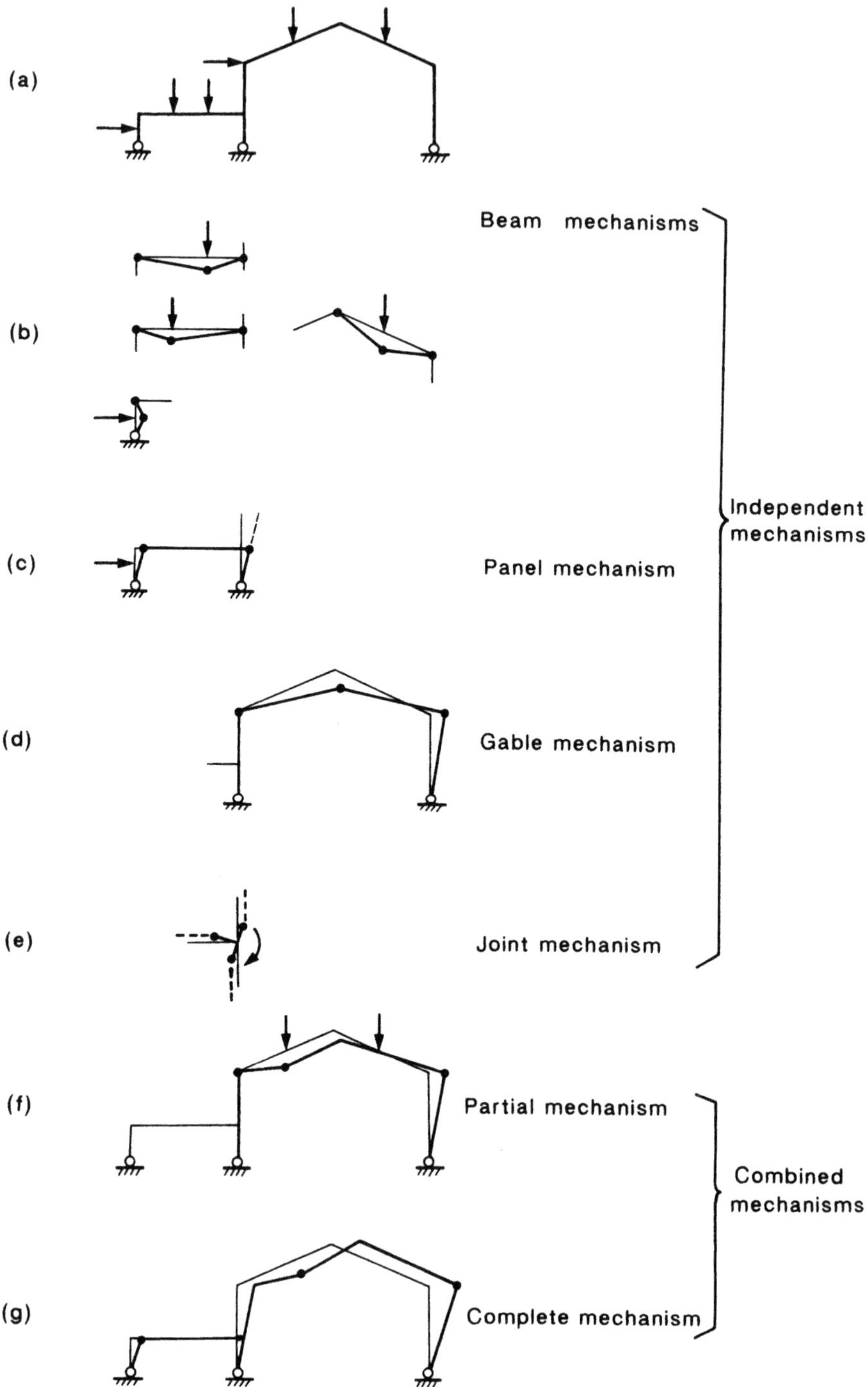

FIGURE 5.22. Types of mechanisms.

nate as many hinges as possible. The method will be explained here with reference to simple rectangular portal frame problems. The analysis and design of multistory and multibay frames subjected to concentrated loads will be presented in the next section. The illustration of a technique for dealing with cases in which the members are subjected to uniformly distributed loads is then followed.

Example 5.8.1. **A pinned-Ended Portal Frame:** Determine the plastic limit load for the pinned-ended frame in Fig. 5.3 by combining the beam and panel mechanisms.

Solution: For the frame in Example 5.3.3, the beam and the panel mechanisms [Fig. 5.3(b) and (c)] have a common hinge at B. Since the rotation at B in the beam mechanism is opposite that in the panel mechanism, the addition of these two mechanisms will lead to a cancellation of the hinge at B. Thus, the plastic limit load of the frame can be determined as follows.

Beam mechanism gives [Fig. 5.3(b)]:

$$\frac{PL}{2}\theta = 4M_p\theta. \tag{5.8.2}$$

Panel mechanism gives [Fig. 5.3(c)]

$$\frac{PL}{4}\theta = 2M_p\theta. \tag{5.8.3}$$

The addition of Eqs. (5.8.2) and (5.8.3) gives:

$$\frac{3}{4}PL\theta = 6M_p\theta. \tag{5.8.4}$$

The cancellation of the hinge at B reduces the internal work by $2M_p\theta$ ($M_p\theta$ from the beam mechanism and another $M_p\theta$ from the panel mechanism). Therefore, the work equation corresponding to the combined mechanism [Fig. 5.3(d)] can be obtained directly from Eq. (5.8.4) by simply reducing $2M_p\theta$ from the right-hand side of this equation

$$\frac{3}{4}PL\theta = 6M_p\theta - 2M_p\theta = 4M_p\theta, \tag{5.8.5}$$

which gives

$$P = \frac{16}{3}\frac{M_p}{L}. \tag{5.8.6}$$

This leads to the same answer as that obtained in Example 5.3.3. The procedure that was just described for deriving the work equation for the combined mechanism is a little shorter for this particular example than the direct derivation from considering the kinematics of the combined mechanism.

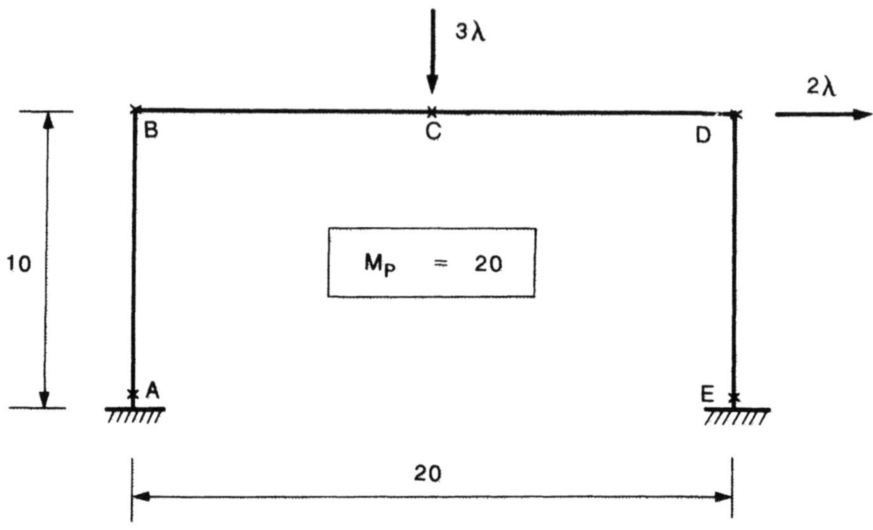

FIGURE 5.23. Fixed-ended portal frame with uniform section (Example 5.8.2).

However, for more complicated problems, this procedure will result in a considerable reduction of computational work. This can be seen clearly in the later examples when more complicated frames are analyzed and designed.

Example 5.8.2. **A fixed-Ended Portal Frame:** A fixed-ended rectangular portal frame has a uniform section with $M_p = 20$ and carries the load as shown in Fig. 5.23. Determine the value of load factor λ at collapse.

Solution: The frame has five critical sections ($N = 5$) and three redundancies ($R = 3$), so the number of independent mechanisms is two ($n = 2$), which will be taken here as those shown in Figs. 5.24(a) and (b). The two independent work equations are therefore

panel mechanism [Fig. 5.24(a)]:

$$20\lambda = 4(20) = 80, \lambda = 4 \qquad (5.8.7)$$

and beam mechanism [Fig. 5.24(b)]:

$$30\lambda = 4(20) = 80, \lambda = 2.67. \qquad (5.8.8)$$

Now the combination of the two independent mechanisms must be examined to see if it will give a value of λ less than 2.67. It can be seen that only one combination of the two mechanisms is possible. This combined mechanism, shown in Fig. 5.24(c), involves the cancellation of the hinge at B.

The calculations leading to the work equation for mode (c) can conveniently be laid out as

FIGURE 5.24. Three possible
mechanisms of collapse for
the frame in Figure 5.23: (a)
panel mechanism, (b) beam
mechanism, and (c) com-
bined mechanism.

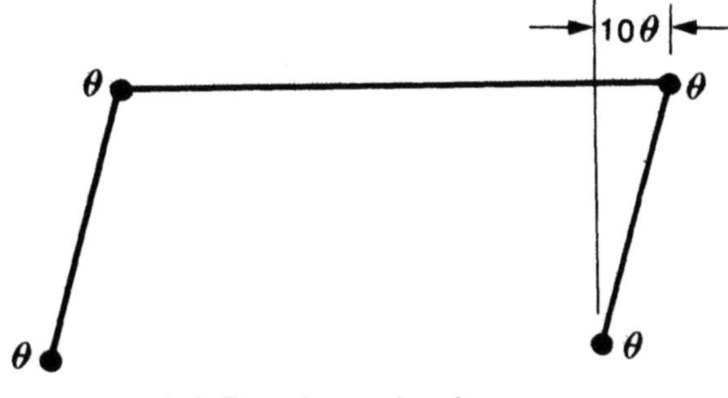

(a) Panel mechanism

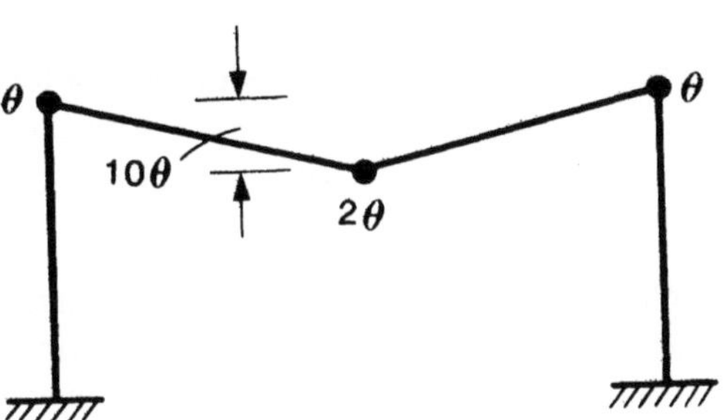

(b) Beam mechanism

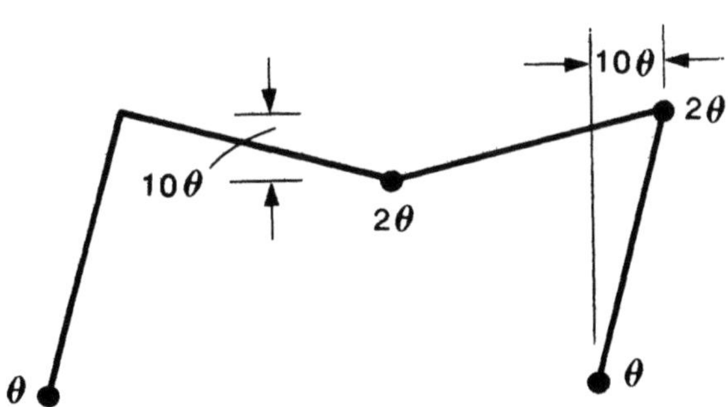

(c) Combined Mechanism

Panel mechanism:	$20\lambda = 4(20)$
Beam mechanism:	$30\lambda = 4(20)$
Addition:	$50\lambda = 8(20)$
Cancel hinge at B:	$-2(20)$
Combined mechanism:	$50\lambda = 6(20) = 120$
	$\lambda = 2.4.$

Note that the cancellation of hinges is the key to the method of combination of mechanisms. If two equations such as (5.8.7) and (5.8.8) are added without any reduction of the right-hand side, a value of λ will result that is between the two original values. Thus, the mechanisms cannot possibly be combined to give a smaller value of λ unless some terms on the right-hand side are cancelled.

Example 5.8.3. **A Portal Frame with Nonuniform Section:** A pinned- and fixed-ended rectangular portal frame is subjected to loads as shown in Fig. 5.25. Determine the load factor λ at collapse using the technique of combining mechanisms.

Solution: Critical sections at which plastic hinges might form are at loading points and joints. The five critical sections for the given frame are marked with crosses in Fig. 5.25 ($N = 5$). The frame has two redundancies ($R = 2$). So there are three independent mechanisms of collapse ($n = 3$). These may be taken as those shown in Fig. 5.26. The work equations corresponding to

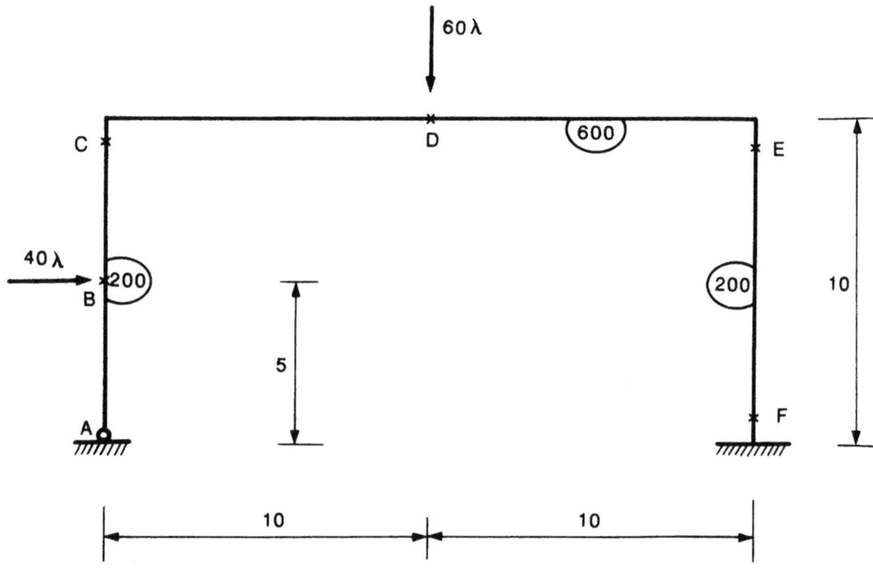

FIGURE 5.25. Rectangular portal frame with nonuniform sections (Example 5.8.3).

FIGURE 5.26. Three independent mechanisms for the frame in Figure 5.25: (a) panel mechansm, (b) column mechanism, and (c) beam mechanism.

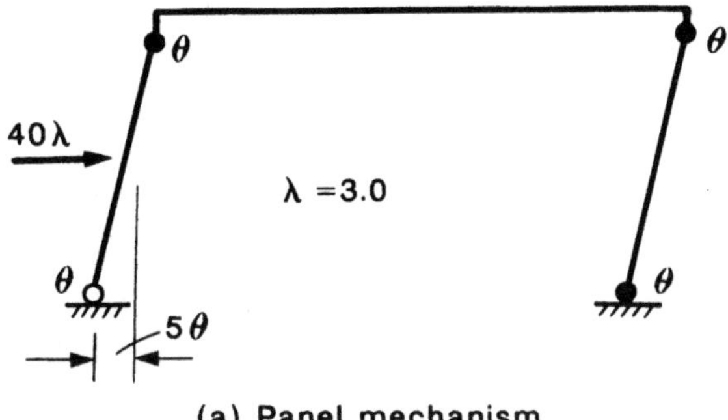

(a) Panel mechanism

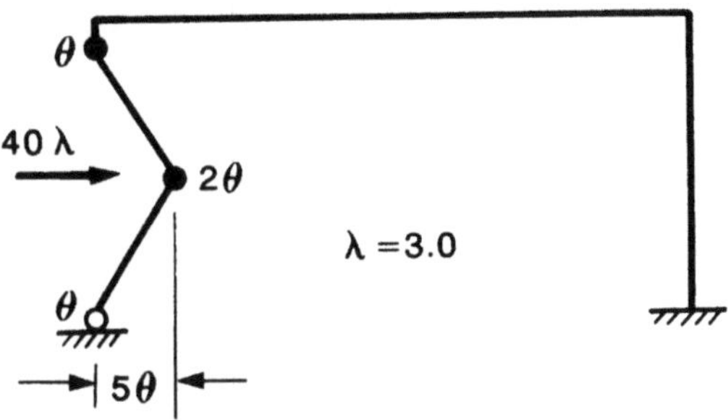

(b) Column mechanism

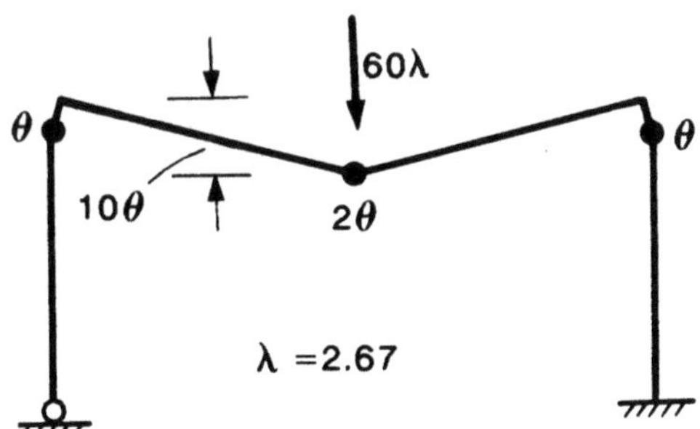

(c) Beam mechanism

these three mechanisms are

Panel mechanism, Fig. 5.26(a): (sides-way)	$200\lambda = 3(200) = 600$ $\lambda = 3$	(5.8.9)
Column mechanism, Fig. 5.26(b): (left-hand column)	$200\lambda = 3(200) = 600$ $\lambda = 3$	(5.8.10)
Beam mechanism, Fig. 5.26(c): (top beam)	$600\lambda = 2(200) + 2(600) = 1,600$ $\lambda = 2.67.$	(5.8.11)

Examination of the three independent mechanisms shows that there are only two critical sections (at the top of each column) at which hinges occur in more than one of the mechanisms. For example, the hinge at the right-hand column foot occurs only in Fig. 5.26(a), and hence cannot be cancelled by combination with any other mechanisms. Only the two hinges at the top of the columns can be eliminated in this way. The hinge at the top of the right-hand column could be eliminated by subtracting the mechanisms of Fig. 5.26(a) and (c). But this would lead to unreasonable mechanisms of the type shown in Fig. 5.27: Fig. 5.27(a) = Fig. 5.26(a) − Fig. 5.26(c), and Fig. 5.27(b) = Fig. 5.26(c) − Fig. 5.26(a).

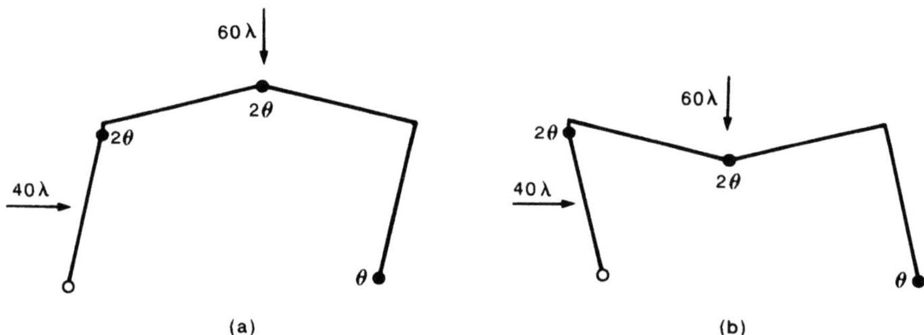

FIGURE 5.27. Unreasonable combined mechanisms from independent mechanisms in Figure 5.26.

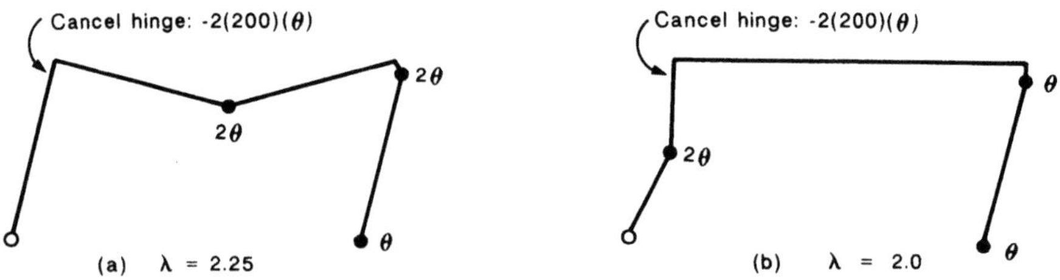

FIGURE 5.28. Reasonable combined mechanisms from independent mechanisms in Figure 5.26.

It seems likely therefore that the only hinge that can possibly be cancelled by any combination of mechanisms is at the top of the left-hand column. The positive rotation $(+\theta)$ at this location of Fig. 5.26(a) can be cancelled either with negative rotation at the same location of Fig. 5.26(c) or with that of Fig. 5.26(b). The two possible combined mechanisms are: Fig. 5.28(a) = Fig. 5.26(a) + Fig. 5.26(c) and Fig. 5.28(b) = Fig. 5.26(a) + Fig. 5.26(b). The calculations leading to work equations for these two modes are laid out in the following:

First Case, Fig. 5.28(a):

Panel mechanism, Fig. 5.26(a): (side-sway)	$200\lambda = 600$
Beam mechanism, Fig. 5.26(c): (top beam)	$600\lambda = 1600$
Addition:	$800\lambda = 2200$
Cancel hinge at C:	$-2(200)$
Combined mechanism, Fig. 5.28(a):	$800\lambda = 1800$ $\lambda = 2.25.$

Second Case, Fig. 5.28(b):

Panel mechanism, Fig. 5.26(a): (side-sway)	$200\lambda = 600$
Column mechanism, Fig. 5.26(b): (left-hand column)	$200\lambda = 600$
Addition:	$400\lambda = 1200$
Cancel hinge at B:	$-2(200)$
Combined mechanism, Fig. 5.28(b):	$400\lambda = 800$ $\lambda = 2.$

By performing a moment check (or a statical analysis) of the combined mechanism of Fig. 5.28(b), which has a lower load factor than that of Fig. 5.28(a), we can show that the moment condition is satisfied everywhere in the structure. Thus, $\lambda = 2$ is the correct answer.

5.9 Multi-Story and Multi-Bay Frames

The basic technique of the method for combining mechanisms was illustrated in the preceding section with reference to a simple one-story and one-bay frame. In this section, we will show its applications to multistory and multi-bay frames with concentrated loads. The analysis of a portal frame to illustrate a technique for dealing with distributed loads will be given in the following section. Applications to more complicated frames such as two-bay pitched-roof portal frames with distributed loads, among other types, will be described in Section. 5.11.

5.9.1 A Two-Story and One-Bay Rectangular Frame

Example 5.9.1. The two-story rectangular frame shown in Fig. 5.29 has a uniform cross section with a full plastic moment capacity of 200 units. Determine the load factor λ at collapse.

Solution: The first step is to decide on the correct number of independent mechanisms. The frame has 6 redundancies and 12 critical sections (marked with crosses in Fig. 5.29). So the number of independent mechanisms is given by the following calculations.

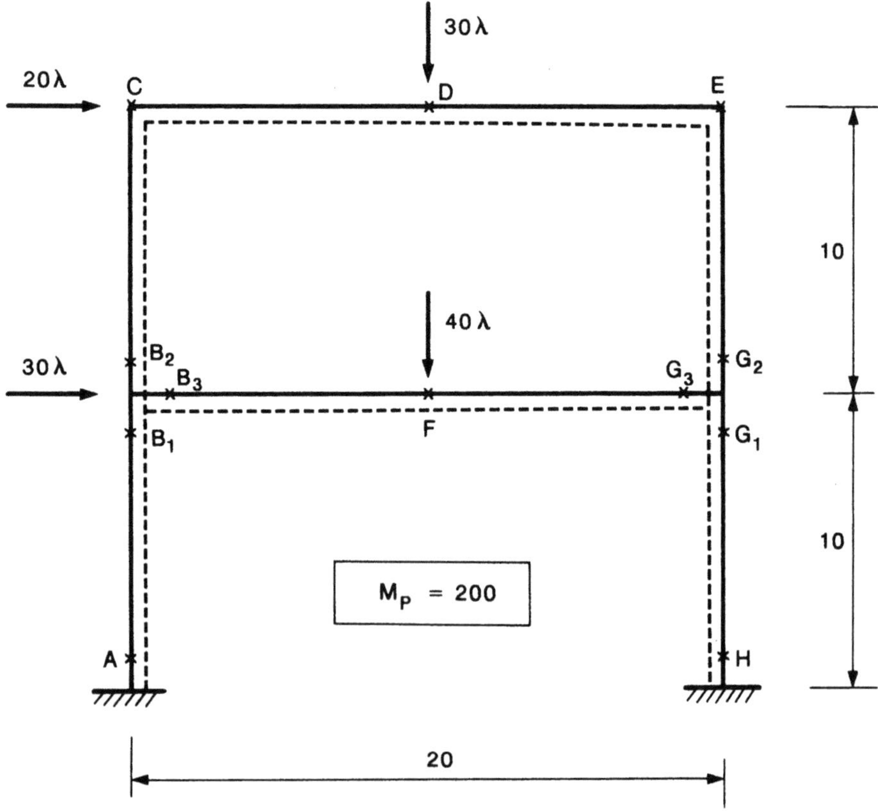

FIGURE 5.29. Two-story and one-bay rectangular frame (Example 5.9.1).

Possible hinge locations:	$N = 12$
(crosses marked in Fig. 5.29)	
Number of redundanices:	$R = 6$
(two cuts, say, at sections D and F)	

Number of independent mechanisms:	$n = 6$
Joint mechanisms at B and G:	$= 2$

Number of true independent mechanisms:	$= 4$

Note that if the two critical sections are taken at each end of the upper beam, then we will have 14 critical sections. However, two extra joint mechanisms are introduced and the number of true independent mechanisms remains four. The four true independent mechanisms are readily identified as shown in Fig. 5.30, consisting of two beam mechanisms and two sway (panel) mechanisms. These four mechanisms are independent, since none of them could be obtained by combining the other three in any way.

Note that for a building frame of m bays and n stories, we will have mn beam mechanisms and n panel (sway) mechainsms. So the total number of true independent mechanisms will be $n(m + 1)$. The work equations corresponding to the independent mechanisms of Fig. 5.30 are as follows:

Beam mechanism, (a):	$30\lambda(10\theta) = M_p[\theta + 2\theta + \theta]$	
(top beam)	$300\lambda = 800, \quad \lambda = 2.67$	(5.9.1)
Beam mechanism, (b):	$40\lambda(10\theta) = M_p(\theta + 2\theta + \theta)$	
(bottom beam)	$400\lambda = 800, \quad \lambda = 2.0$	(5.9.2)
Panel mechanism, (c):	$20\lambda(10\theta) = M_p(\theta + \theta + \theta + \theta)$	
(top story)	$200\lambda = 800, \quad \lambda = 4$	5.9.3
Panel mechanism, (d):	$20\lambda(10\theta) + 30\lambda(10\theta) = M_p(\theta + \theta + \theta + \theta)$	
(bottom story)	$500\lambda = 800, \quad \lambda = 1.6.$	(5.9.4.)

These four work equations, together with the two joint mechanisms, provide the basic equations for obtaining the final solution. The order of combination of mechanisms is, to some extent, arbitrary, although it is common-sense to start with those elementary mechanisms that give the lowest values of λ.

Let us first combine two panel (side-sway) mechanisms (c) and (d). In the resulting mechanism [Fig. 5.31(e)], the clockwise joint rotation at B closes up the two hinges in the columns so that, on this account, $2M_p\theta$ can be subtracted from the right-hand side of the work equations. However, a hinge discontinuity appears in the beam so that a single $M_p\theta$ term reappears in the equations. The net gain at the joint is therefore $M_p\theta$ which is numerically equal to 200θ. The joint rotation mechanism may thus be regarded as meaningless in itself but of significance when it is combined with other mechanisms. To sum up, the total hinge discontinuity of 2θ in columns at joint

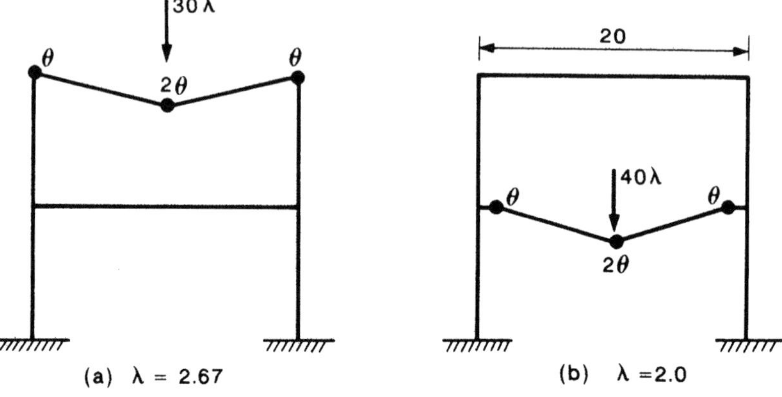

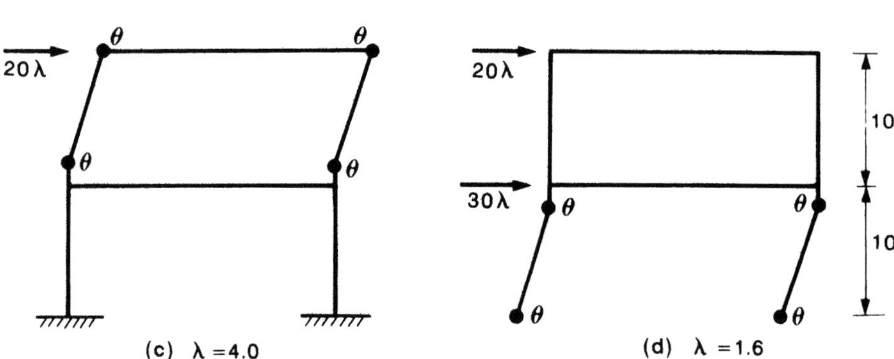

FIGURE 5.30. Four true independent mechanisms for the frame in Figure 5.29.

B is replaced by a discontinuity of θ in beam BG after the joint rotation [Fig. 5.31(e)]. The same reasoning applies to the joint at G. These calculations may be laid out as

Panel mechanism, (c): (top story)	$200\lambda = 800$
Panel mechanism, (d): (bottom story)	$500\lambda = 800$
	$\overline{}$
Addition:	$700\lambda = 1600$
Joint rotation at B and G:	$-2(200)$
	$\overline{}$
Combined Mechanism, (e):	$700\lambda = 1200, \qquad \lambda = 1.714.$

$$(5.9.5)$$

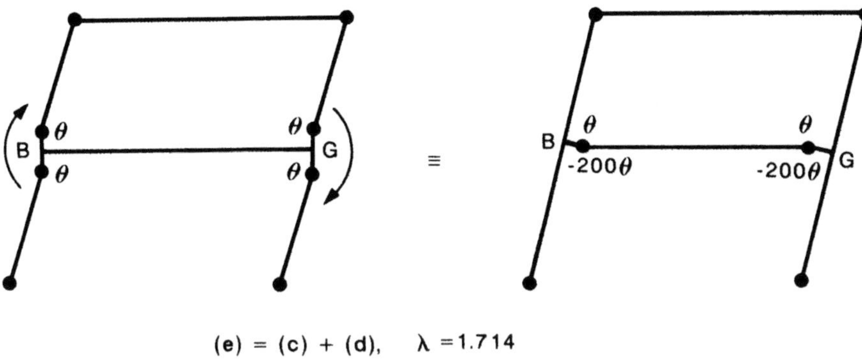

(e) = (c) + (d),   λ =1.714

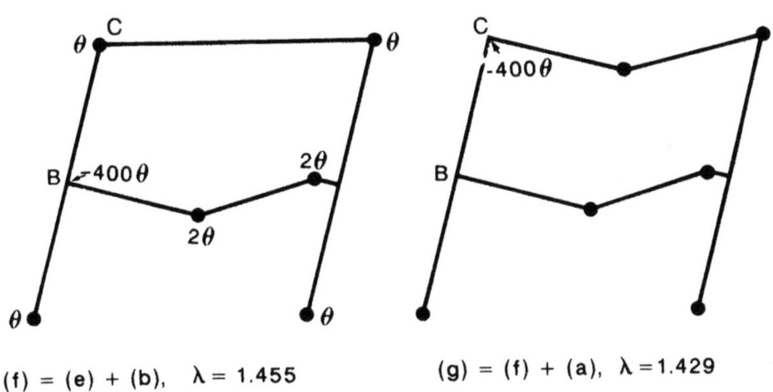

(f) = (e) + (b),   λ = 1.455   (g) = (f) + (a),   λ =1.429

FIGURE 5.31. Combined mechanisms from the independent mechanisms in Figure 5.30.

As already pointed out, no work equation can be written for the joint rotation mechanism. Without the joint rotations at B and G, the straightforward addition of the displacement and hinge rotations of mechanisms (c) and (d) would be pointless, for they have no common hinge whose rotation would be cancelled by the addition and thus reduce the internal work absorbed in the plastic hinges in the combined mechanism.

This solution is worse than that given by the sway of the lower story alone, but it now opens up the possibility of adding in two beam mechanisms, with corresponding cancellation of hinges. The two stages are shown in Fig. 5.31(f) and (g). In each case, the cancelled hinge at the left-hand end of the beam gives rise to a deduction of $2M_p\theta$ numerically equal to 400θ. The calculations for the work equations corresponding to mechanisms (f) and (g) in Fig. 5.31 are laid out in the following:

Combined mechanism, (e):	$700\lambda = 1200$
Beam mechanism, (b):	$400\lambda = 800$
(bottom beam)	

Addition:	$1100\lambda = 2000$
Cancel hinge at B:	-400

Combined mechanism, (f):	$1100\lambda = 1600$
	$\lambda = 1.455$
Beam mechanism, (a):	$300\lambda = 800$
(top beam)	

Addition:	$1400\lambda = 2400$
Cancel hinge at C:	-400

Combined mechanism: (g):	$1400\lambda = 2000$
	$\lambda = 1.429.$

It seems that mechanism (g) is correct, since it gives the lowest value of the load factor. However, to be sure, we need to carry out a moment check for mechanism (g).

Moment Check: Corresponding to mechanism (g), the degree of indeterminacy can be calculated as

Number of plastic hinges in the mechanism:	$M = 6$
Redundancy in the original frame:	$X = 6$
Redundancy at collapse:	$I = X - (M - 1) = 1.$

Assume that the moment at the lower end of column CB at B_2 is the redundant (Fig. 5.29).

Now by using equilibrium, the moments in the frame can be calculated in terms of plastic moment and the redundant moment at B_2. The moment at C is determined by applying the virtual work equation to the equilibrium and geometry sets shown in Fig. 5.32(a). For simplicity, all unknown moments are assumed to be positive in the following computations. The usual sign convention is adopted here for the bending moments.

Top beam CE:

$$M_C(-\theta) + (+200)(+2\theta) + (-200)(-\theta) = (1.429 \times 30)(10\theta),$$

which gives

$$M_C = 600 - 428.7 = 171.3 < 200, \quad \text{okay.} \tag{5.9.6}$$

Similarly, the beam end moment at B_3 is determined by applying the virtual work equation to the equilibrium and geometry sets shown in Fig. 5.32(b).

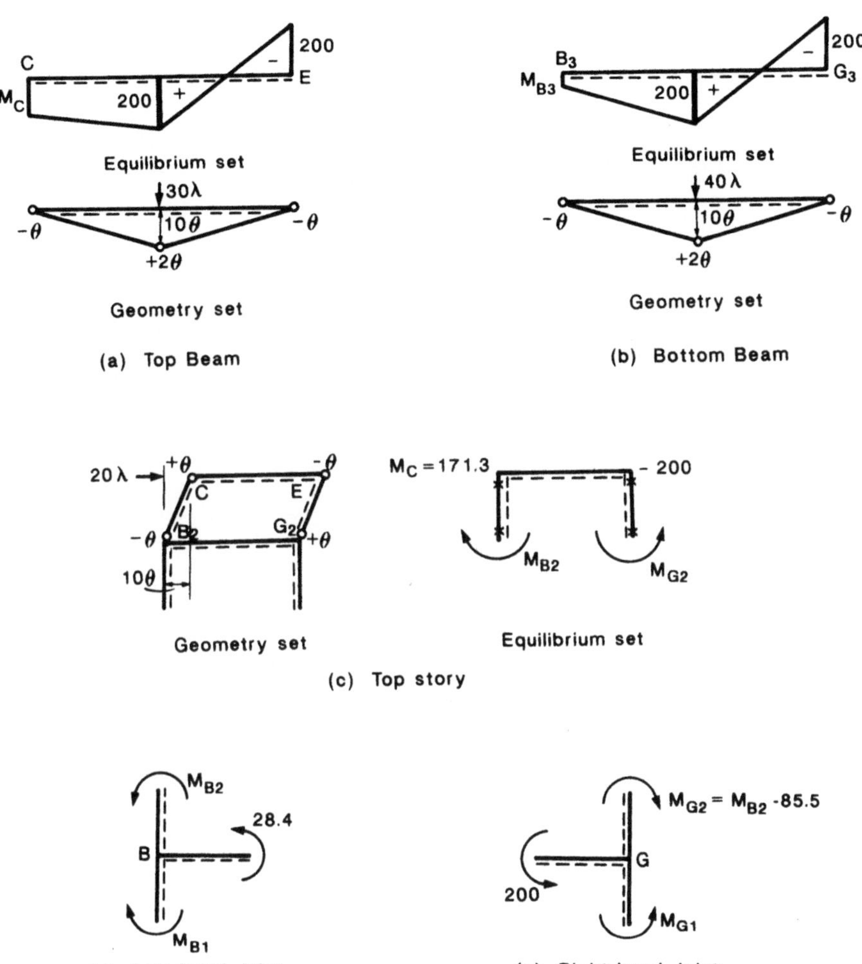

FIGURE 5.32. Moment check for mechanism (g) in Figure 5.31.

Bottom beam BG:

$$M_{B3}(-\theta) + (+200)(+2\theta) + (-200)(-\theta) = (1.429 \times 40)(10\theta),$$

which gives

$$M_{B3} = 600 - 571.6 = 28.4 < 200, \quad \text{okay.} \tag{5.9.7}$$

The column end moment at G_2 is determined by applying the virtual work equation to equilibrium and geometry sets of Fig. 5.32(c).

Top story side-sway:

$$(M_{B2})(-\theta) + (+171.3)(+\theta) + (-200)(-\theta) + (M_{G2})(+\theta) = (20\lambda)(10\theta),$$

which gives

$$M_{G2} = M_{B2} - 171.3 - 200 + 285.5 = M_{B2} - 85.5. \qquad (5.9.8)$$

The column end monent at B_1 is determined by considering the equilibrium of joint B in Fig. 5.32(d).

Left-hand joint B:

$$M_{B1} = M_{B2} + 28.4. \qquad (5.9.9)$$

The column end moment at G_1 is determined by considering the equilibrium of joint G in Fig. 5.32(e).

Right-hand joint G:

$$M_{G1} = M_{B2} - 285.5. \qquad (5.9.10)$$

The complete bending moment diagram for the collapse mechanism in Fig. 5.31(g), expressed in terms of the redundant moment M_{B2}, is shown in Fig. 5.33. The condition that the absolute moment at B_2, G_2, B_1, and G_1 be less than $M_p = 200$ leads to

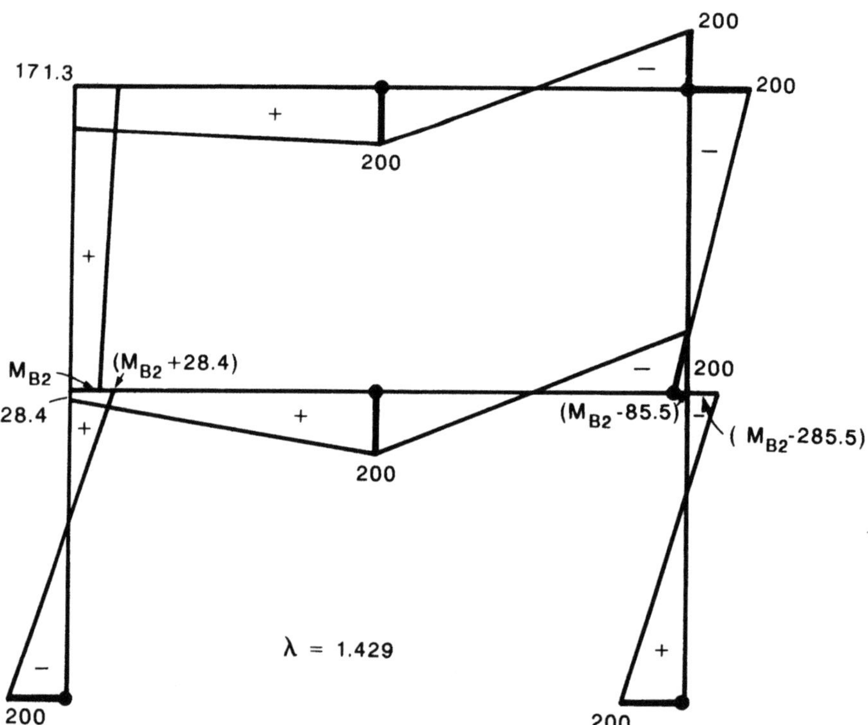

FIGURE 5.33. Moment diagram for mechanism (g) in Figure 5.31.

$$-200 \leq M_{B2} \leq 200 \tag{5.9.11}$$

$$-200 \leq M_{B2} - 85.5 \leq 200 \tag{5.9.12}$$

$$-200 \leq M_{B2} + 28.4 \leq 200 \tag{5.9.13}$$

$$-200 \leq M_{B2} - 285.5 \leq 200. \tag{5.9.14}$$

Inequalities (5.9.11) to (5.9.14) can be expressed as

$$-200 \leq M_{B2} \leq 200 \tag{5.9.15}$$

$$-114.5 \leq N_{B2} \leq 285.5 \tag{5.9.16}$$

$$-228.4 \leq M_{B2} \leq 171.6 \tag{5.9.17}$$

$$85.5 \leq M_{B2} \leq 485.5. \tag{5.9.18}$$

Inequalities (5.9.15) to (5.9.18) are equivalent to

$$85.5 \leq M_{B2} \leq 171.6, \tag{5.9.19}$$

which indicates that there exists an M_{B2} value corresponding to which the absolute moments at B_2, G_2, B_1, and G_1 are less than $M_p = 200$ so that the moment check is complete. Thus, $\lambda = 1.429$ is exact.

5.9.2 A Two-Bay Two-Story Rectangular Frame

Example 5.9.2. The two-bay two-story frame shown in Fig. 5.34 has full-strength connections and carries the loads as shown. The full plastic moments are marked adjacent to the members. Determine the collapse load factor λ according to simple plastic theory.

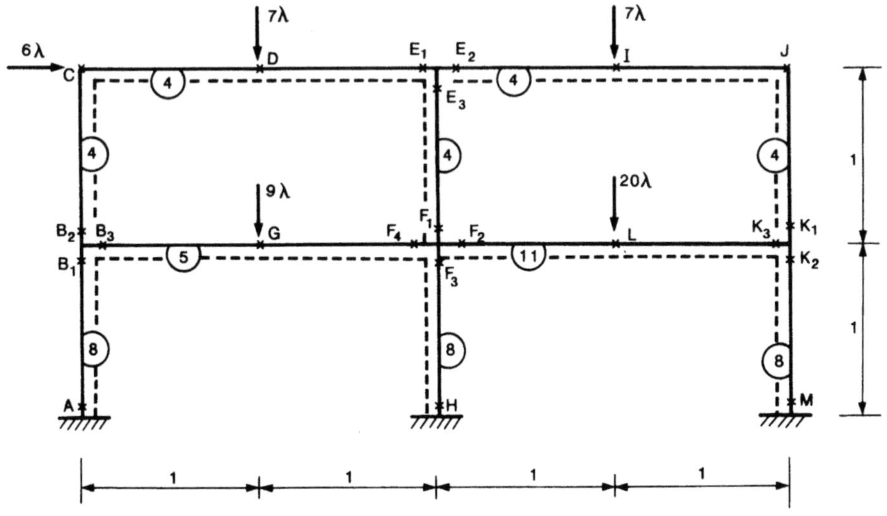

FIGURE 5.34. Two-story two-bay rectangular frame (Example 5.9.2).

Solution: The number of independent mechanisms can be calculated as

Possible hinge locations:	$N = 22$
(crosses marked in Fig. 5.34)	
Redundancies	$R = 12$
(4 cuts, say, at sections D, G, I, L)	

Independent mechanisms:	$n = 10$
Joint mechanisms:	$= 4$
(joint B, E, F, K)	

True independent mechanisms:	$= 6$

The six independent mechanisms are shown in Fig. 5.35 consisting of four beam mechanisms and two side-sway mechanisms. The work equations corresponding to these independent mechanisms are as follows:

Beam mechanism, (a):	$7\lambda = 16,$	$\lambda = 2.286$	(5.9.20)
(top left-hand)			
Beam mechanism, (b):	$7\lambda = 16,$	$\lambda = 2.286$	(5.9.21)
(top right-hand)			
Beam mechanism, (c):	$9\lambda = 20,$	$\lambda = 2.222$	(5.9.22)
(bottom left-hand)			
Beam mechanism, (d):	$20\lambda = 44,$	$\lambda = 2.20$	(5.9.23)
(bottom right-hand)			
Panel mechanism, (e):	$6\lambda = 24,$	$\lambda = 4$	(5.9.24)
(top story)			
Panel mechanism, (f):	$6\lambda = 48,$	$\lambda = 8$	(5.9.25)
(bottom story)			

Combining the top story side-sway mechanism (e) with the four beam mechanisms (a), (b), (c), and (d) [Fig. 5.36], we have

Panel mechanism, (e):	$6\lambda = 24$		
(top story)			
Beam mechanism, (a):	$7\lambda = 16$		
(top left-hand)			
	——————		
Addition:	$13\lambda = 40$		
Cancel hinge at C:	-8		
Combined mechanism, (g):	$13\lambda = 32,$	$\lambda = 2.462$	(5.9.26)
Beam mechanism, (b)	$7\lambda = 16$		
(top right-hand)			
	——————		
Addition:	$20\lambda = 48$		
Rotate joint E:	-4		

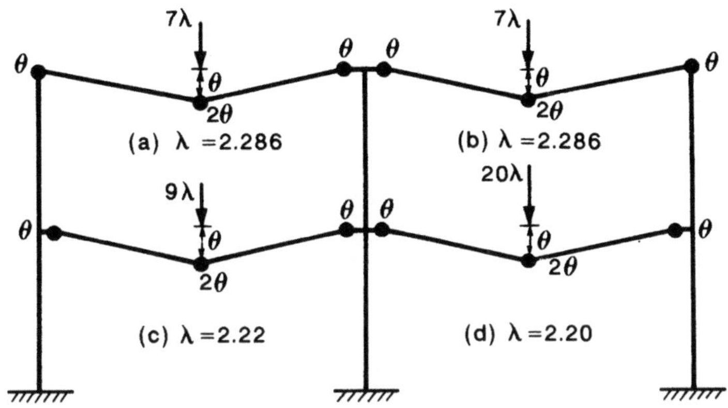

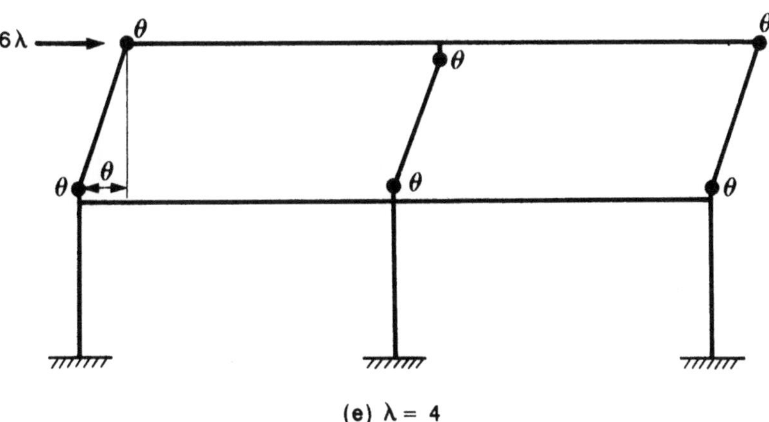

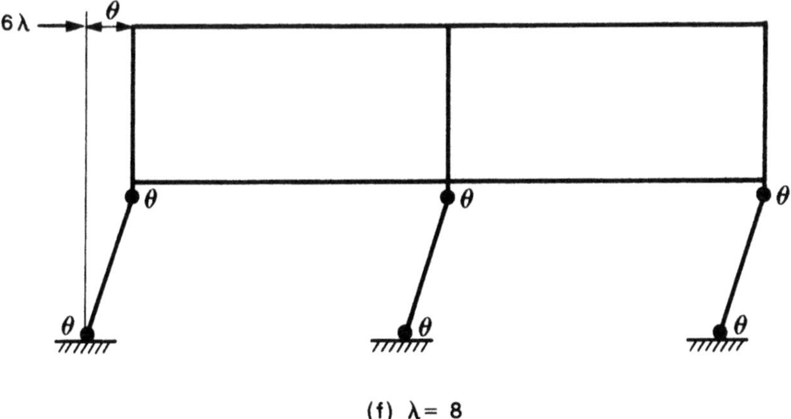

FIGURE 5.35. Six true independent mechanisms for the frame in Figure 5.34.

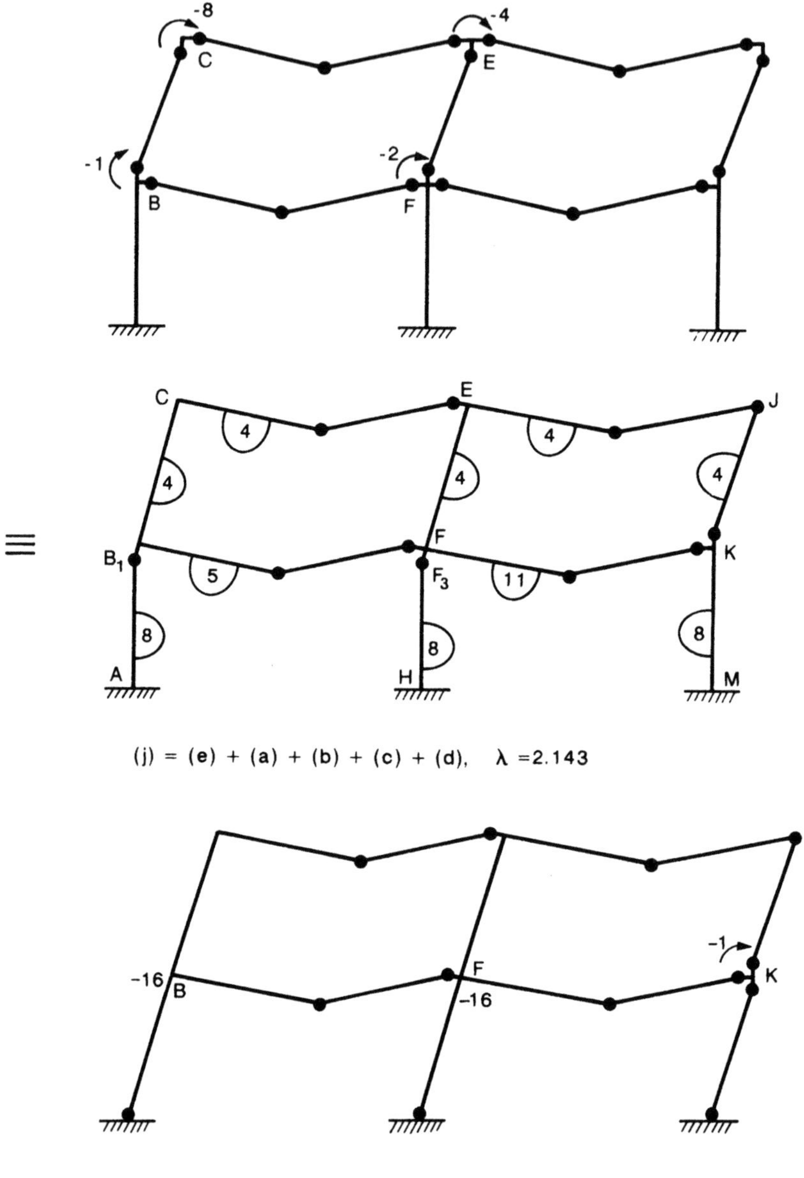

$$(j) = (e) + (a) + (b) + (c) + (d), \quad \lambda = 2.143$$

$$(k) = (j) + (f), \quad \lambda = 2.182$$

FIGURE 5.36. Combined mechanisms from the independent mechanisms in Figure 5.35.

Combined mechanism, (h):	$20\lambda = 44,$	$\lambda = 2.2$	(5.9.27)
Beam mechanism, (c):	$9\lambda = 20$		
(bottom left-hand)			

Addition:	$29\lambda = 64$
Rotate joint B:	-1

Combined mechanism, (i):	$29\lambda = 63,$	$\lambda = 2.172$	(5.9.28)
Beam mechanism, (d):	$20\lambda = 44$		
(bottom right-hand)			

Addition:	$49\lambda = 107$
Rotate joint F:	-2

Combined mechanism, (j):	$49\lambda = 105,$	$\lambda = \dfrac{15}{7} = 2.143.$	(5.9.29)

Now, we shall check the combination of mechanism (j) and panel mechanism (f) to show that mechanism (j) in Fig. 5.36 gives the lowest λ value.

Combined mechanism, (j):	$49\lambda = 105$
Panel mechanism (f):	$6\lambda = 48$
(bottom story)	

Addition:	$55\lambda = 153$
Cancel hinge at B_1:	-16
Cancel hinge at F_3:	-16
Rotate joint K:	-1

Combined mechanism, (k):	$55\lambda = 120,$	$\lambda = 2.182.$	(5.9.30)

Despite the hinge cancellations achieved at joints B_1 and F_3 and by the joint rotation at K this value of λ still exceeds the values of 2.143 obtained in Eq. (5.9.29). It can therefore be concluded that the lowest value of λ (2.143) is corresponding to mechanism (j) in Fig. 5.36. To confirm that $\lambda = 2.143$ is the actual solution, we need to make a moment check for mechanism (j).

Moment Check: The best method for obtaining the moment equilibrium equations for this type of frame is to apply the virtual work equation. The indeterminacy corresponding to mechanism (j) can be calculated as

Number of plastic hinges in the mechanism:	$M = 11$
Redundancy in the original structure:	$X = 12$
Redundancy at collapse:	

$$I = X - (M - 1) = 12 - (11 - 1) = 2$$

Let the fixed-ended moments M_A and M_H be the redundant moments. All other moments can be expressed in terms of these two redundants by use of the virtual work equation and the equilibrium of joints. For simplicity, all known moments are assumed to be positive in the following computations. Due to symmetry, the beam end moments M_C and M_{E2} are equal and therefore can be determined by considering either of the two upper beams. From Fig. 5.37(a), with $\lambda = 2.143 = 15/7$, we have

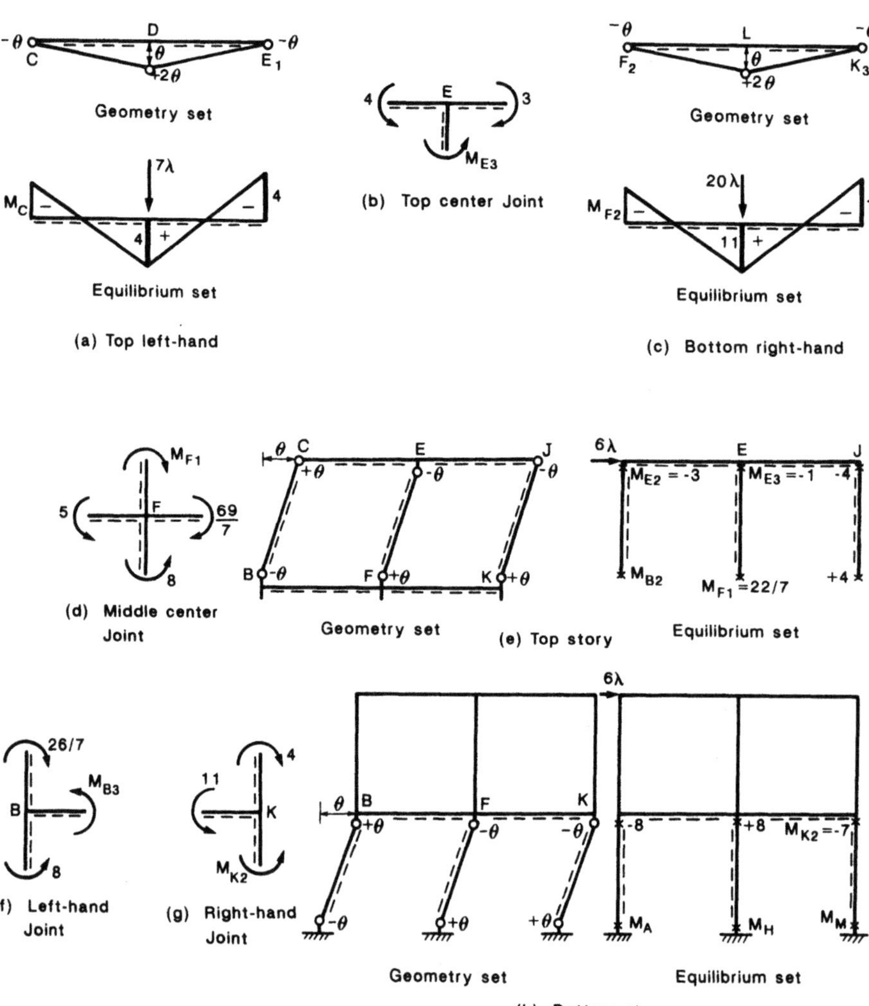

FIGURE 5.37. Moment check for mechanism (j) in Figure 5.36: (a) top left-hand beam, (b) top center joint, (c) bottom right-hand beam, (d) middle center joint, (e) top story, (f) left-hand joint, (g) right-hand joint, and (h) bottom story.

top left-hand beam CE:

$$(M_C)(-\theta) + (+4)(+2\theta) + (-4)(-\theta) = 7\lambda\theta = 15\theta,$$

which gives

$$M_C = M_{E2} = -15 + 12 = -3, \qquad |M_C| \le 4, \quad \text{okay.}$$

The column end moment M_{E3} is determined from equilibrium of joint E in Fig. 5.37(b):

top central joint E:

$$M_{E3} = -4 + 3 = -1, \qquad |M_{E3}| < 4, \quad \text{okay.} \tag{5.9.31}$$

The beam end moment M_{F2} is determined by applying the virtual work equation to beam $F_2 K_3$ as shown in Fig. 5.37(c),

bottom right-hand beam FK:

$$(M_{F2})(-\theta) + (+11)(+2\theta) + (-11)(-\theta) = (20\lambda)(\theta) = \frac{300}{7}\theta,$$

which gives

$$M_{F2} = -\frac{300}{7} + 33 = -\frac{69}{7}, \qquad |M_{F2}| < 11, \quad \text{okay.} \tag{5.9.32}$$

Now, the column end moment M_{F1} can be determined by considering the equilibrium of joint F in Fig. 5.37(d):

middle central joint F:

$$M_{F1} = 5 + 8 - \frac{69}{7} = \frac{22}{7} < 4, \quad \text{okay.} \tag{5.9.33}$$

The column end moment M_{B2} is determined by applying the virtual work equation to panel mechanism in Fig. 5.37(e):

top story side-sway:

$$M_{B2}(-\theta) + (-3)(+\theta) + (-1)(-\theta) + (-4)(-\theta) + \left(\frac{22}{7}\right)(+\theta) + (+4)(+\theta)$$

$$= (6\lambda)(\theta),$$

which gives

$$M_{B2} = -\frac{90}{7} - 3 + 9 + \frac{22}{7} = -\frac{26}{7}, \qquad |M_{B2}| < 4, \quad \text{okay.} \tag{5.9.34}$$

The beam end moment M_{B3} is determined by considering equilibrium of joint B in Fig. 5.37(f):

left-hand joint B:

$$M_{B3} = -8 + \frac{26}{7} = -\frac{30}{7}, \qquad |M_{B3}| < 5, \quad \text{okay.} \qquad (5.9.35)$$

The column end moment M_{K2} is determined by considering equilibrium of joint K in Fig. 5.37(g):

Right-hand joint K:

$$M_{K2} = -11 + 4 = -7, \qquad |M_{K2}| < 8, \quad \text{okay.} \qquad (5.9.36)$$

Now the column end moment M_M can be expressed in terms of the redundants M_A and M_H by considering the panel mechanism of the first story in Fig. 5.37(h):

bottom story side-sway:

$$M_A(-\theta) + M_H(+\theta) + M_M(+\theta) + (-8)(+\theta) + (+8)(-\theta) + (-7)(-\theta)$$
$$= (6\lambda)(\theta),$$

which gives

$$M_A - M_H - M_M = -\frac{90}{7} - 9 = -\frac{153}{7}. \qquad (5.9.37)$$

The moment check will be complete if we can show that there exists a set of redundants M_A and M_H values such that the value of M_M as given in (5.9.37) will satisfy the moment condition $M_M \leq 8$. This can be demonstrated by choosing, say, $M_A = -8$ and $M_H = +8$, which gives

$$M_M = \frac{153}{7} - 16 = \frac{41}{7} < 8 \quad \text{okay.} \qquad (5.9.38)$$

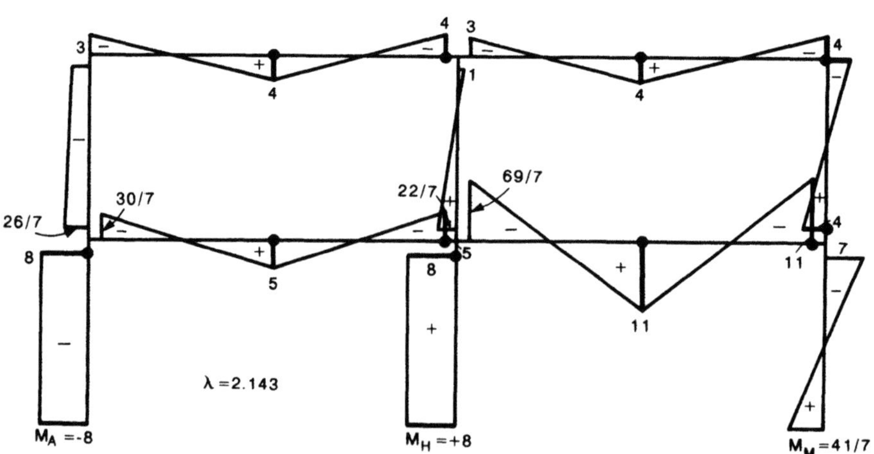

FIGURE 5.38. Moment diagram for mechanism (j) in Figure 5.36.

A complete bending moment diagram for mechanism (j) of Fig. 5.36 with $M_A = -8$ and $M_H = +8$ is shown in Fig. 5.38. The moment condition is not violated anywhere in the frame. The load factor $\lambda = 2.143$ is thus exact. Note that there are many other possible safe and statically admissible bending moment distributions corresponding to this partial collapse mechanism, but as shown here it is only necessary to establish the existence of one such distribution to verify the solution.

5.9.3 A Three-Story Two-Bay Rectangular Frame

Example 5.9.3. Find the load factor at collapse for the frame shown in Fig. 5.39. Note that the members have different plastic moment capacities.

Solution: The number of independent mechanisms are calculated as

Possible hinge locations:	$N = 27$
(crosses marked in Fig. 5.39)	
Redundancies:	$R = 12$
(4 cuts, say, at sections E, G, I, L)	
Independent mechanisms:	$= 15$
Joint mechanisms:	$= 7$
(joints C, D, F, H, J, K, M)	
True independent mechanisms:	$= 8.$

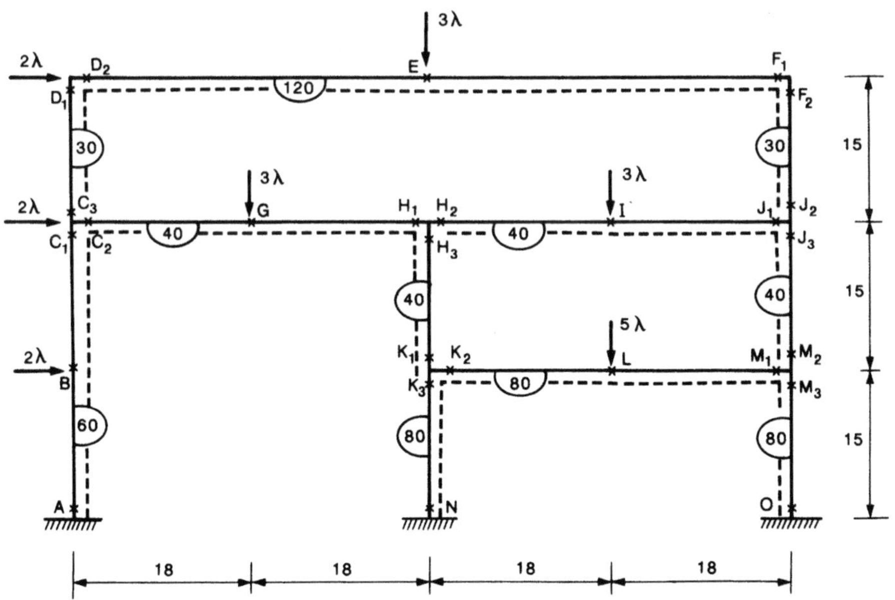

FIGURE 5.39. Three-story two-bay rectangular frame (Example 5.9.3).

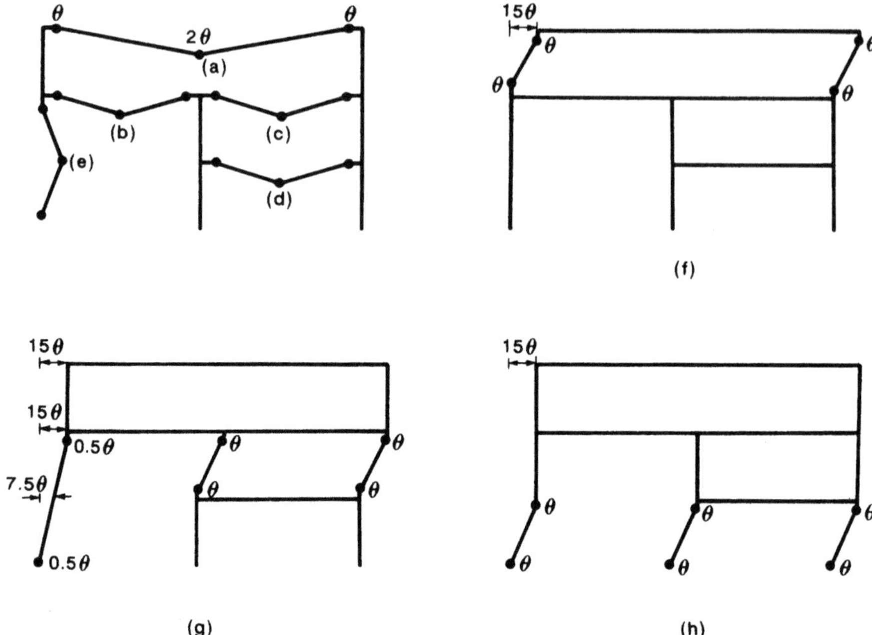

FIGURE 5.40. Eight true independent mechanisms for the frame in Figure 5.39.

The eight independent mechanisms are shown in Fig. 5.40, consisting of five beam-type mechanisms and three side-sway mechanisms. The work equations corresponding to these independent mechanisms are summarized as follows:

Mechanisms (a) to (e) are beam type:

(a) Top beam:
$(3\lambda)(36\theta) = (120)(4\theta)$
$108\lambda\theta = 480\theta,$ $\qquad\qquad\qquad\qquad \lambda = 4.444.$ (5.9.39)

(b) Bottom left-hand beam:
$(3\lambda)(18\theta) = (40)(4\theta)$
$54\lambda\theta = 160\theta,$ $\qquad\qquad\qquad\qquad \lambda = 2.963.$ (5.9.40)

(c) Middle right-hand beam:
$(3\lambda)(18\theta) = (40)(4\theta)$
$54\lambda\theta = 160\theta,$ $\qquad\qquad\qquad\qquad \lambda = 2.963.$ (5.9.41)

(d) Bottom right-hand beam:
$(5\lambda)(18\theta) = (80)(4\theta)$
$90\lambda\theta = 320\theta,$ $\qquad\qquad\qquad\qquad \lambda = 3.556.$ (5.9.42)

(e) Left-hand column:
$(2\lambda)(15\theta) = (60)(4\theta)$
$30\lambda\theta = 240\theta,$ $\qquad\qquad\qquad\qquad\qquad \lambda = 8.$ $\qquad$ (5.9.43)

Mechanisms (f) *to* (h) *are side-sway type:*

(f) Top story:
$(2\lambda)(15\theta) = (30)(4\theta)$
$30\lambda\theta = 120\theta,$ $\qquad\qquad\qquad\qquad\qquad \lambda = 4.$ $\qquad$ (5.9.44)

(g) Middle story:
$(2\lambda)(15\theta + 15\theta + 7.5\theta = (60)(0.5\theta)(2) + (40)(4\theta)$
$75\lambda\theta = 220\theta,$ $\qquad\qquad\qquad\qquad \lambda = 2.933.$ $\quad$ (5.9.45)

(h) Bottom story:
$(2\lambda)(15\theta)(3) = (60)(2\theta) + (80)(4\theta)$
$90\lambda\theta = 440\theta,$ $\qquad\qquad\qquad\qquad \lambda = 4.889.$ $\quad$ (5.9.46)

Combining these independent mechanisms and looking for a minimum value of λ, we have (Fig. 5.41):

The combination: (i) = 0.5(f) + (g)

0.5(f) top side-sway:	$30\lambda(0.5\theta) = 120(0.5\theta)$
(g) middle side-sway:	$75\lambda\theta = 220\theta$
	$90\lambda\theta = 280\theta$
Rotate joint C by 0.5θ:	-25θ
Rotate joint J by 0.5θ:	-15θ
(i) combined:	$90\lambda\theta = 240\theta,$ $\qquad \lambda = 2.667.$

$$(5.9.47)$$

The combination: (j) = (i) + 0.5(a)

(i) combined:	$90\lambda\theta = 240\theta$
0.5(a) top beam DF:	$108\lambda(0.5\theta) = 480(0.5\theta)$
	$144\lambda\theta = 480\theta$
Rotate joint D by 0.5θ:	-75θ
Rotate joint F by 0.5θ:	-45θ
(j) combined:	$144\lambda\theta = 360\theta,$ $\qquad \lambda = 2.5.$

$$(5.4.48)$$

The combination: (k) = (j) + 0.5(b)

(j) combined:	$144\lambda\theta = 360\theta$
0.5(b) bottom left-hand beam CH:	$54\lambda(0.5\theta) = 160(0.5\theta)$
	$171\lambda\theta = 440\theta$
Cancel hinge at C_2:	-40θ
(k) combined:	$171\lambda\theta = 400\theta,$ $\qquad \lambda = 2.339.$

$$(5.9.49)$$

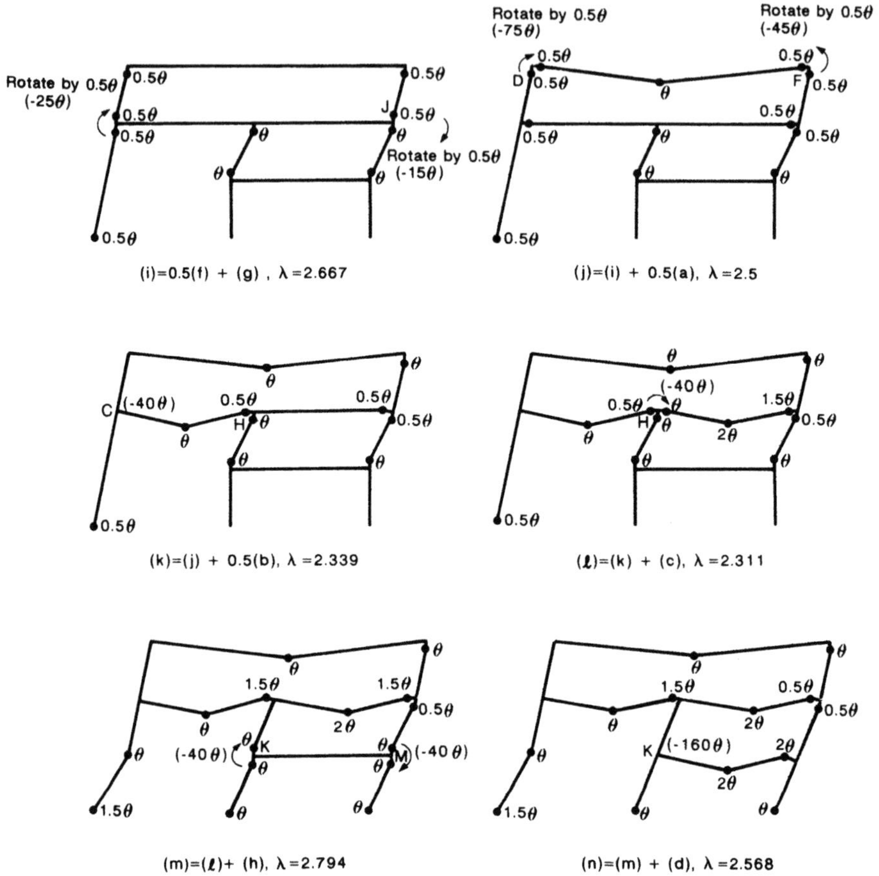

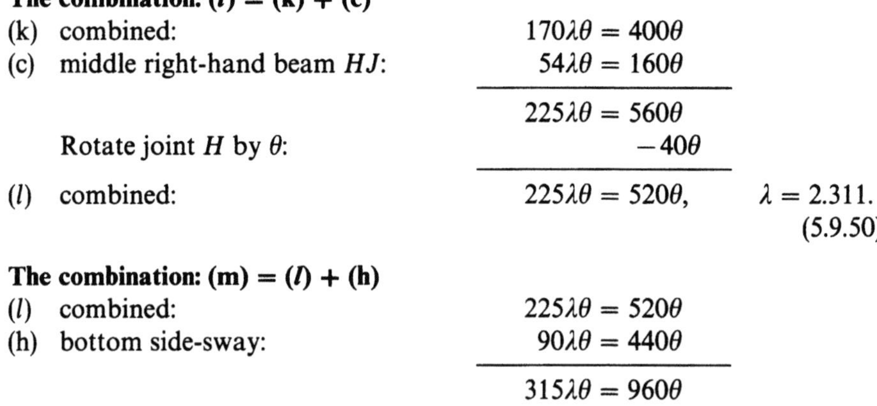

FIGURE 5.41. Combined mechanisms from the independent mechanisms in Figure 5.40.

The combination: (*l*) = (k) + (c)

(k) combined:	$170\lambda\theta = 400\theta$	
(c) middle right-hand beam *HJ*:	$54\lambda\theta = 160\theta$	
	$225\lambda\theta = 560\theta$	
Rotate joint *H* by θ:	-40θ	
(*l*) combined:	$225\lambda\theta = 520\theta,$	$\lambda = 2.311.$

$$(5.9.50)$$

The combination: (m) = (*l*) + (h)

(*l*) combined:	$225\lambda\theta = 520\theta$
(h) bottom side-sway:	$90\lambda\theta = 440\theta$
	$315\lambda\theta = 960\theta$

Rotate joint K by θ:	-40θ	
Rotate joint M by θ:	-40θ	
(m) combined:	$315\lambda\theta = 880\theta,$	$\lambda = 2.794.$

$$(5.9.51)$$

The combination: (n) = (m) + (d)

(m) combined:	$315\lambda\theta = 880\theta$	
(d) bottom right-hand beam KM:	$90\lambda\theta = 320\theta$	

$$405\lambda\theta = 1200\theta$$

Cancel hinge at K_2:	-160θ

(n) combined:	$450\lambda\theta = 1040\theta,$	$\lambda = 2.568.$

$$(5.9.52)$$

The lowest value of λ (2.311) is corresponding to mechanism (l) as given by Eq. (5.9.50). To confirm that $\lambda = 2.311$ is exact, we shall make a moment check for mechanism (l) in Fig. 5.41.

Moment Check: The indeterminacy corresponding to mechanism (l) can be calculated as

Number of plastic hinges in the mechanism:	$M = 10$
Redundancy in the original structure:	$X = 12$
Redundancy at collapse:	$I = X - (M - 1)$
	$= 12 - (10 - 1) = 3.$

Let the column end moment M_{M3} and the fixed-ended moments M_N and M_O be the three chosen redundant moments. All other unknown moments in Fig. 5.42(a), all assumed to be positive, can be expressed in terms of these redundants by the virtual work equation and the equilibrium of joints. Details of these calculations are given in the forthcoming (Fig. 5.42).

The Top Beam Moments $M_{D2} = M_{D1}$: From Fig. 5.42(b), we have

top beam:

$$(3\lambda)(36\theta) = M_{D2}(-\theta) + (+120)(+2\theta) + (-30)(-\theta)$$

$$M_{D2} = -108\lambda + 270 = 20.41, |M_{D2}| \leq 120 \text{ and } |M_{D1}| < 30, \quad \text{okay.}$$

$$(5.9.53)$$

The Middle Beam Moments $M_{C2} = M_{H2}$: From Fig. 5.42(c), we have

bottom left-hand beam:

$$(3\lambda)(18\theta) = M_{C2}(-\theta) + (+40)(+2\theta) + (-40)(-\theta)$$

$$M_{C2} = 120 - 54\lambda = -4.79, |M_{C2}| \text{ and } |M_{H2}| < 40, \quad \text{okay.}$$

$$(5.9.54)$$

The Column End Moment M_{J2}: From the equilibrium of joint J in Fig. 5.42(d), we have

right-hand joint:

$$M_{J2} = 0, |M_{J2}| < 30, \quad \text{okay.} \qquad (5.9.55)$$

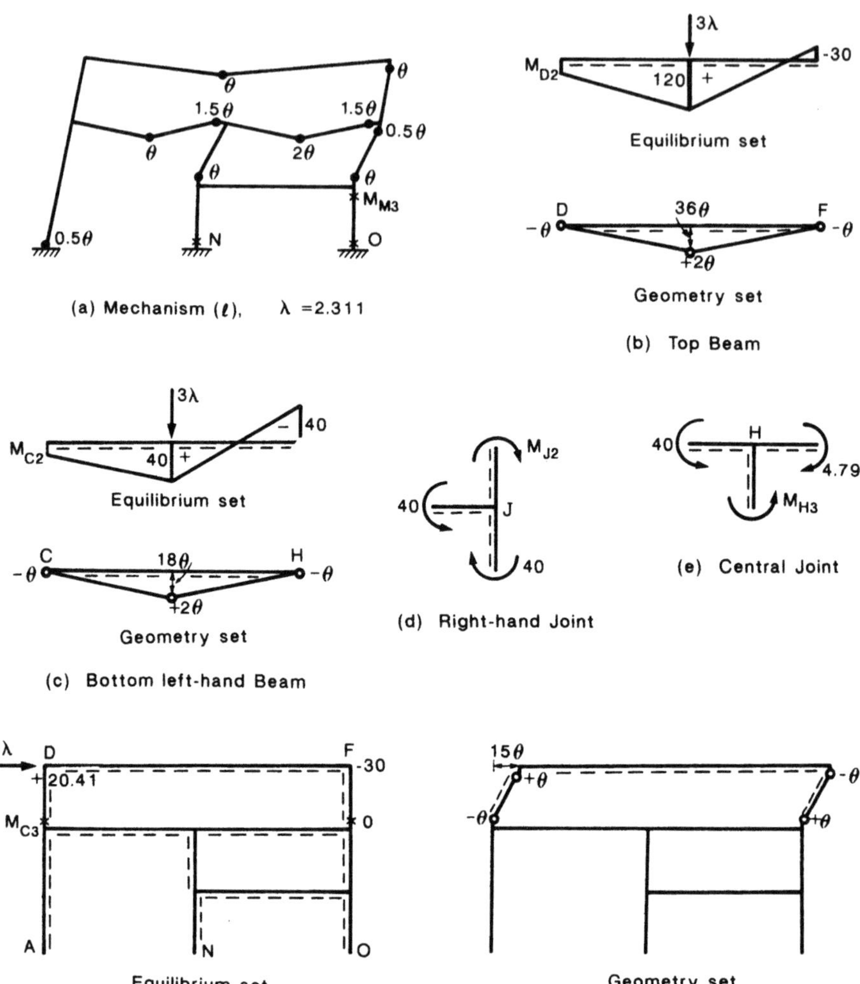

(a) Mechanism (ℓ), $\lambda = 2.311$

(b) Top Beam

(c) Bottom left-hand Beam

(d) Right-hand Joint

(e) Central Joint

(f) Top story sidesway

FIGURE 5.42. Moment check for mechanism (l) in (a).

The Column End Moment M_{H3}: From the equilibrium of joint H in Fig. 5.42(e), we have

central joint:

$$M_{H3} = -40 + 4.79 = -35.21, \quad |M_{H3}| < 40, \quad \text{okay.} \qquad (5.9.56)$$

The Column End Moment M_{C3}: M_{C3} is determined by applying the virtual work equation to the panel mechanism in Fig. 5.42(f):

top story side-sway:

$$(2\lambda)(15\theta) = (M_{C3})(-\theta) + (+20.41)(+\theta) + (-30)(-\theta) + (0)(+\theta),$$

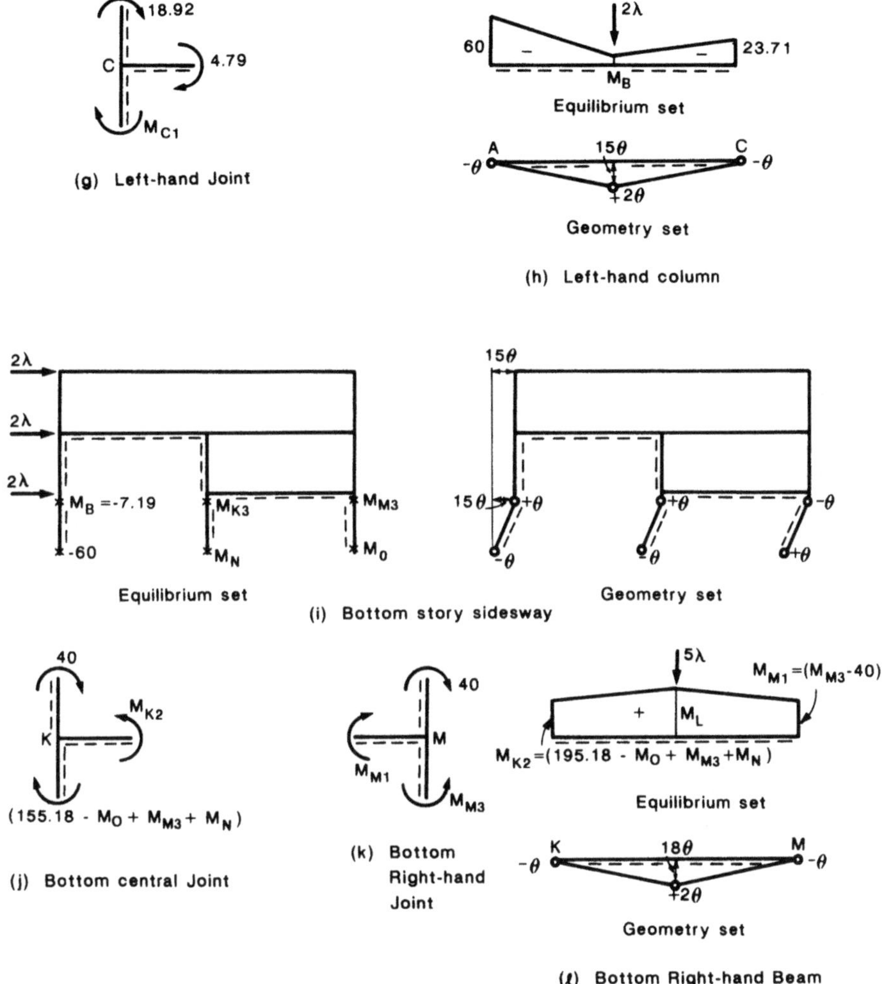

FIGURE 5.42 (*cont.*)

which gives

$$M_{C3} = 20.41 + 30 - 30\lambda = -18.92, \; |M_{C3}| < 30, \quad \text{okay.} \quad (5.9.57)$$

The Column End Moment M_{C1}: From equilibrium of joint C in Fig. 5.42(g), we have

left-hand joint:

$$M_{C1} = -23.71, \; |M_{C1}| < 60, \quad \text{okay.} \quad (5.9.58)$$

The Column Midspan Moment M_B: By applying the virtual work equation to Fig. 5.42(h), we have

left-hand column:

$$(2\lambda)(15\theta) = (-60)(-\theta) + (M_B)(+2\theta) + (-23.71)(-\theta),$$

which gives

$$M_B = 15\lambda - \frac{1}{2}(60 + 23.71)$$

or

$$M_B = -7.19, \; |M_B| < 60, \quad \text{okay.} \tag{5.9.59}$$

Now, we shall express moments M_{K3}, M_{K2}, M_{M1}, and M_L in terms of the redundants M_{M3}, M_N, and M_O.

The Column End Moment M_{K3}: From the panel mechanism shown in Fig. 5.42(i), we have

bottom story side-sway:

$$(2\lambda)(15\theta)(3) = (-60)(-\theta) + (-7.19)(+\theta) + (M_N)(-\theta) + (M_{K3})(+\theta)$$
$$+ (M_{M3})(-\theta) + (M_O)(+\theta),$$

which gives

$$M_{K3} = 90\lambda - 60 + 7.19 + M_N + M_{M3} - M_O$$

or

$$M_{K3} = 155.18 - M_O + M_{M3} + M_N. \tag{5.9.60}$$

The Beam End Moment M_{K2}: From equilibrium of joint K in Fig. 5.42(j), we have

bottom central joint:

$$M_{K2} = 195.18 - M_O + M_{M3} + M_N. \tag{5.9.61}$$

The Beam End Moment M_{M1}: From equilibrium of joint M in Fig. 5.42(k), we have

bottom right-hand joint:

$$M_{M1} = M_{M3} - 40. \tag{5.9.62}$$

The Beam Midspan Moment M_L: By applying the virtual work equation to Fig. 5.42(l), we have

bottom right-hand beam:

$$(5\lambda)(18\theta) = (195.18 - M_O + M_{M3} + M_N)(-\theta)$$
$$+ M_L(+2\theta) + (M_{M3} - 40)(-\theta),$$

which gives

$$M_L = 45\lambda + \frac{1}{2}(195.18 - M_0 + M_{M3} + M_N + M_{M3} - 40)$$

or

$$M_L = 181.59 - \frac{1}{2}M_0 + M_{M3} + \frac{M_N}{2}. \qquad (5.9.63)$$

The moment condition is that the absolute values of the seven unknown moments M_N, M_O, M_{M3}, M_{K3}, M_{K2}, M_{M1}, and M_L be less than the plastic moments. This leads to the following inequalities:

$$-80 \le M_N \le 80 \qquad (5.9.64)$$

$$-80 \le M_O \le 80 \qquad (5.9.65)$$

$$-80 \le M_{M3} \le 80 \qquad (5.9.66)$$

$$-80 \le 155.18 - M_O + M_{M3} + M_N \le 80 \qquad (5.9.67)$$

$$-80 \le 195.18 - M_O + M_{M3} + M_N \le 80 \qquad (5.9.68)$$

$$-80 \le M_{M3} - 40 \le 80 \qquad (5.9.69)$$

$$-80 \le 181.59 - \frac{M_O}{2} + M_{M3} + \frac{M_N}{2} \le 80. \qquad (5.9.70)$$

Inequalities (5.9.67) to (5.9.70) can be written as

$$75.18 \le M_O - M_{M3} - M_N \le 235.18 \qquad (5.9.71)$$

$$115.18 \le M_O - M_{M3} - M_N \le 275.18 \qquad (5.9.72)$$

$$-40 \le M_{M3} \le 120 \qquad (5.9.73)$$

$$101.59 \le \frac{M_O}{2} - M_{M3} - \frac{M_N}{2} \le 261.59. \qquad (5.9.74)$$

Thus, all inequalities (5.9.64) to (5.9.66) and (5.9.71) to (5.9.74) can be arranged to have the form

$$-80 \le M_N \le 80 \qquad (5.9.75)$$

$$-80 \le M_O \le 80 \qquad (5.9.76)$$

$$-40 \le M_{M3} \le 80 \qquad (5.9.77)$$

$$115.18 \le M_O - M_{M3} - M_N \le 235.18 \qquad (5.9.78)$$

$$101.59 \le \frac{M_O}{2} - M_{M3} - \frac{M_N}{2} \le 261.59. \qquad (5.9.79)$$

These inequalities (5.9.75) to (5.9.79) can be satisfied when $M_{M3} = -40$, $M_O = 80$, $M_N = -80$. None of the seven bending moments found in this way exceeds the corresponding fully plastic value, so a distribution of bending moment has been found for the entire frame that is not only statically admissible but also safe. This bending moment diagram for the moment check is

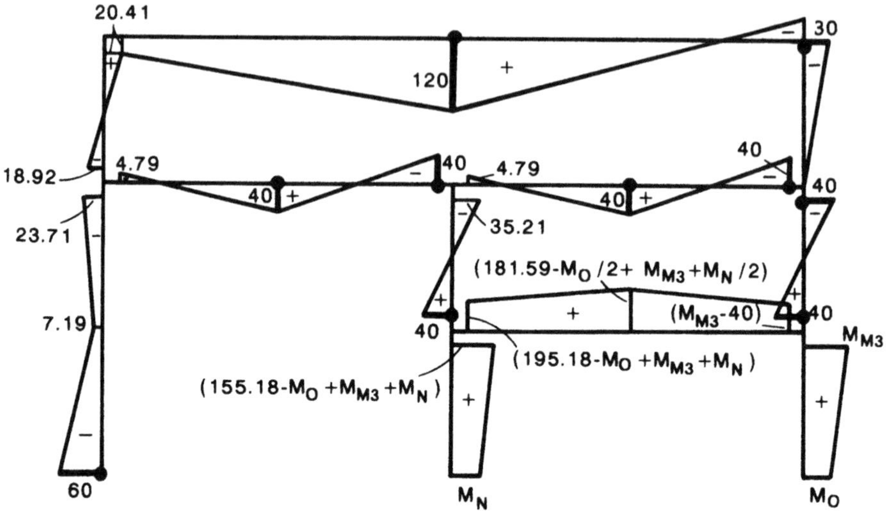

FIGURE 5.43. Moment check for the mechanism (*l*) in Figure 5.42 (a).

shown in Fig. 5.43. As can be seen, moments are all less than or at most equal to the plastic moments in the frame. Therefore, moment check is complete and the load factor $\lambda = 2.311$ is exact.

5.10 Distributed Loads

When a member is subjected to distributed loads (Fig. 5.44), the maximum moment and hence the plastic hinge location is not known in advance. The exact location of the hinge may be determined by first writing the work equation in terms of the unknown distance (for example, z, in Fig. 5.45) and then maximizing the plastic moment (or minimizing the loads) by formal differentiation. In actual practice, however, it is not necessary to make such an exact calculation; instead, it is convenient to substitute the uniform load with a pattern of equivalent concentrated loads. In this way, plastic hinge locations are limited to well-defined load and reaction points. Herein, we shall first solve exactly the single-bay frame of uniform section shown in Fig. 5.44, followed by an approximate analysis assuming the plastic hinge at quarter points as well as at the midpoint of the beam (Fig. 5.46), respectively.

5.10.1 Exact Analysis of Portal Frame

The solution of problems involving frames subject to uniformly distributed vertical loading, without resort to an alternate concentrated load pattern, requires the determination of two unknowns, the M_p value corresponding to

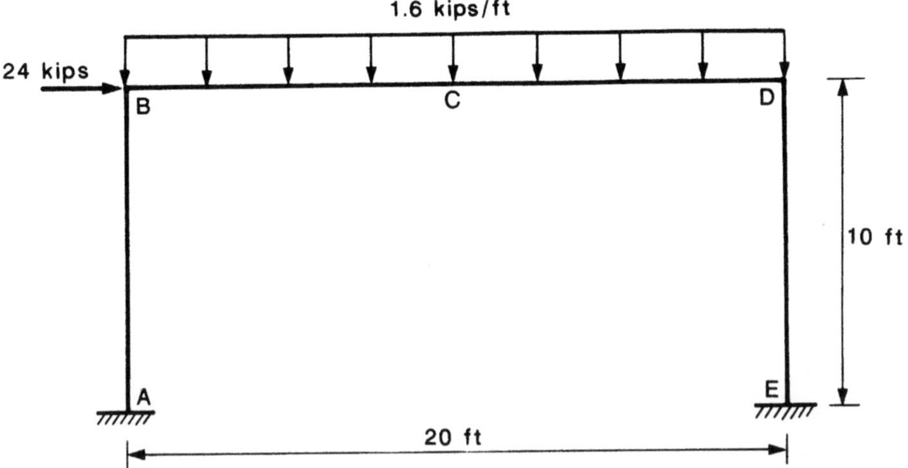

FIGURE 5.44. A portal frame subjected to vertical uniformaly distributed and horizontal concentrated loads.

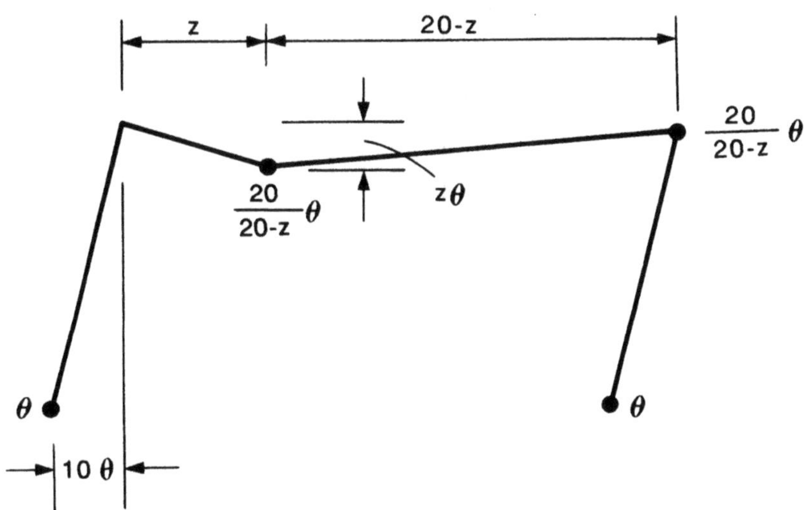

FIGURE 5.45. Exact analysis for portal frame subjected to loads shown in Figure 5.44.

the failure mechanism and the location of any plastic hinges that would have to form at points along the beam in developing that mechanism. Herein, we shall use the combination of mechanisms to derive the expression for the M_p value of the combined mechanism in terms of the unknown hinge location z in the beam. Differentiating this expression for M_p with respect to z and setting the derivative equal to zero, the critical hinge location z is obtained. Finally, this value is substituted in the equation for M_p to obtain the desired solution.

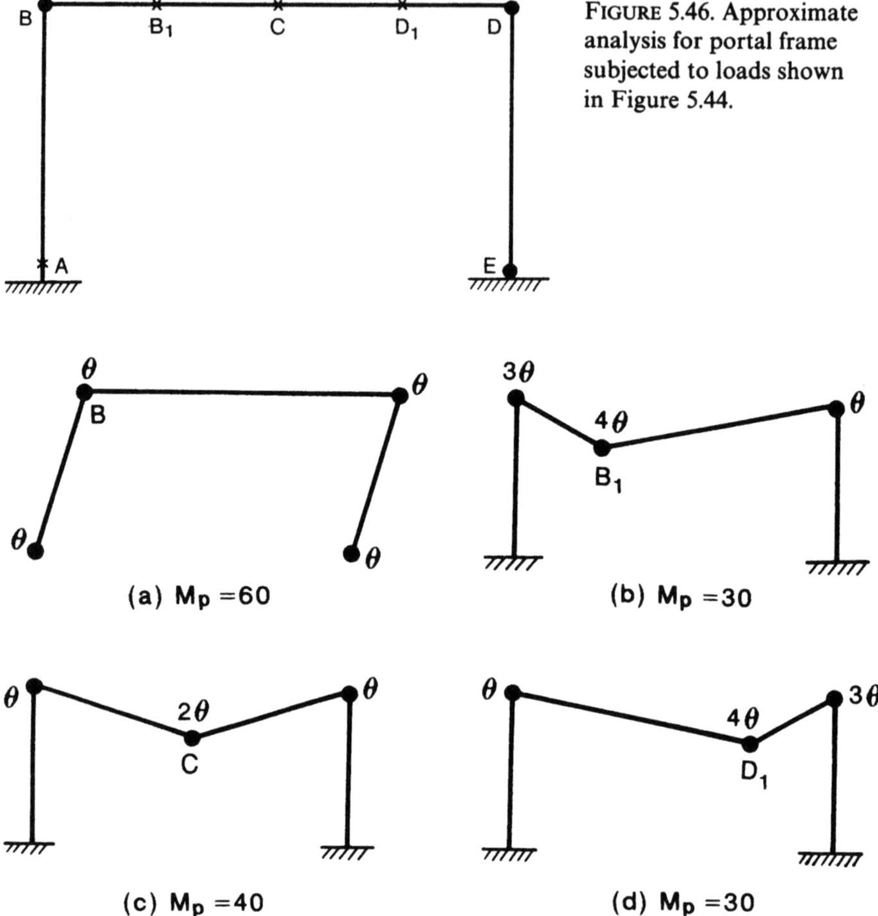

FIGURE 5.46. Approximate analysis for portal frame subjected to loads shown in Figure 5.44.

FIGURE 5.47. Independent mechanisms assuming critical plastic hinge locations shown in Figure 5.46.

The side-sway mode of collapse in Fig. 5.47(a) leads to

side-sway:

$$4M_p\theta = 24(10\theta),$$

which gives

$$M_p = 60 \text{ kip-ft.} \tag{5.10.1}$$

The beam mechanism of Fig. 5.47(c) leads to

beam with midspan hinge:

$$4M_p\theta = \frac{1}{2}(10\theta)\,32,$$

which gives

$$M_p = 40 \text{ kip-ft.} \tag{5.10.2}$$

In fact, the correct mechanism is shown in Fig. 5.45, in which distance z is as yet unknown. The work equation corresponding to this mechanism is

$$24(10\theta) + \frac{1}{2}(1.6)(20)(z\theta) = M_p\left[2 + 2\left(\frac{20}{20 - z}\right)\right]\theta,$$

which gives

$$M_p = \frac{(240 + 16z)(20 - z)}{(80 - 2z)}. \tag{5.10.3}$$

To maximize M_p, the derivative of M_p is set equal to zero, i.e.,

$$\frac{dM_p}{dz} = 0$$

or

$$(80 - 2z)(80 - 32z) - (4{,}800 + 80z - 16z^2)(-2) = 0,$$

which gives

$$z = 40 - \sqrt{1{,}100} = 6.83 \text{ ft,} \tag{5.10.4}$$

and the exact solution is

$$M_p = 69.34 \text{ kip-ft.} \tag{5.10.5}$$

General formulas for the maximum bending moment in a member subject to a uniformly distributed load, together with the position where this maximum occurs, can be derived. This is given as Problem 5.18.

5.10.2 Approximate Analysis of Portal Frame

Approximate solutions are obtained by considering several failure mechanisms. The frame has three degrees of redundancies. If we assume seven critical sections (A, B, B_1, C, D_1, D, and E) as shown in Fig. 5.46, we will have four independent mechanisms. The two usual beam and side-sway mechanisms are shown in Figs. 5.47(a) and (c). The two additional beam mechanisms involving hinges at quarter points B_1 and D_1 are shown in Figs. 5.47(b) and (d). The corresponding work equations for these four mechanisms are as follows:

Mechanism (a), side-sway:

$$240\theta = 4M_p\theta$$
$$M_p = 60 \text{ kip-ft.} \tag{5.10.6}$$

Mechanism (b), beam hinge at B_1:

$$\frac{1}{2}(1.6)(20)(15\theta) = M_p(3\theta) + (M_p)(4\theta) + M_p\theta$$

$$M_p = 30 \text{ kip-ft.}$$

(5.10.7)

Mechanism (c), beam hinge at midspan:

$$\frac{1}{2}(1.6)(20)(10\theta) = M_p\theta + (M_p)(2\theta) + M_p\theta$$

$$M_p = 40 \text{ kip-ft.}$$

(5.10.8)

Mechanism (d), beam hinge at D_1:

$$\frac{1}{2}(1.6)(20)(15\theta) = M_p\theta + (M_p)(4\theta) + M_p(3\theta)$$

$$M_p = 30 \text{ kip-ft.}$$

(5.10.9)

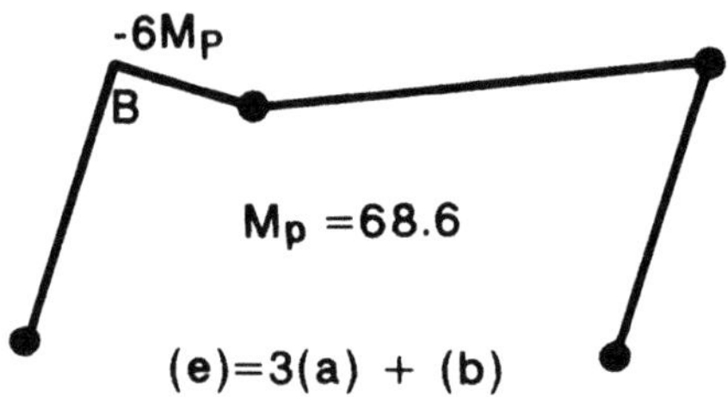

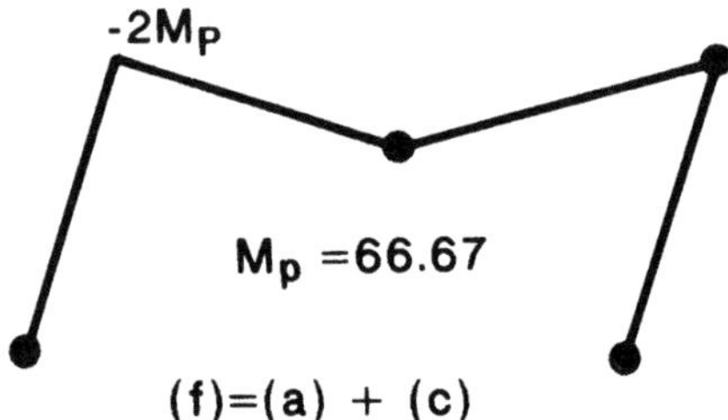

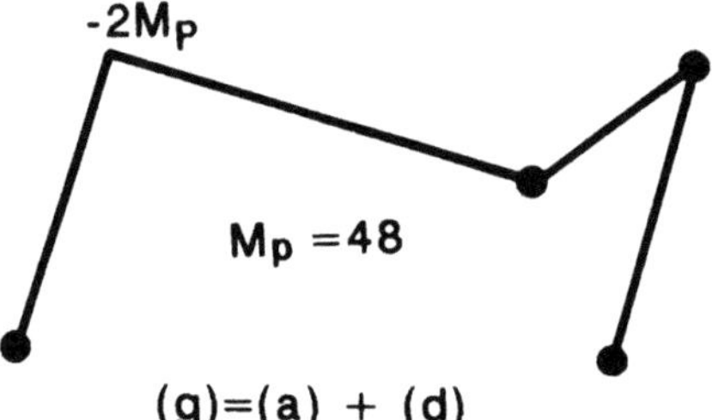

FIGURE 5.48. Combining independent mechanisms of Figure 5.47.

New mechanisms can be obtained by combining these four independent mechanisms. Combining mechanism (a) with mechanisms (b), (c), and (d) results in mechanisms (e), (f), and (g) as shown in Fig. 5.48. Since the rotation at B in mechanism (a) is opposite that in (b), (c), and (d), the combinations result in a cancellation of the hinge at B and thus provide higher values of M_p. Mechanism (a) is combined with (b), (c), and (d) as follows.

Mechanism (e): 3(a) + (b)

3(a) side-sway: $\qquad\qquad 3(240\theta) = 3(4M_p\theta)$

(b) beam at B_1: $\qquad\qquad\quad 240\theta = 8M_p\theta$

Cancel hinge at B: $\qquad\qquad\qquad -6M_p\theta$

(e) combined at B_1: $\qquad\quad 960\theta = 14M_p\theta$

$$M_p = 68.6 \text{ kip-ft.} \qquad\qquad (5.10.10)$$

Mechanism (f): (a) + (c)

(a) side-sway: $\qquad\qquad\quad 240\theta = 4M_p\theta$

(c) beam at C: $\qquad\qquad\quad 160\theta = 4M_p\theta$

Cancel hinge at B: $\qquad\qquad\qquad -2\,M_p\theta$

(f) combined at C: $\qquad\qquad 400\theta = 6M_p\theta$

$$M_p = 66.67 \text{ kip-ft.} \qquad\qquad (5.10.11)$$

Mechanism (g): (a) + (d)

(a) side-sway: $\qquad\qquad\quad 240\theta = 4M_p\theta$

(d) beam at D_1: $\qquad\qquad\quad 240\theta = 8M_p\theta$

Cancel hinge at B: $\qquad\qquad\qquad -2M_p\theta$

(g) combined at D_1: $\qquad\qquad 480\theta = 10M_p$

$$M_p = 48 \text{ kip-ft.} \qquad\qquad (5.10.12)$$

Note that the direct superposition of mechanisms (a) and (b) would not lead to the cancellation of the hinge at B. To cancel this hinge, the mechanism (a) in Eq. (5.10.6) is multiplied by a factor 3 before adding to mechanism (b) in Eq. (5.10.7). Also note that the cancellation of this hinge reduces the internal work by $6M_p\theta$, since the hinge in each mechanism rotates by 3θ.

Among all the mechanisms considered, mechanism (e) gives the highest value of M_p and thus gives the closest approximate solution. Note that the approximate solution differs from the exact one by little more than 1%. If the independent mechanisms (b) and (d) are not considered, then the combined mechanism (f) leads to $M_p = 66.67$ kip-ft, which differs from the exact solution by only 4%.

The approximate analysis of the distributed load can be made by replacing distributed loads by a set of concentrated loads as shown in Fig. 5.49. In the case of vertical loading, the distributed load can be replaced by the concen-

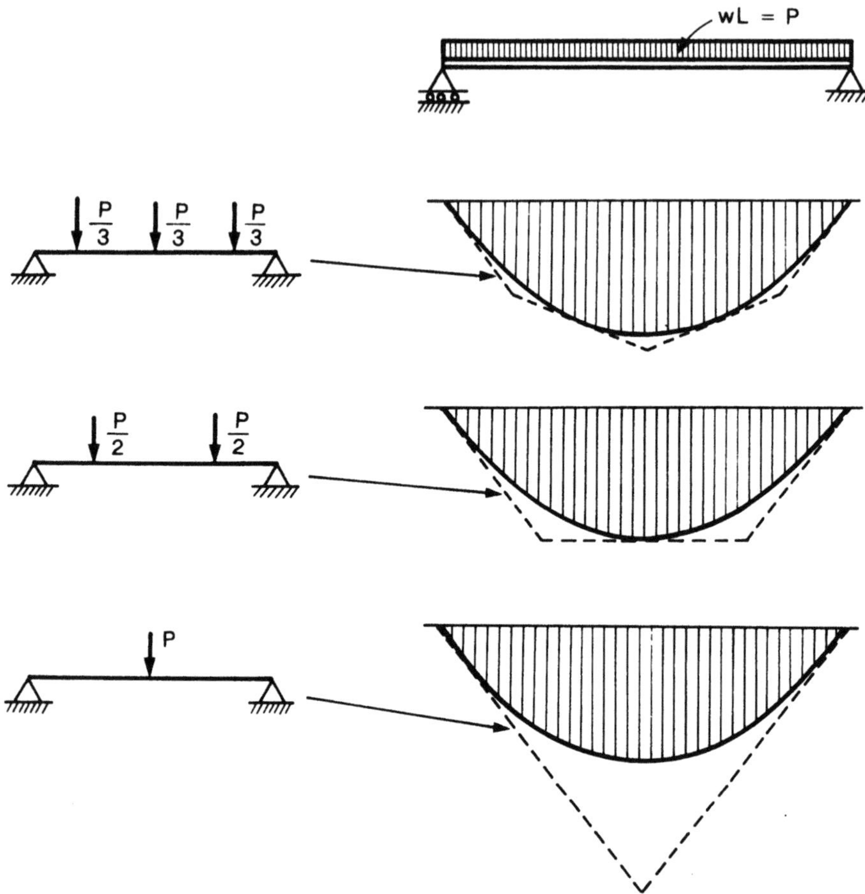

FIGURE 5.49. Replacing a distributed load by an equivalent set of concentrated loads.

trated loads in several ways shown. The uniform load parabola is always circumscribed by the moment diagram due to equivalent concentrated loads. Any consequent error will be small and on the "safe" side when the alternate pattern consists of three or more concentrated loads, evenly spaced as shown in the top of Fig. 5.49. Of course, the more concentrated loads assumed, the closer the approximation is to the real problem. Any resulting error will likewise be small and on the safe side if uniformly distributed horizontal wind loading is replaced by a single concentrated load at the top of the windward column (at the eaves in the case of a gable frame). The computed magnitude of this load is such that its overturning moment, taken about the base of the windward column, is equal to the overturning moment that the wind loading produces about the same point.

5.11 Examples for Distributed Loads

In this section, we will present two realistic examples. The first deals with a
two-bay gable frame subjected to distributed loading, the second deals with
a two-story two-bay rectangular frame subjected to distributed loading.

5.11.1 A Two-Bay Gable Frame

Example 5.11.1. A two-bay gable frame shown in Fig. 5.50 has same cross
section throughout. Use simple plastic theory to determine the plastic moment capacity M_p of the cross section. Note that all the loads are distributed
over the lengths of the members, the magnitude of each load being indicated
against the dotted arrow that shows the direction in which the load acts. The
kinematics of gable frames are more difficult to analyze than those of the
rectangular frames. This example is used to illustrate the technique of dealing
with such frames subjected to distributed loads.

Solution: The number of independent mechanisms is calculated as

Possible hinge locations:	$N = 16$
(crosses marked in Fig. 5.50)	
Redundancies:	$R = 6$
(two cuts, say, sections C and F)	
Independent mechanisms:	$n = 10$
Joint mechanism:	$= 1$
(joint D)	
True independent mechanisms:	$= 9.$

The nine independent mechanisms are identified and shown in Fig. 5.51,
consisting of six beam-type mechanisms and three side-sway-type mechanisms. In deriving the work equations for the beam-type mechanisms, we
shall assume first that the plastic hinges within the spans occur at midspan.
In the actual collapse mechanisms, these plastic hinges may occur anywhere
within the span. However, in the initial analysis, it is convenient to combine
the beam-type mechanisms with the other independent mechanisms while
keeping the plastic hinges at midspans. Only when we obtain the actual
collapse mechanism will an adjustment be made to correct the incorrect
positioning of these hinges.

The work equations corresponding to the nine independent mechanisms
are summarized as follows:

Mechanisms (a) *to* (f) *are beam type*:

(a) column AB:
$$\frac{1}{2}(0.45)(7\theta) = 4M_p\theta$$

$$1.58\theta = 4M_p\theta, \qquad M_p = 0.4. \qquad (5.11.1)$$

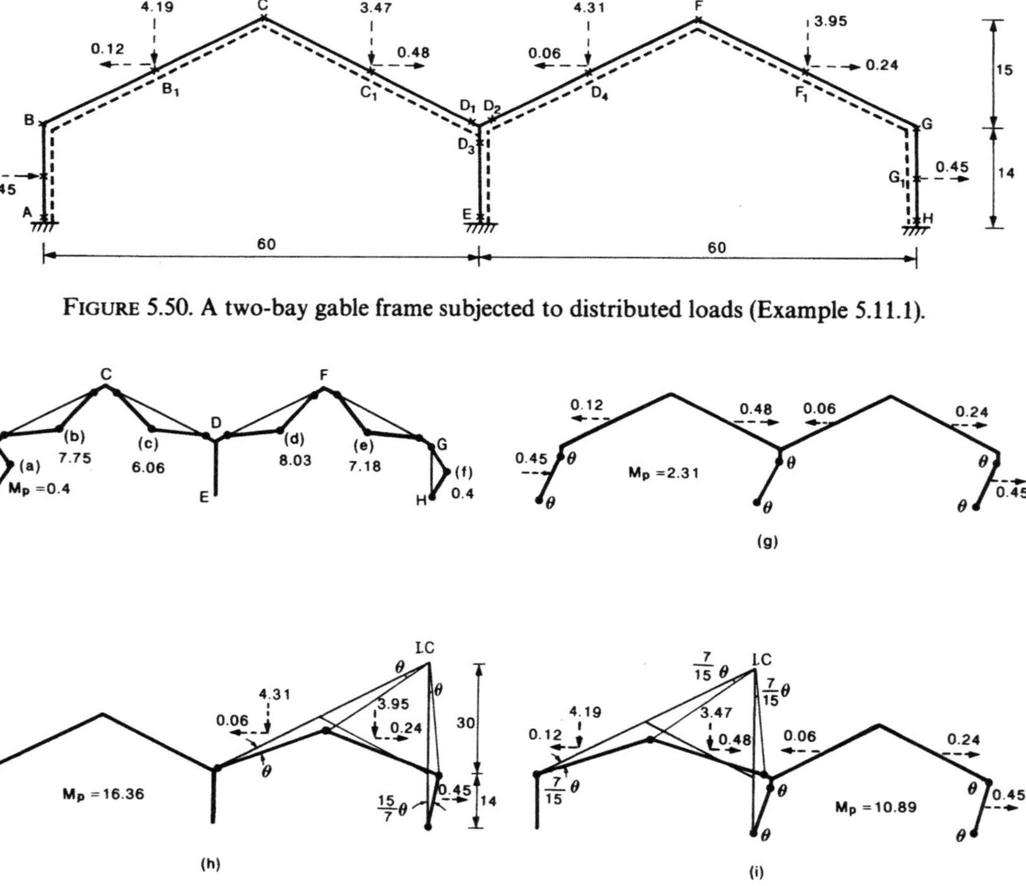

FIGURE 5.50. A two-bay gable frame subjected to distributed loads (Example 5.11.1).

FIGURE 5.51. Nine true independent mechanisms for the frame in Figure 5.50.

(b) beam BC:

$$\frac{1}{2}(4.19)(15\theta) - \frac{1}{2}(0.12)(7.5\theta) = 4M_p\theta$$

$$30.98\theta = 4M_p\theta, \qquad M_p = 7.75. \tag{5.11.2}$$

(c) beam CD:

$$\frac{1}{2}(3.47)(15\theta) - \frac{1}{2}(0.48)(7.5\theta) = 4M_p\theta$$

$$24.23\theta = 4M_p\theta, \qquad M_p = 6.06. \tag{5.11.3}$$

(d) beam DF:

$$\frac{1}{2}(4.31)(15\theta) - \frac{1}{2}(0.06)(7.5\theta) = 4M_p\theta$$

$$32.1\theta = 4M_p\theta, \qquad M_p = 8.03. \tag{5.11.4}$$

(e) beam FG:

$$\frac{1}{2}(3.95)(15\theta) - \frac{1}{2}(0.24)(7.5\theta) = 4M_p$$

$$28.73\theta = 4M_p\theta, \qquad M_p = 7.18. \tag{5.11.5}$$

(f) column GH:

$$\frac{1}{2}(0.45)(7\theta) = 4M_p\theta$$

$$1.58\theta = 4M_p\theta, \qquad M_p = 0.4. \tag{5.11.6}$$

Mechanisms (g) to (i) are side-sway type:

(g) total side-sway:

$$\left[\frac{1}{2}(0.45)(2) - 0.12 + 0.48 - 0.06 + 0.24\right](14\theta) = 6M_p\theta$$

$$13.86\theta = 6M_p\theta, \qquad M_p = 2.31. \tag{5.11.7}$$

(h) right-hand sway:

$$\frac{1}{2}(4.31)(30\theta) - \frac{1}{2}(0.06)(15\theta) + \frac{1}{2}(3.95)(30\theta)$$

$$+ \frac{1}{2}(0.24)(15 + 30)\theta + \frac{1}{2}(0.45)\left(\frac{15}{7}\theta\right)(14)$$

$$= M_p\left(1 + 2 + \frac{22}{7} + \frac{15}{7}\right)\theta$$

$$135.6\theta = 8.29M_p\theta, \qquad M_p = 16.36. \tag{5.11.8}$$

(i) combined sway:

$$\frac{1}{2}(4.19)\left(\frac{7}{15}\theta\right)(30) - \frac{1}{2}(0.12)\left(\frac{7}{15}\theta\right)(15)$$

$$+ \frac{1}{2}(3.47)\left(\frac{7}{15}\theta\right)(30) + \frac{1}{2}(0.48)((15 + 30)\left(\frac{7}{15}\theta\right)$$

$$- (0.06)(14\theta) + (0.24)(14\theta)$$

$$+ \frac{1}{2}(0.45)(14\theta) = M_p\left(\frac{7}{15} + \frac{14}{15} + \frac{7}{15} + 4\right)\theta$$

$$63.91\theta = 5.87M_p\theta, \qquad M_p = 10.89. \tag{5.11.9}$$

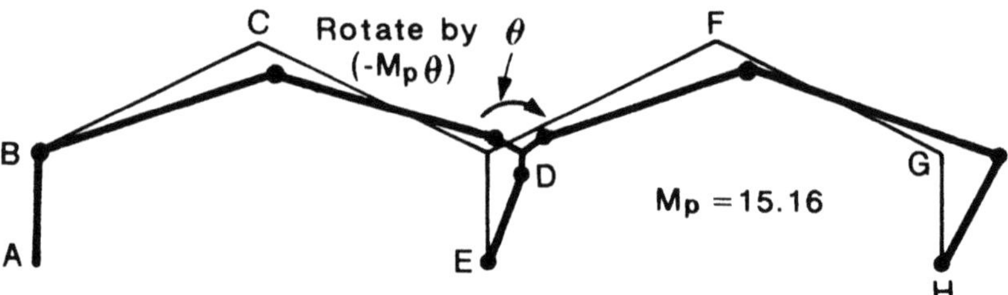

(a) Mechanism (j)=(h) + (i)

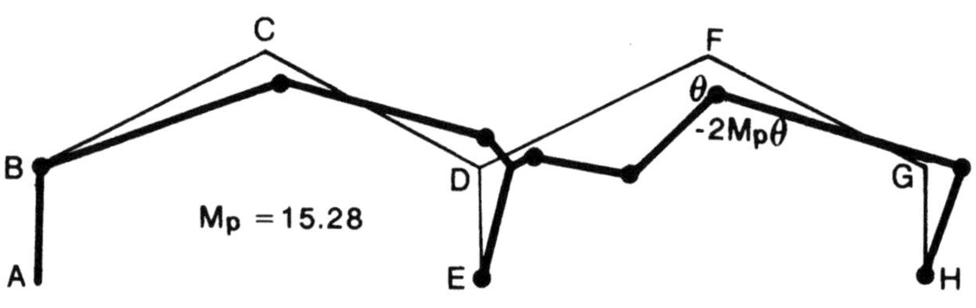

(b) Mechanism (k) = (j) + (d)

FIGURE 5.52. Combining the independent mechanisms in Figure 5.51: (a) mechanism (j) = (h) + (i) and (b) mechanism (k) = (j) + (d).

(A) Determining the "Correct" Mechanism: Combining the mechanisms in Fig. 5.52 and looking for a maximum value of M_p, we note that the beam-type mechanisms cannot be combined with one another in such a way as to achieve the cancellation of any hinges, with a possible consequent increase in the corresponding value of M_p. The analysis therefore takes the form of investigating combinations of the side-sway mechanisms first and then with the beam-type mechanisms. To this end, we choose first to combine the two side-sway mechanisms (h) and (i), and then with the beam mechanism (d). This is outlined in the following.

The combination: (j) = (h) + (i)

(h) right-hand side-sway: $\qquad 135.6\theta = 8.29M_p\theta$

(i) combined side-sway: $\qquad 63.91\theta = 5.87M_p\theta$

$$\overline{\qquad 199.51\theta = 14.16M_p\theta \qquad}$$

Rotation of joint D: $\qquad\qquad -M_p\theta$

(j) combined: $\qquad 199.51\theta = 13.16M_p\theta$

$$M_p = 15.16. \qquad\qquad (5.11.10)$$

The combination: (k) = (j) + (d)

(j) combined: $\qquad\qquad 199.51\theta = 13.16M_p\theta$

(d) beam D_2F $\qquad\qquad\quad 32.1\theta = 4M_p\theta$

$$231.61\theta = 17.16M_p\theta$$

Cancel hinge at F: $\qquad\qquad\qquad\quad -2M_p\theta$

(k) combined: $\qquad\qquad 231.61\theta = 15.16M_p\theta$

$$M_p = 15.28. \qquad\qquad (5.11.11)$$

The addition of beam mechanisms does not seem to improve the solution. So we shall check the moments for side-sway mechanism (h) for which the value of M_p is highest among the mechanisms considered thus far. It is therefore concluded here that this mechanism is the actual mechanism, subject only to possible adjustments of the positions of the plastic hinges within the spans of the beams DF and FG and column GH.

(B) Adjusting Hinges to the "Correct" Locations: Since the structure is subjected to distributed loads, the moments within the members may be critical. Therefore, we shall first check the moments in members DF, FG, and GH and, if necessary, make the adjustments in the location of hinges in the mechanism. The moment check will be carried out by the application of the virtual work equation. Details of this procedure need not be given, since they have already been discussed in previous examples.

Moments in Member DF: The equation for moment M at a horizontal distance x from D is obtained by applying the virtual work equation to equilibrium and geometry sets shown in Fig. 5.53(a)

$$\frac{1}{2}(4.31)(x\theta) - \frac{1}{2}(0.06)\left(\frac{x\theta}{2}\right)$$

$$= (-M_p)(-\theta) + (+M)\left(\frac{+30\theta}{30-x}\right) + (+M_p)\left(\frac{-x\theta}{30-x}\right),$$

which gives

$$M = -0.0713x^2 + 3.231x - 16.36$$

or

$$M = -0.0713(x - 22.66)^2 + 20.25.$$

Thus, $M_{\text{max}} = 20.25$, $|M_{\text{max}}| > M_p = 16.36$ at $x = 22.66$, not okay.

Moments in Member FG: From Fig. 5.53(b), the virtual work equation for moment M is

$$\frac{1}{2}(3.95)(x\theta) - \frac{1}{2}(0.24)\left(\frac{x\theta}{2}\right)$$

$$= (+M_p)(-\theta) + (+M)\left(\frac{+30\theta}{30-x}\right) + (-M_p)\left(\frac{-x\theta}{30-x}\right),$$

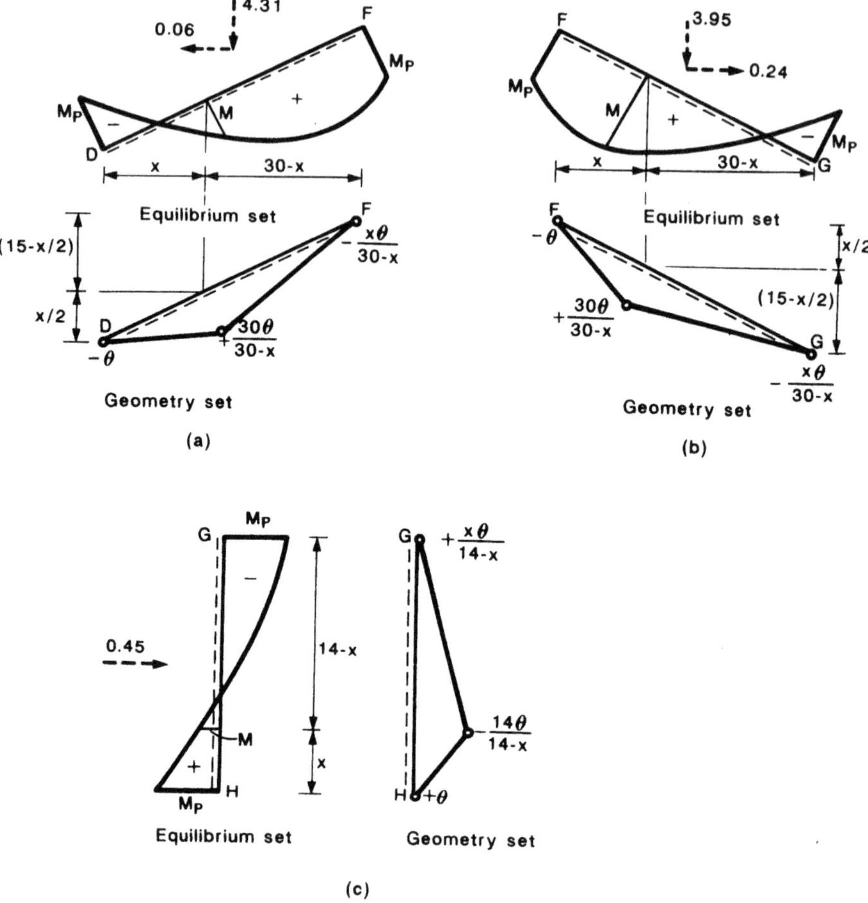

FIGURE 5.53. Moment check for mechanism (h) in Figure 5.51.

which provides

$$M = -0.0638x^2 + 0.824x + 16.36$$

or

$$M = -0.0638(x - 6.46)^2 + 19.02.$$

Thus, $M_{max} = 19.02$, $|M_{max}| > M_p = 16.36$ at $x = 6.46$, not okay.

Moments in Member *GH*: From Fig. 5.53(c), the virtual work equation for moment M is

$$\frac{1}{2}(0.45)(x\theta) = (+M_p)(+\theta) + (+M)\left(\frac{-14\theta}{14 + x}\right) + (-M_p)\left(\frac{+x\theta}{14 - x}\right),$$

which furnishes

$$M = -0.0161x^2 + 2.562x - 16.36$$

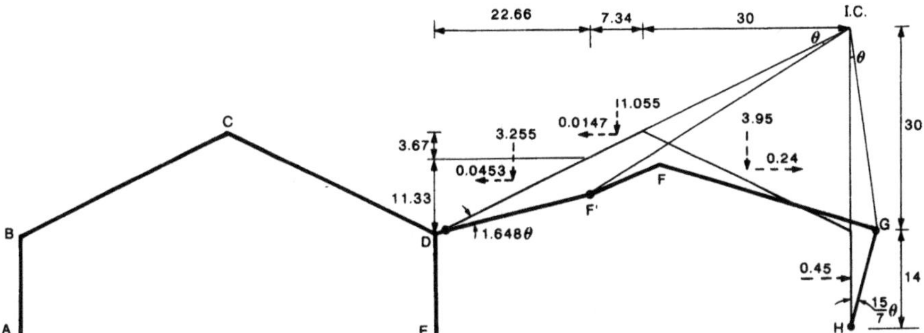

FIGURE 5.54. Mechanism (l) obtained by adjusting mechanism (h) in Figure 5.51.

or

$$M = -0.0161(x - 79.61)^2 + 85.68.$$

Since the value of x found from the equation $dM/dx = 0$ exceeds the member vertical height 14, the bending moment increases continuously from the column end G to the column end H with $M_{\max} = M_G = M_p$. As the maximum moment does not occur within the member, we have $|M| \leq M_p$ along GH.

The moment condition is violated in members DF and FG and the maximum moment in DF is higher than that in FG. The necessary adjustments for the distributed loads will thus be made by moving the plastic hinge at F in mechanism (h) in Fig. 5.51 to F' (Fig. 5.54). The work equation corresponding to this new mechanim (l) is

$$\left[\frac{1}{2}(3.255)(22.66) - \frac{1}{2}(0.0453)(11.33)\right](1.648\theta)$$

$$+ \frac{1}{2}(1.055)(30 + 37.34)\theta - \frac{1}{2}(0.0147)(15 + 18.67)\theta$$

$$+ \frac{1}{2}(3.95)(30\theta) + \frac{1}{2}(0.24)(15 + 30)\theta + \frac{1}{2}(0.45)(14)\left(\frac{15}{7}\theta\right)$$

$$= M_p\left(1.648 + 2.648 + \frac{22}{7} + \frac{15}{7}\right)\theta$$

$$167.03\theta = 9.58M_p\theta, \qquad M_p = 17.44. \qquad (5.11.12)$$

This M_p is an improvement over the previous one ($M_p = 16.36$). Now, we shall check the moment for this new mechanism (l) in Fig. 5.54.

(C) Performing the Moment Check for the Adjusted Mechanism (l): The redundancy corresponding to the mechanism (h) is calculated as

Number of plastic hinges in the mechanism: $M = 4$
Redundancy in the original structure: $X = 6$
Redundancy at collapse: $I = X - (M - 1)$
$$= 6 - (4 - 1) = 3.$$

Assume the moments at B, C, and D_3 as the three redundants. Now all the moments in the frame can be expressed in terms of these redundants and the plastic moment ($M_p = 17.44$). For simplicity, all unknown moments are assumed to be positive in the following computations. The moment distribution in the right-hand collapse portion of this partial collapse frame is statically determinate and can be uniquely determined, while the bending moment distribution in the left-hand frame is statically indeterminate with three redundants. To

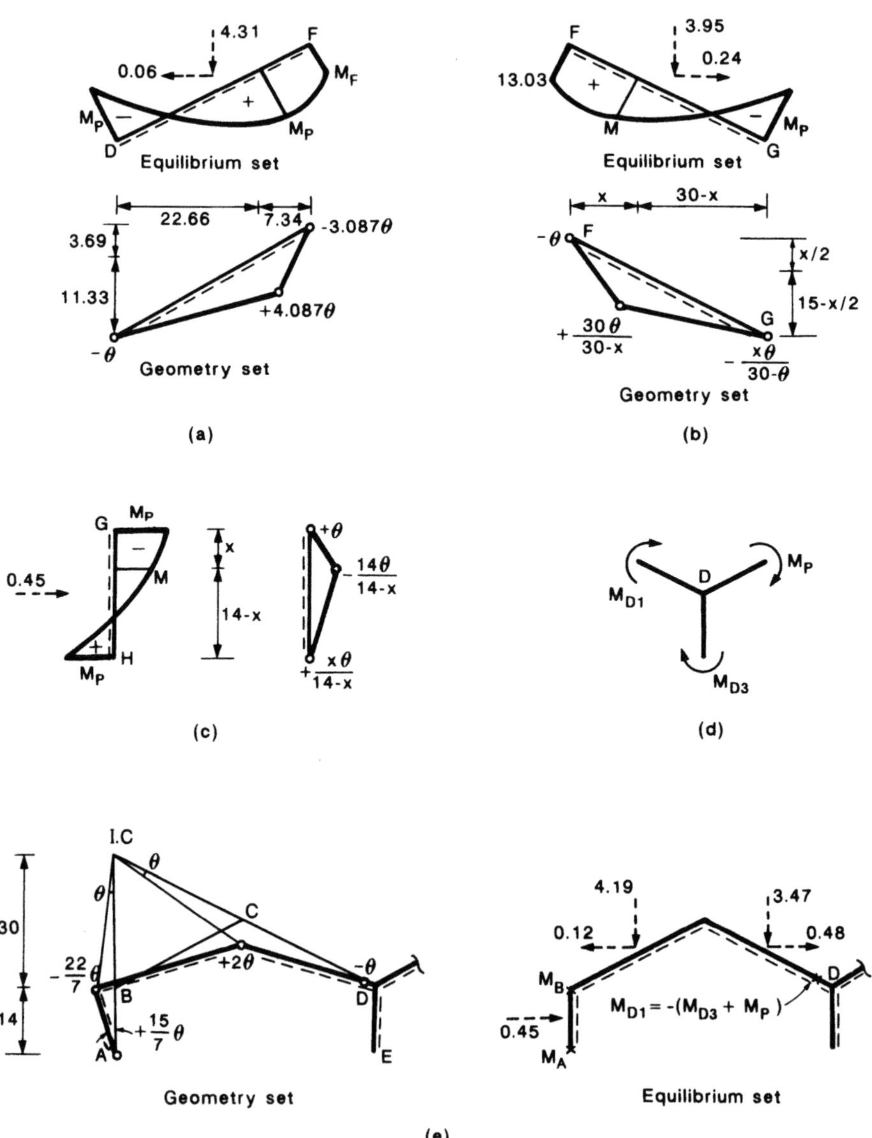

FIGURE 5.55. Moment check for mechanism (l) in Figure 5.54.

complete the moment check, it is necessary to assign values to these three redundants, when all other moments are expressed in terms of these three.

Moment Check for the Statically Determinate Portion of the Frame: Moment M_F: From Fig. 5.55(a), the virtual work equation for M_F can be written as

$$\frac{1}{2}(4.31)(22.66\theta) - \frac{1}{2}(0.06)(11.33\theta)$$

$$= (-M_p)(-\theta) + (+M_p)(+4.087\theta) + (M_F)(-3.087\theta),$$

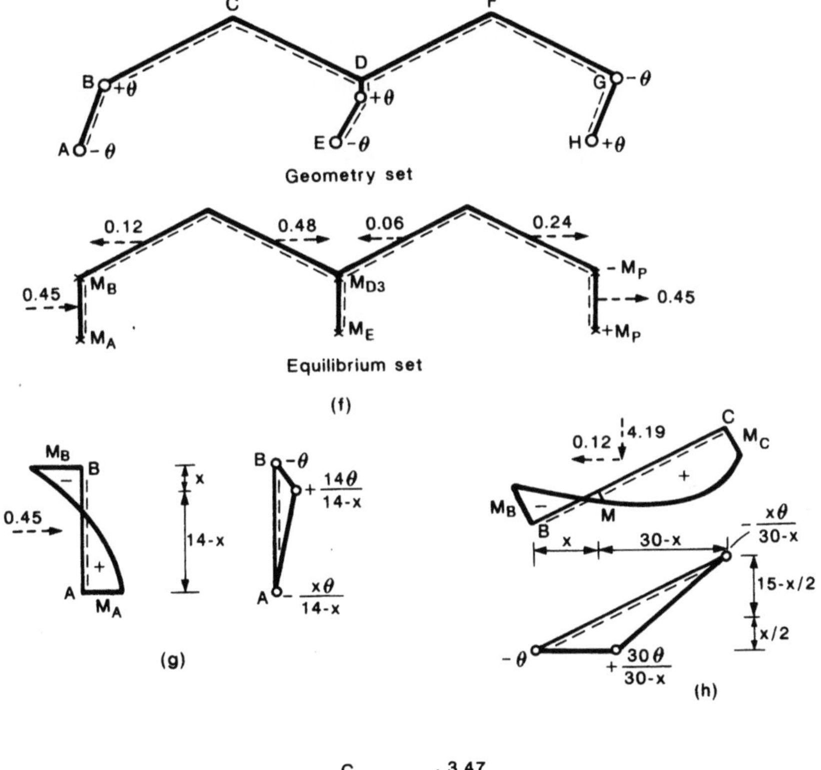

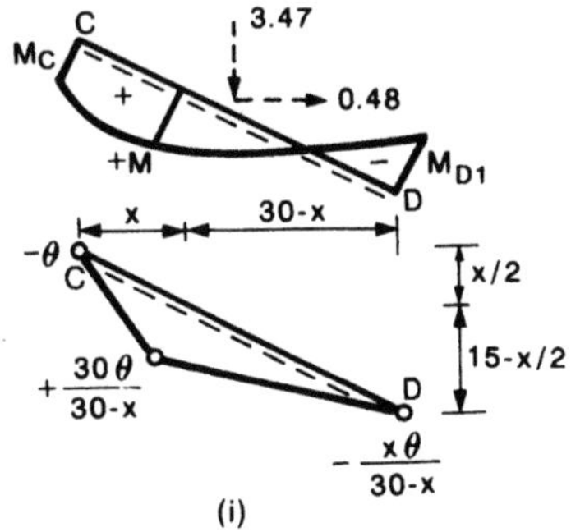

FIGURE 5.55 (cont.)

which gives

$$M_F = 13.03 < M_p = 17.44, \quad \text{okay.}$$

Moments in Member FG: The equation for moment M at a horizontal distance x from F can be written from Fig. 5.55(b) as

$$\frac{1}{2}(3.95)(x\theta) - \frac{1}{2}(0.24)\left(\frac{x}{2}\theta\right)$$

$$= (+13.03)(-\theta) + (M)\left(\frac{+30\theta}{30 - x}\right) + (-M_p)\left(\frac{-x\theta}{30 - x}\right)$$

$$M = -0.0638x^2 + 0.899x + 13.03$$

$$= -0.0638(x - 7.05)^2 + 16.20.$$

Thus, we find $M_{\max} = 16.20$ at the location $x = 7.05$. Since $|M_{\max}| \le M_p = 17.44$ okay.

Moments in Member GH: From Fig. 5.55(c), the virtual work equation for the moment in member GH can be written as

$$\frac{1}{2}(0.45)(\theta x) = (-M_p)(+\theta) + (+M)\left(\frac{-14\theta}{14 - x}\right) + (+M_p)\left(\frac{+x\theta}{14 - x}\right)$$

$$M = 0.0161x^2 + 2.266x - 17.44$$

or

$$M = 0.0161(x + 70.37)^2 - 97.17.$$

It is found that the maximum moment does not occur within GH; we can therefore conclude that $|M| \le M_p = 17.44$, okay.

Moment Check for the Statically Indeterminate Portion of the Frame: We will first carry out a moment check at the ends of the members and then in the members subjected to uniformly distributed loads.

Moment M_{D1}: Since M_{D3} has been chosen as one of the three redundants, it follows from the equilibrium of joint D, Fig. 5.55(d), that the unknown moment M_{D1} has the value

$$M_{D1} = -M_{D3} - M_p.$$

Moment M_A: From the equilibrium and geometry sets shown in Fig. 5.55(e), we have

$$\frac{1}{2}(3.47)(30\theta) - \frac{1}{2}(0.48)(15\theta) + \frac{1}{2}(4.19)(30\theta)$$

$$+ \frac{1}{2}(0.12)(15 + 30)\theta - \frac{1}{2}(0.45)(14)\left(\frac{15}{7}\theta\right)$$

$$= (+M_A)\left(+\frac{15}{7}\theta\right) + (+M_B)\left(-\frac{22}{7}\theta\right) + (+M_C)(+2\theta)$$

$$+ (-M_{D3} - M_p)(-\theta)$$

$$107.25 = \frac{15}{7}M_A - \frac{22}{7}M_B + 2M_C + M_{D3} + 17.44$$

$$M_A = 1.47M_B - 0.93M_C - 0.47M_{D3} + 41.91.$$

Moment M_E: From Fig. 5.55(f), we have

$$\frac{1}{2}(0.45)(14\theta)(2) + (14\theta)(-0.12 + 0.48 - 0.06 + 0.24)$$

$$= (1.47M_B - 0.93M_C - 0.47M_{D3} + 41.91)(-\theta)$$

$$+ (+M_B)(+\theta) + (+M_E)(-\theta) + (+M_{D3})(+\theta)$$

$$+ (-M_p)(-\theta) + (+M_p)(+\theta)$$

$$M_E = -0.47M_B + 0.93M_C + 1.47M_{D3} - 20.89.$$

The condition that the absolute value of the moment at A, B, C, D_1, D_3, and E is less than M_p leads to the following inequalities

$$-17.44 \leq 1.47M_B - 0.93M_C - 0.47M_{D3} + 41.91 \leq 17.44 \quad (5.11.13)$$

$$-17.44 \leq M_B \leq 17.44 \quad (5.11.14)$$

$$-17.44 \leq M_C \leq 17.44 \quad (5.11.15)$$

$$-17.44 \leq -M_{D3} - 17.44 \leq 17.44 \quad (5.11.16)$$

$$-17.44 \leq M_{D3} \leq 17.44 \quad (5.11.17)$$

$$-17.44 \leq -0.47M_B + 0.93M_C + 1.47M_{D3} - 20.89 \leq 17.44. \quad (5.11.18)$$

Inequalities (5.11.13), (5.11.16), and (5.11.18) can be rearranged as

$$-59.35 \leq 1.47M_B - 0.93M_C - 0.47M_{D3} \leq -24.47 \quad (5.11.19)$$

$$-34.88 \leq M_{D3} \leq 0 \quad (5.11.20)$$

$$-3.45 \leq -0.47M_B + 0.93M_C + 1.47M_{D3} \leq 38.33. \quad (5.11.21)$$

Inequalities (5.11.14), (5.11.15), (5.11.17), (5.11.19), (5.11.20), and (5.11.21) can be satisfied if we substitute $M_B = -15$, $M_C = +10$, and $M_{D3} = -1.0$.

So the moment condition is satisfied at all nodal points (A, B, C, D_1, D_2, D_3, E, F, G, and H) and in the members FG and GH. Next, we shall check the maximum moments in the members AB, BC, and CD.

Further Moment Check for Members Carrying Distributed Loads: Moments in Member AB: From Fig. 5.55(g), the equation for moment M in member AB is written as

$$\frac{1}{2}(0.45)(\theta x) = (M_B)(-\theta) + (+M)\left(\frac{+14\theta}{14-x}\right) + (+M_A)\left(\frac{-x\theta}{14-x}\right),$$

which gives

$$M = -0.0161x^2 + \left(0.225 + \frac{M_A - M_B}{14}\right)x + M_B.$$

Substituting $M_B = -15$ [Fig. 5.55(g)] and

$$M_A = 1.47M_B - 0.93M_C - 0.47M_{D3} + 41.91$$

or

$$M_A = 1.47(-15) - 0.93(10) - 0.47(-1.0) + 41.91 = 11.03$$

we have

$$M = -0.0161x^2 + 2.084x - 15$$

$$M = -0.0161(x - 64.72)^2 + 52.44.$$

Since the maximum moment occurs at the top end $M_{max} = M_B = -15$, it follows that $|M| \leq M_p = 17.44$ along the length of member AB.

Moments in Member BC: From Fig. 5.55(h), the virtual work equation for moment M in member BC is

$$\frac{1}{2}(4.19)(\theta x) - \frac{1}{2}(0.12)\left(\frac{x}{2}\right)$$

$$= (M_B)(-\theta) + (+M)\left(\frac{+30\theta}{30 - x}\right) + (+M_C)\left(\frac{-x\theta}{30 - x}\right),$$

which gives

$$M = -0.0688x^2 + \left(2.065 + \frac{M_C - M_B}{30}\right)x + M_B.$$

Substituting $M_B = -15$ and $M_C = +10$, we have

$$M = -0.0688x^2 + 2.898x - 15$$

or

$$M = -0.0688(x - 21.06)^2 + 15.51.$$

So $M_{max} = +15.51$ at $x = 21.06$.

$$|M_{max}| < M_p = 17.44, \quad \text{okay.}$$

Moments in Member CD: From Fig. 5.55(i), the virtual work equation for moment M can be written as

$$\frac{1}{2}(3.47)(\theta x) - \frac{1}{2}(0.48)\left(\frac{\theta x}{2}\right)$$

$$= (+M_C)(-\theta) + (+M)\left(\frac{+30\theta}{30 - x}\right) + (M_{D1})\left(\frac{-x\theta}{30 - x}\right),$$

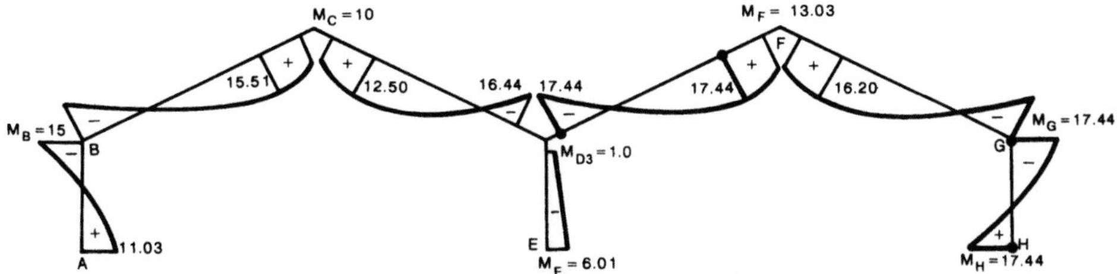

FIGURE 5.56. Final moment diagram for mechanism (l) in Figure 5.51.

which gives

$$M = -0.0538x^2 + \left[1.615 + \frac{M_{D1} - M_C}{30} \right] x + M_C.$$

Substituting $M_C = +10$ and $M_{D1} = -M_{D3}, -17.44 = -16.44$, we have

$$M = -0.0538x^2 + 0.734x + 10$$

or

$$M = -0.0538(x - 6.82)^2 + 12.50.$$

So $M_{max} = 12.50$ at $x = 6.82$. Thus $|M_{max}| < M_p = 17.44$ okay.

The final bending moment diagram for the failure mechanism (l) in Fig. 5.51 is shown in Fig. 5.56. The moment condition is satisfied everywhere in the frame. Therefore, the moment check is complete and the solution $M_p = 17.44$ is exact.

5.11.2 A Two-Story Two-Bay Rectangular Frame

Example 5.11.2. A two-story two-bay frame is loaded as shown in Fig. 5.57. The distributed loads on span *CE* and *IK* are shown by dashed arrows. Determine the plastic limit load *P*.

Solution: The number of independent mechanisms can be calculated as follows:

Possible hinge locations:	$N = 22$
(crosses marked in Fig. 5.57)	
Redundancies:	$R = 12$
(4 cuts at sections *D, H, F, J*)	
Independent mechanisms:	$n = 10$
Joint mechanisms:	$= 4$
(joints *B, E, I, K*)	
True independent mechanisms:	$= 6.$

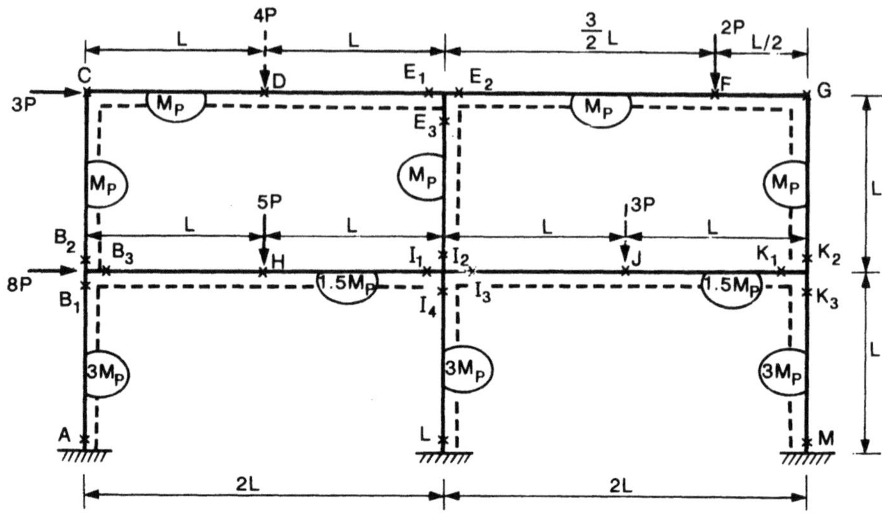

FIGURE 5.57. A two-bay two-story frame with distributed loads (Example 5.11.2).

The six independent mechanisms are shown in Fig. 5.58 consisting of four beam mechanisms and two side-sway mechanisms. The work equations corresponding to these independent mechanisms are summarized as follows, assuming the plastic hinges for the distributed loads on span CE and IK occur at their midspans.

Mechanics (a) *to* (d) *are beam mechanisms*:
(a) Left-hand top beam:

$$\frac{1}{2}(4P)(L\theta) = M_p\theta + (M_p)(2\theta) + M_p\theta$$

$$2PL\theta = 4M_p\theta, \qquad P = 2\frac{M_p}{L}.$$

$$(5.11.22)$$

(b) Right-hand top beam:

$$2P\left(\frac{3}{2}L\theta\right) = M_p\theta + (M_p)(4\theta) + M_p(3\theta)$$

$$3PL\theta = 8M_p\theta, \qquad P = 2.67\frac{M_p}{L}.$$

$$(5.11.23)$$

(c) Left-hand bottom beam:

$$5P(L\theta) = 1.5M_p\theta + (1.5M_p)(2\theta) + 1.5M_p\theta$$

$$5PL\theta = 6M_p\theta, \qquad P = 1.2\frac{M_p}{L}.$$

$$(5.11.24)$$

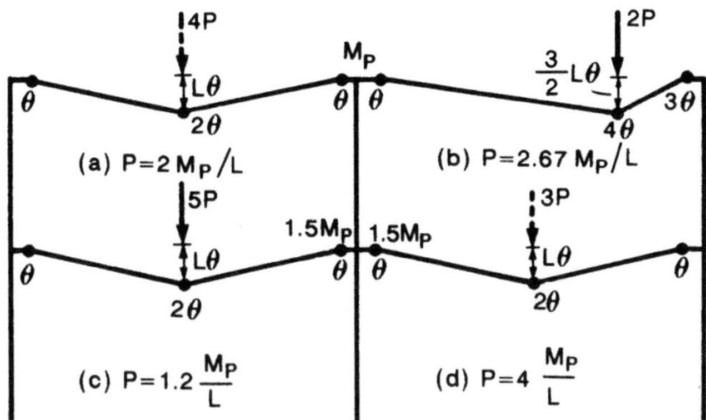

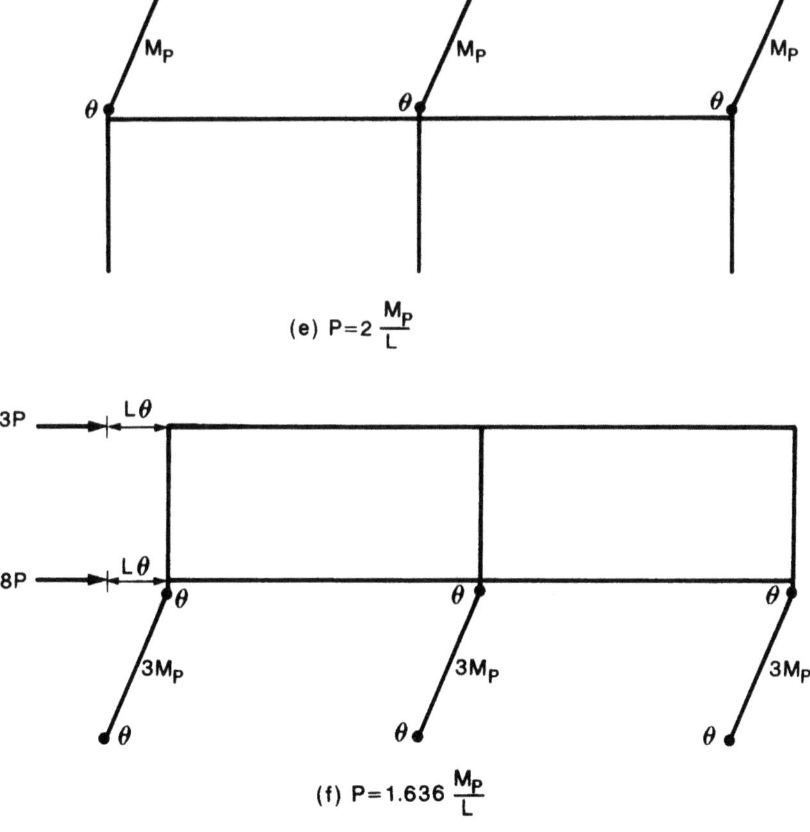

FIGURE 5.58. Six true independent mechanisms for the frame in Figure 5.57.

(d) Right-hand bottom beam:

$$\frac{1}{2}(3P)(L\theta) = 1.5M_p\theta + (1.5M_p)(2\theta) + 1.5M_p\theta$$

$$1.5PL\theta = 6M_p\theta, \qquad P = 4\frac{M_p}{L}.$$

(5.11.25)

Mechanism (e) and (f) are side-sway mechanisms:
(e) Top story:

$$3P(L\theta) = M_p\theta + M_p\theta + M_p\theta + M_p\theta + M_p\theta + M_p\theta$$

$$3PL\theta = 6M_p\theta, \qquad P = 2\frac{M_p}{L}.$$

(5.11.26)

(f) Bottom story:

$$3P(L\theta) + 8P(L\theta) = 3M_p\theta + 3M_p\theta + 3M_p\theta + 3M_p\theta + 3M_p\theta + 3M_p\theta$$

$$11PL\theta = 18M_p\theta, \qquad P = 1.636\frac{M_p}{L}.$$

(5.11.27)

Combining these mechanisms and looking for a minimum value of P, we have (Fig. 5.59):

The combination: (g) = (e) + (f)

(e) Top side-sway:	$3PL\theta = 6M_p\theta$
(f) Bottom side-sway:	$11PL\theta = 18M_p\theta$
	$14PL\theta = 24M_p\theta$
Rotate joint B by θ:	$-2.5M_p\theta$
Rotate joint I by θ:	$-M_p\theta$
Rotate joint K by θ:	$-2.5M_p\theta$
(g) Combined:	$14PL\theta = 18M_p\theta$
	$P = 1.286(M_p/L)$ (5.11.28)

The combination: (h) = (g) + (c)

(g) Combined:	$14PL\theta = 18M_p\theta$
(c) Left-hand bottom beam BI:	$5PL\theta = 6M_p\theta$
	$19PL\theta = 24M_p\theta$
Cancel hinge at B_3:	$-3M_p\theta$
(h) Combined:	$19PL\theta = 21M_p\theta$
	$P = 1.105(M_p/L)$ (5.11.29)

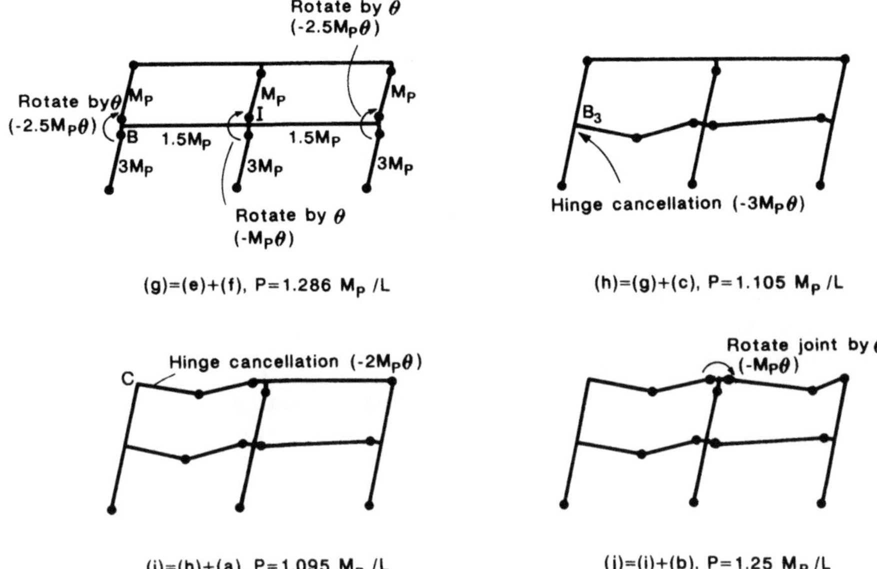

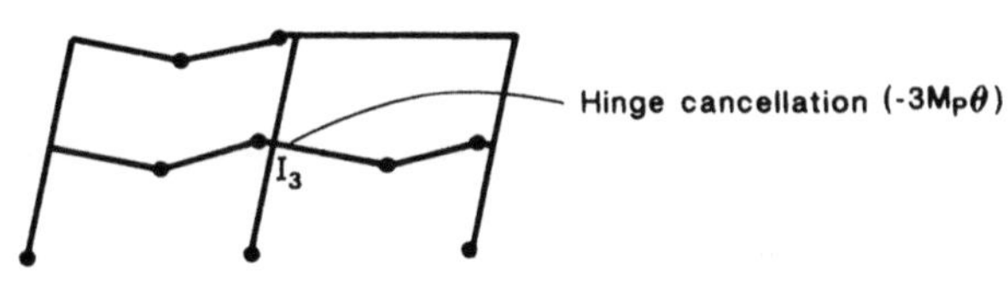

FIGURE 5.59. Combined mechanisms from the independent mechanisms in Figure 5.58.

The combination: (i) = (h) + (a)

(h) Combined: $\qquad\qquad 19PL\theta = 21M_p\theta$

(a) Left-hand top beam CE: $\qquad 2PL\theta = 4M_p\theta$

$$21PL\theta = 25M_p\theta$$

Cancel hinge at C: $\qquad\qquad\qquad -2M_p\theta$

(i) Combined: $\qquad\qquad 21PL\theta = 23M_p\theta$

$$P = 1.095(M_p/L) \qquad (5.11.30)$$

The combination: (j) = (i) + (b)
(i) Combined:
(b) Right-hand top beam *EG*:

$$21PL\theta = 23M_p\theta$$
$$3PL\theta = 8M_p\theta$$

$$\overline{24PL\theta = 31M_p\theta}$$

Rotate joint *E* by θ:

$$-M_p\theta$$

$$\overline{24PL\theta = 30M_p\theta}$$
$$P = 1.25(M_p/L) \qquad (5.11.31)$$

The combination: (k) = (i) + (d)
(j) Combined:
(d) Right-hand bottom beam *IK*:

$$21PL\theta = 23M_p\theta$$
$$1.5PL\theta = 6M_p\theta$$

$$\overline{22.5PL\theta = 29M_p\theta}$$

Cancel hinge at I_3:

$$-3M_p\theta$$

(k) Combined:

$$\overline{22.5PL\theta = 26M_p\theta}$$
$$P = 1.156(M_p/L) \qquad (5.11.32)$$

Mechanism (i) gives the lowest value of P. Therefore, we shall make a moment check for this mechanism. However, before proceeding to a moment check, we note that the left-hand top beam CE_1 carries a distributed load, and the assumption that the hinge develops in the middle of this beam should be adjusted.

Adjustment of the Hinge in the Left-Hand Top Beam CE_1: Assume that the hinge is located a distance $\alpha 2L$ from C (Fig. 5.60). Then the external work W_E done by the distributed load is

$$W_E = \frac{1}{2}(4P)(2\alpha L\theta) = 4\alpha PL\theta.$$

The internal work W_I is

$$W_I = M_p(\theta) + (M_p)\left(\frac{\theta}{1-\alpha}\right) + M_p\left(\frac{\alpha}{1-\alpha}\theta\right) = M_p\left(\frac{2}{1-\alpha}\right)\theta.$$

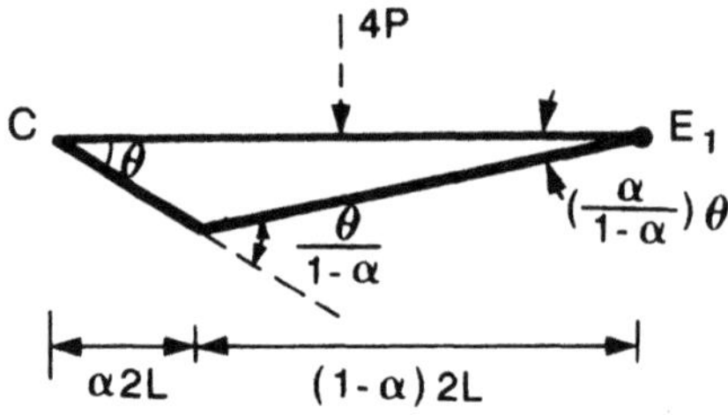

FIGURE 5.60. Adjustment of hinge location in member CE_1.

Recall that we have $W_I = 4M_p\theta$ and $W_E = 2PL\theta$ for mechanism (a). Now the work equation corresponding to mechanism (i) can be modified to be

$$21PL\theta - 2PL\theta + 4\alpha PL\theta = 23M_p\theta - 4M_p\theta + M_p\left(\frac{2}{1-\alpha}\right)\theta,$$

which gives

$$P = \frac{M_p}{L}\frac{21 - 19\alpha}{(1-\alpha)(19 + 4\alpha)}.$$

For P to be minimum, we set $dP/d\alpha$ to zero

$$\frac{dP}{d\alpha} = \frac{(4\alpha^2 + 15\alpha - 19)(19) - (19\alpha - 21)(8\alpha + 15)}{(4\alpha^2 + 15\alpha - 19)^2} = 0,$$

which gives

$$\alpha = 0.320.$$

So P is

$$P = 1.082\frac{M_p}{L}. \tag{5.11.33}$$

Moment Check for Adjusted Mechanism (i): The redundancy corresponding to the mechanism (i) can be calculated as follows:

Number of plastic hinges in the mechanism: $M = 11$
Redundancy in the original structure: $X = 12$
Redundancy at collapse: $I = X - (M - 1)$
$= 12 - 10 = 2.$

Let moments at B_2 and K_2 be the two redundants. We will first check the moments in member I_3K, which carries distributed loads.

Moment in Member I_3K_1: From Fig. 5.61(d), the equation for moment M in the right-hand bottom beam I_3K_1 can be written as

$$\frac{1}{2}(3P)(x\theta) = (+1.5M_p)(-\theta) + (+M)\left(\frac{+2L\theta}{2L - x}\right) + (-1.5M_p)\left(\frac{-x\theta}{2L - x}\right),$$

which gives

$$M = \frac{M_p}{2L^2}[-1.623x^2 + 0.246xL + 3L^2]$$

or

$$= -0.812\frac{M_p}{L^2}[(x - 0.076L)^2 - 1.854L^2].$$

318   5. Work Method

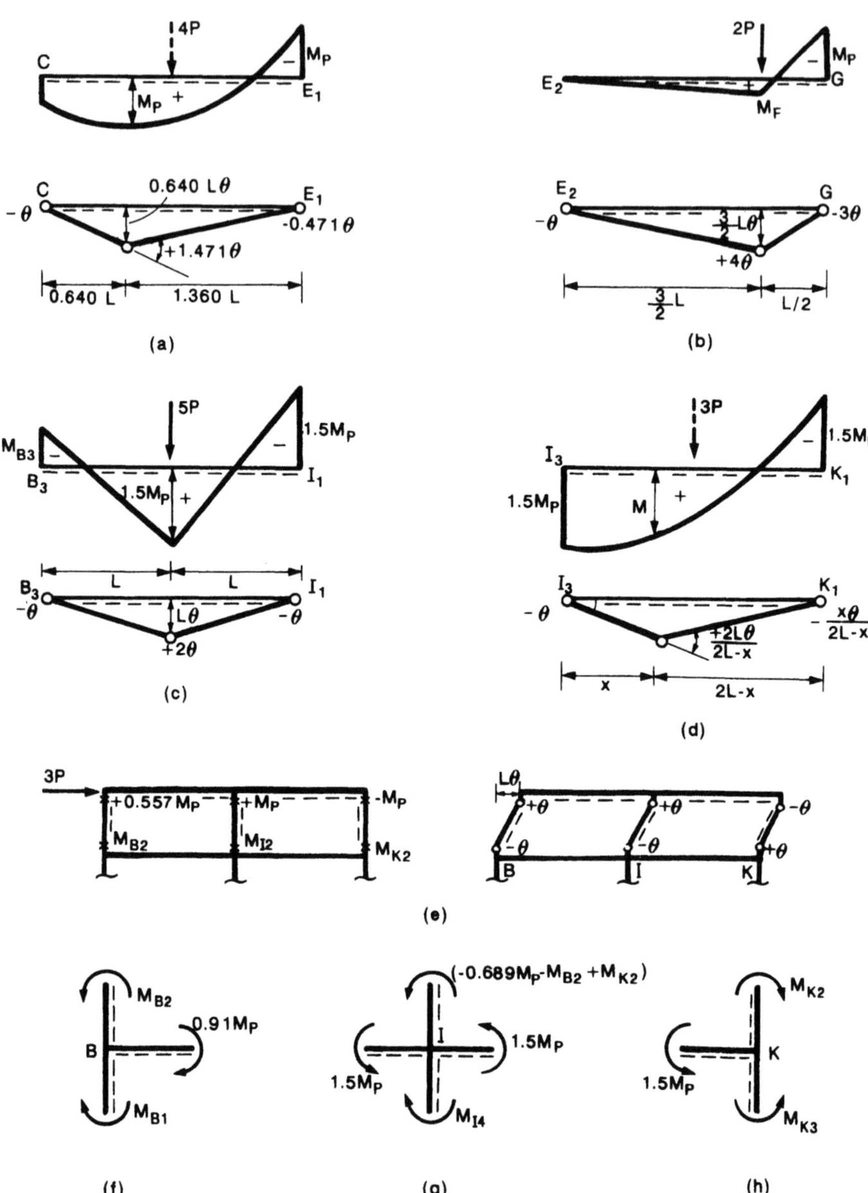

FIGURE 5.61. Moment check for mechanism (i) in Figure 5.59.

The maximum moment in member I_3K_1 is

$$M_{\max} = \left(0.812\frac{M_p}{L^2}\right)(1.854L^2) = 1.505M_p \text{ at } x = 0.076L$$

or

$$|M_{\max}| = 1.5\ M_p, \quad \text{okay.}$$

Moment at C: By applying the virtual work equation to the left-hand top beam CE_1 in Fig. 5.61(a), we have

$$\frac{1}{2}(4P)(0.640L\theta) = (+M_C)(-\theta) + (+M_p)(+1.471\theta) + (-M_p)(-0.471\theta),$$

which gives

$$M_C = M_p[-1.385 + 1.471 + 0.471] = 0.557M_p$$

or

$$|M_C| < M_p, \quad \text{okay.}$$

Moment at F: From the equilibrium and geometry sets for the right-hand top beam E_2G shown in Fig. 5.61(b), we have

$$(2P)\left(\frac{3}{2}L\theta\right) = (+M_F)(+4\theta) + (-M_p)(-3\theta),$$

which gives

$$M_F = \frac{1}{4}[3.246 - 3]M_p = 0.062M_p$$

or

$$|M_F| < M_p, \quad \text{okay.}$$

Note that from equilibrium of joint E, the top beam end moment at E_2 is zero.

Moment at B_3: Applying the virtual work equation to the left-hand bottom beam B_3I_1, in Fig. 5.61(c), we have

$$5P(L\theta) = (M_{B3})(-\theta) + (+1.5M_p)(+2\theta) + (-1.5M_p)(-\theta),$$

which gives

$$M_{B3} = -M_p[5.41 - 3 - 1.5] = -0.91M_p$$

or

$$|M_{B3}| < 1.5M_p, \quad \text{okay.}$$

Moment M_{I2}: Applying the virtual work equation to the equilibrium and geometry sets shown in Fig. 5.61(e), we have

$$3PL\theta = (M_{B2})(-\theta) + (+0.557M_p)(+\theta) + (M_{I2})(-\theta) + (+M_p)(+\theta)$$
$$+ (M_{K2})(+\theta) + (-M_p)(-\theta),$$

which gives the beam end moment:

$$M_{I2} = -0.689M_p - M_{B2} + M_{K2}.$$

Moment M_{B1}: From the equilibrium of joint B shown in Fig. 5.61(f), we have the left column end moment:

$$M_{B1} = 0.91M_p + M_{B2}.$$

Moment M_{I4}: From the equilibrium of joint I shown in Fig. 5.61(g), we have the middle column end moment:

$$M_{I4} = 2.311M_p - M_{B2} + M_{K2}.$$

Moment M_{K3}: From the equilibrium of joint K shown in Fig. 5.61(h), we have the right column end moment:

$$M_{K3} = -1.5M_p + M_{K2}.$$

The conditions that $|M_{B2}|$, $|M_{K2}|$, and $|M_{I2}|$ be less than M_p and $|M_{K3}|$, $|M_{B1}|$, and $|M_{I4}|$ be less than $3M_p$ lead to

$$-M_p \le M_{B2} \le M_p \tag{5.11.34}$$

$$-M_p \le M_{K2} \le M_p \tag{5.11.35}$$

$$-M_p \le -0.689M_p - M_{B2} + M_{K2} \le M_p \tag{5.11.36}$$

$$-3M_p \le -0.91M_p + M_{B2} \le 3M_p \tag{5.11.37}$$

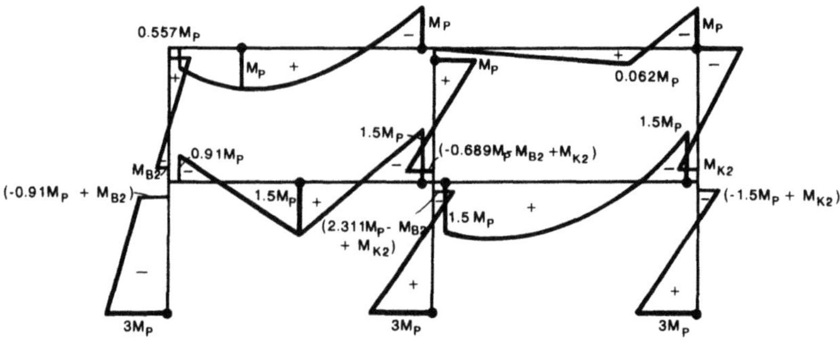

FIGURE 5.62. Moment diagram for failure mechanism (i) in Figure 5.59.

$$-3M_p \leq 2.311M_p - M_{B2} + M_{K2} \leq 3M_p \qquad (5.11.38)$$

$$-3M_p \leq -1.5M_p + M_{K2} \leq 3M_p. \qquad (5.11.39)$$

Inequalities (5.11.34) to (5.11.39) can be satisfied by choosing the redundant moments $M_{B2} = M_{K2} = 0$. The final bending moment diagram for failure mechanism (i) in Fig. 5.59 is shown in Fig. 5.62. With $M_{B2} = M_{K2} = 0$, the moment condition is satisfied everywhere in the frame. So the moment check is complete and the solution $P = 1.082(M_p/L)$ is exact.

5.11.3 Analysis Procedures for Distributed Loads

Thus, in general, the procedure for dealing with distributed loads can be summarized as follows:

i. Determine the "correct" collapse mechanism by the usual method of combining mechanisms, assuming that plastic hinges occur at midspan for those members carrying uniformly distributed loads.
ii. Perform a moment check and determine the moments at the ends of the members and at the midspan in the loaded members.
iii. Determine the maximum moments in the members carrying uniformly distributed loads and their respective locations.
iv. Adjust the plastic hinges in the "correct" mechanism under (i) to the positions of maximum moments under (iii) and analyze the resulting mechanism by the work method. The corresponding value of M_p will then be the required value for the actual distributed loads.
v. Perform a moment check for the adjusted mechanism.

5.12 Summary

Out of the two basic methods of plastic analysis and design, the equilibrium method was described in Chapter 4. The work method, based on the upper-bound theorem of limit analysis, was presented herein. Like the equilibrium method, the work method consists of two stages of operation. In the first stage, a valid mechanism is assumed or obtained through a combination of independent mechanisms and the corresponding work equation is formed by equating the external work to internal dissipation of energy at the plastic hinge locations, which is then solved to determine the plastic limit load or the required plastic moment capacity of the structure.

In the second stage, a moment check is performed to show that for the postulated mechanism, the moments everywhere in the structure shall not exceed the plastic moment capacities of the corresponding members. If the structure at collapse is statically determinate, then the moment check can be performed by simple statics or the virtual work equation. However, difficulties arise when the actual collapse mechanism is of the partial type, in which only a portion of the frame is statically determinate at collapse. To carry out a moment check for this kind of mechanism, it is a simple matter to construct a uniquely determined bending moment diagram for that part of the frame that is statically determinate at collapse, by the usual procedure and see whether the fully plastic moment is exceeded anywhere within this part of the distribution. This is followed by examining the remainder of the frame, in which the distribution of bending moment is not uniquely determined. To this end, the virtual work equation can be used conveniently to express the unknown moments in terms of the redundants, from which we can check and see whether there exists at least one set of values of redundants such that the resulting moment diagram will not violate the plastic moment condition. Such an investigation may prove to be of great difficulty, if the noncollapse portion of the frame is highly redundant. The solution of the work equation is exact only when the plastic moment condition is not violated anywhere in the structure; otherwise the solution is unsafe, i.e., the computed plastic limit load is higher than the exact plastic limit load or the computed required plastic moment is lower than the exact required plastic moment.

The work method is applied to the analysis and design of rectangular and gable frames. The application of the work method to gable frames involves complex geometrical calculations for determining the displacements in the direction of applied loads and rotation at plastic hinges. The methods of instantaneous center and the virtual work equation simplifies these geometrical calculations and are used in examples of analysis and design of complex gable frames.

The method of combining mechanisms is described and applied to the analysis and design of frames including gable and multistory and multibay frames. The basic concept underlying the method of combining mechanisms is that for a given frame and loading, every possible collapse mechanism can be obtained as some combination of a certain number of independent mechanisms. Once these independent mechanisms are identified, the work equation for each combination can be obtained and solved to determine the corresponding collapse load or required plastic moment. The basic aim in combining mechanisms is to maximize the external work and minimize the internal dissipation of energy. In this way, the lowest possible loads or highest plastic moment is obtained.

For structures having members with distributed loads, the exact location of the plastic hinge is not known in advance and its determination requires the use of differential calculus. However, the final solution for these structures is not very sensitive to the location of the plastic hinges, and in most practical cases, a safe solution with small error can be obtained by substituting the uniform load with a pattern of equivalent concentrated loads.

References

5.1. Chen, W.F., "A Rapid Method of Computing Geometric Relations in Structural Analysis," *Civil Engineering Magazine*, ASCE, New York, May 1983.

5.2. Drucker, D.C., *Introduction to Mechanics of Deformable Solids*, McGraw-Hill, New York, 1967.

5.3. Hodge, P.G., Jr., *Plastic Analysis of Structures*, McGraw-Hill, New York, 1959.

5.4. Massonnet, C.E., and Save, M.A., *Plastic Analysis and Design, Vol. 1, Beams and Frames*, Ginn., New York, 1965.

5.5. Neal, B.G., *The Plastic Methods of Structural Analysis*, third edition, Chapman and Hall, London, 1977.

Problems

5.1. The two-bay frame as shown in Fig. P5.1 has uniform cross section with a full plastic moment of 100 units. Find the collapse load factor [16/9].

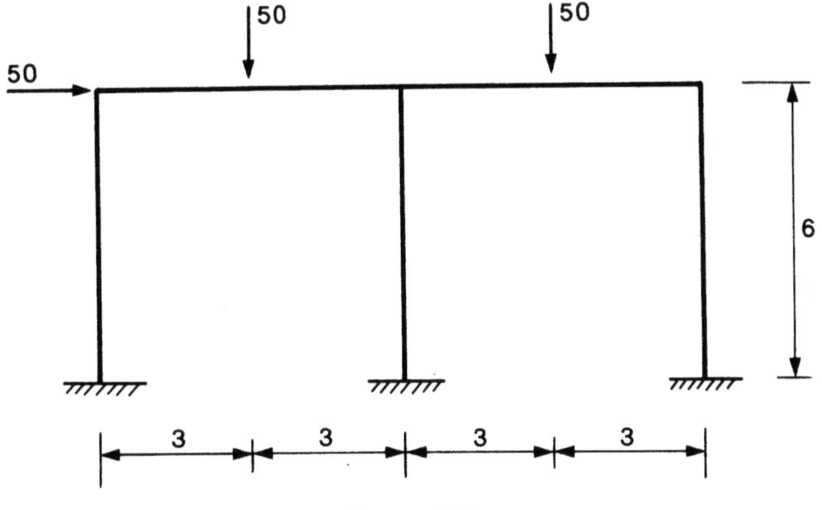

FIGURE P5.1

5.2. The two-story frame as shown in Fig. P5.2 has uniform section with a full plastic moment of 45 units. Determine the collapse load factor [2.00].

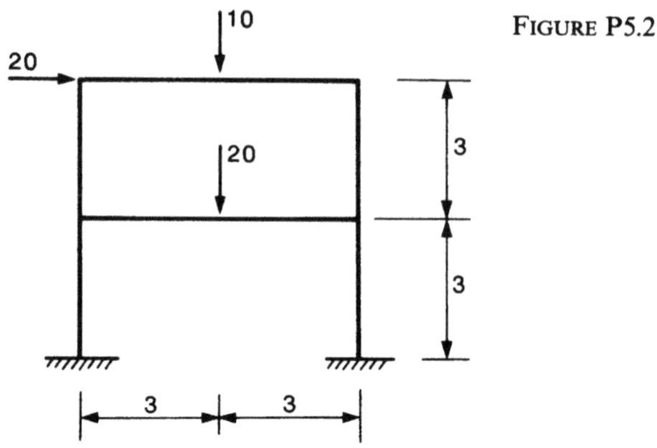

5.3. Repeat Example 5.2 if the load of 10 units on the top beam is replaced by a load of (a) 15 units and (b) 20 units [2.00, 1.875].

5.4. A two-story frame has full plastic moment capacity of its members as shown in Fig. P5.4. Find the collapse load factor [19/9].

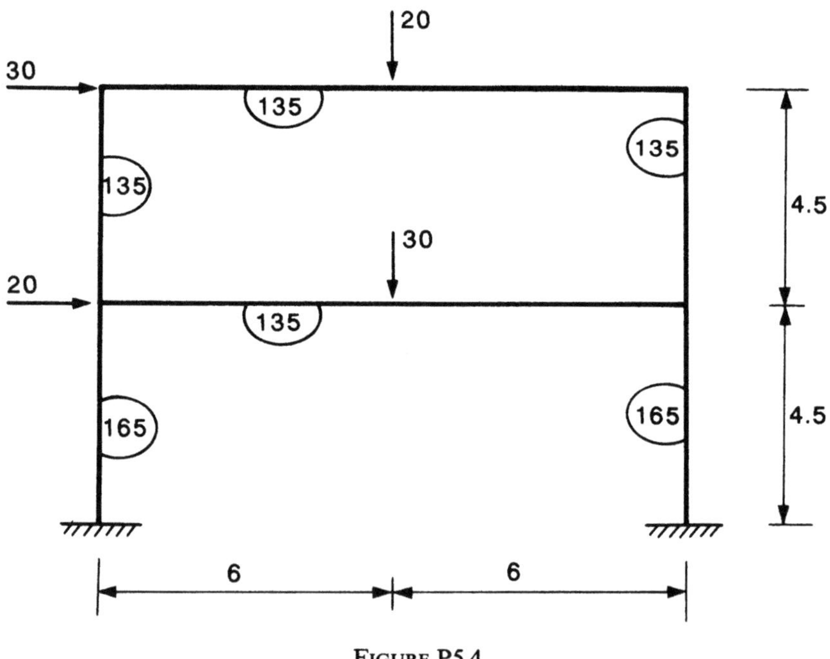

Figure P5.4

5.5. The two-story two-bay frame shown in Fig. P5.5 has uniform section with a full plastic moment of 10 units. Find the collapse load factor [2.50].

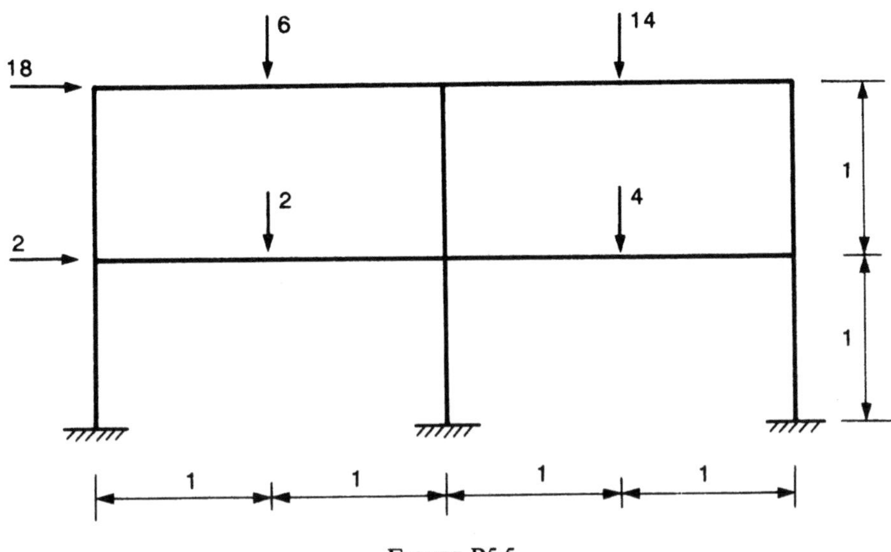

FIGURE P5.5

5.6. Find the value of W to cause collapse of the two-story tow-bay frame as shown in Fig. P5.6. The vertical loads act as midpoints of the beams and the full plastic moments are marked in the figure against each member $[3.60\ M_p/l]$.

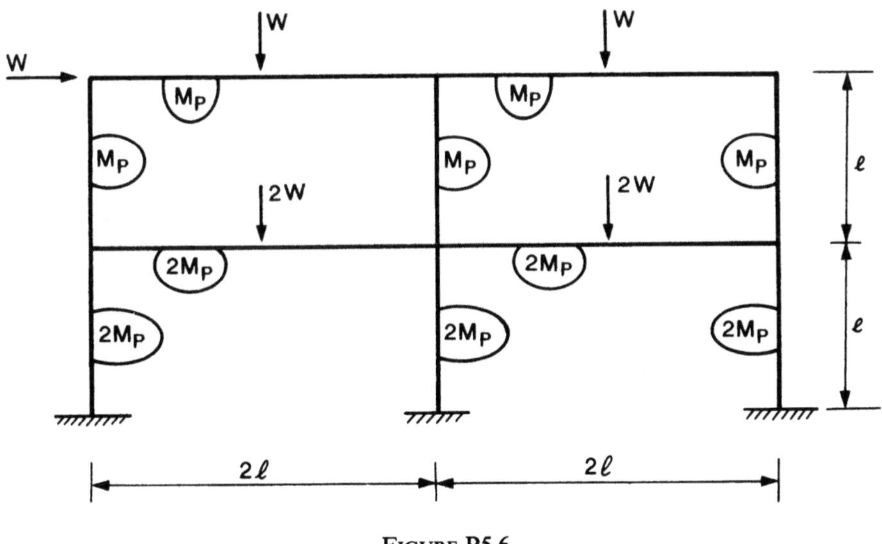

FIGURE P5.6

5.7. The frame as shown in Fig. P5.7 has uniform cross section with a full plastic moment M_p. Find the value of W that will just cause collapse $[2.5\ M_p/l]$.

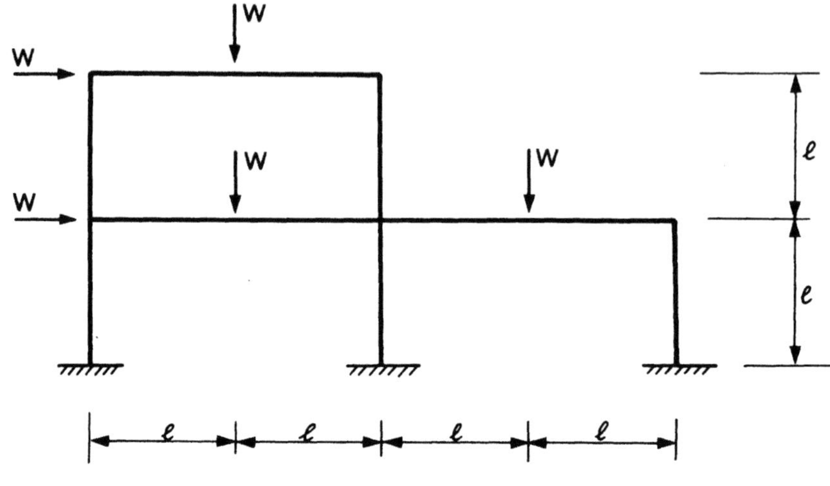

FIGURE P5.7

5.8. A three-story two-bay frame is subjected to vertical distributed and horizontal concentrated loads as shown in Fig. P5.8. The full plastic moments are marked in the figure against each member. Determine the load factor λ [1.976].

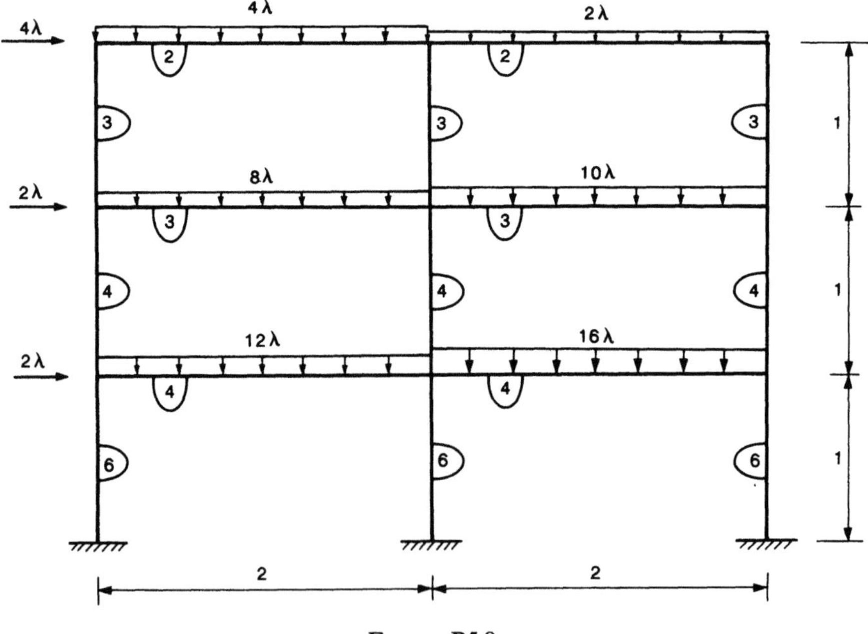

FIGURE P5.8

5.9. The two-bay frame shown in Fig. P5.9 is to be designed to have a uniform section with full plastic moment M_p. The loads W act at midspan of the beams. Construct an interaction diagram with axes Ph/M_p and Wl/M_p from which the mode of collapse and the value of required M_p may be determined for all ratios of Ph/Wl. Assume that P and W are always positive. For given $h = 3$ units, $l = 6$ units, $P = 50$ units, and $W = 30$ units, determine M_p [30].

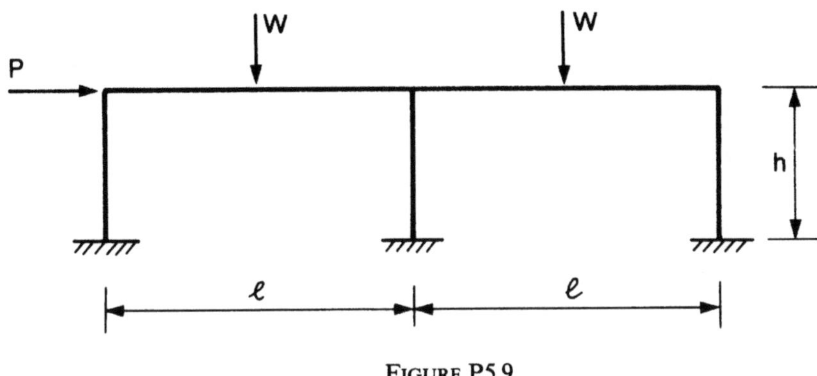

FIGURE P5.9

5.10. For the two-story frame shown in Fig. P5.10, the ratio of full plastic moment of columns to that of beams is 0.8. Investigate the plastic collapse behavior of the frame for various ratios of W to H. Present the results in graphical form.

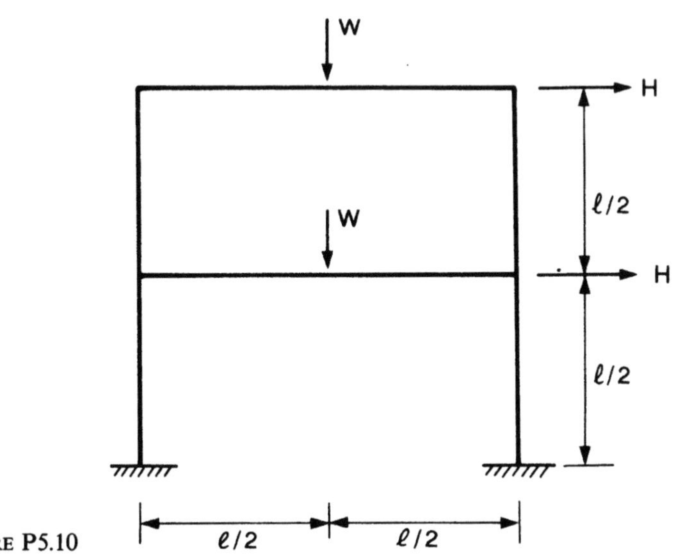

FIGURE P5.10

5.11. A gable frame is subjected to concentrated vertical and horizontal loads as shown in Fig. P5.11. Assuming that the frame has a uniform cross section, determine the required plastic moment M_p [30.75].

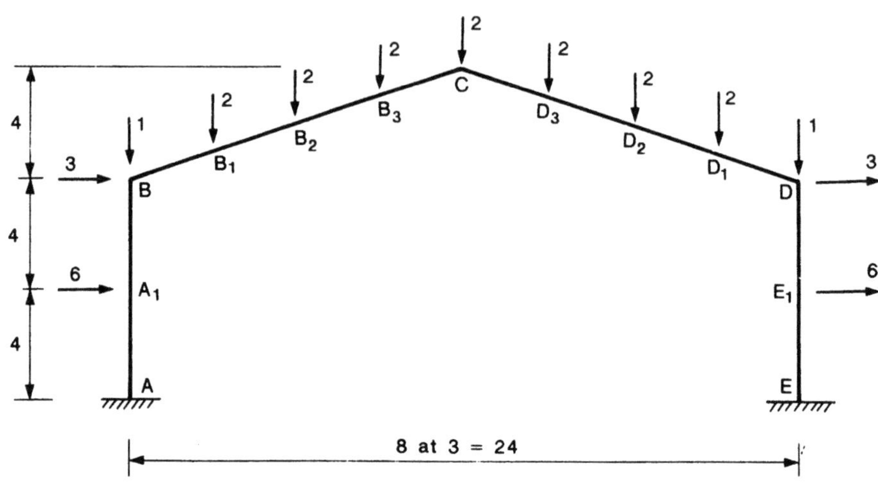

FIGURE P5.11

5.12. A two-bay gable frame with fixed feet is subjected to concentrated loads as shown in Fig. P5.12. Determine the required plastic moment for $W = 28$ kips and $H = 22.4$ kips. Assume that the frame has uniform cross section [312 kip-ft].

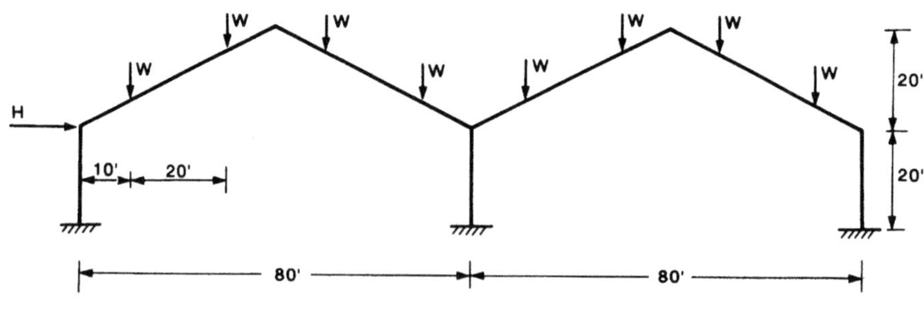

FIGURE P5.12

5.13. Repeat Problem 5.12 for $W = 37$ kips and $H = 0$. Compare the results with that from Problem 5.12 [423 kip-ft].

5.14. A three-bay gable frame is subjected to the loading as shown in Fig. P5.14. The full plastic moments are marked in the figure against each member in terms of M_p. If $P = 18.5$ kips and $H = 0$,
  (a) find M_p by using k_1, k_2, and k_3 proportional to the corresponding spans, i.e., $k_1 = (80/60)^2 = 1.78$; $k_2 = (100/60)^2 = 2.78$; and $k_3 = 1$.
  (b) revise k_1, k_2, and k_3 to have simultaneous collapse in all three spans [370 kip-ft; 1.74, 2.5, and 1].

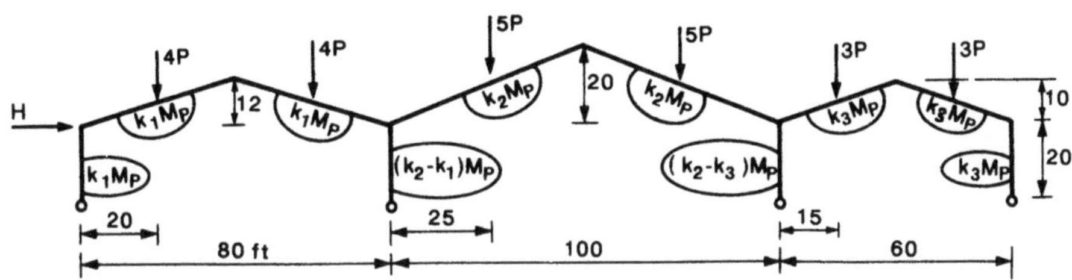

FIGURE P5.14

5.15. Determine the value of required M_p for the three-bay gable frame of Problem 5.14, given that $P = 14.0$ kips and $H = 28.0$ kips. Use k_1, k_2, and K_3 revised in Problem 5.14 [304 kip-ft].

5.16. Repeat Problem 5.16 when the direction of horizontal load is reversed [301 kip-ft].

5.17. A uniform beam of length L and full plastic moment M_p is simply supported at one end and fixed at the other end. A concentrated load Q is applied anywhere within the span. Use the work method to
  (a) find the smallest value of M_p that will carry the load Q in its most unfavorable position.
  (b) construct the bending moment diagram for the beam assuming that plastic hinges occur at the midspan and the fixed end. Find the greatest bending moment in this distribution for the load Q applied anywhere in the span, and hence determine a safe value of M_p.

5.18. The maximum bending moment in a member of length L subjected to a uniformly distributed load W is given by

$$M_{\max} = M_C + \frac{Wx_0^2}{2L}, \quad x_0 = \frac{M_L - M_R}{W}$$

in which M_R, M_L, and M_C are the bending moments on the right-hand and left-hand ends and the central span of the beam, respectively. The maximum bending moment within the span occurs at a distance x_0 to the left of the center bending moment within the span occurs at a distance x_0 to the left of the center of the beam. Thus, if M_L, M_C, and M_R have been calculated in a moment check, the value of x_0 and M_{max} can be found from these general formulas.

(i) Derive these general formulas using the virtual work equation.

(ii) What is the value of M_{max} when the value of x_0 exceeds $L/2$ in magnitude?

(iii) Apply these results to the particular portal frame example shown in Fig. 5.44.

6
Estimate of Deflections

6.1 Introduction

In most structures, the primary concern is that the structure should have adequate strength. Deflections are mostly of secondary concern. The calculations of deflections are important, however, because excessive deflections may hinder the operation of moving parts or the closing of doors or they may cause the cracking of a plaster-finished ceiling. Also, deflection control is needed to limit the lateral deflection (drift) of tall buildings. In fact, the simple plastic theory is valid only when the deflections at the collapse load (P-Δ effect) are small.

The hinge-by-hinge method of tracing the load-deflection curve and computing deflections at collapse load was introduced in Chapter 1. For determining deflections at the collapse load by this method, a series of sequential elastic analyses is carried out. In this chapter, we will first describe and illustrate the slope-deflection and virtual work methods by which the deflections at collapse load can be determined in one-step analysis, provided that the location of the last plastic hinge is known in advance. Then, we will present the deflection theorem, which is very useful in the estimation of collapse deflections, if the location of the last plastic hinge is not known in advance. Finally, we will illustrate the computations of deflections for simple beams, simple frames, and multistory and multibay frames by the use of the virtual work method [1.8, 1.10, 6.1].

6.2 Deflections at Collapse and Working Loads

After the collapse mechanism is formed, the structure becomes unstable and its deflections are unrestricted as shown in the load-deflection curve in Fig. 1.11 of a fixed-ended beam in Chapter 1. However, just before the formation of the plastic mechanism, elastic continuity still exists at the point of the last plastic hinge formation and the structure is stable. The deflections at this point can thus be determined by any method suitable for computing the deflections in the elastic range. However, to compute deflections at the collapse load in a one-step analysis, the location of the last plastic hinge must be known in advance.

The determination of exact deflection at working load may involve elastic-plastic or elastic analysis and therefore require relatively more calculation steps. However, an upper limit on the deflection δ_w at working load can be determined from the deflection at collapse load by assuming a linear relationship between the deflection and the load, i.e., by dividing the calculated deflection δ_c at collapse load by the factor λ, that is,

$$\delta_w = \frac{\delta_c}{\lambda}. \tag{6.2.1}$$

The error in the estimated deflection at working load by this approach will increase with an increase in the number of statical indeterminacy. The procedure, however, serves its purpose when the estimated deflections are less than the prescribed deflection limit.

6.3 Slope Deflection Method

The slope at an end of a beam subjected to end moments and lateral loads (Fig. 6.1) can be expressed as [1.8, 1.10]

$$\theta_A = \theta_A' + \frac{\Delta}{L} + \frac{L}{3EI}\left(M_{AB} - \frac{1}{2}M_{BA}\right) \tag{6.3.1}$$

in which M_{AB} and M_{BA} are end moments, L and EI are, respectively, length and bending stiffness of the beam, Δ is lateral translation between the two ends of the beam, and θ_A' is the slope at end A of a simply supported beam due to the given lateral loading on the segment. The end moments M_{AB} and M_{BA} at collapse load are determined from the plastic analysis. Note that in this method, moments and rotations are positive when clockwise and vice-versa. The deflections are then determined by solving equations formed by using the compatibility condition as illustrated in the following examples.

Example 6.3.1. The collapse mechanism and its corresponding bending moment diagram of the fixed-ended beam of Example 1.8.2 in Chapter 1 are shown in Fig. 6.2. Determine the vertical deflection at B at the collapse load by the slope deflection method. Assume the last plastic hinge to form, in turn, at A, B, and C.

Solution: Last Hinge at A: For the last plastic hinge to form at A, θ_A must be equal to zero. The angle θ_A can also be expressed in terms of deflection at B by applying the slope-deflection equation to segment AB [Fig. 6.2(c)].

$$\theta_A = \theta_A' + \frac{\delta_{BA}}{L/3} + \frac{L/3}{3EI}\left(M_{AB} - \frac{M_{BA}}{2}\right) = 0 \tag{6.3.2}$$

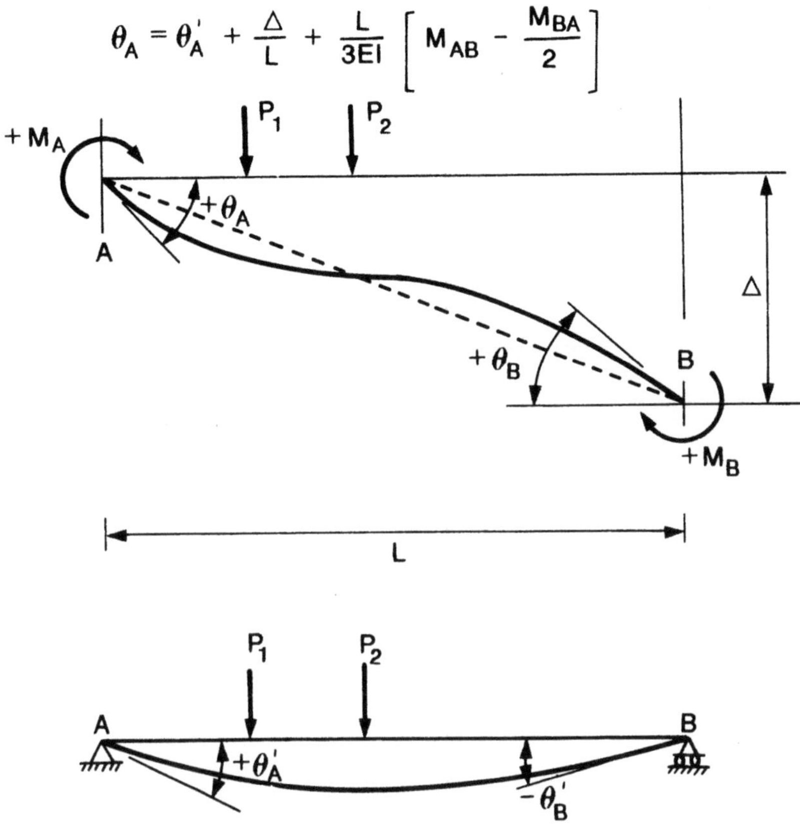

$$\theta_A = \theta'_A + \frac{\Delta}{L} + \frac{L}{3EI}\left[M_{AB} - \frac{M_{BA}}{2}\right]$$

FIGURE 6.1. Sign convention and nomenclature for slope-deflection equation.

where δ_{BA} is the deflection at B corresponding to the last hinge at A. Since there is no lateral load on segment AB, $\theta'_A = 0$. So by substituting $\theta'_A = 0$, $M_{AB} = -M_p$, and $M_{BA} = -M_p$ [Fig. 6.2(c)], we have

$$\frac{\delta_{BA}}{L/3} + \frac{L/3}{3EI}\left(-M_p + \frac{M_p}{2}\right) = 0, \tag{6.3.3}$$

which gives

$$\delta_{BA} = \frac{M_p L^2}{54EI}. \tag{6.3.4}$$

Last Hinge at B: For the last plastic hinge to form at B, we have the continuity at B, that is, $\theta_{BA} = \theta_{BC}$. The slope-deflection equations of segments AB and BC [Fig. 6.2(d)] provide θ_{BA} and θ_{BC}, respectively, as

$$\theta_{BA} = \frac{\delta_{BB}}{L/3} + \frac{L/3}{3EI}\left(-M_p + \frac{M_p}{2}\right) = \frac{3\delta_{BB}}{L} - \frac{M_p L}{18EI} \tag{6.3.5}$$

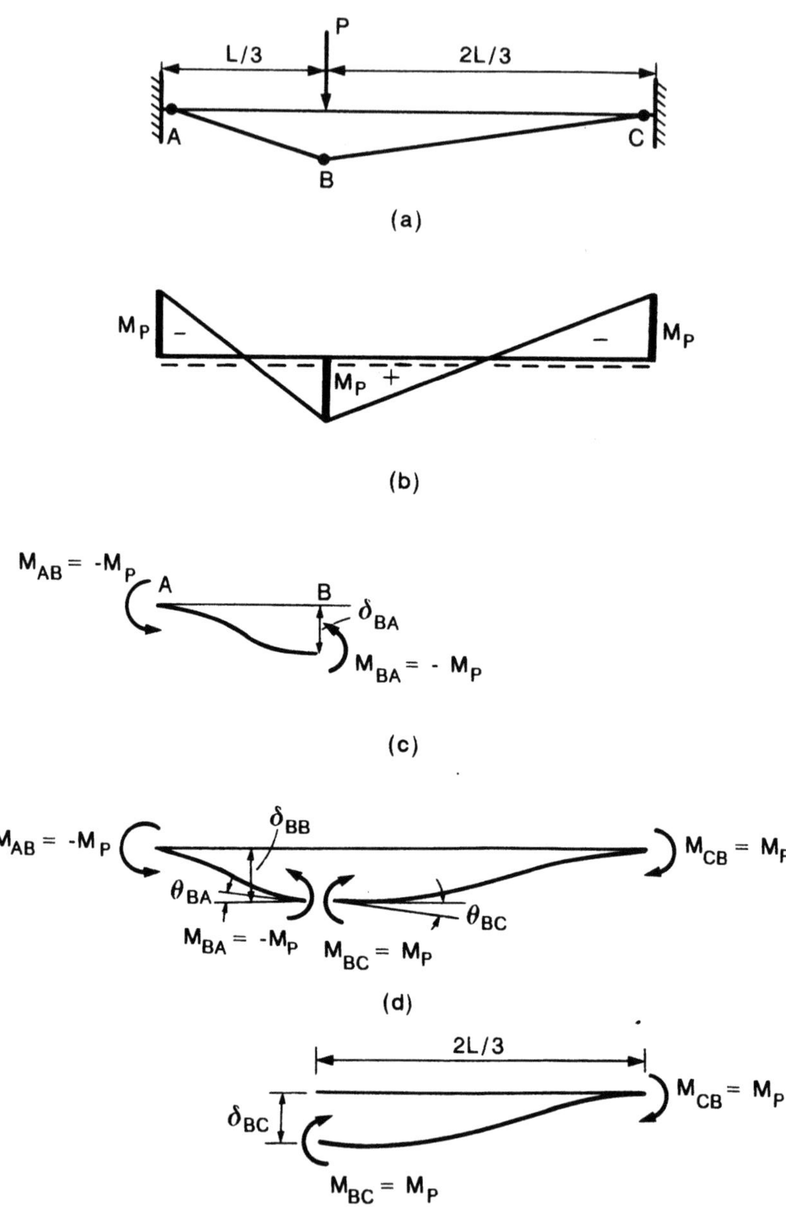

FIGURE 6.2. Deflection calculation of fixed-ended beam with concentrated load at the third point by slope-deflection method.

$$\theta_{BC} = -\frac{\delta_{BB}}{2L/3} + \frac{2L/3}{3EI}\left(M_p - \frac{M_p}{2}\right) = -\frac{3\delta_{BB}}{2L} + \frac{M_pL}{9EI}. \qquad (6.3.6)$$

By equating θ_{BA} and θ_{BC} and simplifying, we have

$$\delta_{BB} = \frac{M_pL^2}{27EI}. \qquad (6.3.7)$$

Last Hinge at C: For this case, we must have $\theta_C = 0$. The slope-deflection equation for segment CB [Fig. 6.2(e)] gives θ_C as

$$\theta_C = -\frac{\delta_{BC}}{2L/3} + \frac{2L/3}{3EI}\left(M_p - \frac{M_p}{2}\right) = 0, \qquad (6.3.8)$$

which gives

$$\delta_{BC} = \frac{2}{27}\frac{M_pL^2}{EI}. \qquad (6.3.9)$$

Example 6.3.2. The collapse mechanism and the bending moment diagram for a fixed-ended beam of Example 4.4.1 in Chapter 4 are shown in Fig. 6.3. Determine the vertical deflection at midspan at the collapse load. Assume the last plastic hinge to form, in turn, at A and B.

Solution: Last Hinge at A: For this case, we must have $\theta_A = 0$. From the slope-deflection equation for segment AB [Fig. 6.3(c)], θ_A, can be expressed as

$$\theta_A = \theta'_A + \frac{\delta_{BA}}{L/2} + \frac{L/2}{3EI}\left(-M_p + \frac{M_p}{2}\right) = 0 \qquad (6.3.10)$$

where θ'_A in this case is the end slope of a simply supported beam of length $L/2$ subjected to a uniformly distributed load w. Using $w = 16M_p/L^2$ (Example 4.4.1), θ'_A can be expressed as

$$\theta'_A = \frac{M_pL}{12EI}. \qquad (6.3.11)$$

So Eq. (6.3.10) becomes

$$\frac{M_pL}{12EI} + \frac{2\delta_{BA}}{L} - \frac{L}{6EI}\frac{M_p}{2} = 0, \qquad (6.3.12)$$

which gives

$$\delta_{BA} = 0. \qquad (6.3.13)$$

Last Hinge at B: For this case, we must have continuity at B, and that leads to $\theta_{BA} = \theta_{BC} = 0$. Applying the slope-deflection equation to segment AB [Fig. 6.3(d)], θ_{BA} can be written as

$$\theta_{BA} = \theta'_B + \frac{\delta_{BB}}{L/2} + \frac{L/2}{3EI}\left(-M_p + \frac{M_p}{2}\right) = 0 \qquad (6.3.14)$$

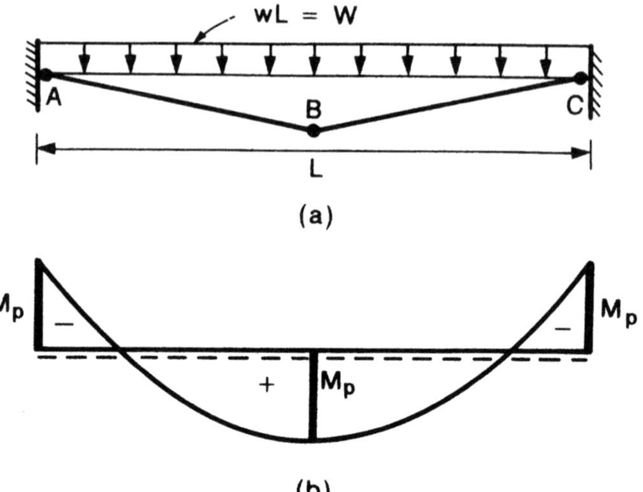

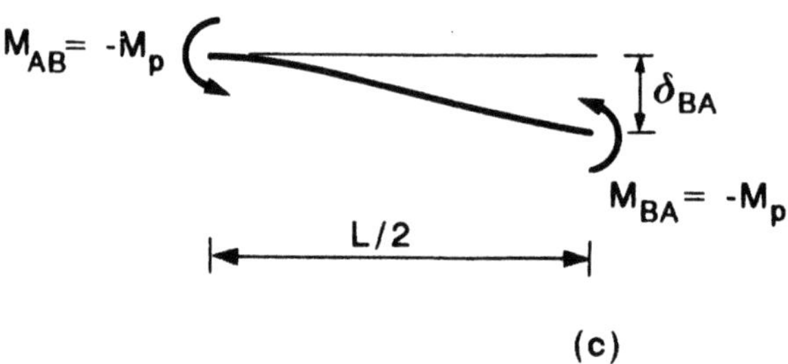

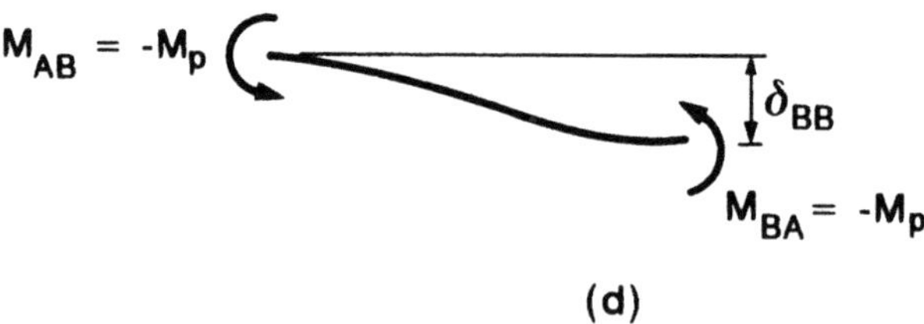

FIGURE 6.3. Deflection calculation of fixed-ended beam with uniformly distributed load by slope-deflection method.

in which θ'_B is

$$\theta'_B = -\frac{M_pL}{12EI}. \tag{6.3.15}$$

Eq. (6.3.14) becomes

$$-\frac{M_pL}{12EI} + \frac{2\delta_{BB}}{L} - \frac{M_pL}{12EI} = 0, \tag{6.3.16}$$

which gives

$$\delta_{BB} = \frac{M_pL^2}{12EI}. \tag{6.3.17}$$

6.4 Dummy Load Method

In this method, the deflections can be obtained by the virtual work equation involving the integration of the product of two moment diagrams. One moment diagram is of the actual structure under given loads and the other is of an auxiliary structure under a dummy load. The integral is often conveniently evaluated by a graphical procedure. The derivation of the virtual work equation and the corresponding graphical procedure of integration are presented in the forthcoming.

6.4.1 Virtual Work Equation

The virtual work equation is derived by considering an auxiliary structure obtained from the actual structure by eliminating the last plastic hinge and replacing the rest of them by real frictionless hinges. A dummy unit load is applied to this structure at the point and in the direction of desired deflection. Let m be a set of internal moments in equilibrium with this unit load.

Let δ be the desired deflection and Φ be the set of internal deformations (curvatures) in the actual structure under the actual loads. Since the auxiliary structure differs from the actual structure only by nature of the hinges, the displacement δ and the curvature Φ in the actual structure are also compatible on the auxiliary structure. So δ and Φ, and the unit load and the corresponding moment m in the auxiliary structure, respectively, provide the compatible set and the equilibrium set in the auxiliary structure. The external and internal virtual work done by the equilibrium set in the auxiliary structure will thus be

$$\text{external virtual work: } (1)\delta = \delta \tag{6.4.1}$$

$$\text{internal virtual work: } \int m\Phi\, ds \tag{6.4.2}$$

in which ds is the distance along a member of the structure. Since we assume a elastic–perfectly plastic moment-curvature relationship, the members are fully elastic between the plastic hinges. The curvature Φ in Eq. (6.4.2) can thus be replaced by M/EI. Then by equating external virtual work to internal virtual work, we have

$$\delta = \int \Phi m \, ds = \int \frac{M}{EI} m \, ds \tag{6.4.3}$$

in which δ = desired deflection; M = moments in the actual structure under actual loads; m = moments in the auxiliary structure due to the unit load at the point and direction of the desired deflection; ds = distance along the member; E = Young's modulus; and I = moment of inertia of cross section of the member.

The integration in Eq. (6.4.3) is over the whole structure and is carried out through the intervals where all functions exist and are piecewise continuous. By defining limits at joints, concentrated loads, and hinges, discontinuities in the curvature functions are readily handled.

6.4.2 Graphical Procedures

If we have specific expressions for M and m, the deflection δ can be obtained by substituting these expressions in Eq. (6.4.3) and carrying out the necessary integrations. However, in most cases, it is convenient to obtain this deflection δ directly by feeding the coordinates of m and M diagrams into the algebraic expressions derived by integrating various shapes of moment diagram. Note that the m-diagram will always be linear since it corresponds to a concentrated unit load.

For example, assume that both m and M diagrams of a structural member are linear as shown in Fig. 6.4(a). The value of this integral is in fact the volume of the three-dimensional figure shown and is

$$\int_0^L Mm \, dx = \frac{L}{6}[M_A(2m_A + m_B) + M_B(m_A + 2m_B)] \tag{6.4.4}$$

where m_A, m_B, M_A, and M_B are coordinates of m- and M-diagrams as shown in Fig. 6.4(a).

When the applied loads are distributed, the M-diagram is nonlinear. When the loads are uniformly distributed, the M-diagram is a second-degree parabola as shown in Figs. 6.4(b) and (c), the values of the integral for these two types of second-degree parabola are

$$\int_0^L Mm \, dx = \frac{L}{3} M_k(m_A + m_B) \tag{6.4.5}$$

and

$$\int_0^L Mm \, dx = \frac{L}{12} M_k(3m_A + 5m_B) \tag{6.4.6}$$

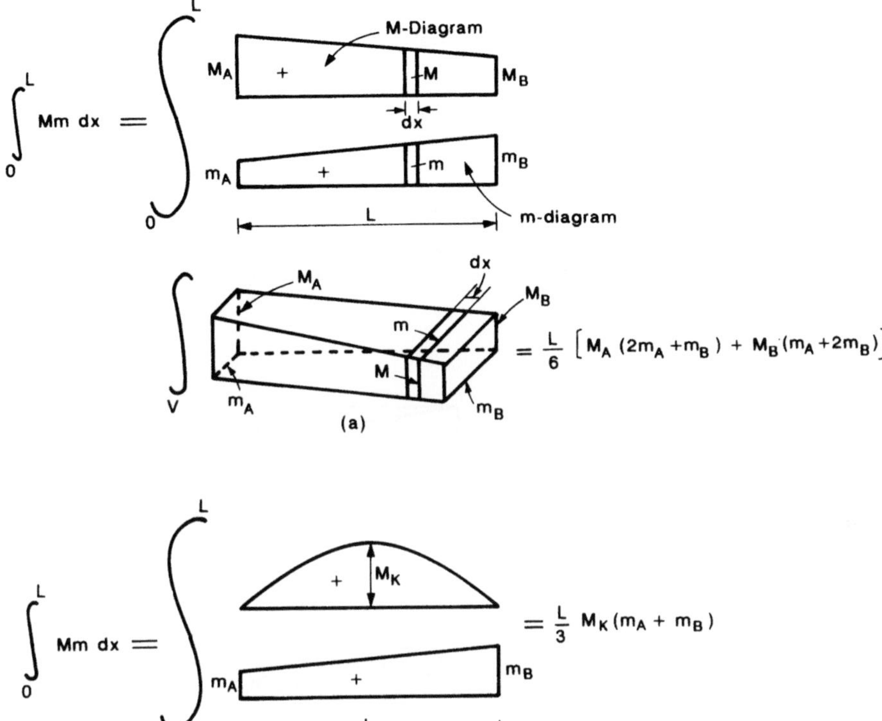

$$\int_0^L Mm\,dx = \frac{L}{6}\left[M_A(2m_A+m_B)+M_B(m_A+2m_B)\right]$$

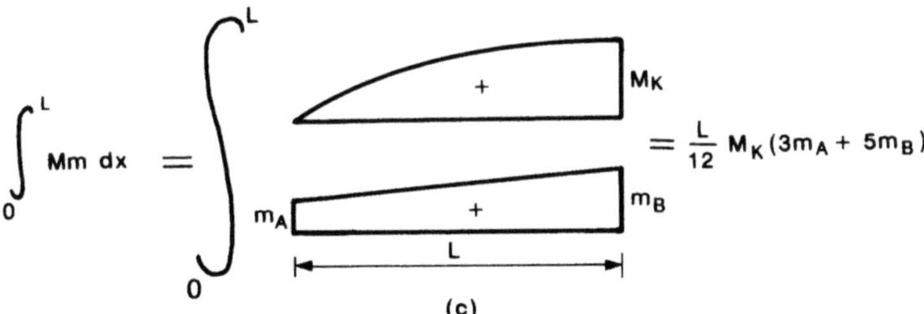

$$\int_0^L Mm\,dx = \frac{L}{3}M_K(m_A+m_B)$$

$$\int_0^L Mm\,dx = \frac{L}{12}M_K(3m_A+5m_B)$$

FIGURE 6.4(a, b, c). Graphical integration for various shapes of M-diagram.

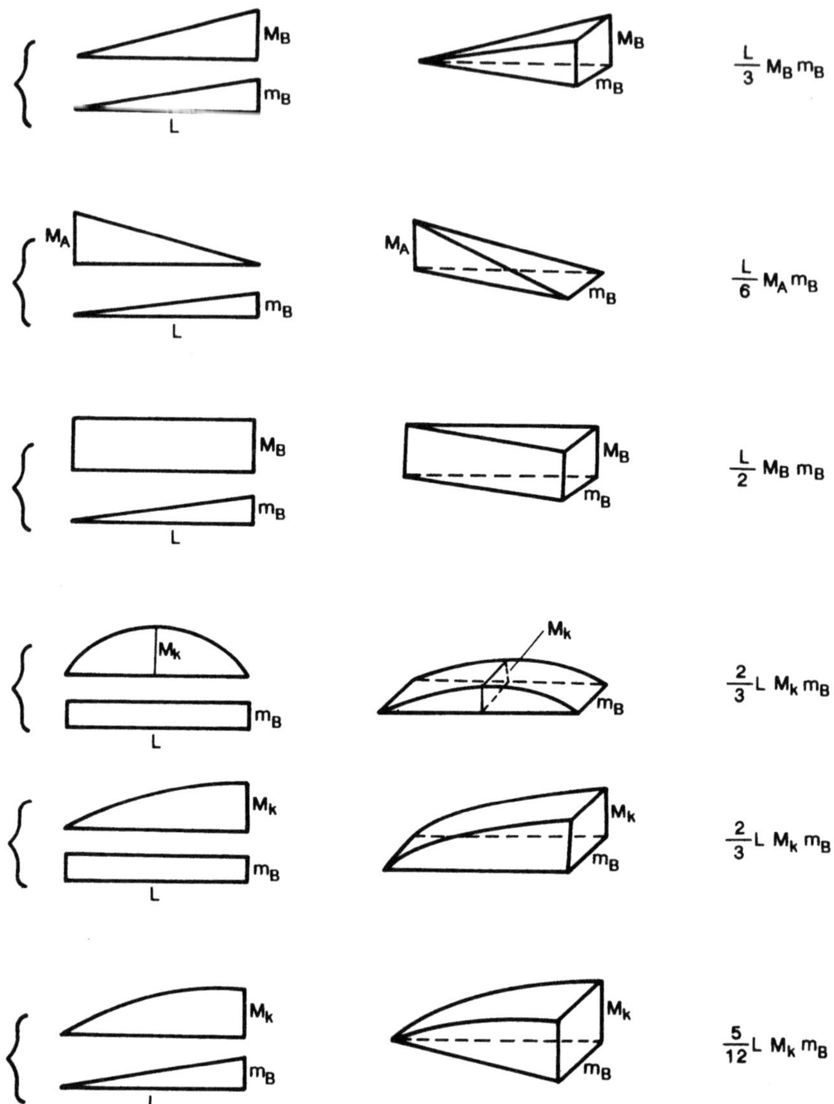

FIGURE 6.4(d). Volumetric formulas for integrating simple shapes of M and m diagrams.

where M_k is shown in Fig. 6.4(b, c). Thus, Eqs. (6.4.4), (6.4.5), and (6.4.6) can be used to determine deflections of a structure directly without going through the actual integration process. Volumetric formulas for integrating several simple shapes of M and m diagrams are summarized in Fig. 6.4(d).

The applications of the dummy load method to beams, simple frames, and multistory frames will be illustrated, respectively, in Sections 6.6, 6.7, and 6.8.

6.5 The Deflection Theorem

As discussed in Sections 6.2 and 6.3, the deflection at collapse load is determined by assuming continuity at the last plastic hinge. However, the plastic analysis using either the equilibrium method or the work method provides us only the position of the plastic hinges. But it does not provide any information regarding the sequence of formation of the hinges and thus the location of the last hinge. In the absence of this information, the deflection theorem assists in determining the deflection of a structure at its collapse load. It states that: If the deflections are calculated on the assumption that each hinge, in turn, is last to form, then the correct deflection at the collapes load is the maximum value obtained from various trials, and the corresponding hinge will in fact be the last to form in the collapse mechanism.

This theorem is based on the reasoning that every solution computed by making an incorrect assumption may be obtained from the correct solution by simply applying a rigid body motion to the structure in a direction opposite to the collapse mechanism.

For instance, consider the fixed-ended beam of Example 6.2.1, shown again in Fig. 6.5. If we assume that the last hinge is at section C (the right assumption), then the deflections are shown in Fig. 6.5(b). However, if we assume that the last plastic hinge is at section B (the wrong assumption), a "kink" or mechanism angle that should have been formed at section B will be removed and a negative slope discontinuity will be created at section C as shown in Fig. 6.5(c), where continuity should have existed. Note that the plastic moment condition will be satisfied only when we have negative slope at C.

To obtain deflections of Fig. 6.5(c) from the deflections of Fig. 6.5(b) the beam must be given a motion shown in Fig. 6.5(d). This motion, however, is opposite to the failure mechanism of the beam. Therefore, it will reduce all

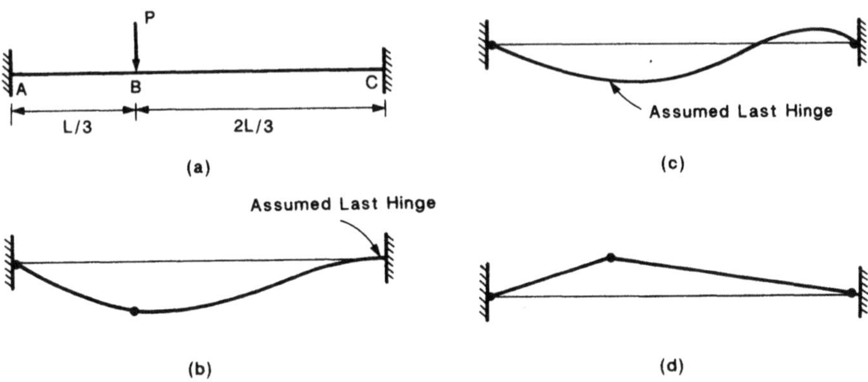

FIGURE 6.5. Logic behind the deflection theorem.

the displacements and rotations. Similarly, it can be shown that the deflections obtained by assuming the last plastic hinge at A will be less than the correct ones. Thus, it follows that the displacements computed by making a wrong assumption regarding the last plastic hinge will always be less than the correct deflections.

The theorem is very useful in the computation of deflections at collapse load. For example, by applying the *deflection theorem* to the deflections for these beams computed in Examples 6.3.1 and 6.3.2, the correct deflections for these beams at collapse loads are, respectively, chosen as

$$\delta_{BC} = \frac{2}{27} \frac{M_p L^2}{EI} \tag{6.5.1}$$

and

$$\delta_{BB} = \frac{M_p L^2}{12EI}. \tag{6.5.2}$$

Moreover, when deflections are being computed to ensure that the deflections at ultimate load are less than the permissible deflections, then the deflection theorem may be used to reduce the required number of trial calculations. For example, if the deflection from the first trial calculation is greater than the permissible deflection, then it is established that the correct deflection at ultimate load will be greater than the permissible deflection and no further trials are required.

6.6 Simple Beams

The dummy load method was presented in Section 6.4. Herein, we shall apply this method to two beam examples.

Example 6.6.1.   Redo Example 6.3.1 by the dummy load method.

Solution: The collapse mechanism for the beam is shown in Fig. 6.6(a). Herein, we will calculate the vertical deflection at B by successively assuming the last hinge to form at A, B, and C. The correct deflection will be the largest value out of these three calculations (the *deflection theorem*).

Last Hinge at A: The moment diagram (M-diagram) due to the applied load on the actual beam is shown in Fig. 6.6(b). Assume the last plastic hinge forms at A and apply a dummy unit load to the auxiliary structure at B in the vertical direction as shown in Fig. 6.6(c). The corresponding m-diagram is shown in Fig. 6.6(d). Using the M- and m-diagrams, the deflection at B corresponding to the last hinge at A, δ_{BA}, can graphically be expressed as in Fig. 6.6(e). Applying Eqs. (6.4.4) to these diagrams, δ_{BA} is reduced to

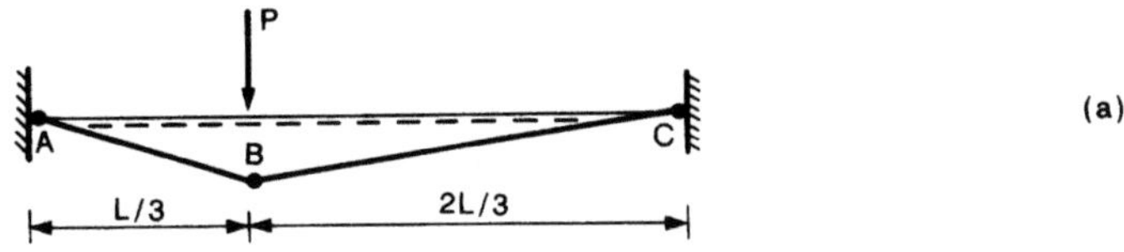

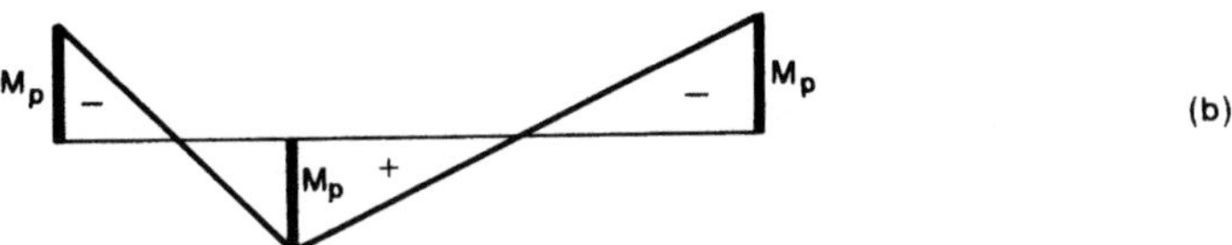

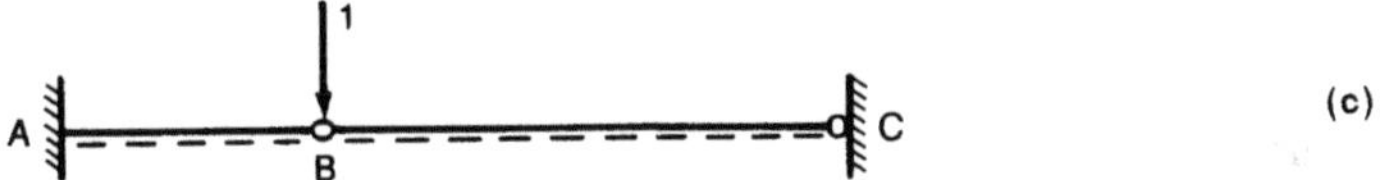

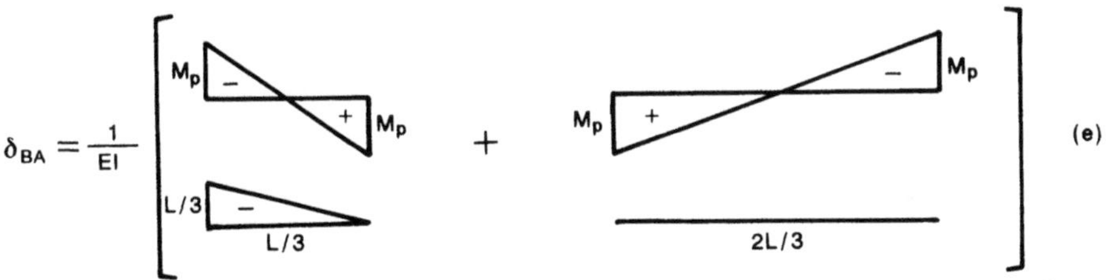

FIGURE 6.6. Deflection calculation of fixed-ended beam with concentrated load at the third point by dummy load method.

$$\delta_{BA} = \frac{L/3}{6EI}\left[-(M_p)\left(-\frac{2L}{3}\right) + (M_p)\left(-\frac{L}{3}\right)\right] = \frac{M_p L^2}{54EI}. \qquad (6.6.1)$$

Last Hinge at B: For this case, the auxiliary structure and the corresponding m-diagram are shown, respectively, in Fig. 6.6(f) and (g). Applying Eq. (6.4.4) to the diagram in, Fig. 6.6(h), δ_{BB} can be written as

$$\delta_{BB} = \frac{1}{6EI}\left[\left(\frac{L}{3}\right)(-M_p)\left(\frac{2}{9}L\right) + \left(\frac{L}{3}\right)(M_p)\left(\frac{4}{9}L\right)\right.$$
$$\left. + \left(\frac{2}{3}L\right)(M_p)\left(\frac{4}{9}L\right) + \left(\frac{2L}{3}\right)(-M_p)\left(\frac{2}{9}L\right)\right] \qquad (6.6.2)$$

or

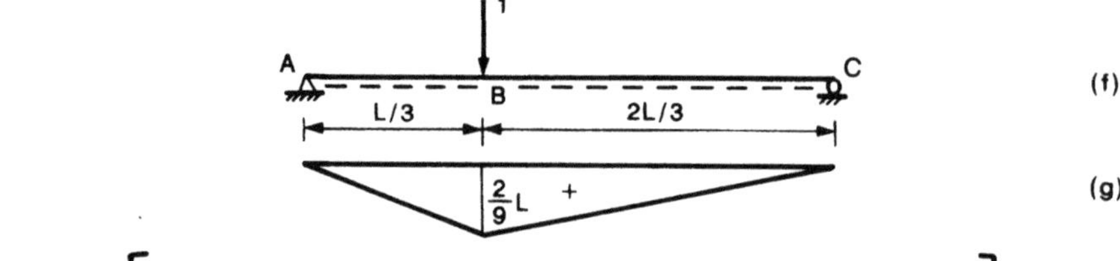

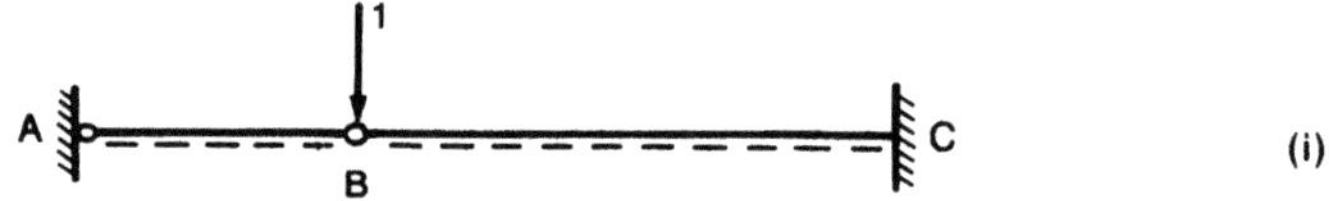

$$\delta_{BB} = \frac{M_p L^2}{27EI}. \tag{6.6.3}$$

Last Hinge at C: The auxiliary structure and the m-diagram corresponding to this case are shown, respectively, in Fig. 6.6(i) and (j). Using M- and m-diagrams, δ_{BC} can graphically be expressed as shown in Fig. 6.6(k). Applying Eq. (6.4.4) to these diagrams, δ_{BC} is reduced to

$$\delta_{BC} = \frac{1}{6EI}\left(\frac{2L}{3}\right)\left[M_p\left(0 - \frac{2L}{3}\right) - M_p\left(-\frac{4L}{3} + 0\right)\right] \tag{6.6.4}$$

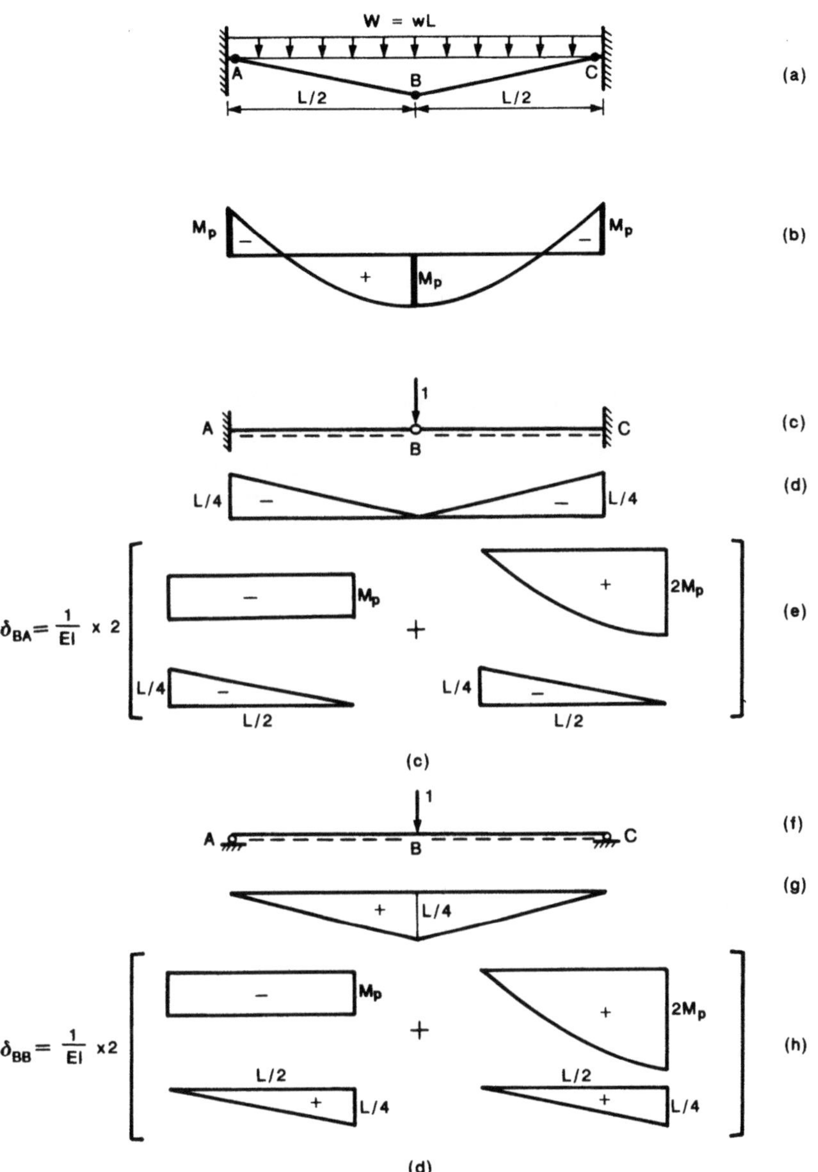

FIGURE 6.7. Deflection calculation of fixed-ended beam with uniformly distributed load by dummy load method.

or

$$\delta_{BC} = \frac{2}{27} \frac{M_p L^2}{EI}.$$

(6.6.5)

Using the deflection theorem, we can conclude that deflection at B is $\delta_B = \delta_{BC}$ as given by Eq. (6.6.5). Note that the results are the same as those in Example 6.3.1.

Example 6.6.2. Redo Example 6.3.2 by the dummy load method.

Solution: The collapse mechanism for the beam and the loading is shown in Fig. 6.7(a). The deflection at midspan will be determined by assuming the last plastic hinge to form first at A and then at B.

Last Hinge at A or C: The M-diagram is shown in Fig. 6.7(b). Corresponding to the last plastic hinge at A, the auxiliary structure and the m-diagram are shown, respectively, in Figs. 6.7(c) and (d). Using the M- and m-diagrams, the deflection at B corresponding to the last plastic hinge at A, δ_{BA}, can graphically be expressed in Fig. 6.7(e). Applying Eqs. (6.4.4) and (6.4.6) to these diagrams, we have

$$\delta_{BA} = \frac{2(L/2)}{6EI}\left\{(-M_p)\left(-2\frac{L}{4}\right) + (-M_p)\left(-\frac{L}{4}\right)\right\}$$
$$+ \frac{2(L/2)}{12EI}(2M_p)\left(-3\frac{L}{4}\right)$$

(6.6.6)

or

$$\delta_{BA} = 0.$$

(6.6.7)

Last Hinge at B: For this case, the auxiliary structure and the m-diagram are shown, respectively, in Figs. 6.7(f) and (g). Using the M- and m-diagrams, the deflection at B corresponding to the last hinge at B, δ_{BB}, can graphically be expressed in Fig. 6.7(h). Applying Eqs. (6.4.4) to these diagrams, we have

$$\delta_{BB} = \frac{2(L/2)}{6EI}\left\{(-M_p)\left(\frac{L}{4}\right) + (-M_p)\left(\frac{L}{2}\right)\right\} + \frac{2(L/2)}{12EI}(2M_p)\left(\frac{5L}{4}\right)$$

(6.6.8)

or

$$\delta_{BB} = \frac{M_p L^2}{12EI}.$$

(6.6.9)

Since $\delta_{BB} > \delta_{BA}$, we conclude that the deflection at B is $\delta_B = \delta_{BB}$ as given by Eq. (6.6.9). The results are the same as those in Example 6.3.2.

6.7 Simple Frames

The dummy load method has been applied to simple beams in the previous section. Herein, we shall illustrate its use for simple frames.

Example 6.7.1. The frame shown in Fig. 6.8 was analyzed previously in

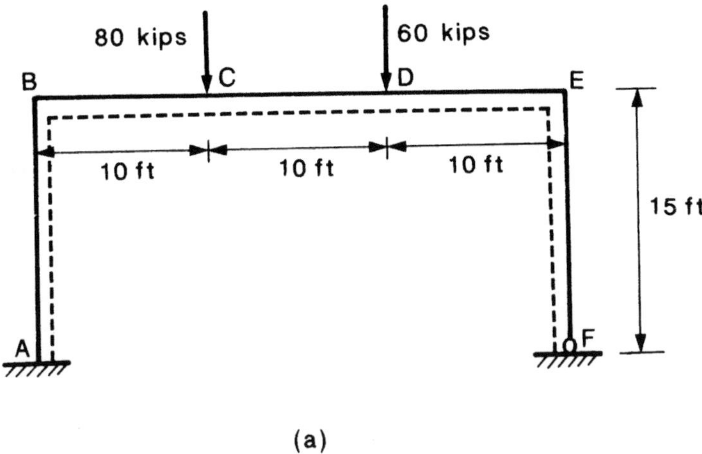

(a)

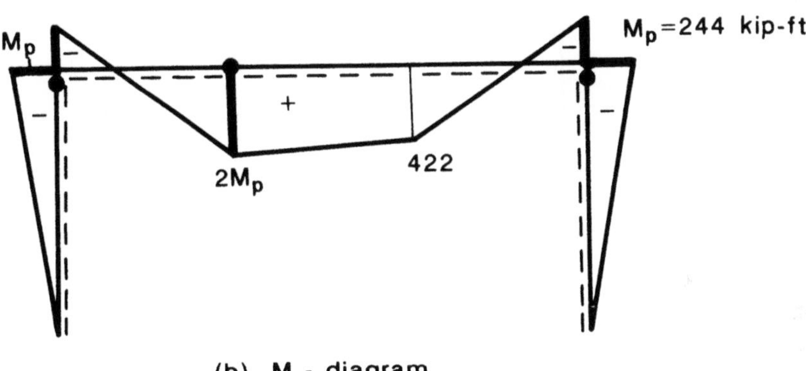

(b) M - diagram

FIGURE 6.8. Rectangular frame under vertical loads and the corresponding collapse moment diagram.

Example 5.5.1 in Chapter 5 and its collapse moment diagram (M-diagram) is shown in Fig. 6.8. Determine the vertical deflection at C by the dummy load method.

Solution: The plastic mechanism has plastic hinges at B, C, and E. We shall determine the vertical deflection at C by successively assuming the last hinge to form at C, B, and E. The lagrest of the three deflections gives the correct deflection (the *deflection theorem*).

Last Hinge at C: The auxiliary structure with a vertical unit load at C and the corresponding m-diagram are shown, respectively, in Fig. 6.9(a) and (b). The deflection δ_{CC} can graphically be expressed as in Fig. 6.9(c). Applying Eq. (6.4.4) to these diagrams, δ_{CC} is reduced to

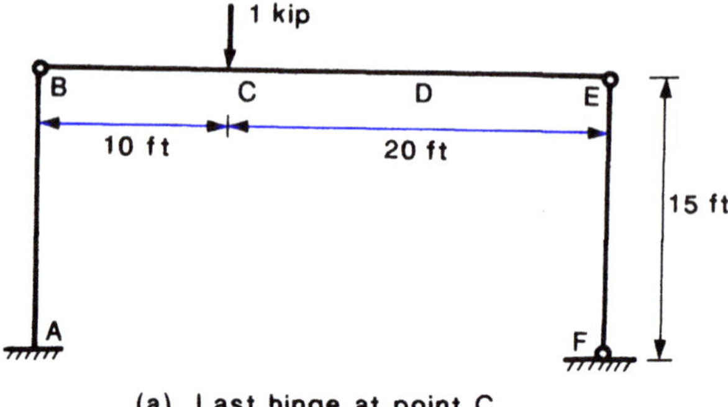

(a) Last hinge at point C

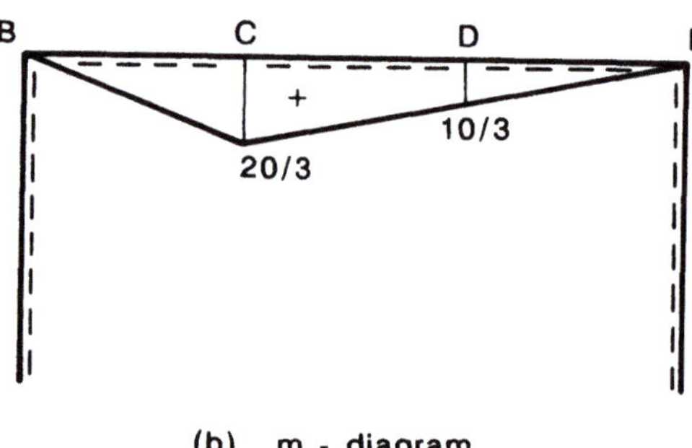

(b)   m - diagram

$$EI\,\delta_{CC} =$$ 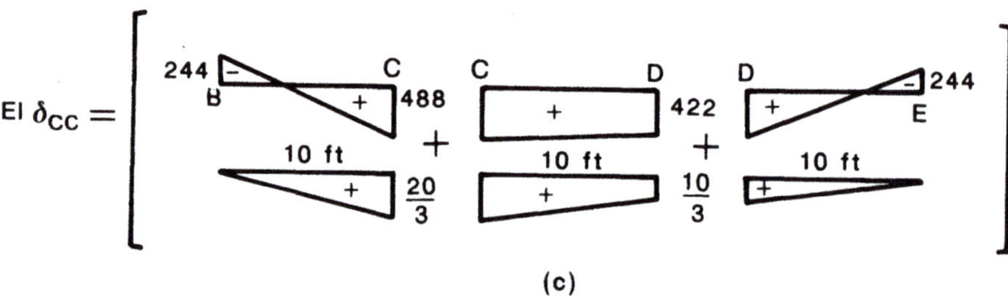

(c)

FIGURE 6.9. Auxiliary structure, m-diagram, and deflection δ_{CC} of the frame of Fig. 6.8(a) corresponding to the assumption of the last hinge at point C.

$$\delta_{CC} = \frac{10 \times 12^3}{6EI}\left[\begin{array}{l}(-244)\left(0 + \dfrac{20}{3}\right) + (488)\left(0 + \dfrac{40}{3}\right) + (488)\left(\dfrac{40}{3} + \dfrac{10}{3}\right) \\[2mm] + (422)\left(\dfrac{20}{3} + \dfrac{20}{3}\right) + (422)\left(\dfrac{20}{3} + 0\right) + (-244)\left(\dfrac{10}{3} + 0\right)\end{array}\right]$$

$$(6.7.1)$$

or

$$\delta_{CC} = \frac{59.4 \times 10^6}{EI}\ \text{in.,} \qquad (6.7.2)$$

where the units of E and I are, respectively, ksi and in^4.

Last Hinge at E: The auxiliary structure and the m-diagram are shown, respectively, in Figs. 6.10(a) and (b). Using the M- and m-diagrams, δ_{CE} can graphically be expressed as in Fig. 6.10(c). Applying Eq. (6.4.4) to these diagrams, δ_{CE} has the value

$$\delta_{CE}$$
$$= \frac{12^3}{EI}\left[\begin{array}{l}\dfrac{15}{6}(-244)(20) + \dfrac{10}{6}\{(488)(-10) + (422)(-20)\} \\[2mm] + \dfrac{10}{6}\{(422)(-20 - 20) + (-244)(-40 - 10)\} + \dfrac{15}{6}(-244)(-40)\end{array}\right]$$

$$(6.7.3)$$

or

$$\delta_{CE} = -\frac{30.8 \times 10^6}{EI}\ \text{in.} \qquad (6.7.4)$$

Last Hinge at B: For this case, the auxiliary structure and the m-diagram are shown, respectively, in Figs. 6.11(a) and (b) of Fig. 6.11. Using the m- and M-diagrams, δ_{CB} can graphically be expressed as in Fig. 6.11(c). Applying Eq. (6.4.4) to these diagrams, we have

$$\delta_{CB} = \frac{12^3}{EI}\left[\frac{15}{6}(-244)(-20 - 10) + \frac{10}{6}\{(-244)(-20) + (488)(-10)\}\right]$$

$$(6.7.5)$$

or

$$\delta_{CB} = \frac{31.6 \times 10^6}{EI}\ \text{in.} \qquad (6.7.6)$$

Comparing δ_{CC}, δ_{CE}, and δ_{CB}, we conclude that the deflection at C is $\delta_C = \delta_{CC}$ as given by Eq. (6.7.2) and the last hinge forms at C.

Example 6.7.2. The frame in Fig. 6.12 has the same plastic moment capacity for all its members. Determine the collapse load P and the moment diagram

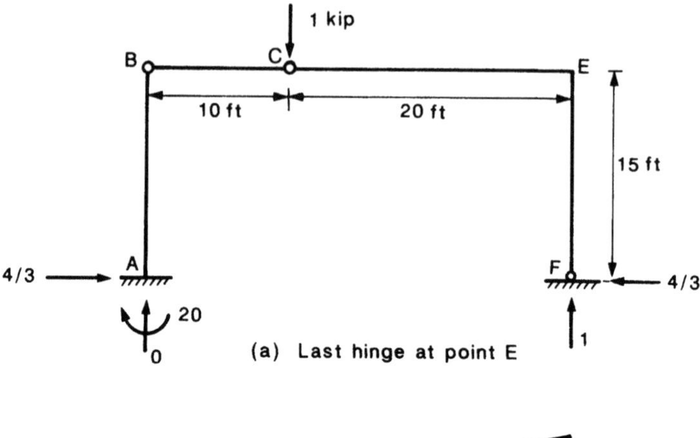

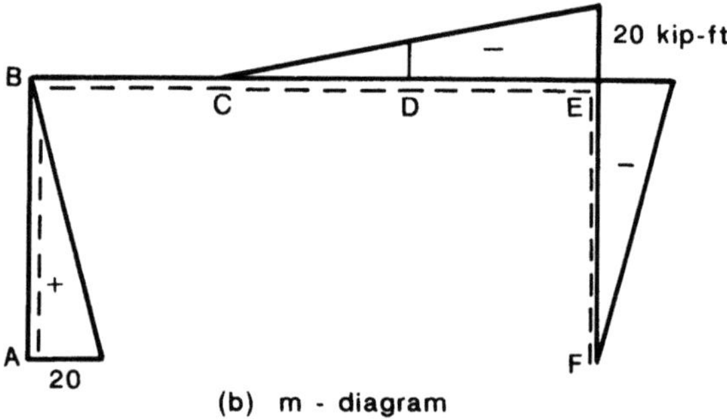

$$EI\,\delta_{CE} =$$

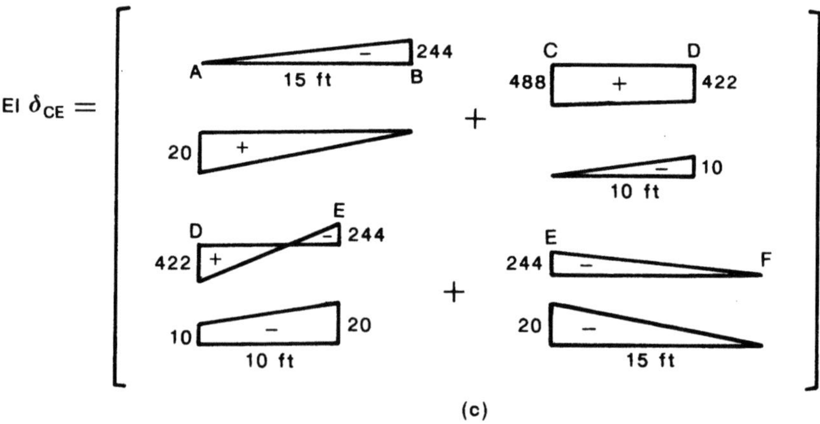

FIGURE 6.10. Auxiliary structure, *m*-diagram, and deflection δ_{CE} of the frame of Fig. 6.8(a) corresponding to the assumption of the last hinge at point *E*.

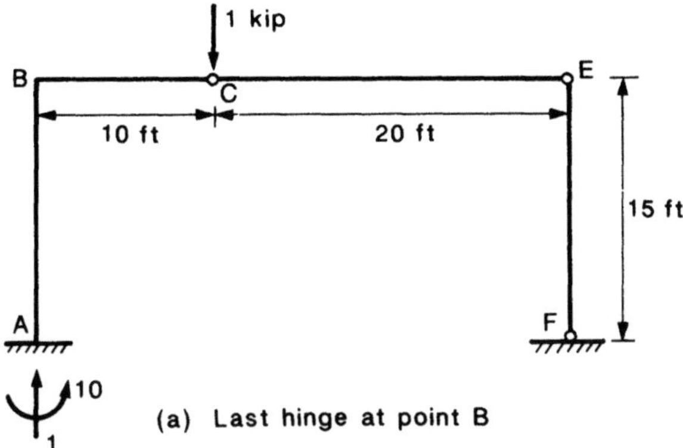

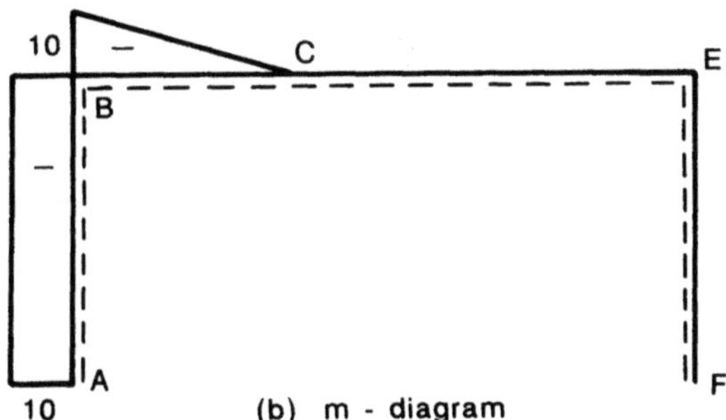

$$EI\,\delta_{CB} =$$

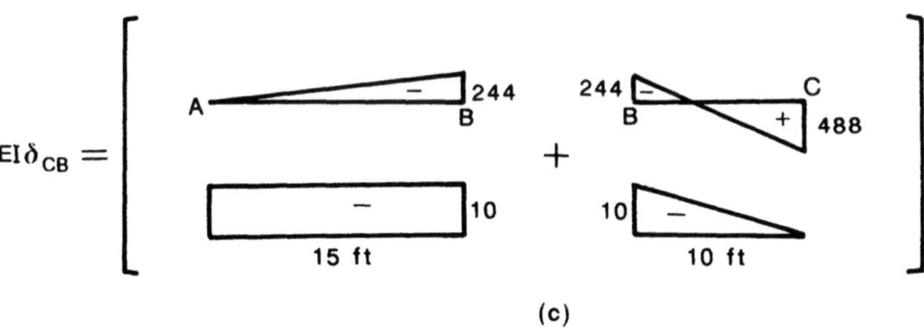

FIGURE 6.11. Auxiliary structure, m-diagram, and deflection δ_{CB} of the frame of Fig. 6.8(a) corresponding to the assumption of the last hinge at point B.

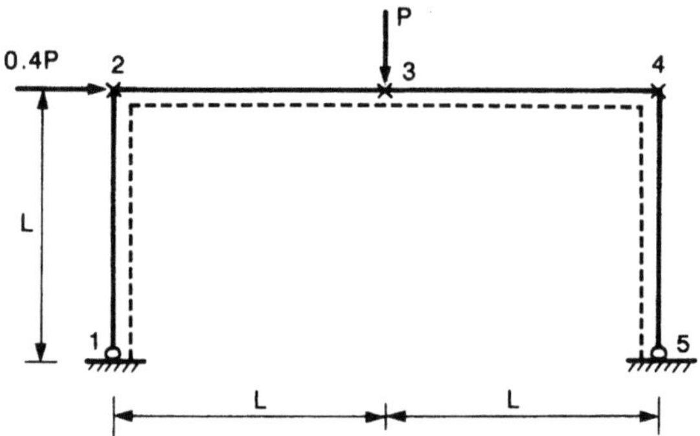

FIGURE 6.12. Rectangular frame with a horizontal and a vertical load.

(M-diagram). Using the dummy load method, calculate the vertical deflection of point 3 and the horizontal deflection of point 4 at the collapse load.

Solution: (A) Collapse Load and Moment Diagram: The frame has one redundancy and three critical sections marked with crosses. The number of independent mechanisms are:

Possible hinge locations	$= 3$
Number of redundancy	$= 1$
Independent mechanisms	$= 2$
Joint mechanisms	$= 0$
True independent mechanisms	$= 2$

The work equations for each independent mechanism as shown in Figs. 6.13(a) and (b) are

$$\text{(a)} \quad PL\theta = M_p(\theta + 2\theta + \theta), \quad P = 4M_p/L \tag{6.7.7}$$

$$\text{(b)} \quad 0.4PL\theta = M_p(\theta + \theta), \quad P = 5M_p/L. \tag{6.7.8}$$

Combining the two independent mechanisms, we have

$$\text{(a)} \quad PL\theta = 4M_p\theta$$
$$\text{(b)} \quad 0.4PL\theta = 2M_p\theta$$

$$1.4PL\theta = 6M_p\theta$$
$$\text{Cancel hinge at } 2 = -2M_p\theta$$

$$\text{(c)} \quad 1.4PL\theta = 4M_p\theta, \quad P = 2.857M_p/L. \tag{6.79}$$

Since there are no more mechanisms to combine, we find that $P = 2.857$ M_p/L is the lowest. This load probably is the correct answer. However, to be sure, we need to carry out a moment check.

Moment Check: Only the moment at 2 is unknown and can be determined by the use of the virtual work equation for the equilibrium and geometry sets shown in Fig. 6.14:

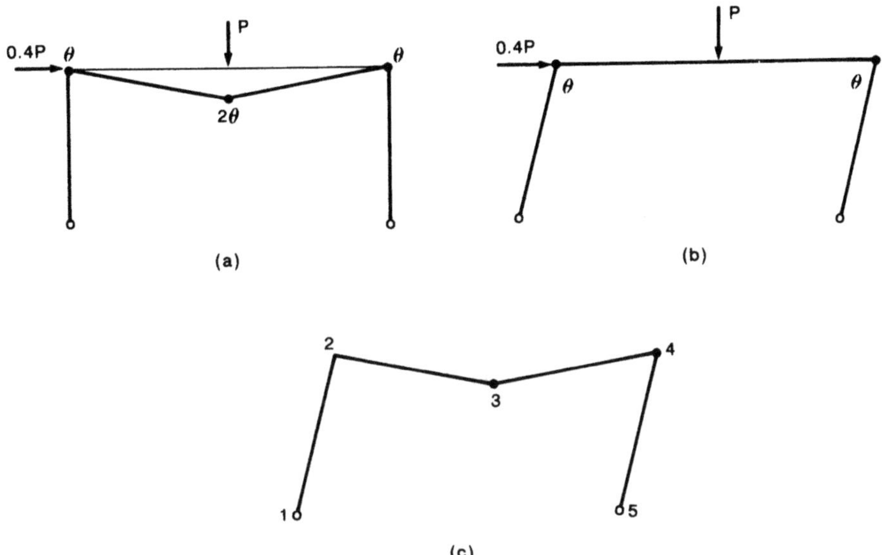

FIGURE 6.13. Possible mechanisms of frame of Fig. 6.12: (a) beam mechanism, (b) panel mechanism, and (c) combined mechanism.

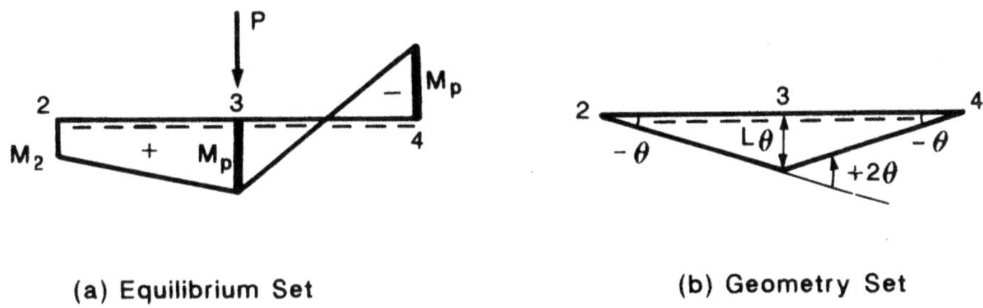

(a) Equilibrium Set   (b) Geometry Set

FIGURE 6.14. Equilibrium and geometry set for determining M_2: (a) equilibrium set and (b) geometry set.

$$PL\theta = M_2(-\theta) + (+M_p)(+2\theta) + (-M_p)(-\theta). \qquad (6.7.10)$$

Substituting P from Eq. (6.7.9) and simplifying, we have

$$M_2 = +0.143M_p. \qquad (6.7.11)$$

The complete moment diagram (M-diagram) of the frame at the collapse is shown in Fig. 6.15

(B) Deflections: The collapse mechanism [Fig. 6.13(c)] has plastic hinges at 3 and 4. Here, as in previous cases, we must determine the deflections by suc-

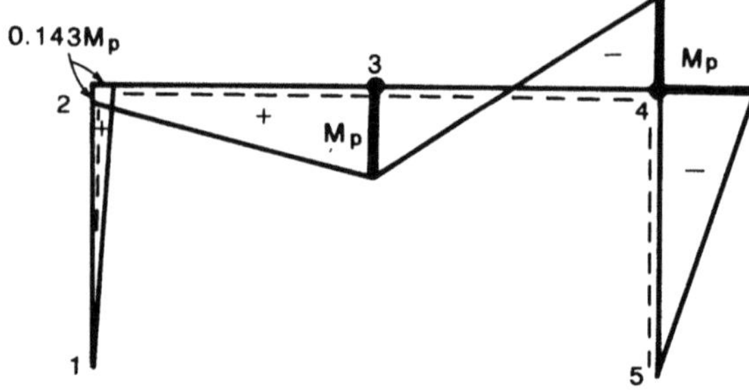

FIGURE 6.15. Moment diagram (M-diagram) for frame of Fig. 6.12.

cessively assuming the last hinge to form at 3 and 4. The correct solution corresponds to the largest deflection so obtained (deflecton theorem).

Last Hinge at 3: For the vertical deflection at point 3, the m-diagram corresponds to a unit vertical load applied to the auxiliary structure at point 3 as shown in Fig. 6.16(a). The m-diagram is shown in Fig. 6.16(b). Using the M- and m-diagrams, the vertical deflection at 3, δ_{3v}, can be graphically expressed as in Fig. 6.16(c). Applying Eq. (6.4.4) to these graphs, δ_{3v} is reduced to

$$\delta_{3v} = \frac{1}{EI}\left[\begin{array}{l} \frac{L}{6}\left\{(0.143M_p)\left(0 + \frac{L}{2}\right) + (M_p)(L + 0)\right\} \\ + \frac{L}{6}\left\{(M_p)(L + 0) + (-M_p)\left(0 + \frac{L}{2}\right)\right\} \end{array}\right]$$

or

$$\delta_{3v} = \frac{0.262M_pL^2}{EI}. \tag{6.7.12}$$

As for the horizontal deflection at point 4, a horizontal unit load is applied to the auxiliary structure at point 4 in Fig. 6.17(a). The m-diagram is shown in Fig. 6.17(b). Using the M- and m-diagrams, the horizontal deflection δ_{4h} can be graphically expressed as in Fig. 6.17(c). Applying Eq. (6.4.4) to these diagrams, δ_{4h} is reduced to

$$\delta_{4h} = \frac{L}{6EI}\left[\begin{array}{l} 0 + (0.143M_p)(2L + 0) + (0.143M_p)\left(2L + \frac{L}{2}\right) + (M_p)(L + L) \\ + (M_p)(L + 0) + (-M_p)\left(0 + \frac{L}{2}\right) \end{array}\right]$$

or

$$\delta_{4h} = 0.524\frac{M_pL^2}{EI}. \tag{6.7.13}$$

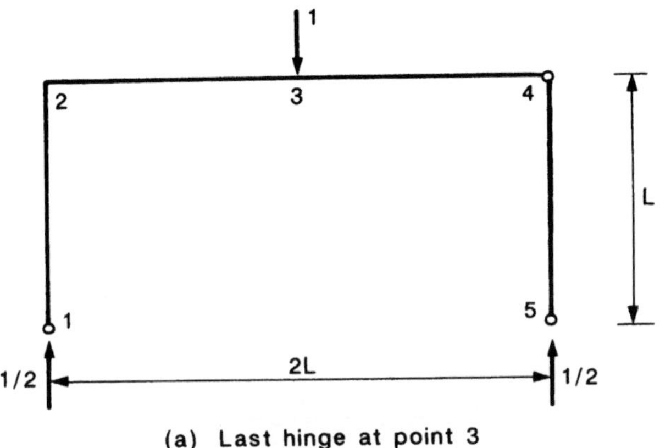

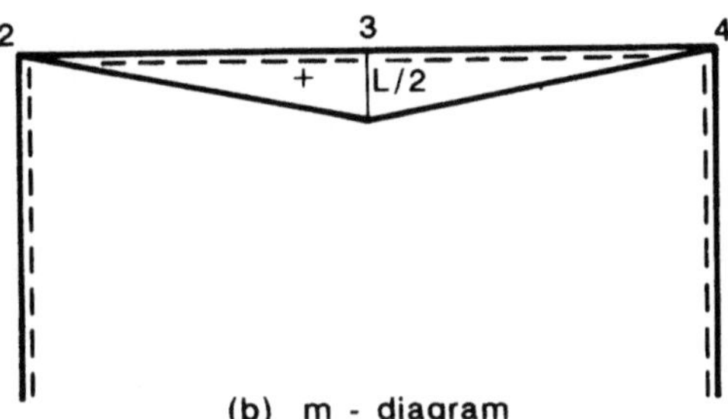

$$\delta_{3v} = \frac{1}{EI}$$

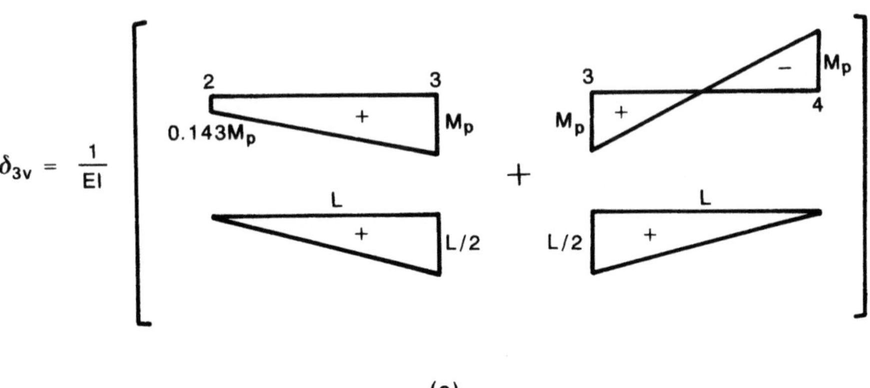

FIGURE 6.16. Auxiliary structure, *m*-diagram, and deflection δ_{3v} of the frame of Fig. 6.12 corresponding to the assumption of the last hinge at point 3.

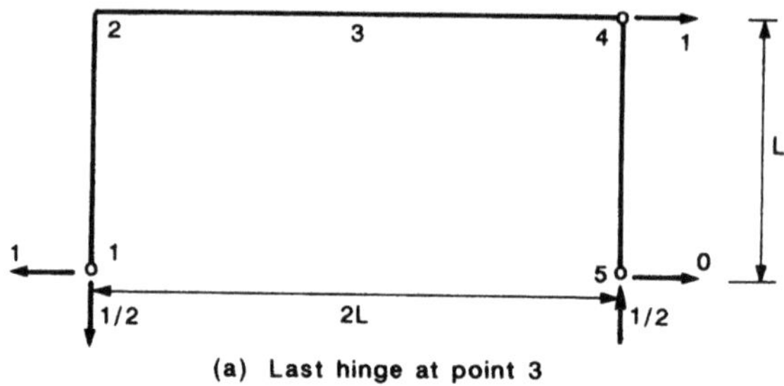

(a) Last hinge at point 3

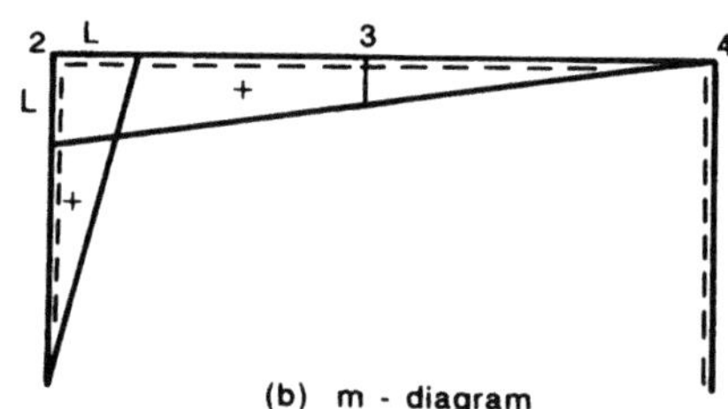

(b) m - diagram

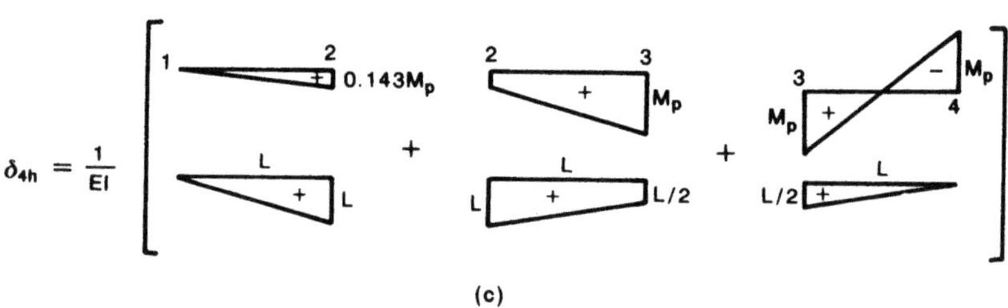

(c)

FIGURE 6.17. Auxiliary structure, m-diagram, and deflection δ_{4h} of the frame of Fig. 6.12 corresponding to the assumption of the last hinge at point 3.

Last Hinge at 4: The auxiliary structure with a unit vertical load at point 3 is shown in Fig. 6.18(a). The m-diagram is shown in Fig. 6.18(b). Combining M- and m-diagrams, δ_{3v} can graphically be expressed as in Fig. 6.19(c). Applying Eq. (6.4.4) to these diagrams, we have

$$\delta_{3v} = \frac{L}{6EI}\left[\begin{array}{l} 0 + (0.143M_p)(-L+0) + (0.143M_p)(-L+0) + (M_p)\left(0 - \frac{L}{2}\right) \\ + (M_p)\left(0 - \frac{L}{2}\right) + (-M_p)(-L+0) + (-M_p)(-L+0) + 0 \end{array}\right]$$

or

$$\delta_{3v} = \frac{0.119M_pL^2}{EI}. \tag{6.7.14}$$

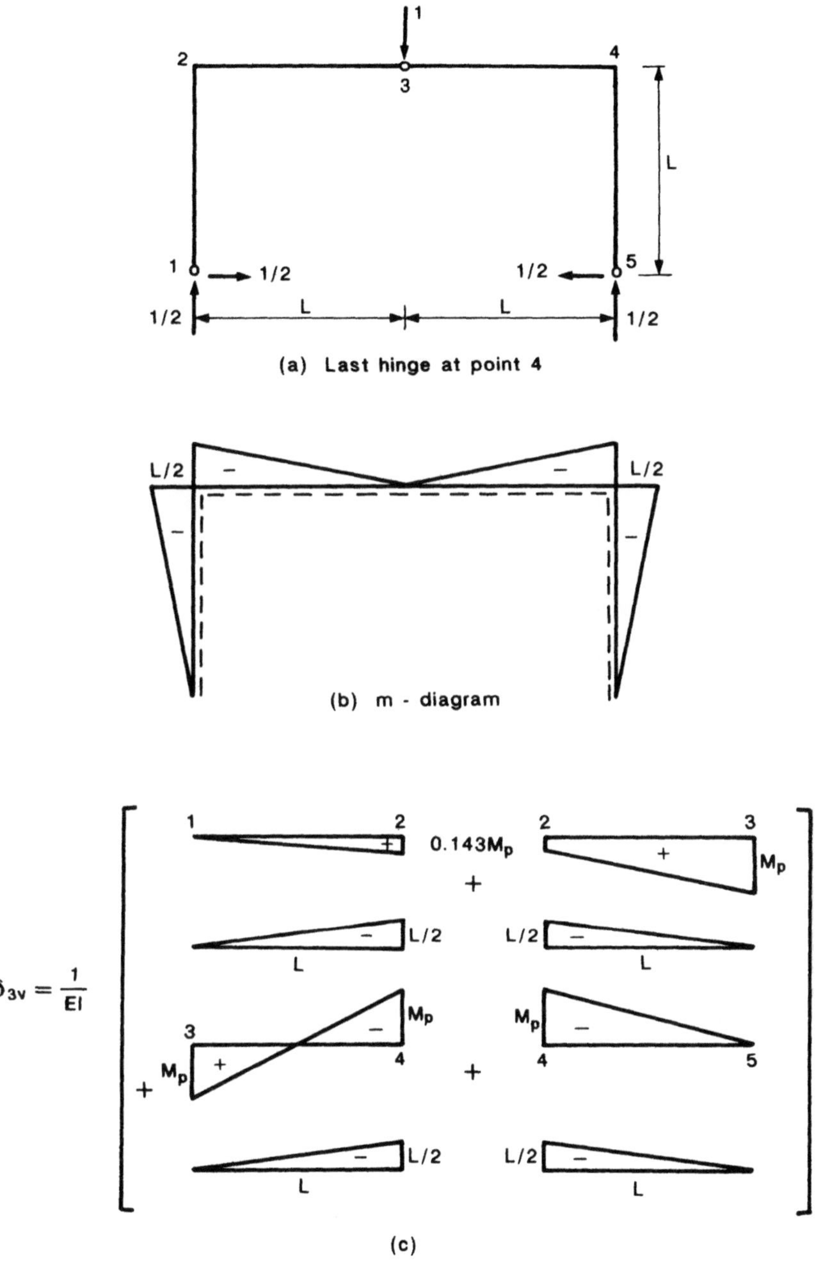

$$\delta_{3v} = \frac{1}{EI}$$

FIGURE 6.18. Auxiliary structure, *m*-diagram, and deflection δ_{3v} of the frame of Fig. 6.12 corresponding to the assumption of the last hinge at point 4.

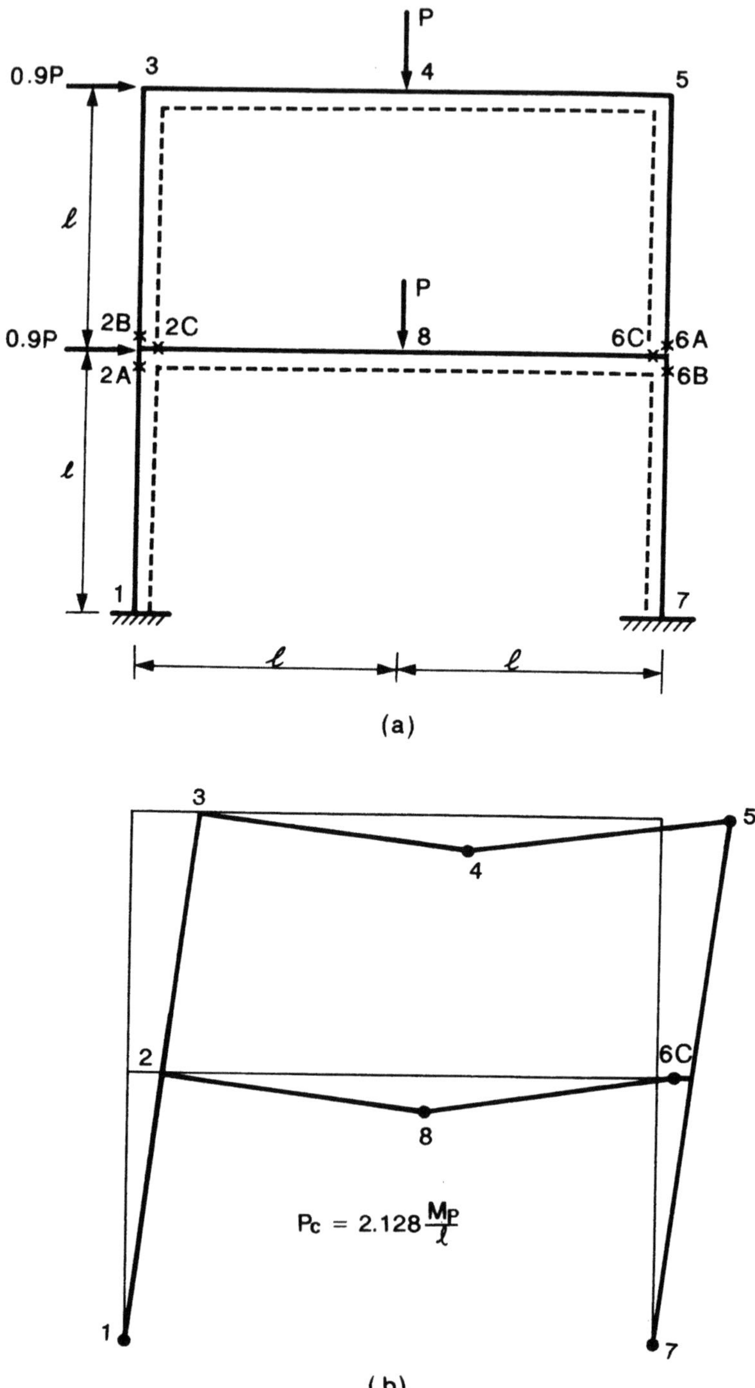

FIGURE 6.19. Collapse mechanism for a two-story frame:
a) two-story frame and
b) collapse mechanism.

Since δ_{3v} is lower than that from the previous case by the deflection theorem, the previous case (last hinge at 3) is correct and therefore δ_{3v} and δ_{4h} are, respectively, given by Eqs. (6.7.12) and (6.7.13).

6.8 Multi-Story and Multi-Bay Frames

The calculations of deflections of multistory and multibay frames are more complicated. First, because, these frames are more likely to be redundant at the collapse state, which complicates the computation of moment diagrams of actual and auxiliary structures. Second, the collapse mechanism of such frames involves more plastic hinges, thus requiring either better intuition or more trials for selecting the correct last hinge. Herein, we shall use the dummy load method (virtual work method) to calculate the deflection of a two-story frame.

Note that the calculations of deflections of high-rise frames are further complicated by the interaction of stability and inelasticity. Example calculations of these frames will therefore not be presented here. The computer-based methods for analyzing such frames will be presented in Chapters 7 and 8.

Example 6.8.1. A two-story frame is shown in Fig. 6.19(a). The collapse load is $P = 2.128\, M_p/l$ and its corresponding mechanism is shown in Fig. 6.19(b). Determine the horizontal deflection δ_{h5} of point 5 at collapse load, assuming the last plastic hinge forms at point 4.

Solution: The use of the virtual work method needs the constructions of the M-diagram of the actual structure at collapse load and the m-diagram of the auxiliary structure under dummy unit load. The first step toward obtaining these moment diagrams is to examine the frame [Fig. 6.19(a)] and determine its redundancy at collapse [Fig. 6.19(b)]. The redundancy of the collapse mechanism and the auxiliary structure is:

Redundancy in the original structure: $X = 6$

No. of plastic hinges in the mechanism: $M = 6$

Redundancy at collapse: $I = X - (M - 1) = 1.$

The redundancy of the auxiliary struture [Fig. 6.21(b)] is also equal to 1. Since the collapse mechanism has a redundant, it is a partial collapse mechanism. For constructing the M-diagram of this mechanism, take the moment at point $6A$ as redundant. The moments at points 1, 4, 5, $6C$, 7, and 8 [Fig. 6.19(b)] are known to be M_p and the moment at points $2A$, $2B$, $2C$, 3, and $6B$ can be obtained by the use of the virtual work equation as follows.

Moment at 3: Applying the virtual work equation to the equilibrium and geometry sets of beam 3-4-5 shown in Fig. 6.20(a), we have

$$Pl\theta = M_3(-\theta) + M_p(2\theta) + (-M_p)(-\theta).$$

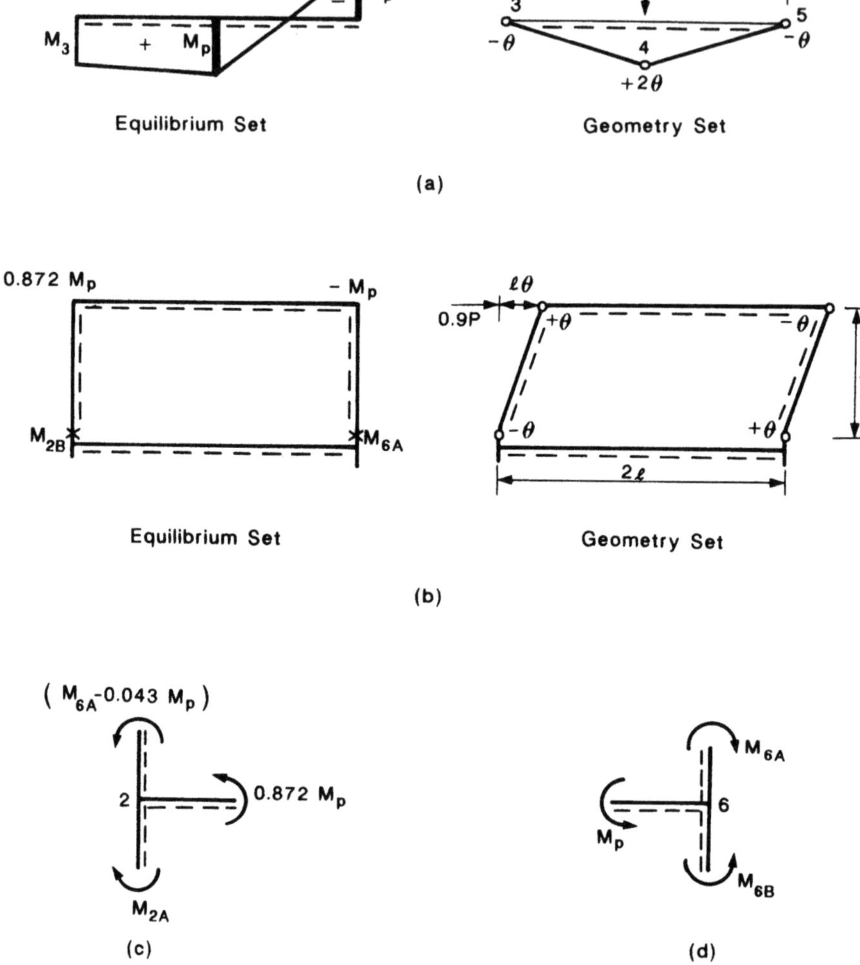

FIGURE 6.20. Computation of moments corresponding to the given mechanism by the virtual work method.

Substituting the value of $P = 2.128M_p/l$, we have

$$M_3 = 0.872M_p.$$

Moment at 2B: M_{2B} can be obtained by applying the virtual work equation to the equilibrium and geometry sets shown in Fig. 6.20(b) as

$$0.9Pl\theta = (M_{2B})(-\theta) + (0.872M_p)(\theta) + (-M_p)(-\theta) + M_{6A}(\theta),$$

which gives

$$M_{2B} = M_p(0.872 + 1 - 0.9 \times 2.128) + M_{6A} = M_{6A} - 0.043M_p.$$

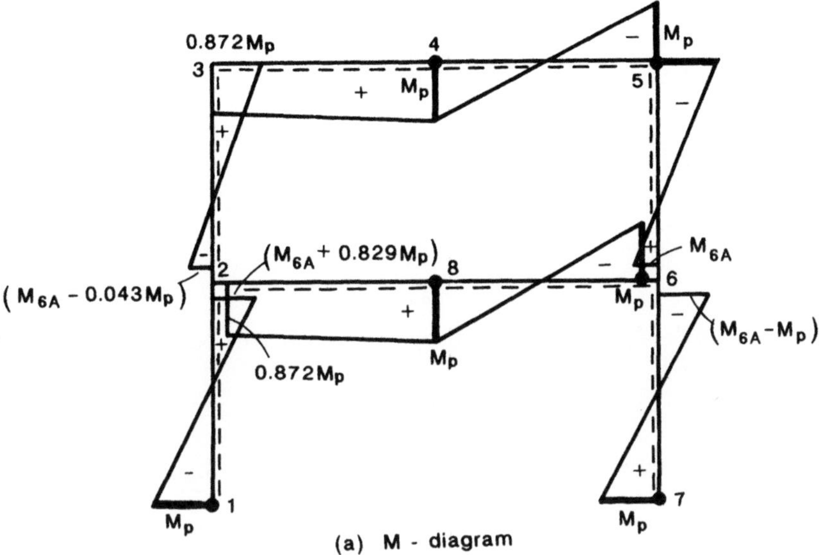

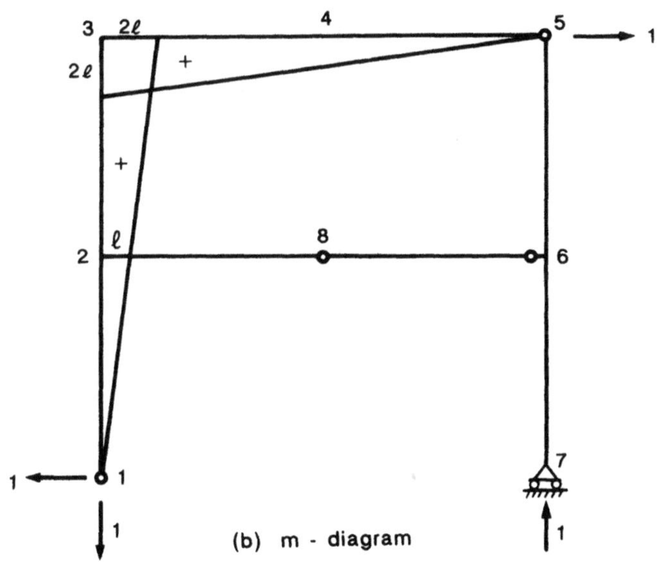

FIGURE 6.21. Moment diagrams of actual and auxiliary frame: (a) actual frame (*M*-diagram) and (b) auxiliary frame (*m*-diagram).

Moment at 2C: Since the moment diagram of member 2-6 is the same as that for member 3-5, the application of the virtual work equation at beam 2-6 gives moment M_{2C} equal to $M_3 = 0.872M_p$.

Moment at 2A: M_{2A} is determined by considering the equilibrium of joint 2 as [Fig. 6.20(c)]

$$M_{2A} = M_{6A} - 0.043M_p + 0.872M_p$$

or

$$M_{2A} = M_{6A} + 0.829M_p.$$

Moment at 6B: M_{6B} is determined by considering the equilibrium of joint 6 as [Fig. 6.20(d)]

$$M_{6B} = M_{6A} - M_p.$$

The bending moment diagram expressed in terms of M_{6A} is shown in Fig. 6.21(a).

The conditions that $|M_{6A}|$, $|M_{2B}|$, $|M_{2A}|$, and $|M_{6B}|$ be less than M_p lead to

$$(1) \quad -M_p \leq M_{6A} \leq M_p \tag{6.8.1}$$

$$(2) \quad -M_p \leq M_{2B} \leq M_p \quad \text{or} \quad -M_p \leq (M_{6A} - 0.043M_p) \leq M_p$$

$$\text{or} \quad -0.957M_p \leq M_{6A} \leq 1.043M_p \tag{6.8.2}$$

$$(3) \quad -M_p \leq M_{2A} \leq M_p \quad \text{or} \quad -M_p \leq (M_{6A} + 0.829M_p) \leq M_p$$

$$\text{or} \quad -1.829M_p \leq M_{6A} \leq 0.171M_p \tag{6.8.3}$$

$$(4) \quad -M_p \leq M_{6B} \leq M_p \quad \text{or} \quad -M_p \leq (M_{6A} - M_p) \leq M_p$$

$$\text{or} \quad 0 \leq M_{6A} \leq 2M_p. \tag{6.8.4}$$

The four inequalities (6.8.1) to (6.8.4) can be expressed as

$$0 \leq M_{6A} \leq 0.171M_p. \tag{6.8.5}$$

The inequality (6.8.5) implies that M_{6A} can be assigned any value between 0 and $0.171M_p$. For simplicity, we take

$$M_{6A} = 0.$$

It follows that

$$M_{2B} = M_{6A} - 0.043M_p = -0.043M_p$$

$$M_{2A} = M_{6A} + 0.829M_p = 0.829M_p$$

$$M_{6B} = M_{6A} - M_p = -M_p.$$

Next, the auxiliary structure [Fig. 6.21(b)] is obtained by assuming continuity at the plastic hinge location formed last (point 4) and by replacing the rest of the plastic hinges by real hinges and then applying a unit load at point 5 in the horizontal direction (location and direction of the desired deflection).

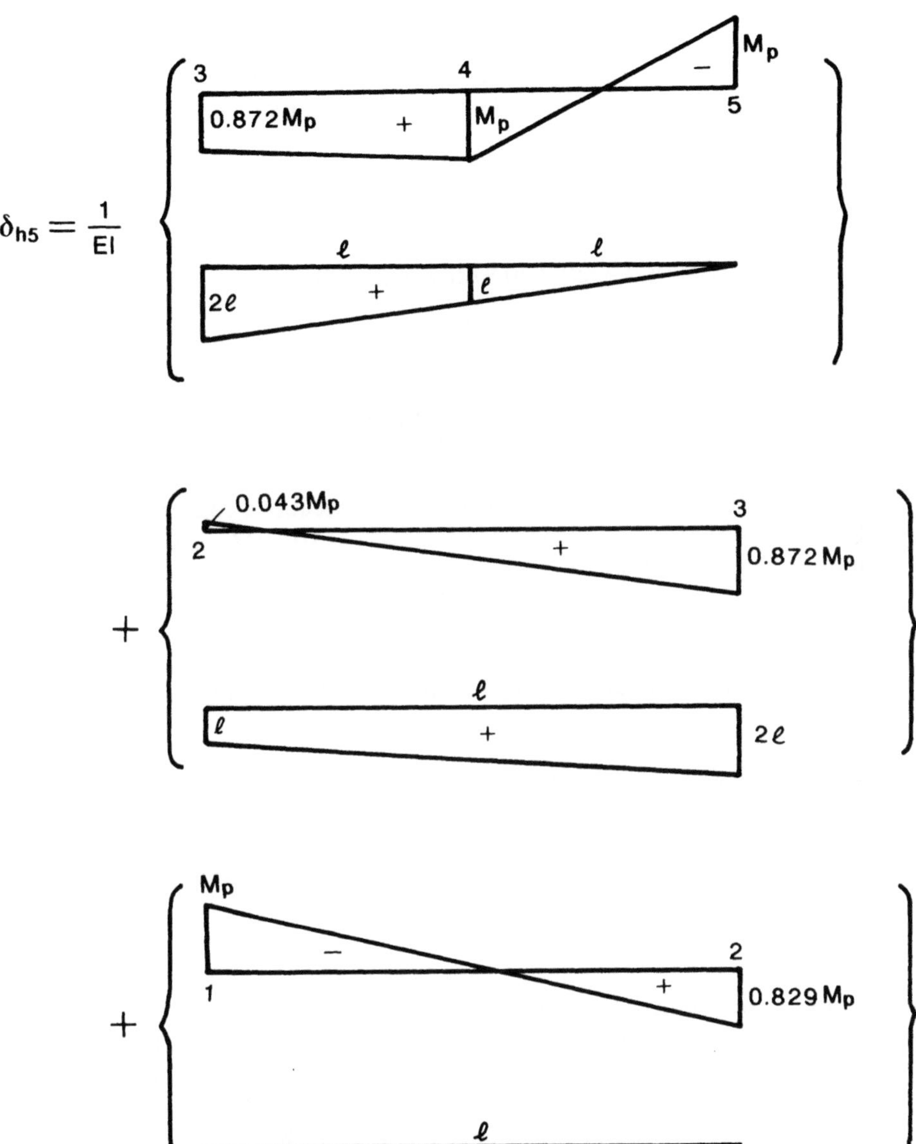

FIGURE 6.22. Graphical integration for determining δ_{h5} horizontal deflection of point 5 of the frame of Fig. 6.19(a).

This auxiliary frame has one redundant. To simplify its m-diagram, the redundant is eliminated by releasing an axial restraint in segment 8-6. The resulting moment diagram is shown in Fig. 6.21(b). Now the deflection δ_{h5} can graphically be expressed as in Fig. 6.22. Applying Eq. (6.4.4) to these diagrams, we have

$$\delta_{h5} = \frac{M_p l}{6EI}[(0.872)(2 \times 2l + l) + (1)(2l + 2l) + (1)(2l + 0) + (-1)(0 + l)$$

$$+ (-0.043)(2l + 2l) + (0.872)(2 \times 2l + l)$$

$$+ (-1)(0 + l) + (0.829)(2l + 0)],$$

which gives

$$\delta_{h5} = 14.206 \frac{M_p l^2}{6EI}. \tag{6.8.6}$$

Note that the above value of δ_{h5} is not exact, because a convenient value of $M_{6A} = 0$ was chosen. The exact value of M_{6A} is $0.069 M_p$. Using this exact solution [Fig. 6.21(a)], we obtain the exact value of δ_{h5} as

$$\delta_{h5} = 14.626 \frac{M_p l^2}{6EI}.$$

The approximate solution of δ_{h5} has an error of about 2.9%.

6.9 Rotational Capacity Requirement

The plastic design considers the reserve strength of structures by allowing redistribution of moments. This process of redistribution of moments is possible only if the first developed plastic hinges are capable of going through the necessary rotations without much loss in moment capacity while additional hinges are being developed elsewhere in the structure. The excessive rotation of these hinges may cause local or lateral buckling of the members. In the plastic design, it is therefore of great interest to determine the required rotation capacity of hinges in structures. Once the moment diagram and deflection at collapse load are known, the calculation of the required hinge rotation capacities (also known as hinge angles) is rather simple and straightforward. The calculated hinge angles must not exceed the rotation capacity of the section [6.1, 6.2, 6.3, 6.4].

In the multistory frames, the hinge angles computed on the basis of the load determined from the first-order analysis are sometimes very large. However, the second-order analysis, which is more appropriate for such frames, mostly gives a significantly lower load-carrying capacity corresponding to which hinge angles are much smaller than those necessary to develop a complete mechanism. Thus, the rotation capacity requirements for the members of these frames will be less strict than those computed by using the loads obtained from the first-order analysis.

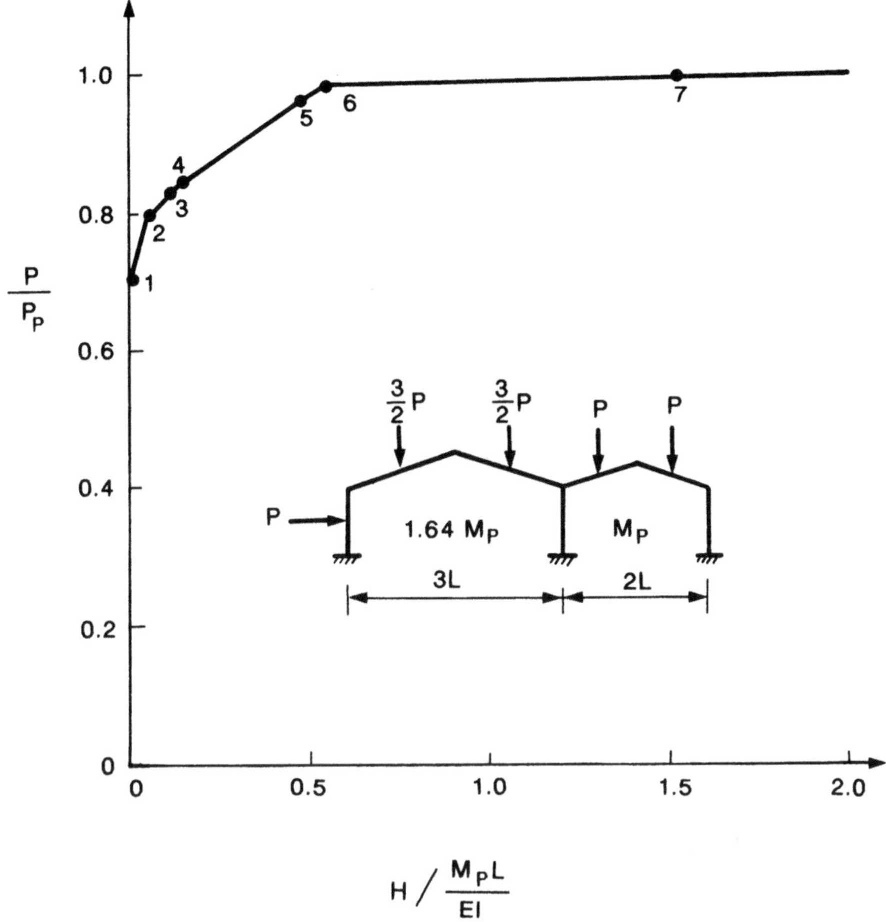

FIGURE 6.23. Required rotation capacity for a two-bay gable frame (H = hinge angle).

For some structures, the calculated hinge angle required to develop a mechanism is very large, thus putting a limitation on the design. However, the hinge rotation required for the next to last plastic hinge to form is a fraction of that required at first to develop the complete mechanism, but the load corresponding to the formation of the next to last plastic hinge is almost the same as that at the collapse (Fig. 6.23). Such structures are therefore appropriate for the plastic design. But their load factor will be slightly less, about 2% for the two-bay gable frame shown in the inset of Fig. 6.23, than the one determined by the plastic analysis.

For carrying out the plastic analysis for design of structures, LRFD requires hinge rotation capacity of three times the elastic rotation capacity. In areas of high seismicity, the required rotation capacity is 7 to 9 times the elastic rotation capacity (page 6-175 of the LRFD specifications).

Example 6.9.1. Determine the required rotation capacity of the fixed-ended beam with a concentrated load at the one-third span [Fig. 6.2(a)]. Its collapse moment diagram is shown in Fig. 6.2(b) and the vertical deflection at point B at collapse load was calculated previously in Example 6.3.1.

Solution: The correct deflection δ_B is the largest of the three deflections calculated in Example 6.3.1, i.e.,

$$\delta_B = \delta_{BC} = \frac{2}{27}\frac{M_p L^2}{EI}. \tag{6.9.1}$$

At A, the hinge angle H_A is determined by applying Eq. (6.3.1) to segment AB [Fig. 6.2(c)] as

$$H_A = \theta_A = \theta_A' + \frac{\Delta}{3/L} + \frac{3/L}{3EI}\left(M_{AB} - \frac{1}{2}M_{BA}\right) \tag{6.9.2}$$

where

$$\theta_A' = 0, \quad \Delta = \delta_B, \quad M_{AB} = -M_p \quad \text{and} \quad M_{BA} = -M_p.$$

Thus

$$H_A = \frac{2}{9}\frac{M_p L}{EI} - \frac{M_p L}{18EI} = \frac{1}{6}\frac{M_p L}{EI}. \tag{6.9.3}$$

Similarly, at C [Fig. 6.2(e)], we have

$$H_C = -\frac{2}{27}\frac{M_p L^2}{EI}\frac{1}{2L/3} + \frac{2L/3}{3EI}(M_p - M_p/2) = 0. \tag{6.9.4}$$

This confirms that C is the last hinge to form. And at C, the rotation capacity required to develop the collapse mechanism is theoretically zero. At B, the hinge angle H_B is

$$H_B = H_{BA} - H_{BC} \tag{6.9.5}$$

where H_{BA} and H_{BC} can be written from Fig. 6.2(d). Substitution of H_{BA} and H_{BC} in Eq. (6.9.5) and simplification lead to

$$H_B = \frac{1}{6}\frac{M_p L}{EI}. \tag{6.9.6}$$

Note that the rotation capacity requirement depends on the structure geometry and loading type. For example, if the fixed-ended beam is subjected to a concentrated load at its midspan, the rotation capacity requirement for all hinges is theoretically zero, because all plastic hinges form simultaneously. However, if the fixed-ended beam is subjected to a uniformly distributed load, the required rotation capacity is the same as that of concentrated load at one-third of the span.

Example 6.9.2. The rectangular frame shown in Fig. 6.12 was analyzed in Example 6.7.2. The collapse mechanism has hinges at points 3 and 4. The

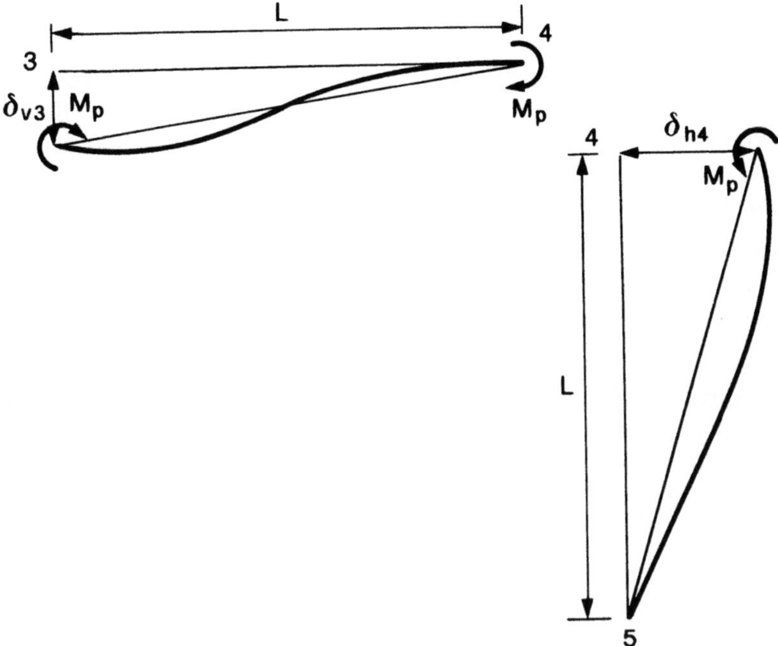

FIGURE 6.24. Hinge rotation requirement at point 4 of the frame shown in Fig. 6.12.

moment diagram is shown in Fig. 6.15. The vertical deflection of points 3 and the horizontal deflection of point 4 are, respectively, calculated as $\delta_{3v} = 0.262 M_p L^2/EI$ and $\delta_{4h} = 0.524 M_p L^2/EI$. Determine the hinge rotations at plastic hinges.

Solution: The rotation of the plastic hinge at 4 can be expressed as

$$H_4 = H_{43} - H_{45}$$

where H_{43} and H_{45} are determined by applying Eq. (6.3.1) to segments 3-4 and 4-5 in Fig. 6.24. H_4 thus becomes

$$H_4 = -0.262 \frac{M_p L^2}{EI} \frac{1}{L} + \frac{L}{3EI}\left(M_p - \frac{M_p}{2}\right)$$

$$+ 0.524 \frac{M_p L^2}{EI} \frac{1}{L} + \frac{L}{3EI}(-M_p - 0)$$

or

$$H_4 = 0.095 \frac{M_p L}{EI}.$$

Since the plastic hinge at 3 is the last to form, the required rotation capacity at this point is zero.

6.10 Examples

In this section, we will present three frame examples. In the first example, we use the *slope deflection method* for deflection calculation. We also determine the required rotation capacity and the increase in the required moment capacity of the frame due to the P-Δ effect. In the second and third examples, we calculate deflections by the *dummy load method*.

Example 6.10.1. A rectangular frame shown in Figs. 6.25(a) is analyzed and its collapse load is found to be $P = (80/11)M_p/L$. The corresponding collapse mechanism and moment diagram are shown, respectively, in Figs. 6.25(b) and (c). Determine

(a) the vertical deflection of point 3 (δ_{v3}) and the horizontal deflection of point 4 (δ_{h4}) at collapse by the slope-deflection method.
(b) the required rotation capacity of the frame.
(c) the precentage increase in the required moment capacity of the frame due to the P-Δ effect.

Solution: (a) *Deflection:* To determine δ_{v3} and δ_{h4}, we need two equations. One is obtained by the use of the continuity condition at point 2, i.e., $\theta_{21} = \theta_{23}$ (Fig. 6.26). The second is by the use of the continuity condition at a point where the last plastic hinge has just formed. Since we do not know in advance whether the last plastic hinge forms at point 3 or 4, we calculate two sets of deflections and select the larger one because of the *deflection theorem*.

The continuity condition at point 2 is obtained by applying the slope-deflection equation (6.3.1) to segments 2-1 and 2-3 (Fig. 6.26), and then by equating slopes of these two segments, i.e., $\theta_{21} = \theta_{23}$. Considering segment 2-1, we have

$$\theta_{21} = 0 + \frac{\delta_{h4}}{L/2} + \frac{L/2}{3EI}\left(\frac{7}{11}M_p - 0\right)$$

or

$$\theta_{21} = \frac{2\delta_{h4}}{L} + \frac{7}{66}\frac{M_pL}{EI}. \tag{6.10.1}$$

Now, considering segment 2-3, we have

$$\theta_{23} = 0 + \frac{\delta_{v3}}{L/2} + \frac{L/2}{3EI}\left(-\frac{7}{11}M_p + \frac{1}{2}M_p\right)$$

or

$$\theta_{23} = \frac{2\delta_{v3}}{L} - \frac{1}{44}\frac{M_pL}{EI}. \tag{6.10.2}$$

Equating θ_{21} and θ_{23}, we have

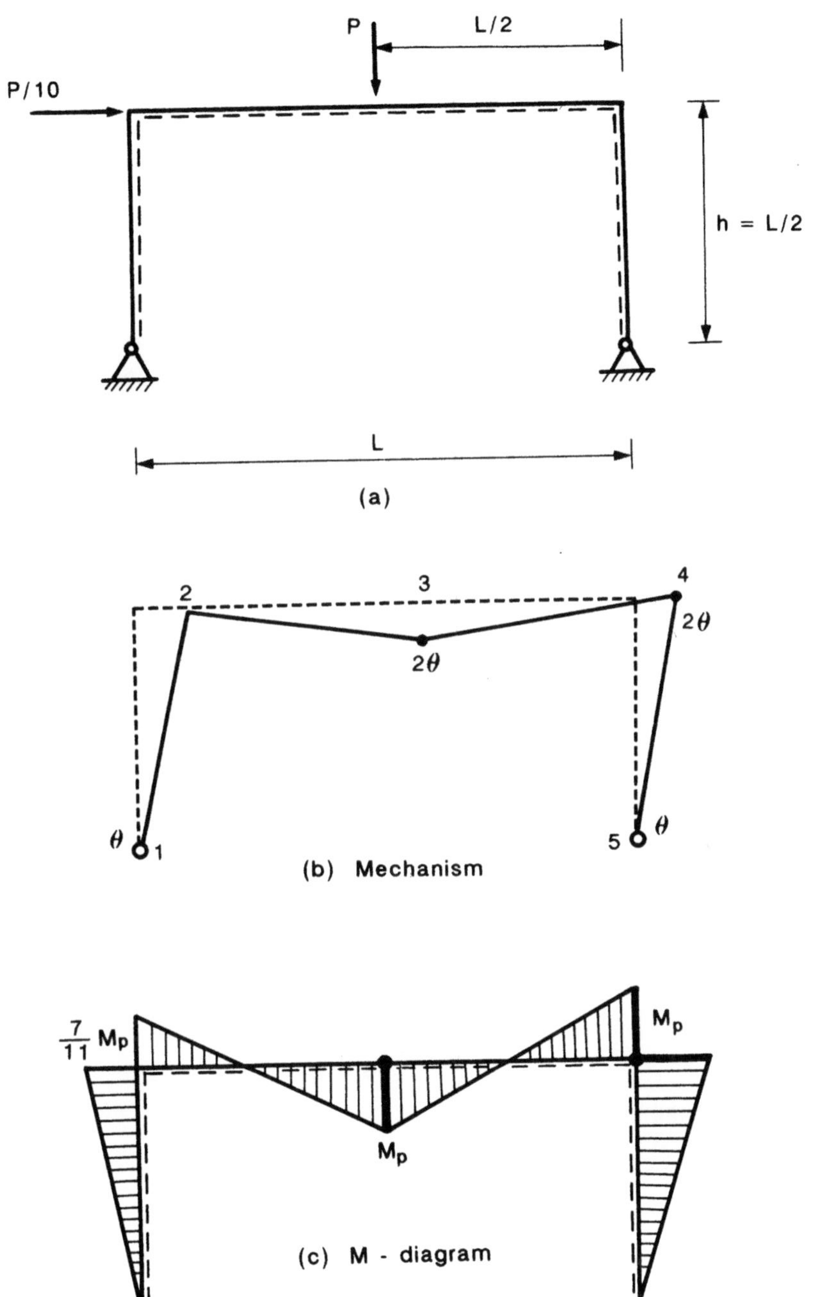

FIGURE 6.25. A rectangular frame, its failure mechanism, and corresponding moment diagram.

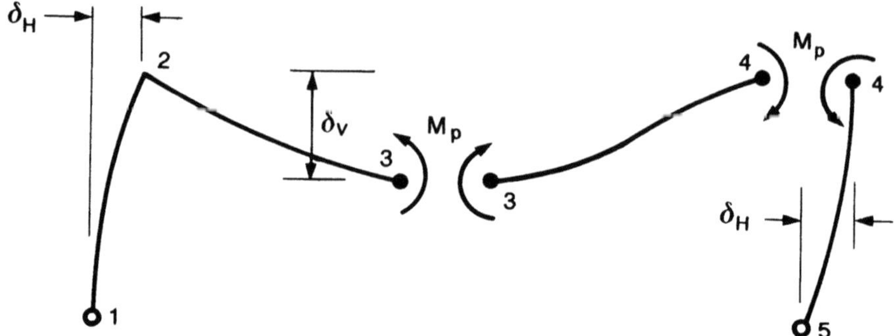

FIGURE 6.26. Free body diagrams of the frame of Fig. 6.26.

$$\frac{2\delta_{h4}}{L} + \frac{7}{66}\frac{M_pL}{EI} = \frac{2\delta_{v3}}{L} - \frac{1}{44}\frac{M_pL}{EI},$$

which gives

$$\delta_{v3} = \frac{17}{264}\frac{M_pL^2}{EI} + \delta_{h4}. \tag{6.10.3}$$

This is one of the two desired relationships between δ_{v3} and δ_{h4}. The second relationship depends on whether the last plastic hinge forms at point 3 or point 4.

Last Hinge at Point 3: If the last plastic hinge forms at point 3, then we have the continuity condition $\theta_{32} = \theta_{34}$ (Fig. 6.26). Applying the slope-deflection equation (6.3.1) to segment 3-2, θ_{32} can be expressed as

$$\theta_{32} = 0 + \frac{\delta_{v3}}{L/2} + \frac{L/2}{3EI}\left(-M_p + \frac{7}{22}M_p\right)$$

or

$$\theta_{32} = \frac{2\delta_{v3}}{L} - \frac{5}{44}\frac{M_pL}{EI}. \tag{6.10.4}$$

Similarly, for segment 3-4, we have

$$\theta_{34} = 0 - \frac{\delta_{v3}}{L/2} + \frac{L/2}{3EI}\left(M_p - \frac{M_p}{2}\right)$$

or

$$\theta_{34} = -\frac{2\delta_{v3}}{L} + \frac{1}{12}\frac{M_pL}{EI}. \tag{6.10.5}$$

Equating θ_{32} and θ_{34}, we have

$$\frac{2\delta_{v3}}{L} - \frac{5}{44}\frac{M_pL}{EI} = -\frac{2\delta_{v3}}{L} + \frac{1}{12}\frac{M_pL}{EI},$$

which gives

$$\delta_{v3} = \frac{13}{264} \frac{M_p L^2}{EI}. \tag{6.10.6}$$

Substituting this δ_{v3} in Eq. (6.10.3), we obtain

$$\delta_{h4} = -\frac{1}{66} \frac{M_p L^2}{EI}. \tag{6.10.7}$$

Last Hinge at Point 4: Applying the slope-deflection equation (6.3.1) to segment 4-3, θ_{43} can be expressed as

$$\theta_{43} = 0 - \frac{\delta_{v3}}{L/2} + \frac{L/2}{3EI}\left(M_p - \frac{M_p}{2}\right)$$

or

$$\theta_{43} = -\frac{2\delta_{v3}}{L} + \frac{1}{12} \frac{M_p L}{EI}. \tag{6.10.8}$$

Similarly, for segment 4-5, we have

$$\theta_{45} = 0 + \frac{\delta_{h4}}{L/2} + \frac{L/2}{3EI}(-M_p - 0)$$

or

$$\theta_{45} = \frac{2\delta_{h4}}{L} - \frac{1}{6} \frac{M_p L}{EI}. \tag{6.10.9}$$

Equating θ_{43} and θ_{45}, we have

$$-\frac{2\delta_{v3}}{L} + \frac{1}{12} \frac{M_p L}{EI} = \frac{2\delta_{h4}}{L} - \frac{1}{6} \frac{M_p L}{EI}$$

$$\delta_{h4} + \delta_{v3} = \frac{1}{8} \frac{M_p L^2}{EI}. \tag{6.10.10}$$

Solving the simultaneous Eqs. (6.10.3) and (6.10.10), we obtain

$$\delta_{v3} = \frac{25}{264} \frac{M_p L^2}{EI} \tag{6.10.11}$$

$$\delta_{h4} = \frac{1}{33} \frac{M_p L^2}{EI}. \tag{6.10.12}$$

Since δ_{v3} and δ_{h4} given by Eqs. (6.10.11) and (6.10.12) are larger than those given by Eqs. (6.10.6) and (6.10.7), by the deflection theorem, the δ_{v3} and δ_{h4} given by Eqs. (6.10.11) and (6.10.12) are the correct solutions and the hinge is formed last at point 4.

Recall that in Example 6.7.2, the geometry of the frame (Fig. 6.12) is identical to the present example. However, the sequence of plastic hinge formation is different. This difference is due to the fact that the ratio of the vertical to horizontal loads is not the same in these two examples.

(b) *Required Rotation Capacity:* The mechanism has two plastic hinges. At point 4, the plastic hinge forms last, so there is no need of rotation capacity. The hinge rotation at point 3 can be expressed as

$$H_3 = \theta_{32} - \theta_{34}. \tag{6.10.13}$$

Substituting θ_{32} and θ_{34}, respectively, from Eq. (6.10.4) and Eq. (6.10.5), we have

$$H_3 = \frac{2\delta_{v3}}{L} - \frac{5}{44}\frac{M_pL}{EI} + \frac{2\delta_{v3}}{L} - \frac{1}{12}\frac{M_pL}{EI}.$$

Substituting δ_{v3} from Eq. (6.10.11) and simplifying, we obtain

$$H_3 = \frac{2}{11}\frac{M_pL}{EI}. \tag{6.10.14}$$

The required rotation capacity of the member 2-4, at least at point 3, should be $H_3 = (2/11)/(M_pL/EI)$

(c) *P-Δ Moment:* In the frame, the maximum increase in moment due to the P-Δ effect is at point 4 and is obtained approximately as follows. By considering the equilibrium of segments 4-5 and 3-4, along with the moment diagram of the frame [Fig. 6.25(c)], the first-order vertical force at point 5 is obtained as

$$V_5 = 4\frac{M_p}{L}. \tag{6.10.15}$$

Now, the P-Δ moment at point 4 is

$$M_{P\Delta} = V_5\delta_{h4}. \tag{6.10.16}$$

Substituting δ_{h4} from Eq. (6.10.12) and V_5 from Eq. (6.10.15), we have

$$M_{P\Delta} = \left(4\frac{M_p}{L}\right)\left(\frac{1}{33}\frac{M_pL^2}{EI}\right) \tag{6.10.17}$$

and

$$\frac{M_{P\Delta}}{M_p} = \frac{4}{33}\frac{M_pL}{EI} = \frac{4}{33}\left(\frac{F_y}{E}\right)\left(\frac{Z}{I}\right)L. \tag{6.10.18}$$

Using Young's modulus $E = 29,000$ ksi and yield stress $F_y = 36$ ksi and taking the shape factor $f = Z/S$ at an average value of 1.14, we have

$$\frac{M_{P\Delta}}{M_p} = \frac{4}{33}\left(\frac{36}{29,000}\right)(1.14)\left(\frac{L}{d/2}\right) = 1.72 \times 10^{-4}\left(\frac{L}{d/2}\right).$$

Taking L as 20 ft and the depth of the member d as 14 in., the percentage increase in moment becomes:

$$\% \text{ increase in moment} = 1.72 \times 10^{-4} \times \left(\frac{20 \times 12}{7}\right) = 0.59\%.$$

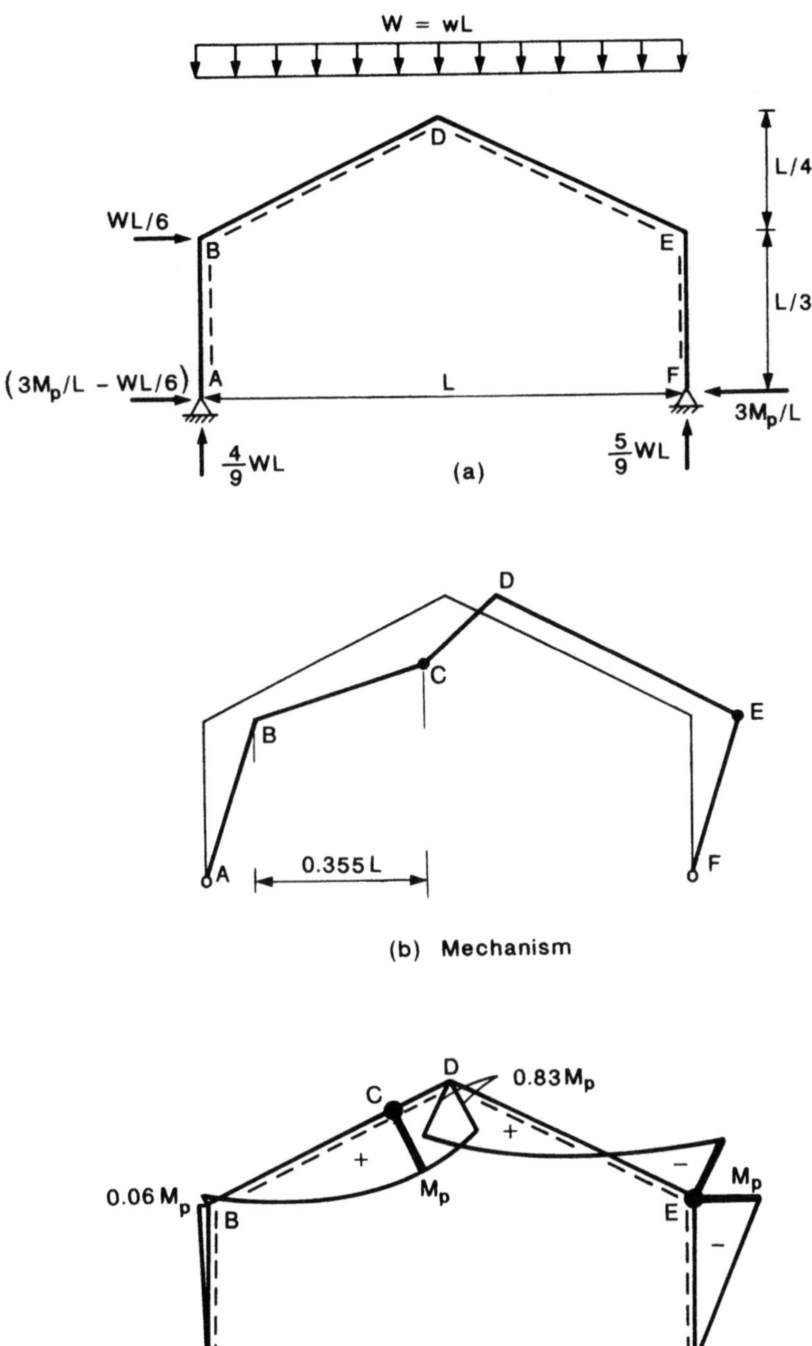

FIGURE 6.27. A gable frame, its failure mechanism, and corresponding moment diagram (*M*-diagram).

The collapse load based on the first-order plastic analysis, $P = (80/11)/(M_p/L)$, has been reduced at least by 0.59%.

Example 6.10.2. The gable frame shown in Fig. 6.27(a) is analyzed and its collapse mechanism and moment diagram are shown, respectively, in Figs. 6.27(b) and (c). Determine the vertical deflection of the ridge of this frame (point D) at collapse load by the dummy load method.

Solution: The collapse mechanism [Fig. 6.27(b)] has plastic hinges at C and E. Since the sequence of the plastic hinge formation is not known in advance, the deflection at collapse load can only be determined by successively assuming the last hinge to be at C and E. The correct deflection is the larger one according to the deflection theorem.

Last Hinge at C: For this case, the m-diagram is obtained by applying a unit vertical load at point D of the auxiliary structure (obtianed by removing the hinge from point C and replacing the plastic hinge at point E by a real frictionless hinge). The auxiliary structure and its m-diagram are shown in Fig. 6.28(a). Combining the M- and m-diagrams, δ_{vD} is graphically expressed as in Fig. 6.28(b). Applying Eqs. (6.4.4) and (6.4.5) to these diagrams, we obtain

$$EI\delta_{vD} = \frac{1}{6}\frac{\sqrt{5}}{4}L\left[(-0.06M_p)\left(0 + \frac{L}{4}\right) + 0.83M_p\left(\frac{L}{2}\right) + 0\right]$$
$$+ \frac{1}{3}\frac{\sqrt{5}}{4}L(0.526M_p)\left(0 + \frac{L}{4}\right)$$
$$+ \frac{1}{6}\frac{\sqrt{5}}{4}L\left[(0.83M_p)\left(\frac{L}{2} + 0\right) + (-M_p)\left(0 + \frac{L}{4}\right)\right]$$
$$+ \frac{1}{3}\frac{\sqrt{5}}{4}L(0.526M_p)\left(\frac{L}{4} + 0\right)\right]$$

or

$$\delta_{vD} = \frac{M_pL^2}{EI}[0.03727 + 0.02450 + 0.01537 + 0.02450]$$

or

$$\delta_{vD} = 0.1016\frac{M_pL^2}{EI}. \tag{6.10.19}$$

Last Hinge at E: For this case, the auxiliary structure with a unit vertical load at D and its m-diagram are shown in Fig. 6.29(a). The m- and M-diagrams combined to express δ_{vD} graphically are shown in Fig. 6.29(b). Applying Eqs. (6.4.4) and (6.4.5) to these diagrams, we obtain

$$EI\delta_{vD} = \frac{1}{6}\frac{L}{3}[0 + (-0.06M_p)(-0.232L + 0)]$$
$$+ \frac{1}{6}\frac{\sqrt{5}}{4}L[(-0.06M_p)(-0.232L + 0.047L)$$

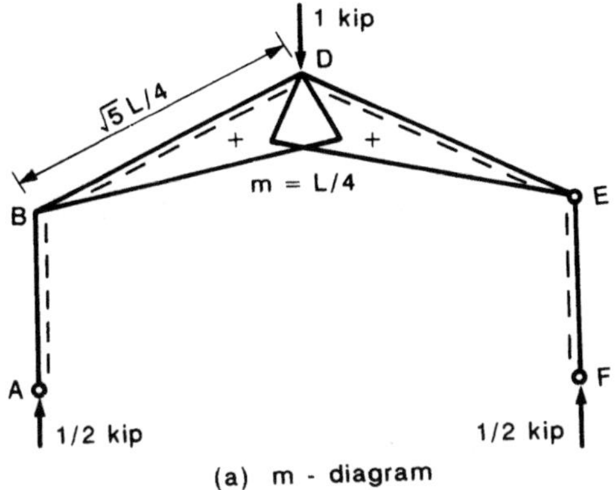

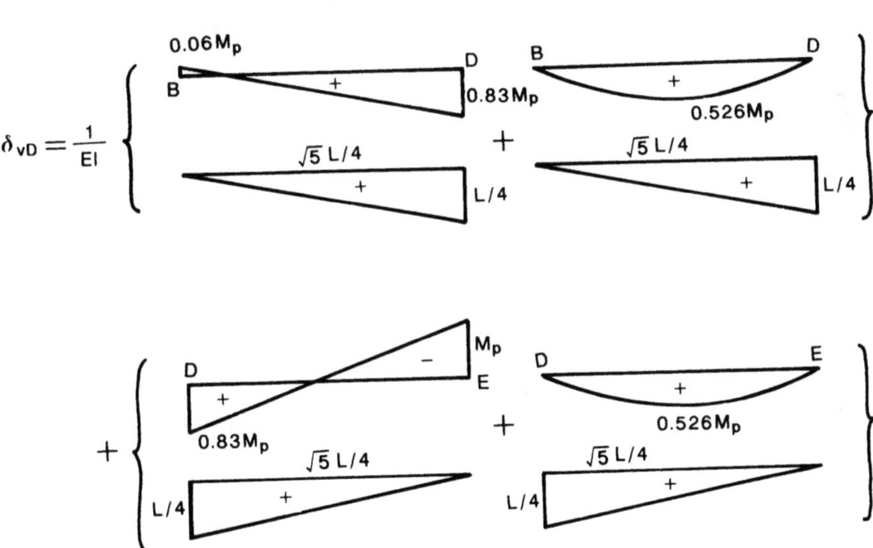

$$\delta_{vD} = \frac{1}{EI} \{ \cdots \}$$

FIGURE 6.28. Auxiliary structure, m-diagram, and graphical representation of integral for δ_{vD} corresponding to the last hinge at point C.

$$+ (0.83M_p)(0.094L - 0.116L)]$$

$$+ \frac{1}{3}\frac{\sqrt{5}L}{4}[(0.526M_p)(-0.116L + 0.047L)]$$

$$+ \frac{1}{6}\frac{\sqrt{5}}{4}L[0.83M_p(0.094L - 0.116L) + (-M_p)(-0.232L + 0.047L)]$$

$$+ \frac{1}{3}\frac{\sqrt{5}L}{4}[(0.526M_p)(0.047L - 0.116L)]$$

$$+ \frac{1}{6}\frac{L}{3}[(-M_p)(-0.232L + 0) + 0)]$$

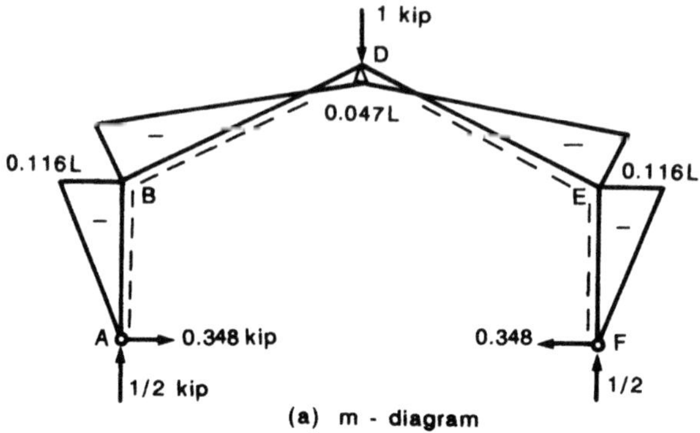

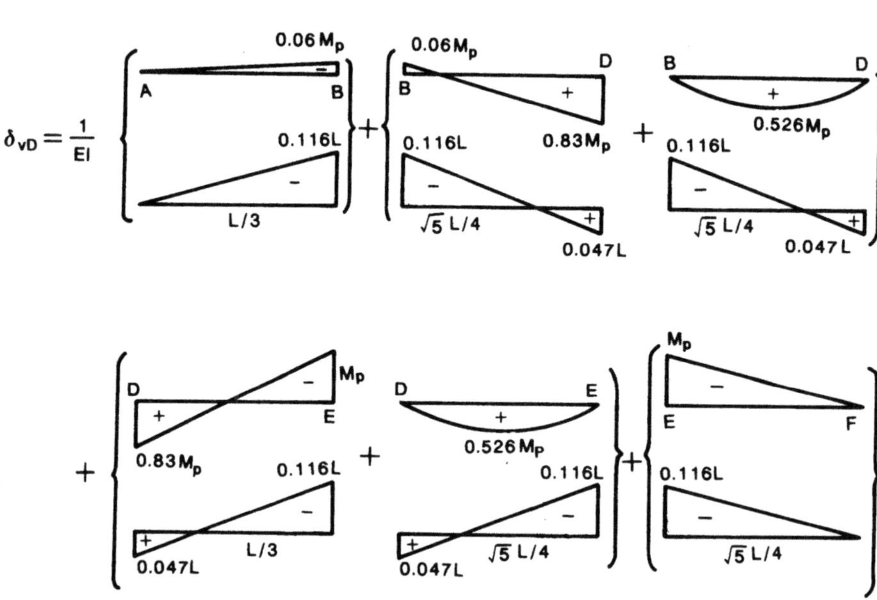

$$\delta_{vD} = \frac{1}{EI}$$

(b)

FIGURE 6.29. Auxiliary structure, m-diagram, and graphical representation of integral for δ_{vD} corresponding to the last hinge at point E.

or

$$\delta_{vD} = \frac{M_p L^2}{EI} [0.00077 - 0.00067 - 0.00676 + 0.01554 - 0.00676 + 0.01289]$$

or

$$\delta_{vD} = 0.015 \frac{M_p L^2}{EI}. \tag{6.10.20}$$

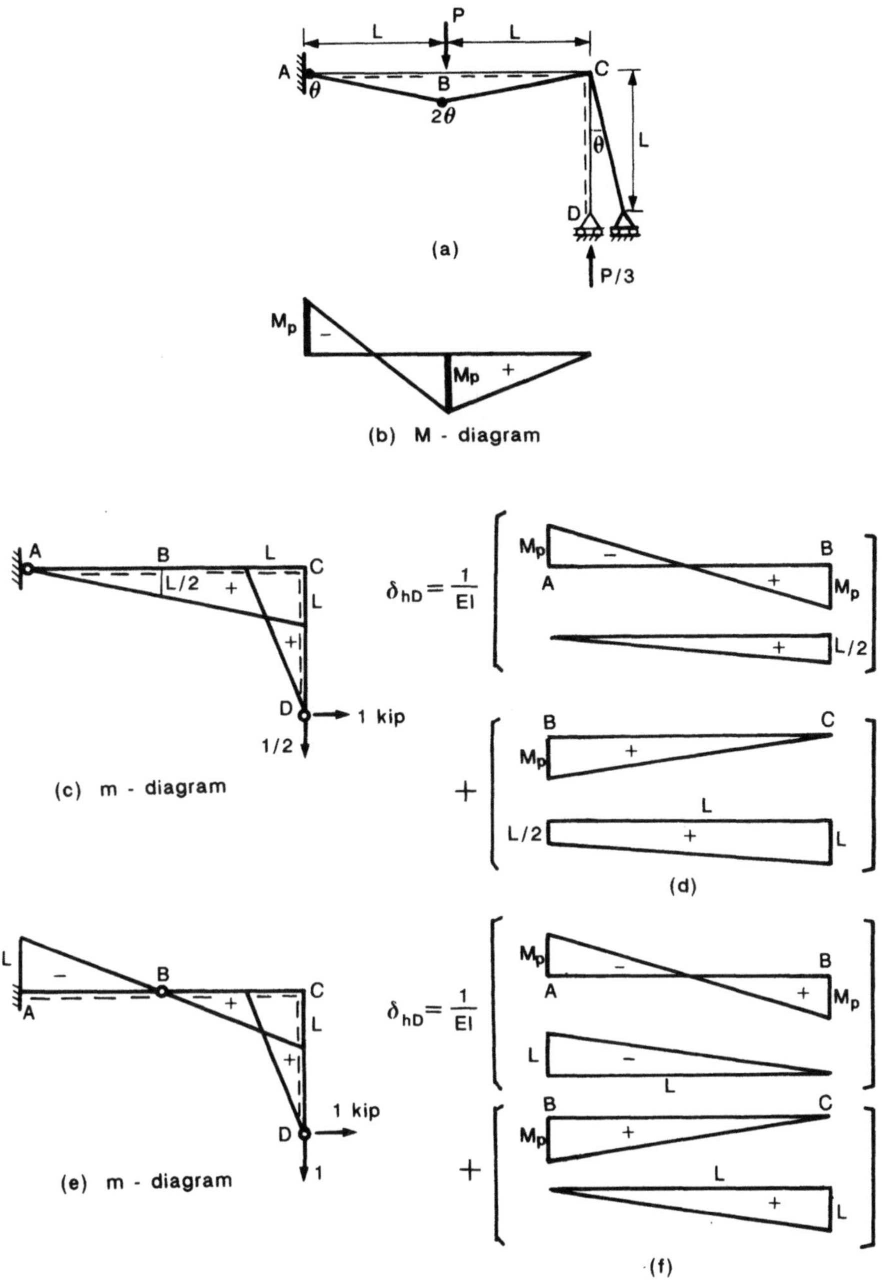

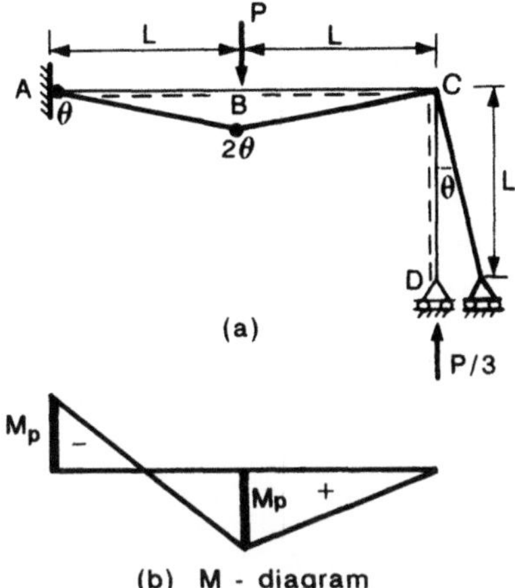

FIGURE 6.30. Computation of deflection for Example 6.10.3:
(a) frame and its collapse mechanism;
(b) M-diagram;
(c) auxiliary structure & m-diagram for the last hinge at point B;
(d) graphical representation of integral for δ_{hD} for the last hinge at point B;
(e) auxiliary structure & m-diagram for the last hinge at point A; and
(f) graphical representation of integral for δ_{hD} for the last hinge at point A.

Since the deflection δ_{vD} for the last hinge at C (Eq. 6.10.19) is higher than that of the last hinge at E (Eq. 6.10.20), the correct solution for δ_{vD} is given by Eq. (6.10.19).

Example 6.10.3. A frame with its collapse mechanism is shown in Fig. 6.30(a). Its corresponding M-diagram is given in Fig. 6.30(b). Determine the horizontal deflection at point D by the dummy load method.

Solution: Assuming in turn the last hinge to form at points B and A in Fig. 6.30(a), we make two deflection calculations in the following.

Last Hinge at B: For this case, the m-diagram is obtained by applying a unit horizontal load at point D of the auxiliary structure. The auxiliary structure and m-diagram are shown in Fig. 6.30(c). Combining the M- and m-diagrams, δ_{hD} is graphically expressed in Fig. 6.30(d). Applying Eq. (6.4.4) to these diagrams, we have

$$EI\delta_{hD} = \frac{1}{6}L\left[(-M_p)\left(0 + \frac{L}{2}\right) + (M_p)(L + 0)\right] + \frac{1}{6}L[(M_p)(L + L) + 0]$$

or

$$\delta_{hD} = \frac{5}{12}\frac{M_pL^2}{EI}. \tag{6.10.21}$$

Last Hinge at A: For this case, the auxiliary structure with a unit horizontal load at D and its m-diagram are shown in Fig. 6.30(e). This m-diagram is combined with the M-diagram to express δ_{hD} graphically in Fig. 6.30(f). Applying Eq. (6.4.4) to these diagrams, we have

$$EI\delta_{hD} = \frac{1}{6}L[(-M_p)(-2L + 0) + M_p(0 - L)]$$

$$+ \frac{1}{6}L[M_p(0 + L) + 0]$$

or

$$\delta_{hD} = \frac{M_pL^2}{3EI}. \tag{6.10.22}$$

Since the deflection δ_{hD} with the last hinge at B (Eq. 6.10.21) is greater than that with the last hinge at A (Eq. 6.10.22), the correct solution is Eq. (6.10.21).

6.11 Summary

The computation of deflections of plastically designed structures must be made to check that the deflections at the plastic limit load do not exceed certain specified limits. The second-order moments introduced by the lateral deflection of the frames, if significant, must be accunted for in the design.

Also, it must be ensured that the developed plastic hinges are capable of going through the rotations necessary to form the collapse mechanism without significant loss in their moment-carrying capacity.

The deflections at the plastic limit load can be estimated by performing a series of sequential elastic analyses. But in order to compute plastic limit deflections in a one-step analysis, the location of the last plastic hinge to form must be known in advance. In the absence of this information, the deflections are calculated by assuming that each hinge, in turn, is the last to form. Then, by the deflection theorem, the maximum value obtained from various trials is the correct deflection.

The plastic limit deflections can be determined by any method suitable for computing the deflections in the elastic range. In this chapter, the slope-deflection equation method and dummy load or virtual work equation method are described and applied to determine the plastic limit displacements, the second-order moments introduced by lateral drift of frames, and the required rotation capacity of plastic hinges of simple beams, simple frames, and multistory and multibay frames.

In the slope-deflection equation method, the slopes of the members are first expressed in terms of their deflections, end moments, and lateral loads. The deflections are then determined by solving equations formed by using compatibility conditions. The method usually involves solution of simultaneous equations. The dummy load or virtual work equation method does not require the solution of simultaneous equations. In this mehtod, deflections are obtained from equations formed by using the virtual work equation. The computational procedure, however, involves an evaluation of an integral of the product of two moment diagrams over the structure. One moment diagram is of the actual structure subjected to actual applied loads. The other is of an auxiliary structure subjected to a unit dummy load. The auxiliary structure is obtained from the actual structure by eliminating the last plastic hinge and replacing the rest of them by real frictionless hinges. A unit dummy load is applied to this structure at the point and in the direction of the desired deflection. The integration of the product of these two moment diagrams is obtained conveniently be substituting the coordinates of these two diagrams into the algebraic expressions derived by integrating various shapes of moment diagrams.

References

6.1. Knudsen, K.E., Yang, C.H., Johnston, B.G., and Beedle, L.S., "Plastic Strength and Deflections of Continuous Beams," *The Welding Journal*, 32, 5, p. 240, 1953.

6.2. Lukey, A.F., and Adams, P.F., "Rotation Capacity of Wide Flange Beams Under Moment Gradient," *Journal of Structural Division*, ASCE, 95, ST6, p. 1173, 1969.

6.3. Lay, M.G., and Galambos, T.V., "Inelastic Beams Under Moment Gradient, *Journal of Structural Division*, ASCE, 93, ST1, p. 381, 1967.

6.4. Driscoll, G.C. Jr., "Rotation Capacity Requirements for Beams and Frames of Structural Steel," *Dissertation*, presented at Lehigh University, Bethlehem, PA, in partial fulfillment of the requirements for the degree of Doctor of Philosophy, 1958.

Problems

6.1. Determine the maximum vertical deflection of the two-span continuous beam shown in Fig. 4.2(a) at plastic limit load by the dummy load method.

6.2. Determine the required rotation capacity of plastic hinges developed at the plastic limit load of the two-span continuous beam of Problem 6.1.

6.3. Repeat Problem 6.1 for a two-span continuous beam shown in Fig. 4.8(a).

6.4. Repeat Problem 6.2 for the two-span continuous beam of Problem 6.3.

6.5. Repeat Problem 6.1 for a two-span continuous beam shown in Fig. 4.9(a).

6.6. Repeat Problem 6.2 for the two-span continuous beam of Problem 6.5.

6.7. Repeat Problem 6.1 for a two-span continuous beam shown in Fig. 5.2(a).

6.8. Repeat Problem 6.2 for the two-span continuous beam of Problem 6.7.

6.9. Repeat Problem 6.1 for a three-span continuous beam shown in Fig. 5.4(a).

6.10. Repeat Problem 6.2 for the three-span continuous beam of Problem 6.9.

6.11. The moment diagram for the gable frame of Fig. 4.32(a) is shown in Fig. 4.34. Determine the vertical deflection of point C and the horizontal deflection of point D by the dummy load method.

6.12. Compute the required rotation capacity of plastic hinges developed at the plastic limit load of the gable frame of Problem 6.11.

6.13. Calculate the vertical deflection of point C and horizontal deflection of point D of the rectangular frame of Fig. 5.3(a) by the dummy load method.

6.14. Repeat Problem 6.12 for the frame of Problem 6.13.

6.15. Determine the vertical deflection of point D and the horizontal deflection of point E of the rectangular frame of Fig. 5.25 by the dummy load method.

6.16. Repeat Problem 6.12 for the frame of Problem 6.15.

6.17. A fixed-based rectangular frame with span $2L$ and height L is subjected to a horizontal load P acting to the right at the top of the left column and a concentrated load P at the center of the girder. Calculate the vertical deflection at the center of the girder and the horizontal deflection of the top of the frame by the dummy load method:

$$\left(P_u = \frac{3M_p}{L}, \delta_v = \delta_h = M_p L^2/3EI \right).$$

6.18. Repeat Problem 6.12 for the frame of Problem 6.17.

6.19. Estimate the horizontal deflection of point E of the two-story frame of Fig. 5.29 by the dummy load method. Assume that the last plastic hinge forms at point D.

6.20. Repeat Problem 6.12 for the frame of Problem 6.19.

7
First-Order Hinge-by-Hinge Analysis

7.1 Introduction

As described in Chapter 1, the first-order hinge-by-hinge analysis is a series of elastic analyses. First, an elastic analysis is performed on the original structure. When the maximum moment in the structure reaches the plastic moment capacity of a member, a plastic hinge is formed at the point of maximum moment. For further loading, the plastic hinge in this member is replaced by a real hinge leading to a new but simpler structure. Next, an elastic analysis is again performed on this new structure. The process of performing an elastic analysis on the current structure to locate a new plastic hinge and then obtaining a newer structure by replacing this new plastic hinge with a real hinge is continued until a sufficient number of plastic hinges are formed to transform the structure into a failure mechanism.

In Chapter 1, the hinge-by-hinge analysis procedure was carried out manually for a fixed-ended beam to obtain its load-deflection curve and plastic collapse load. Herein, we shall computerize this hinge-by-hinge analysis procedure so that it can be conveniently applied to larger structures such as multistory building frames. In this chapter, we will focus our attention on the first-order analysis and ignore the second-order stability effects. Without stability effects, the first-order analysis will obviously overestimate the actual load-carrying capacity, but the difference will be small for most low-rise building frames.

Note that in Chapters 4 and 5 the moment-curvature behavior of members was assumed to be rigid–perfectly plastic, and lower and upper bounds on the plastic collapse load were determined by the plastic limit analysis. Herein, the moment-curvature relationship is taken to be elastic–perfectly plastic, and the exact plastic collapse load along with its load-deflection curve is obtained for a given structure.

Since the matrix method is more appropriate for computerized structural analysis, herein we will first derive the member stiffness matrix for the following five cases: (1) a member without hinges; (2) a member with one end hinge; (3) a member with two end hinges; (4) a member with one intermediate hinge and; (5) a member with one intermediate and one end hinge. Then we will describe the numerical procedure used for the development of a computer program. Finally, we will present results of three illustrative examples. The first example shows the step-by-step computations that have taken place in the computer for a fixed-ended beam. The readers who are interested in learning only about how to run the computer program may skip the first example and directly go to Examples 7.2 and 7.3. The second example compares the computer solution of a multistory and multibay frame with the plastic limit analysis solution described in Chapter 4. The third example compares the deflections of a frame with those computed by the approximate methods described in Chapter 6. In this last example we also compute the required rotation capacity. In the appendix, four additional examples using the first-order plastic analysis (FOPA) computer program are presented.

7.2 Stiffness Matrix of Elastic Beam Element

Considering a beam element subjected to end moments M_A and M_B as shown in Fig. 7.1(a), and using the free body diagram shown in Fig. 7.1(b), the external moment at an arbitrary section x can be expressed as

$$M_{\text{ext}} = M_A - \frac{M_A + M_B}{L} x. \tag{7.2.1}$$

Equating this external moment to internal moment $M_{\text{int}} = -EIy''$ and rearranging, we obtain

$$y'' = -\frac{M_A}{EI} + \frac{M_A + M_B}{EIL} x. \tag{7.2.2}$$

Integrating once and twice to obtain, respectively, the slope and deflection of the member as

$$y' = -\frac{M_A}{EI} x + \frac{M_A + M_B}{EIL} \frac{x^2}{2} + C_1 \tag{7.2.3}$$

$$y = -\frac{M_A}{EI} \frac{x^2}{2} + \frac{M_A + M_B}{EIL} \frac{x^2}{6} + C_1 x + C_2 \tag{7.2.4}$$

Using the boundary conditions $y(0) = y(L) = 0$, we can determine the integration constants C_1 and C_2. The boundary condition $y(0) = 0$ gives

$$C_2 = 0 \tag{7.2.5}$$

and the boundary condition $y(L) = 0$ gives

$$C_1 = \frac{M_A}{EI} \frac{L}{2} - \frac{M_A + M_B}{EI} \frac{L}{6}. \tag{7.2.6}$$

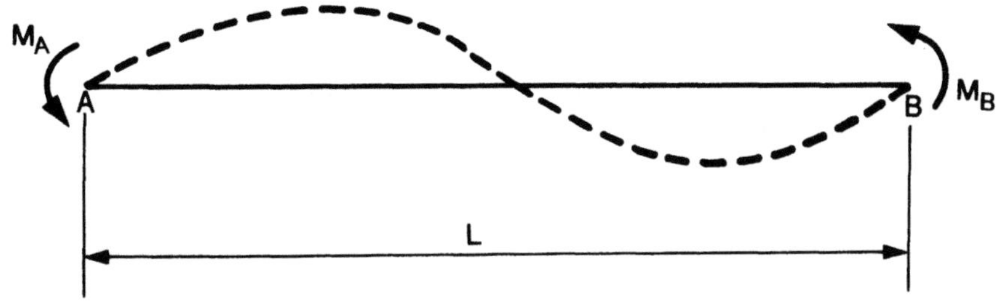

(a) Element

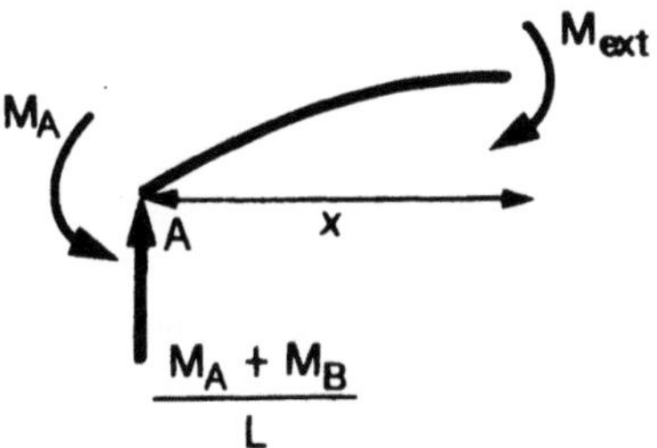

(b) Free Body

FIGURE 7.1. Beam element subjected to end moments: (a) element and (b) free body.

Substituting the value of C_1 into Eq. (7.2.3), we have

$$y' = -\frac{M_A}{EI}x + \frac{M_A + M_B}{EIL}\frac{x^2}{2} + \frac{2M_AL - M_BL}{6EI}. \tag{7.2.7}$$

Since $\theta_A = y'(0)$ and $\theta_B = y'(L)$, we have

$$\theta_A = \frac{M_A}{EI}\frac{L}{3} - \frac{M_B}{EI}\frac{L}{6} \tag{7.2.8}$$

$$\theta_B = -\frac{M_A}{EI}\frac{L}{6} + \frac{M_B}{EI}\frac{L}{3}. \tag{7.2.9}$$

Inverting Eqs. (7.2.8) and (7.2.9), we obtain

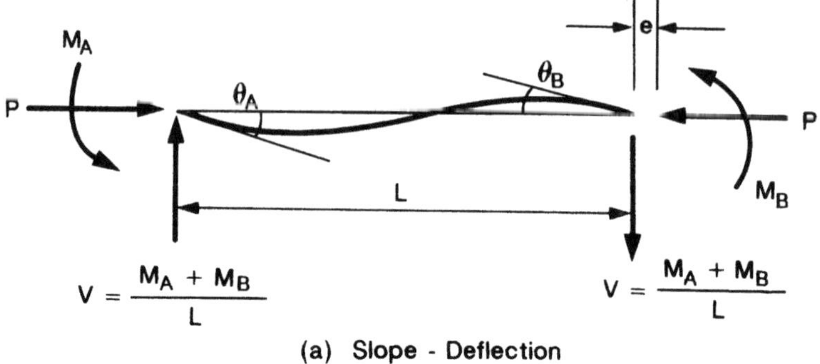

(a) Slope - Deflection

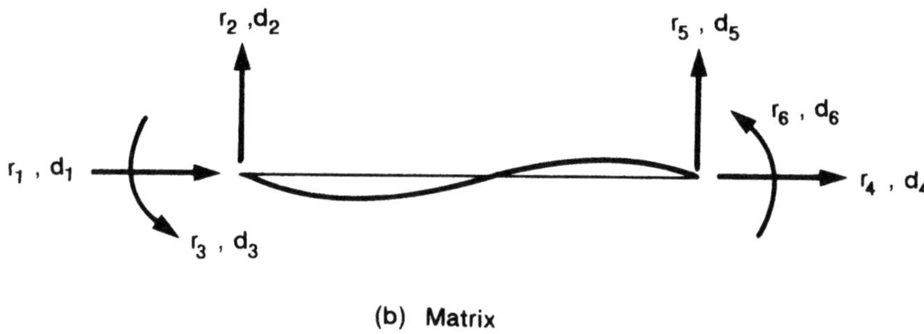

(b) Matrix

FIGURE 7.2. (a) Slope-deflection and (b) matrix analysis coordinates.

$$\begin{Bmatrix} M_A \\ M_B \end{Bmatrix} = \frac{EI}{L} \begin{bmatrix} 4 & 2 \\ 2 & 4 \end{bmatrix} \begin{Bmatrix} \theta_A \\ \theta_B \end{Bmatrix}. \tag{7.2.10}$$

These are the well-known *slope-deflection* equations. However, in a usual structural analysis, it is more convenient to use a matrix analysis coordinate. This transformation can be achieved by simply comparing member end forces and end displacement in the slope-deflection coordinate with the matrix analysis coordinate as shown in Fig. 7.2. The equilibrium relationship between the end forces in these two systems has the form

$$\begin{Bmatrix} r_1 \\ r_2 \\ r_3 \\ r_4 \\ r_5 \\ r_6 \end{Bmatrix} = \begin{bmatrix} 1 & 0 & 0 \\ 0 & \dfrac{1}{L} & \dfrac{1}{L} \\ 0 & 1 & 1 \\ -1 & 0 & 0 \\ 0 & -\dfrac{1}{L} & -\dfrac{1}{L} \\ 0 & 0 & 1 \end{bmatrix} \begin{Bmatrix} P \\ M_A \\ M_B \end{Bmatrix}. \tag{7.2.11}$$

The kinematic relationship between the end displacements in these two systems is

$$\begin{Bmatrix} e \\ \theta_A \\ \theta_B \end{Bmatrix} = \begin{bmatrix} 1 & 0 & 0 & -1 & 0 & 0 \\ 0 & \dfrac{1}{L} & 1 & 0 & -\dfrac{1}{L} & 0 \\ 0 & \dfrac{1}{L} & 0 & 0 & -\dfrac{1}{L} & 1 \end{bmatrix} \begin{Bmatrix} d_1 \\ d_2 \\ d_3 \\ d_4 \\ d_5 \\ d_6 \end{Bmatrix}. \tag{7.2.12}$$

By including the axial force and axial deformation relationship in the slope-deflection equation (7.2.10), the new slope-deflection equation can be expressed as

$$\begin{Bmatrix} P \\ M_A \\ M_B \end{Bmatrix} = \frac{EI}{L} \begin{bmatrix} \dfrac{A}{I} & 0 & 0 \\ 0 & 4 & 2 \\ 0 & 2 & 4 \end{bmatrix} \begin{Bmatrix} e \\ \theta_A \\ \theta_B \end{Bmatrix}. \tag{7.2.13}$$

The relationship between the end forces $\{r\}$ and the end displacements $\{d\}$ can now be obtained by substituting Eq. (7.2.13) into Eq. (7.2.11) and substituting Eq. (7.2.12) in the resulting equation as

$$\begin{Bmatrix} r_1 \\ r_2 \\ r_3 \\ r_4 \\ r_5 \\ r_6 \end{Bmatrix} = \begin{bmatrix} \dfrac{EA}{L} & 0 & 0 & -\dfrac{EA}{L} & 0 & 0 \\ 0 & \dfrac{12EI}{L^3} & \dfrac{6EI}{L^2} & 0 & \dfrac{-12EI}{L^3} & \dfrac{6EI}{L^2} \\ 0 & \dfrac{6EI}{L^2} & \dfrac{4EI}{L} & 0 & -\dfrac{6EI}{L^2} & \dfrac{2EI}{L} \\ -\dfrac{EA}{L} & 0 & 0 & \dfrac{EA}{L} & 0 & 0 \\ 0 & -\dfrac{12EI}{L^3} & -\dfrac{6EI}{L^2} & 0 & \dfrac{12EI}{L^3} & -\dfrac{6EI}{L^2} \\ 0 & \dfrac{6EI}{L^2} & \dfrac{2EI}{L} & 0 & -\dfrac{6EI}{L^2} & \dfrac{4EI}{L} \end{bmatrix} \begin{Bmatrix} d_1 \\ d_2 \\ d_3 \\ d_4 \\ d_5 \\ d_6 \end{Bmatrix}. \tag{7.2.14}$$

7.3 Stiffness Matrix for a Beam Element with a Plastic Hinge at End A

When one plastic hinge is formed at end A, then the incremental moment at end A is zero. Using the condition that $\dot{M}_A = 0$ in the first matrix Eq. (7.2.10), we obtain the kinematic relation in the incremental form as $\dot{\theta}_A = -\dot{\theta}_B/2$.

Substituting this $\dot{\theta}_A$ into the second equation of matrix Eq. (7.2.10) and then rewriting it in matrix form, we have

$$\begin{Bmatrix} \dot{M}_A \\ \dot{M}_B \end{Bmatrix} = \frac{EI}{L} \begin{bmatrix} 0 & 0 \\ 0 & 3 \end{bmatrix} \begin{Bmatrix} \dot{\theta}_A \\ \dot{\theta}_B \end{Bmatrix}. \tag{7.3.1}$$

Including the incremental axial load and axial-deformation relationship in Eq. (7.3.1), we have

$$\begin{Bmatrix} \dot{P} \\ \dot{M}_A \\ \dot{M}_B \end{Bmatrix} = \frac{EI}{L} \begin{bmatrix} \dfrac{A}{I} & 0 & 0 \\ 0 & 0 & 0 \\ 0 & 0 & 3 \end{bmatrix} \begin{Bmatrix} \dot{e} \\ \dot{\theta}_A \\ \dot{\theta}_B \end{Bmatrix}. \tag{7.3.2}$$

This relationship can now be transformed into the matrix analysis coordinate by using the end forces relation (7.2.11) and the kinematic relation (7.2.12) in incremental form as

$$\begin{Bmatrix} \dot{r}_1 \\ \dot{r}_2 \\ \dot{r}_3 \\ \dot{r}_4 \\ \dot{r}_5 \\ \dot{r}_6 \end{Bmatrix} = \begin{bmatrix} \dfrac{EA}{L} & 0 & 0 & -\dfrac{EA}{L} & 0 & 0 \\ 0 & \dfrac{3EI}{L^3} & 0 & 0 & -\dfrac{3EI}{L^3} & \dfrac{3EI}{L^2} \\ 0 & 0 & 0 & 0 & 0 & 0 \\ -\dfrac{EA}{L} & 0 & 0 & \dfrac{EA}{L} & 0 & 0 \\ 0 & -\dfrac{3EI}{L^3} & 0 & 0 & \dfrac{3EI}{L^3} & -\dfrac{3EI}{L^2} \\ 0 & \dfrac{3EI}{L^2} & 0 & 0 & -\dfrac{3EI}{L^2} & \dfrac{3EI}{L} \end{bmatrix} \begin{Bmatrix} \dot{d}_1 \\ \dot{d}_2 \\ \dot{d}_3 \\ \dot{d}_4 \\ \dot{d}_5 \\ \dot{d}_6 \end{Bmatrix}. \tag{7.3.3}$$

7.4 Stiffness Matrix for a Beam Element with a Plastic Hinge at End B

The incremental slope-deflection relation for this case can be derived in a similar way to that of the previous case as

$$\begin{Bmatrix} \dot{M}_A \\ \dot{M}_B \end{Bmatrix} = \frac{EI}{L} \begin{bmatrix} 3 & 0 \\ 0 & 0 \end{bmatrix} \begin{Bmatrix} \dot{\theta}_A \\ \dot{\theta}_B \end{Bmatrix}. \tag{7.4.1}$$

Including the incremental axial load–axial deformation relation in Eq. (7.4.1), we have

$$\begin{Bmatrix} \dot{P} \\ \dot{M}_A \\ \dot{M}_B \end{Bmatrix} = \frac{EI}{L} \begin{bmatrix} \dfrac{A}{I} & 0 & 0 \\ 0 & 3 & 0 \\ 0 & 0 & 0 \end{bmatrix} \begin{Bmatrix} \dot{e} \\ \dot{\theta}_A \\ \dot{\theta}_B \end{Bmatrix}. \tag{7.4.2}$$

Using Eqs. (7.2.11) and (7.2.12) in an incremental form for the transformation to a matrix analysis coordinate, we have

$$
\begin{Bmatrix} \dot{r}_1 \\ \dot{r}_2 \\ \dot{r}_3 \\ \dot{r}_4 \\ \dot{r}_5 \\ \dot{r}_6 \end{Bmatrix} =
\begin{bmatrix}
\dfrac{AE}{L} & 0 & 0 & -\dfrac{AE}{L} & 0 & 0 \\[6pt]
0 & \dfrac{3EI}{L^3} & \dfrac{3EI}{L^2} & 0 & -\dfrac{3EI}{L^3} & 0 \\[6pt]
0 & \dfrac{3EI}{L^2} & \dfrac{3EI}{L^2} & 0 & -\dfrac{3EI}{L^2} & 0 \\[6pt]
-\dfrac{AE}{L} & 0 & 0 & \dfrac{AE}{L} & 0 & 0 \\[6pt]
0 & -\dfrac{3EI}{L^3} & -\dfrac{3EI}{L^2} & 0 & \dfrac{3EI}{L^3} & 0 \\[6pt]
0 & 0 & 0 & 0 & 0 & 0
\end{bmatrix}
\begin{Bmatrix} \dot{d}_1 \\ \dot{d}_2 \\ \dot{d}_3 \\ \dot{d}_4 \\ \dot{d}_5 \\ \dot{d}_6 \end{Bmatrix} . \qquad (7.4.3)
$$

7.5 Plastic Hinges at Both Ends A and B

When plastic hinges are formed at both ends, no additional moment can be applied at these ends, and the incremental slope-deflection relation becomes

$$
\begin{Bmatrix} \dot{M}_A \\ \dot{M}_B \end{Bmatrix} = \begin{bmatrix} 0 & 0 \\ 0 & 0 \end{bmatrix} \begin{Bmatrix} \dot{\theta}_A \\ \dot{\theta}_B \end{Bmatrix} . \qquad (7.5.1)
$$

After transformation to a matrix analysis coordinate, the incremental force-displacement relation becomes

$$
\begin{Bmatrix} \dot{r}_1 \\ \dot{r}_2 \\ \dot{r}_3 \\ \dot{r}_4 \\ \dot{r}_5 \\ \dot{r}_6 \end{Bmatrix} =
\begin{bmatrix}
\dfrac{EA}{L} & 0 & 0 & -\dfrac{EA}{L} & 0 & 0 \\[6pt]
0 & 0 & 0 & 0 & 0 & 0 \\[6pt]
0 & 0 & 0 & 0 & 0 & 0 \\[6pt]
-\dfrac{EA}{L} & 0 & 0 & \dfrac{EA}{L} & 0 & 0 \\[6pt]
0 & 0 & 0 & 0 & 0 & 0 \\[6pt]
0 & 0 & 0 & 0 & 0 & 0
\end{bmatrix}
\begin{Bmatrix} \dot{d}_1 \\ \dot{d}_2 \\ \dot{d}_3 \\ \dot{d}_4 \\ \dot{d}_5 \\ \dot{d}_6 \end{Bmatrix} . \qquad (7.5.2)
$$

7.6 Stiffness Matrix for a Beam with an Intermediate Plastic Hinge

To derive the stiffness matrix for a beam with an intermediate plastic hinge (Fig. 7.3), we will use a slightly different approach. The stiffness matrix is obtained directly in the matrix analysis coordinates by applying a unit displacement along the specified degree of freedom. The corresponding stiffness

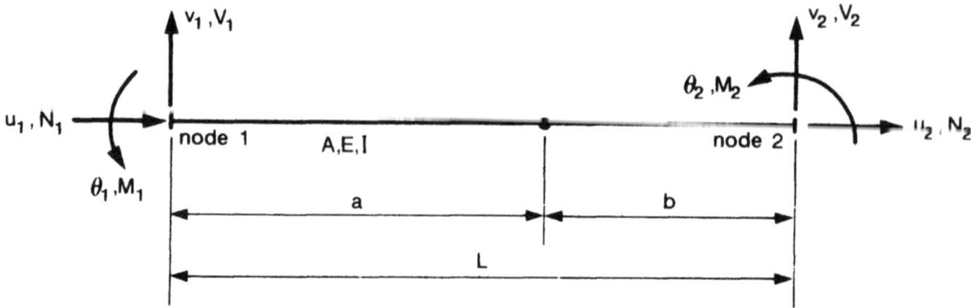

FIGURE 7.3. Nodal forces and degrees of freedom for a frame member with an intermediate plastic hinge.

coefficients, along with constants of integration, are determined from the known boundary conditions.

As an example, consider a beam with a unit displacement along coordinate 1 as shown in Fig. 7.4(a) [7.1]. The external moment at any point a distance x from end A can be expressed as

$$M_{\text{ext}} = k_{11}x - k_{21} \tag{7.6.1}$$

where k_{11} and k_{21} are, respectively, the forces induced along coordinates 1 and 2 by a unit displacement along coordinate 1. Equating the external moment M_{ext} to the internal moment $M_{\text{int}} = EIy''$, we have

$$EIy'' = k_{11}x - k_{21}. \tag{7.6.2}$$

Integrating Eq. (7.6.2) once and twice, we obtain, respectively, the slope and deflection as

for $0 \leq x \leq a$

$$EIy_1' = k_{11}\frac{x^2}{2} - k_{21}x + C_1 \tag{7.6.3}$$

$$EIy_1 = k_{11}\frac{x^3}{6} - k_{21}\frac{x^2}{2} + C_1x + C_2, \tag{7.6.4}$$

and for $a \leq x \leq L$

$$EIy_2' = k_{11}\frac{x^2}{2} - k_{21}x + D_1 \tag{7.6.5}$$

$$EIy_2 = k_{11}\frac{x^3}{6} - k_{21}\frac{x^2}{2} + D_1x + D_2. \tag{7.6.6}$$

In Eqs. (7.6.3) to (7.6.6), there are six unknowns: four constants of integration C_1, C_2, D_1, and D_2; and two stiffness coefficients k_{11} and k_{21}. These unknowns are determined by using the following six boundary conditions: $y_1(0) = 1$, $y_2(L) = 0$, $y''(a) = 0$, and $y_1(a) = y_2(a)$. The solutions for these six equations are

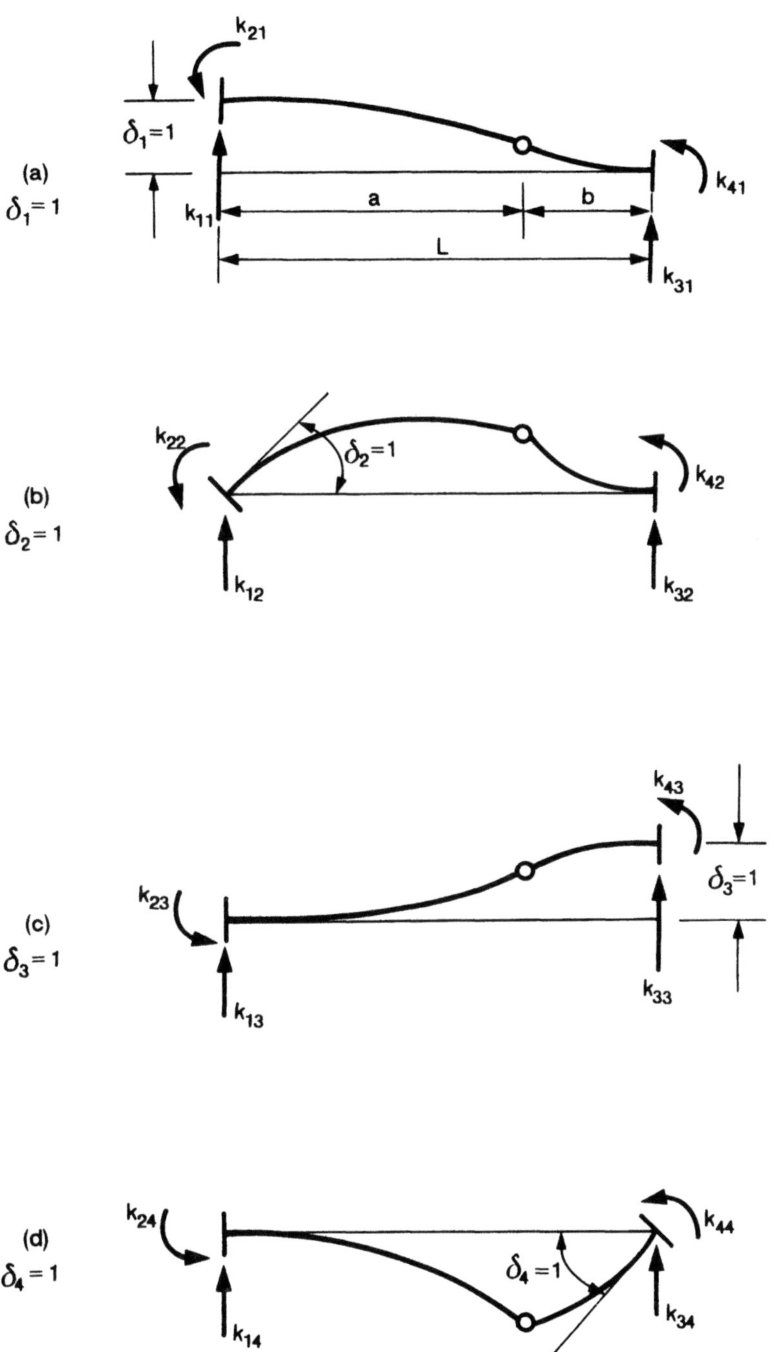

FIGURE 7.4. Unit displacements along coordinates 1, 2, 3, and 4 of a member with an intermediate hinge.

$$C_1 = 0 \tag{7.6.7}$$

$$C_2 = EI \tag{7.6.8}$$

$$D_1 = \frac{3EI(a^2 - b^2)}{(2a^3 + 2b^3)} \tag{7.6.9}$$

$$D_2 = \frac{EI(b^3 - 2a^3 + 3ab^2)}{(a^3 + b^3)} \tag{7.6.10}$$

$$k_{11} = \frac{3EI}{a^3 + b^3} \tag{7.6.11}$$

$$k_{21} = \frac{3EIa}{a^3 + b^3} \tag{7.6.12}$$

By considering the equilibrium of the beam in Fig. 7.4(a), the stiffness coefficients k_{31} and k_{41} can be written as

$$k_{31} = -k_{11} = -\frac{3EI}{a^3 + b^3} \tag{7.6.13}$$

$$k_{41} = k_{11}b = \frac{3EIb}{a^3 + b^3}. \tag{7.6.14}$$

All the stiffness coefficients in column 1 of the stiffness matrix are now known. The coefficients in other columns can be obtained by considering unit displacements along coordinates 2, 3, and 4 as shown in Figs. 7.4(b), (c), and (d) respectively. The resulting stiffness matrix, including the axial force and axial deformation relationship, has the form

$$
\begin{Bmatrix} r_1 \\ r_2 \\ r_3 \\ r_4 \\ r_5 \\ r_6 \end{Bmatrix} = \frac{3EI}{a^3 + b^3}
\begin{bmatrix}
\dfrac{A(a^3 + b^3)}{3I(a + b)} & 0 & 0 & \dfrac{-A(a^3 + b^3)}{3I(a + b)} & 0 & 0 \\
0 & 1 & a & 0 & -1 & b \\
0 & a & a & 0 & -a & ab \\
\dfrac{-A(a^3 + b^3)}{3I(a + b)} & 0 & 0 & \dfrac{A(a^3 + b^3)}{3I(a + b)} & 0 & 0 \\
0 & -1 & -a & 0 & 1 & -b \\
0 & b & ab & 0 & -b & b
\end{bmatrix}
\times
\begin{Bmatrix} d_1 \\ d_2 \\ d_3 \\ d_4 \\ d_5 \\ d_6 \end{Bmatrix}. \tag{7.6.15}
$$

7.7 Numerical Procedure for a First-Order Plastic Analysis

The first-order plastic analysis is a series of first-order elastic analysis performed on the original and subsequent structures modified by the formation of plastic hinges. The plastic hinges are assumed to form only at the member ends. For the members supporting laterally distributed loads, a plastic hinge may form within the member. In the present computer program, this is handled by dividing the member into several elements and replacing the distributed load by a number of concentrated loads, one at each node. The procedure adopted in the present computer program consists of the following steps:

1. An elastic analysis is performed on the original structure subjected to a given set of loads. These loads have relative magnitudes in proportion to those at the ultimate state. This set of loads is referred to as the "reference loads."
2. The *load factor* at each nodal point is computed by dividing the plastic moment capacity at that nodal point by the moment at that nodal point due to the reference loads. The location of the smallest load factor in the whole structure is the location of the first plastic hinge (or hinges).
3. Displacements and internal forces at all nodes of the original structure, subjected to a set of loads obtained by multiplying the lowest load factor with the reference loads, are computed by merely scaling up the results of the first-order elastic analysis of the original structure by the lowest load factor. This is based on the fact that the behavior of the structure is linearly elastic between the formation of any two plastic hinges. With this step, the first stage of the hinge-by-hinge matrix-analysis procedure has been completed.
4. The original structure is now modified by inserting fictitious hinge at the location of the first plastic hinge.
5. An elastic analysis is again performed on this modified structure subjected to the reference loads.
6. The load factor at each node is again calculated by dividing the "remaining" moment capacity (see Example 7.8.1) at each node by the moment due to the reference load applied on the modified structure. The second plastic hinge can now be located where the smallest load factor is.
7. The displacements and internal forces are again scaled up by the new load factor.
8. The cumulative load factor, displacements, and internal forces at all nodal points are then computed. This completes the second stage of the matrix structural analysis.
9. The structure is again modified by inserting another fictitious hinge at the location as determined in step 6.

10. Steps 5 to 9 are repeated until an unstable structure is realized. The unstable structure is recognized when the stiffness matrix ceases to be positive definite or the deformations due to the reference loads become excessively large.

7.8 Numerical Examples

Three numerical examples will be presented herein. The first example is a fixed-ended beam with a concentrated lateral load applied at one-third point as shown in Fig. 7.5(a). This example is solved here manually by the hinge-by-hinge matrix-analysis method. The aim is to show the steps the computer program will go through later. In the second example, a portal frame taken from Chapter 4 is solved by the computer program and the results are compared with the plastic limit load analysis in Chapter 4. In the third example, a frame in Chapter 6 is analyzed by the computer program and the deflections are compared with those obtained by the approximate methods described in Chapter 6.

Example 7.8.1. A 12-foot-long fixed-ended beam with a concentrated load at one-third point is shown in Fig. 7.5(a). Determine the displacements and internal forces in this beam by the hinge-by-hinge matrix-analysis procedure described in Section 7.7. Compare the results with those obtained in Example 1.8.2. Assume $E = 29,000$ ksi, $I = 1000$ in.^{4}, and $Z_x = 157$ in.^{3} (W14 × 90).

Solution: Divide the beam into two elements. The first element has 1, 2, 3, and 4 as global coordinates [Fig. 7.5(b)], and the second element has 3, 4, 5, and 6 as global coordinates [Fig. 7.5(c)]. The beam overall has two degrees of freedom as 3 and 4. Corresponding to these degrees of freedom, the reference load vector is chosen as

$$R = \left\{ \begin{matrix} 1 \\ 0 \end{matrix} \right\}. \tag{7.8.1}$$

Step 1: The first step is to carry out the elastic analysis on the original beam. To this end, the element stiffness matrices are written and then assembled. The stiffness matrix for the first element is (Eq. 7.2.14)

$$k_1 = EI \begin{bmatrix} \dfrac{324}{L^3} & \dfrac{54}{L^2} & -\dfrac{324}{L^3} & \dfrac{54}{L^2} \\ & \dfrac{12}{L} & -\dfrac{54}{L^2} & \dfrac{6}{L} \\ & & \dfrac{324}{L^3} & -\dfrac{54}{L^2} \\ & & & \dfrac{12}{L} \end{bmatrix}$$

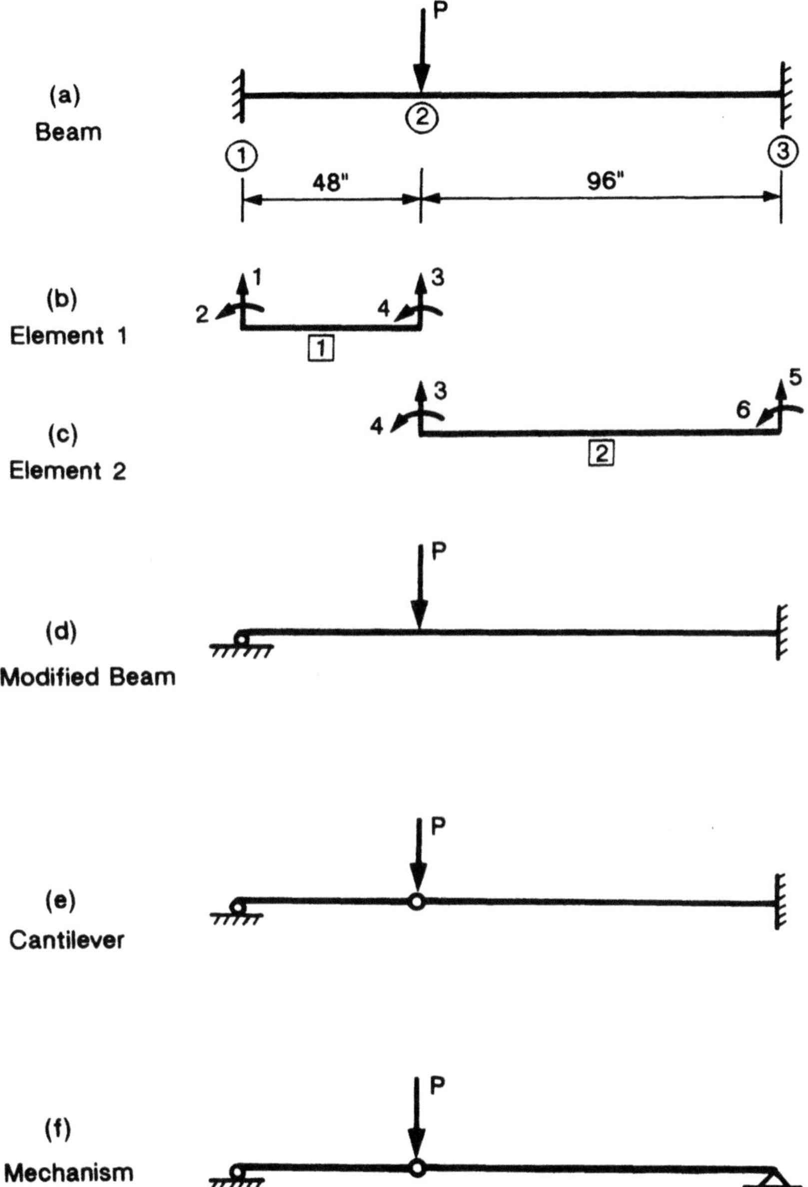

FIGURE 7.5. Hinge-by-hinge matrix analysis of a fixed-ended beam with a concentrated load applied at one-third point: (a) beam, (b) element 1, (c) element 2, (d) modified beam, (e) cantilever, and (f) mechanism.

$$
= \begin{bmatrix} 3147 & 75{,}521 & -3147 & 75{,}521 \\ & 2{,}416{,}667 & -75{,}521 & 1{,}208{,}333 \\ & & 3147 & -75{,}521 \\ & & & 2{,}416{,}617 \end{bmatrix}. \tag{7.8.2}
$$

Similarly, for the second element, we have

$$
k_2 = EI \begin{bmatrix} \dfrac{40.5}{L^3} & \dfrac{13.5}{L^2} & -\dfrac{40.5}{L^3} & \dfrac{13.5}{L^2} \\[2mm] & \dfrac{6}{L} & -\dfrac{13.5}{L^2} & \dfrac{3}{L} \\[2mm] & & \dfrac{40.5}{L^3} & -\dfrac{13.5}{L^2} \\[2mm] & & & \dfrac{6}{L} \end{bmatrix}
$$

$$
= \begin{bmatrix} 393 & 18{,}880 & -393 & 18{,}880 \\ & 1{,}208{,}333 & -18{,}880 & 604{,}167 \\ & & 393 & -18{,}880 \\ & & & 1{,}208{,}333 \end{bmatrix}. \tag{7.8.3}
$$

Assemble these two matrices to form a global stiffness matrix $[k]$ corresponding to the chosen degrees of freedom as

$$
k = \begin{bmatrix} 3540 & -56{,}641 \\ & 3{,}625{,}000 \end{bmatrix}. \tag{7.8.4}
$$

Now the force-displacement relationship can be written as

$$
\begin{Bmatrix} -1 \\ 0 \end{Bmatrix} = \begin{bmatrix} 3540 & -56{,}641 \\ & 3{,}625{,}000 \end{bmatrix} \begin{Bmatrix} V_3 \\ V_4 \end{Bmatrix}. \tag{7.8.5}
$$

Solving for displacements, we have

$$
\begin{Bmatrix} V_3 \\ V_4 \end{Bmatrix} = \begin{Bmatrix} -3.766 \times 10^{-4} \\ -5.885 \times 10^{-6} \end{Bmatrix}. \tag{7.8.6}
$$

Note that V_3 obtained here is the same as Δ_1 calculated from Eq. 1.8.23 in Chapter 1.

Step 2: Herein, the element internal forces expressed as load factors are determined. For the first element, the internal forces are determined as

$$
\begin{Bmatrix} F_1 \\ F_2 \\ F_3 \\ F_4 \end{Bmatrix} = \begin{bmatrix} -- & -3147 & 75{,}521 \\ -- & -75{,}521 & 1{,}208{,}333 \\ -- & 3147 & -75{,}521 \\ -- & -75{,}521 & 2{,}416{,}617 \end{bmatrix} \begin{Bmatrix} 0 \\ 0 \\ -3.766 \times 10^{-4} \\ -5.885 \times 10^{-6} \end{Bmatrix}
$$

$$= \begin{Bmatrix} 0.74 \\ 21.33 \\ -0.74 \\ 14.21 \end{Bmatrix}. \tag{7.8.7}$$

For the second element, the internal forces are calculated as

$$\begin{Bmatrix} F_3 \\ F_4 \\ F_5 \\ F_6 \end{Bmatrix} = \begin{bmatrix} 393 & 18,880 & -- \\ 18,880 & 1,208,333 & -- \\ -393 & -18,880 & -- \\ 18,880 & 604,167 & -- \end{bmatrix} \begin{Bmatrix} -3.776 \times 10^{-4} \\ -5.885 \times 10^{-6} \\ 0 \\ 0 \end{Bmatrix}$$

$$= \begin{Bmatrix} -0.26 \\ -14.21 \\ +0.26 \\ -10.68 \end{Bmatrix}. \tag{7.8.8}$$

The load factors corresponding to moments at these three nodes are

$$\lambda_1 = \frac{5,652}{21.33} = 265 \tag{7.8.9}$$

$$\lambda_2 = \frac{5,652}{14.21} = 397.7 \tag{7.8.10}$$

$$\lambda_3 = \frac{5,652}{10.68} = 529.2 \tag{7.8.11}$$

where $M_p = \sigma_y z_x = 36 \times 157 = 5,652$ has been used. The lowest value of λ occurs at node 1 with $\lambda_1 = 265$. Thus, the first hinge will form at node 1.

Step 3: The displacements and internal forces at this stage are determined by multiplying those determined in steps 1 and 2 by the load factor $\lambda = 265$. The displacements are amplified to be:

$$\begin{Bmatrix} V_3 \\ V_4 \end{Bmatrix} = \begin{Bmatrix} -0.0998 \\ -0.00156 \end{Bmatrix}. \tag{7.8.12}$$

The displacement V_3 in Eq. (7.8.12) is the same as Δ_1 calculated from Eq. (1.8.23) in Chapter 1. The amplified internal forces in the first element become

$$\begin{Bmatrix} F_1 \\ F_2 \\ F_3 \\ F_4 \end{Bmatrix} = \begin{Bmatrix} 196 \\ 5652 \\ -196 \\ 3766 \end{Bmatrix}. \tag{7.8.13}$$

The internal forces in the second element are calculated to be

$$\begin{Bmatrix} F_3 \\ F_4 \\ F_5 \\ F_6 \end{Bmatrix} = \begin{Bmatrix} -69 \\ -3766 \\ +69 \\ -2830 \end{Bmatrix}. \tag{7.8.14}$$

The load vector at this stage can be written as

$$\begin{Bmatrix} F_3 \\ F_4 \end{Bmatrix} = \begin{Bmatrix} 265 \\ 0 \end{Bmatrix}. \tag{7.8.15}$$

Note that the load F_3 in Eq. (7.8.15) is the same as P_1 calculated from Eq. (1.8.22) in Chapter 1.

Step 4: The hinge has formed at the first node and the modified structure is shown in Fig. 7.5(d).

Step 5: An elastic analysis is again performed on the modified structure subjected to the reference load shown in Fig. 7.5(d). Now the first element has a hinge at the left-hand end. Therefore, Eq. (7.3.3) is its stiffness matrix as

$$k_1 = EI \begin{bmatrix} \dfrac{81}{L^3} & 0 & -\dfrac{81}{L^3} & \dfrac{27}{L^2} \\ & 0 & 0 & 0 \\ & & \dfrac{81}{L^3} & -\dfrac{27}{L^2} \\ & & & \dfrac{9}{L} \end{bmatrix} = \begin{bmatrix} 787 & 0 & -787 & 37{,}760 \\ & 0 & 0 & 0 \\ & & 787 & -37{,}760 \\ & & & 1{,}812{,}500 \end{bmatrix}. \tag{7.8.16}$$

The stiffness matrix for the second element remains the same as that given by Eq. (7.8.3). The assembled structural stiffness matrix corresponding to the third and fourth degree of freedom has the value

$$k = \begin{bmatrix} 1180 & -18{,}880 \\ & 3{,}020{,}833 \end{bmatrix}. \tag{7.8.17}$$

The displacements $\dot{V}_3$ and $\dot{V}_4$ are calculated to be

$$\begin{Bmatrix} \dot{V}_3 \\ \dot{V}_4 \end{Bmatrix} = -\begin{Bmatrix} 9.416 \times 10^{-4} \\ 5.884 \times 10^{-6} \end{Bmatrix}. \tag{7.8.18}$$

Step 6: The internal forces in the first element are calculated as

$$\begin{Bmatrix} \dot{F}_1 \\ \dot{F}_2 \\ \dot{F}_3 \\ \dot{F}_4 \end{Bmatrix} = \begin{bmatrix} -0 & -787 & 37{,}760 \\ -0 & 0 & 0 \\ -0 & 787 & -37{,}760 \\ -0 & -37{,}760 & 1{,}812{,}500 \end{bmatrix} \begin{Bmatrix} 0 \\ -- \\ -9.416 \times 10^{-4} \\ -5.884 \times 10^{-6} \end{Bmatrix}$$

$$= \left\{ \begin{array}{c} 0.519 \\ 0 \\ -0.519 \\ +24.89 \end{array} \right\}. \qquad (7.8.19)$$

In the second element, the internal forces come out to be

$$\left\{ \begin{array}{c} \dot{F_3} \\ \dot{F_4} \\ \dot{F_5} \\ \dot{F_6} \end{array} \right\} = \left[\begin{array}{cccc} 393 & 18,880 & - & - \\ 18,880 & 1,208,333 & - & - \\ -393 & -18,880 & - & - \\ 18,880 & 604,167 & - & - \end{array} \right] \left\{ \begin{array}{c} -9.416 \times 10^{-4} \\ -5.884 \times 10^{-6} \\ 0 \\ 0 \end{array} \right\}$$

$$= \left\{ \begin{array}{c} -0.48 \\ -24.89 \\ 0.48 \\ -21.33 \end{array} \right\}. \qquad (7.8.20)$$

In Eqs. (7.8.18 to 7.8.20) the dot denotes the incremental displacement or internal forces that take place after the latest modification to the given structure.

Since a plastic hinge has already been formed at node 1, calculation of the load factor at this node is not needed. At nodes 2 and 3, the load factors are calculated by dividing the "remaining" plastic moment capacity by the moment at that node due to the reference loads applied to the modified structure as ($M_p = \sigma_y Z_x = 36 \times 157 = 5652$)

$$\lambda_2 = \frac{5652 - 3766}{24.89} = 75.7 \qquad (7.8.21)$$

$$\lambda_3 = \frac{5652 - 2830}{21.33} = 132.3 \qquad (7.8.22)$$

The lowest λ occurs at node 2 with $\lambda = 75.7$. Thus, the second hinge forms at node 2.

Step 7: The deflections amplified by $\lambda = 75.7$ are

$$\left\{ \begin{array}{c} \dot{V_3} \\ \dot{V_4} \end{array} \right\} = - \left\{ \begin{array}{c} 9.416 \times 10^{-4} \\ 5.884 \times 10^{-6} \end{array} \right\} (75.7) = - \left\{ \begin{array}{c} 0.0712 \\ 0.000445 \end{array} \right\}. \qquad (7.8.23)$$

The internal forces in the first element are magnified to be

$$\left\{ \begin{array}{c} \dot{F_1} \\ \dot{F_2} \\ \dot{F_3} \\ \dot{F_4} \end{array} \right\} = \left\{ \begin{array}{c} 39 \\ 0 \\ -39 \\ 1882 \end{array} \right\}. \qquad (7.8.24)$$

Similarly, the internal forces in the second element are scaled up by $\lambda = 75.7$ to be

$$\begin{Bmatrix} \dot{F}_3 \\ \dot{F}_4 \\ \dot{F}_5 \\ \dot{F}_6 \end{Bmatrix} = \begin{Bmatrix} -36 \\ -1882 \\ 36 \\ -1613 \end{Bmatrix}. \tag{7.8.25}$$

Step 8: Herein, the cumulative displacements and internal forces are determined. The cumulative displacements are found to be:

$$\begin{Bmatrix} V_3 \\ V_4 \end{Bmatrix} = \begin{Bmatrix} -0.0998 \\ -0.00156 \end{Bmatrix} + \begin{Bmatrix} -0.0712 \\ -0.000445 \end{Bmatrix} = \begin{Bmatrix} -0.1710 \\ -0.0020 \end{Bmatrix}. \tag{7.8.26}$$

The displacement V_3 in Eq. (7.8.26) is the same as Δ_2 calculated from Eq. (1.8.28) in Chapter 1.

The cumulative internal forces in the first element are determined as

$$\begin{Bmatrix} F_1 \\ F_2 \\ F_3 \\ F_4 \end{Bmatrix} = \begin{Bmatrix} 196 \\ 5652 \\ -196 \\ 3766 \end{Bmatrix} + \begin{Bmatrix} 39 \\ 0 \\ -39 \\ 1882 \end{Bmatrix} = \begin{Bmatrix} 235 \\ 5652 \\ -235 \\ 5648 \end{Bmatrix}. \tag{7.8.27}$$

In the second element, the cumulative internal forces are computed as

$$\begin{Bmatrix} F_3 \\ F_4 \\ F_5 \\ F_6 \end{Bmatrix} = \begin{Bmatrix} -69 \\ -3766 \\ +69 \\ -2830 \end{Bmatrix} + \begin{Bmatrix} -36 \\ -1882 \\ 36 \\ -1613 \end{Bmatrix} = \begin{Bmatrix} -105 \\ -5648 \\ +105 \\ -4443 \end{Bmatrix}. \tag{7.8.28}$$

The load vector at this stage can be calculated as

$$\begin{Bmatrix} F_3 \\ F_4 \end{Bmatrix} = \begin{Bmatrix} 265 \\ 0 \end{Bmatrix} + 75.7 \begin{Bmatrix} 1 \\ 0 \end{Bmatrix} = \begin{Bmatrix} 341 \\ 0 \end{Bmatrix}. \tag{7.8.29}$$

The force F_3 calculated in Eq. (7.8.29) is the same as P_2 calculated from Eq. (1.8.26) in Chapter 1.

Step 9: After the second hinge is formed at the second node, the beam reduces to a simple cantilever as shown in Fig. 7.5(e).

Step 10: Again, an elastic analysis is performed on the modified structure shown in Fig. 7.5(e). The first element has hinges on both ends. The stiffness matrix of this element as given by Eq. (7.5.2) is null and can be expressed as

$$k_1 = \begin{Bmatrix} 0 & 0 & 0 & 0 \\ & 0 & 0 & 0 \\ & & 0 & 0 \\ & & & 0 \end{Bmatrix}. \tag{7.8.30}$$

The second element has a hinge at its left end. Thus, its stiffness is obtained from Eq. (7.3.3) as

$$k_2 = EI \begin{bmatrix} \dfrac{10.13}{L^3} & 0 & -\dfrac{10.13}{L^3} & \dfrac{6.75}{L^2} \\ & 0 & 0 & 0 \\ & & \dfrac{10.13}{L^3} & -\dfrac{6.75}{L^2} \\ & & & \dfrac{4.5}{L} \end{bmatrix} = \begin{bmatrix} 98 & 0 & -98 & 9440 \\ & 0 & 0 & 0 \\ & & 98 & -9440 \\ & & & 906{,}250 \end{bmatrix}.$$

$$(7.8.31)$$

The structural stiffness matrix corresponding to the third and fourth degree of freedom is written as

$$k = \begin{bmatrix} 98 & 0 \\ 0 & 0 \end{bmatrix}. \tag{7.8.32}$$

This stiffness matrix is now used to obtain the increment of V_3 as

$$\dot{V}_3 = -0.01017. \tag{7.8.33}$$

The incremental internal forces in the first element are zero. The incremental forces in the second element come out to be

$$\begin{Bmatrix} \dot{F}_3 \\ \dot{F}_4 \\ \dot{F}_5 \\ \dot{F}_6 \end{Bmatrix} = \begin{Bmatrix} -1.0 \\ 0 \\ 1.0 \\ -96 \end{Bmatrix}. \tag{7.8.34}$$

Since plastic hinges have already been formed at nodes 1 and 2, the load factors λ_1 and λ_2 need not be calculated. The load factor λ_3 is calculated as

$$\lambda_3 = \frac{5652 - 4443}{96} = 12.59. \tag{7.8.35}$$

So the controlling load factor is $\lambda = \lambda_3 = 12.59$.

Step 11: Amplifying the displacement $\dot{V}_3$ by the controlling load factor, we have

$$\dot{V}_3 = (12.59)(-0.01017) = -0.1280. \tag{7.8.36}$$

Applying the load factor to the internal forces in the second element, we obtain

$$\begin{Bmatrix} \dot{F}_1 \\ \dot{F}_2 \\ \dot{F}_3 \\ \dot{F}_4 \end{Bmatrix} = \begin{Bmatrix} -12.6 \\ 0 \\ 12.6 \\ -1209 \end{Bmatrix}. \tag{7.8.37}$$

Step 12: The cumulative displacement comes out to be

$$V_3 = -0.1710 - 0.1280 = 0.2990 \text{ inch.} \tag{7.8.38}$$

This value of V_3 is the same as Δ_3 obtained in Eq. (1.8.31). The cumulative internal forces for the second element can be calculated as

$$\begin{Bmatrix} F_3 \\ F_4 \\ F_5 \\ F_6 \end{Bmatrix} = \begin{Bmatrix} -105 \\ -5648 \\ +105 \\ -4443 \end{Bmatrix} + \begin{Bmatrix} -12.6 \\ 0 \\ 12.6 \\ -1209 \end{Bmatrix} = \begin{Bmatrix} -118 \\ -5648 \\ 118 \\ 5652 \end{Bmatrix}. \tag{7.8.39}$$

The cumulative load vector at this stage is calculated as

$$\begin{Bmatrix} F_3 \\ F_4 \end{Bmatrix} = \begin{Bmatrix} 341 \\ 0 \end{Bmatrix} + 12.59 \begin{Bmatrix} 1 \\ 0 \end{Bmatrix} = \begin{Bmatrix} 353 \\ 0 \end{Bmatrix}. \tag{7.8.40}$$

The applied load F_3 in Eq. (7.8.39) is the same as P_3 calculated from Eq. (1.8.30) in Chapter 1. Since plastic hinges are now formed at all three nodes, a mechanism as shown in Fig. 7.5(f) has developed, and the load cannot be further increased.

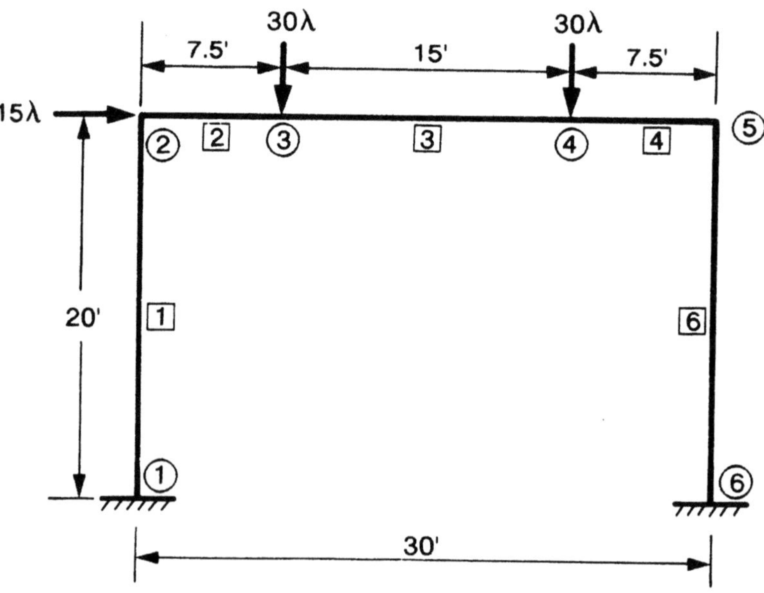

FIGURE 7.6. Node and element numbers for matrix analysis of the portal frame of Example 4.8.1.

Example 7.8.2. Using the computer program First-Order Plastic Analysis described in this chapter and given in the appendix, compute the load factor for the frame subjected to the loads shown in Fig. 7.6. All members are made of W16 × 45 section.

Solution: The elements and node numbers used in the present computer analysis are shown in Fig. 7.6. The input and ouput of the analysis from the computer program are shown in the following.

```
Prob. 7.8.2--Plastic Limit Load
 6 5 3 2 1
 29000 5 2
 0  0
 0  240
 90  240
 270  240
 360  240
 360  0
 1 2  1  1 1
 2 3  1  1 1
 3 4  1  1 1
 4 5  1  1 1
 5 6  1  1 1
 13.3 586 2963
 2 15  0  0
 3  0 -30 0
 4  0 -30 0
 1  1 1 1
 6  1 1 1
```

```
==============================================================================
                    First Order Plastic Analysis "FOPA"
                                    By
                         Dr. Maheeb Abdel-Ghaffar
                            Cairo University
                                  1992
==============================================================================
                              INPUT DATA
                              ==========
```

Prob. 7.8.2--Plastic Limit Load

# Nds.	# Elems.	# Ldd.Nds	# B.C.Nds	E	# Prp.Grps
6	5	3	2	.29E+05	1

JOINT COORDINATES

Joint	X-Coord.	Y-coord.	u-fix	v-fix	r-fix
1	.000	.000	1	1	1
2	.000	240.000	0	0	0
3	90.000	240.000	0	0	0
4	270.000	240.000	0	0	0
5	360.000	240.000	0	0	0
6	360.000	.000	1	1	1

LOAD VECTOR

Joint	Fx	Fy	M
1	.00	.00	.00
2	15.00	.00	.00
3	.00	-30.00	.00
4	.00	-30.00	.00
5	.00	.00	.00
6	.00	.00	.00

MATERIAL GROUP PROPERTIES

Group	Area (A)	Inertia (I)	Plastic Moment (Mp)
1	.1330E+02	.5860E+03	.2963E+04

Maximum Displacement occurs at Joint No.:   2
In the Horizontal Direction

```
==============================================================================
                              OUTPUT DATA
==============================================================================
                          INCREMENT NO.    1
                          ==================
Incr. Load Factor =      1.326
Total Load Factor =      1.326
Condition Number  = 1/  .647E-06
```

NODAL DISPLACEMENTS

```
--------------------
                    INCREMENTAL                          TOTAL
jt       hrz.         vrt.         rot.        hrz.        vrt.         rot.

  1    .000E+00     .000E+00     .000E+00    .000E+00    .000E+00     .000E+00
  2    .109E+01    -.215E-01    -.105E-01    .109E+01   -.215E-01    -.105E-01
  3    .109E+01    -.976E+00    -.792E-02    .109E+01   -.976E+00    -.792E-02
  4    .108E+01    -.752E+00     .872E-02    .108E+01   -.752E+00     .872E-02
  5    .107E+01    -.281E-01     .377E-02    .107E+01   -.281E-01     .377E-02
  6    .000E+00     .000E+00     .000E+00    .000E+00    .000E+00     .000E+00
```

END FORCES

```
                      INCREMENTAL END FORCES
Mem j1  j2      N1          V1          M1          N2          V2          M2
  1  1   2   .345E+02    -.257E+01    .438E+03    -.345E+02    .257E+01    -.106E+04
  2  2   3   .225E+02     .345E+02    .106E+04    -.225E+02   -.345E+02     .205E+04
  3  3   4   .225E+02    -.530E+01   -.205E+04    -.225E+02    .530E+01     .109E+04
  4  4   5   .225E+02    -.451E+02   -.109E+04    -.225E+02    .451E+02    -.296E+04
  5  5   6   .451E+02     .225E+02    .296E+04    -.451E+02   -.225E+02     .243E+04
                        TOTAL END FORCES
  1  1   2   .345E+02    -.257E+01    .438E+03    -.345E+02    .257E+01    -.106E+04
  2  2   3   .225E+02     .345E+02    .106E+04    -.225E+02   -.345E+02     .205E+04
  3  3   4   .225E+02    -.530E+01   -.205E+04    -.225E+02    .530E+01     .109E+04
  4  4   5   .225E+02    -.451E+02   -.109E+04    -.225E+02    .451E+02    -.296E+04
  5  5   6   .451E+02     .225E+02    .296E+04    -.451E+02   -.225E+02     .243E+04
```

Plastic Hinge Locations for Load Level = 1.326
```
At Joint    In Member
     5          4
     5          5
```

```
================================================================================
                          INCREMENT NO.    2
                          ==================
Incr. Load Factor =          .242
Total Load Factor =         1.568
Condition Number  = 1/  .668E-06
```

NODAL DISPLACEMENTS

```
                    INCREMENTAL                          TOTAL
jt       hrz.         vrt.         rot.        hrz.        vrt.         rot.

  1    .000E+00     .000E+00     .000E+00    .000E+00    .000E+00     .000E+00
  2    .606E+00    -.475E-02    -.426E-02    .170E+01   -.262E-01    -.148E-01
  3    .605E+00    -.365E+00    -.314E-02    .169E+01   -.134E+01    -.111E-01
  4    .604E+00    -.380E+00     .308E-02    .168E+01   -.113E+01     .118E-01
  5    .604E+00    -.429E-02     .000E+00    .167E+01   -.323E-01     .377E-02
  6    .000E+00     .000E+00     .000E+00    .000E+00    .000E+00     .000E+00
```

END FORCES

```
                      INCREMENTAL END FORCES
Mem j1  j2      N1          V1          M1          N2          V2          M2
  1  1   2   .763E+01     .141E+01    .470E+03    -.763E+01   -.141E+01    -.133E+03
  2  2   3   .223E+01     .763E+01    .133E+03    -.223E+01   -.763E+01     .554E+03
  3  3   4   .223E+01     .369E+00   -.554E+03    -.223E+01   -.369E+00     .621E+03
  4  4   5   .223E+01    -.690E+01   -.621E+03    -.223E+01    .690E+01     .000E+00
  5  5   6   .690E+01     .223E+01    .000E+00    -.690E+01   -.223E+01     .534E+03
                        TOTAL END FORCES
  1  1   2   .421E+02    -.117E+01    .908E+03    -.421E+02    .117E+01    -.119E+04
  2  2   3   .247E+02     .421E+02    .119E+04    -.247E+02   -.421E+02     .260E+04
  3  3   4   .247E+02    -.493E+01   -.260E+04    -.247E+02    .493E+01     .171E+04
```

```
4    4    5    .247E+02   -.520E+02   -.171E+04   -.247E+02    .520E+02   -.296E+04
5    5    6    .520E+02    .247E+02    .296E+04   -.520E+02   -.247E+02    .296E+04
```

```
Plastic Hinge Locations for Load Level =  1.568
At Joint    In Member
      6           5
```

```
=====================================================================================
                              INCREMENT NO.    3
                              ==================
Incr. Load Factor  =          .127
Total Load Factor  =         1.695
Condition Number   = 1/  .652E-06
```

NODAL DISPLACEMENTS

```
                   INCREMENTAL                              TOTAL
jt       hrz.         vrt.         rot.         hrz.         vrt.         rot.

 1    .000E+00     .000E+00     .000E+00     .000E+00     .000E+00     .000E+00
 2    .476E+00    -.233E-02    -.289E-02     .217E+01    -.285E-01    -.177E-01
 3    .476E+00    -.230E+00    -.187E-02     .217E+01    -.157E+01    -.129E-01
 4    .476E+00    -.227E+00     .188E-02     .216E+01    -.136E+01     .137E-01
 5    .476E+00    -.241E-02     .000E+00     .215E+01    -.348E-01     .377E-02
 6    .000E+00     .000E+00    -.198E-02     .000E+00     .000E+00    -.198E-02
```

END FORCES

```
                        INCREMENTAL END FORCES
Mem  j1   j2      N1           V1           M1           N2           V2           M2
 1    1    2    .374E+01     .191E+01     .434E+03    -.374E+01    -.191E+01     .238E+02
 2    2    3   -.124E-04     .374E+01    -.238E+02     .124E-04    -.374E+01     .361E+03
 3    3    4   -.618E-05    -.661E-01    -.361E+03     .618E-05     .661E-01     .349E+03
 4    4    5   -.124E-04    -.388E+01    -.349E+03     .124E-04     .388E+01     .000E+00
 5    5    6    .388E+01    -.252E-06     .000E+00    -.388E+01     .252E-06     .000E+00
                          TOTAL END FORCES
 1    1    2    .459E+02     .737E+00     .134E+04    -.459E+02    -.737E+00    -.116E+04
 2    2    3    .247E+02     .459E+02     .116E+04    -.247E+02    -.459E+02     .296E+04
 3    3    4    .247E+02    -.500E+01    -.296E+04    -.247E+02     .500E+01     .206E+04
 4    4    5    .247E+02    -.559E+02    -.206E+04    -.247E+02     .559E+02    -.296E+04
 5    5    6    .559E+02     .247E+02     .296E+04    -.559E+02    -.247E+02     .296E+04
```

```
Plastic Hinge Locations for Load Level =  1.695
At Joint    In Member
      3           2
      3           3
```

```
=====================================================================================
                              INCREMENT NO.    4
                              ==================
Incr. Load Factor  =          .225
Total Load Factor  =         1.920
Condition Number   = 1/  .107E-05
```

NODAL DISPLACEMENTS

```
                   INCREMENTAL                              TOTAL
jt       hrz.         vrt.         rot.         hrz.         vrt.         rot.

 1    .000E+00     .000E+00     .000E+00     .000E+00     .000E+00     .000E+00
 2    .229E+01    -.561E-02    -.172E-01     .446E+01    -.341E-01    -.349E-01
 3    .229E+01    -.168E+01     .000E+00     .446E+01    -.325E+01    -.129E-01
 4    .229E+01    -.691E+00     .693E-02     .445E+01    -.205E+01     .206E-01
 5    .229E+01    -.280E-02     .000E+00     .444E+01    -.376E-01     .377E-02
 6    .000E+00     .000E+00    -.954E-02     .000E+00     .000E+00    -.115E-01
```

```
END FORCES
-----------
                              INCREMENTAL END FORCES
 Mem j1  j2      N1           V1          M1            N2          V2          M2
  1   1   2    .901E+01    .338E+01    .162E+04    -.901E+01   -.338E+01   -.811E+03
  2   2   3   -.185E-03    .901E+01    .811E+03     .185E-03   -.901E+01    .000E+00
  3   3   4   -.926E-04    .225E+01    .000E+00     .926E-04   -.225E+01    .405E+03
  4   4   5   -.185E-03   -.451E+01   -.405E+03     .185E-03    .451E+01    .000E+00
  5   5   6    .451E+01   -.143E-05    .000E+00    -.451E+01    .143E-05    .000E+00
                                TOTAL END FORCES
  1   1   2    .549E+02    .412E+01    .296E+04    -.549E+02   -.412E+01   -.198E+04
  2   2   3    .247E+02    .549E+02    .198E+04    -.247E+02   -.549E+02    .296E+04
  3   3   4    .247E+02   -.274E+01   -.296E+04    -.247E+02    .274E+01    .247E+04
  4   4   5    .247E+02   -.604E+02   -.247E+04    -.247E+02    .604E+02   -.296E+04
  5   5   6    .604E+02    .247E+02    .296E+04    -.604E+02   -.247E+02    .296E+04


Plastic Hinge Locations for Load Level =  1.920
At Joint    In Member
     1           1
===============================================================================
Incremental Displacements up to       .1860E+06
Excessive Deformations.... Must be a mechanism
===============================================================================
```

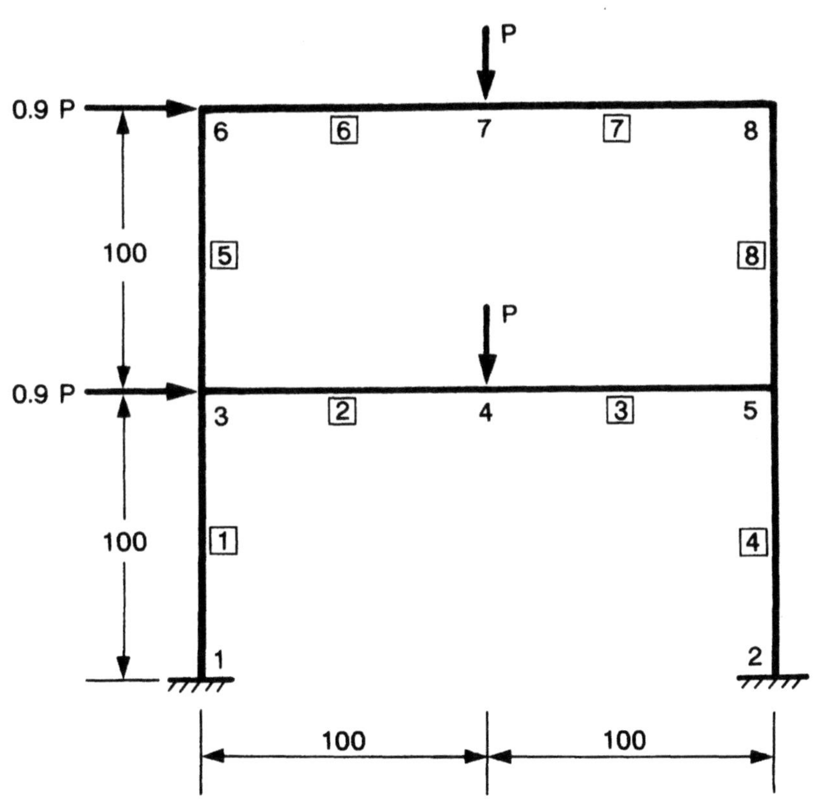

FIGURE 7.7. A two-story frame for deflection computation.

The resulting load factor $\lambda = 1.92$ is the same as that calculated in Example 4.8.1 of Chapter 4.

Example 7.8.3. A two-story rectangular frame is subjected to the loads shown in Fig. 7.7. Analyze the frame by the First-Order Plastic Hinge Analysis computer program. Assume all members are made of W16 × 45.

(a) Determine the load factor and horizontal deflection of point 8, and compare with those determined in Example 6.8.1.
(b) Calculate the required rotation capacity at points 1 and 8.

Solution: The input data for the analysis of the structure shown in Fig. 7.7 and the resulting output data are listed in following.

```
Prob.7.8.3   Deflection of Two Story Frame
8 8 4 2 1
29000 8 2
0 0
200 0
0 100
100 100
200 100
0 200
100 200
200 200
1 3 1 1 1
3 4 1 1 1
4 5 1 1 1
2 5 1 1 1
3 6 1 1 1
6 7 1 1 1
7 8 1 1 1
5 8 1 1 1
13.3 586 2963
3 0.9 0 0
4 0 -1.0 0
6 0.9 0 0
7 0 -1.0 0
1 1 1 1
2 1 1 1
```

```
=====================================================================
                First Order Plastic Analysis "FOPA"
                               by
                     Dr. Maheeb Abdel-Ghaffar
                        Cairo University
                             1992
=====================================================================
                          INPUT DATA
                          ==========
```

Prob.7.8.3 Deflection of Two Story Frame

```
 # Nds.    # Elems.   # Ldd.Nds   # B.C.Nds       E          # Prp.Grps
   8          8           4           2         .29E+05           1
```

JOINT COORDINATES

Joint	X-Coord.	Y-coord.	u-fix	v-fix	r-fix
1	.000	.000	1	1	1
2	200.000	.000	1	1	1
3	.000	100.000	0	0	0
4	100.000	100.000	0	0	0
5	200.000	100.000	0	0	0
6	.000	200.000	0	0	0
7	100.000	200.000	0	0	0
8	200.000	200.000	0	0	0

LOAD VECTOR

Joint	Fx	Fy	M
1	.00	.00	.00
2	.00	.00	.00
3	.90	.00	.00
4	.00	-1.00	.00
5	.00	.00	.00
6	.90	.00	.00
7	.00	-1.00	.00
8	.00	.00	.00

MATERIAL GROUP PROPERTIES

Group	Area (A)	Inertia (I)	Plastic Moment (Mp)
1	.1330E+02	.5860E+03	.2963E+04

Maximum Displacement occurs at Joint No.:   6
In the Horizontal Direction

```
=====================================================================
                          OUTPUT DATA
=====================================================================
                       INCREMENT NO.    1
                       ===================
```

```
Incr. Load Factor  =        42.927
Total Load Factor  =        42.927
Condition Number   = 1/  .377E-06
```

NODAL DISPLACEMENTS
----

jt		INCREMENTAL			TOTAL	
	hrz.	vrt.	rot.	hrz.	vrt.	rot.
1	.000E+00	.000E+00	.000E+00	.000E+00	.000E+00	.000E+00
2	.000E+00	.000E+00	.000E+00	.000E+00	.000E+00	.000E+00
3	.390E+00	-.286E-02	-.447E-02	.390E+00	-.286E-02	-.447E-02
4	.387E+00	-.142E+00	.186E-02	.387E+00	-.142E+00	.186E-02
5	.385E+00	-.194E-01	-.345E-02	.385E+00	-.194E-01	-.345E-02
6	.814E+00	-.527E-02	-.364E-02	.814E+00	-.527E-02	-.364E-02
7	.805E+00	-.179E+00	.108E-02	.805E+00	-.179E+00	.108E-02
8	.796E+00	-.281E-01	-.136E-02	.796E+00	-.281E-01	-.136E-02

END FORCES

INCREMENTAL END FORCES

Mem	j1	j2	N1	V1	M1	N2	V2	M2
1	1	3	.110E+02	.339E+02	.245E+04	-.110E+02	-.339E+02	.936E+03
2	3	4	.861E+01	.170E+01	-.989E+03	-.861E+01	-.170E+01	.116E+04
3	4	5	.861E+01	-.412E+02	-.116E+04	-.861E+01	.412E+02	-.296E+04
4	2	5	.748E+02	.434E+02	.275E+04	-.748E+02	-.434E+02	.158E+04
5	3	6	.932E+01	.388E+01	.531E+02	-.932E+01	-.388E+01	.335E+03
6	6	7	.348E+02	.932E+01	-.335E+03	-.348E+02	-.932E+01	.127E+04
7	7	8	.348E+02	-.336E+02	-.127E+04	-.348E+02	.336E+02	-.209E+04
8	5	8	.336E+02	.348E+02	.138E+04	-.336E+02	-.348E+02	.209E+04

TOTAL END FORCES

Mem	j1	j2	N1	V1	M1	N2	V2	M2
1	1	3	.110E+02	.339E+02	.245E+04	-.110E+02	-.339E+02	.936E+03
2	3	4	.861E+01	.170E+01	-.989E+03	-.861E+01	-.170E+01	.116E+04
3	4	5	.861E+01	-.412E+02	-.116E+04	-.861E+01	.412E+02	-.296E+04
4	2	5	.748E+02	.434E+02	.275E+04	-.748E+02	-.434E+02	.158E+04
5	3	6	.932E+01	.388E+01	.531E+02	-.932E+01	-.388E+01	.335E+03
6	6	7	.348E+02	.932E+01	-.335E+03	-.348E+02	-.932E+01	.127E+04
7	7	8	.348E+02	-.336E+02	-.127E+04	-.348E+02	.336E+02	-.209E+04
8	5	8	.336E+02	.348E+02	.138E+04	-.336E+02	-.348E+02	.209E+04

```
Plastic Hinge Locations for Load Level = 42.927
At Joint    In Member
     5           3
```

===
INCREMENT NO.   2
====================

```
Incr. Load Factor  =         2.681
Total Load Factor  =        45.608
Condition Number   = 1/  .385E-06
```

NODAL DISPLACEMENTS
----

jt		INCREMENTAL			TOTAL	
	hrz.	vrt.	rot.	hrz.	vrt.	rot.
1	.000E+00	.000E+00	.000E+00	.000E+00	.000E+00	.000E+00
2	.000E+00	.000E+00	.000E+00	.000E+00	.000E+00	.000E+00
3	.416E-01	-.331E-03	-.558E-03	.431E+00	-.319E-02	-.502E-02
4	.415E-01	-.330E-01	.163E-04	.429E+00	-.175E+00	.187E-02
5	.414E-01	-.106E-02	-.630E-03	.427E+00	-.205E-01	-.408E-02
6	.897E-01	-.367E-03	-.340E-03	.904E+00	-.564E-02	-.398E-02
7	.892E-01	-.125E-01	.111E-03	.894E+00	-.192E+00	.119E-02
8	.888E-01	-.172E-02	-.144E-03	.885E+00	-.298E-01	-.150E-02

END FORCES

INCREMENTAL END FORCES

Mem	j1	j2	N1	V1	M1	N2	V2	M2
1	1	3	.128E+01	.279E+01	.234E+03	-.128E+01	-.279E+01	.450E+02
2	3	4	.272E+00	.114E+01	-.407E+02	-.272E+00	-.114E+01	.154E+03
3	4	5	.272E+00	-.154E+01	-.154E+03	-.272E+00	.154E+01	.000E+00
4	2	5	.408E+01	.203E+01	.209E+03	-.408E+01	-.203E+01	-.546E+01
5	3	6	.140E+00	.654E+00	-.427E+01	-.140E+00	-.654E+00	.697E+02
6	6	7	.176E+01	.140E+00	-.697E+02	-.176E+01	-.140E+00	.837E+02
7	7	8	.176E+01	-.254E+01	-.837E+02	-.176E+01	.254E+01	-.170E+03
8	5	8	.254E+01	.176E+01	.546E+01	-.254E+01	-.176E+01	.170E+03

TOTAL END FORCES

Mem	j1	j2	N1	V1	M1	N2	V2	M2
1	1	3	.123E+02	.367E+02	.269E+04	-.123E+02	-.367E+02	.981E+03
2	3	4	.888E+01	.284E+01	-.103E+04	-.888E+01	-.284E+01	.131E+04
3	4	5	.888E+01	-.428E+02	-.131E+04	-.888E+01	.428E+02	-.296E+04
4	2	5	.789E+02	.454E+02	.296E+04	-.789E+02	-.454E+02	.158E+04
5	3	6	.946E+01	.454E+01	.489E+02	-.946E+01	-.454E+01	.405E+03
6	6	7	.365E+02	.946E+01	-.405E+03	-.365E+02	-.946E+01	.135E+04
7	7	8	.365E+02	-.361E+02	-.135E+04	-.365E+02	.361E+02	-.226E+04
8	5	8	.361E+02	.365E+02	.139E+04	-.361E+02	-.365E+02	.226E+04

Plastic Hinge Locations for Load Level = 45.608
At Joint   In Member
     2   4

===

INCREMENT NO.   3

    Incr. Load Factor =   1.955
    Total Load Factor =   47.563
    Condition Number   = 1/   .378E-06

NODAL DISPLACEMENTS
----

jt	INCREMENTAL hrz.	vrt.	rot.	TOTAL hrz.	vrt.	rot.
1	.000E+00	.000E+00	.000E+00	.000E+00	.000E+00	.000E+00
2	.000E+00	.000E+00	-.437E-03	.000E+00	.000E+00	-.437E-03
3	.444E-01	-.178E-03	-.527E-03	.476E+00	-.337E-02	-.555E-02
4	.449E-01	-.285E-01	.263E-04	.474E+00	-.203E+00	.190E-02
5	.453E-01	-.836E-03	-.486E-03	.472E+00	-.213E-01	-.457E-02
6	.847E-01	-.180E-03	-.255E-03	.988E+00	-.582E-02	-.423E-02
7	.843E-01	-.852E-02	.892E-04	.978E+00	-.200E+00	.128E-02
8	.839E-01	-.134E-02	-.136E-03	.969E+00	-.312E-01	-.164E-02

END FORCES

INCREMENTAL END FORCES

Mem	j1	j2	N1	V1	M1	N2	V2	M2
1	1	3	.686E+00	.369E+01	.274E+03	-.686E+00	-.369E+01	.948E+02
2	3	4	-.169E+01	.676E+00	-.603E+02	.169E+01	-.676E+00	.128E+03
3	4	5	-.169E+01	-.128E+01	-.128E+03	.169E+01	.128E+01	.000E+00
4	2	5	.323E+01	-.168E+00	.000E+00	-.323E+01	.168E+00	-.168E+02
5	3	6	.922E-02	.235E+00	-.345E+02	-.922E-02	-.235E+00	.580E+02
6	6	7	.152E+01	.922E-02	-.580E+02	-.152E+01	-.922E-02	.590E+02
7	7	8	.152E+01	-.195E+01	-.590E+02	-.152E+01	.195E+01	-.136E+03
8	5	8	.195E+01	.152E+01	.168E+02	-.195E+01	-.152E+01	.136E+03

TOTAL END FORCES

Mem	j1	j2	N1	V1	M1	N2	V2	M2
1	1	3	.130E+02	.404E+02	.296E+04	-.130E+02	-.404E+02	.108E+04
2	3	4	.719E+01	.351E+01	-.109E+04	-.719E+01	-.351E+01	.144E+04
3	4	5	.719E+01	-.440E+02	-.144E+04	-.719E+01	.440E+02	-.296E+04

```
4    2    5    .821E+02    .452E+02    .296E+04   -.821E+02   -.452E+02    .156E+04
5    3    6    .947E+01    .477E+01    .144E+02   -.947E+01   -.477E+01    .463E+03
6    6    7    .380E+02    .947E+01   -.463E+03   -.380E+02   -.947E+01    .141E+04
7    7    8    .380E+02   -.381E+02   -.141E+04   -.380E+02    .381E+02   -.240E+04
8    5    8    .381E+02    .380E+02    .140E+04   -.381E+02   -.380E+02    .240E+04
```

```
Plastic Hinge Locations for Load Level = 47.563
At Joint    In Member
     1            1
```

===

```
                          INCREMENT NO.     4
                          ==================
Incr. Load Factor  =          5.373
Total Load Factor  =         52.936
Condition Number   = 1/   .348E-06
```

NODAL DISPLACEMENTS

```
                 INCREMENTAL                            TOTAL
jt       hrz.          vrt.         rot.        hrz.          vrt.         rot.

 1    .000E+00    .000E+00   -.462E-02    .000E+00    .000E+00   -.462E-02
 2    .000E+00    .000E+00   -.433E-02    .000E+00    .000E+00   -.477E-02
 3    .400E+00    .488E-03   -.277E-02    .876E+00   -.288E-02   -.832E-02
 4    .400E+00   -.128E+00    .226E-03    .874E+00   -.331E+00    .212E-02
 5    .400E+00   -.327E-02   -.334E-02    .872E+00   -.246E-01   -.790E-02
 6    .613E+00    .102E-02   -.124E-02    .160E+01   -.480E-02   -.547E-02
 7    .613E+00   -.290E-01    .435E-03    .159E+01   -.229E+00    .171E-02
 8    .612E+00   -.520E-02   -.689E-03    .158E+01   -.364E-01   -.232E-02
```

END FORCES

```
                          INCREMENTAL END FORCES
Mem j1   j2      N1           V1           M1           N2           V2           M2
 1    1    3   -.188E+01    .630E+01    .000E+00    .188E+01   -.630E+01    .630E+03
 2    3    4    .111E+01    .191E+00   -.499E+03   -.111E+01   -.191E+00    .518E+03
 3    4    5    .111E+01   -.518E+01   -.518E+03   -.111E+01    .518E+01    .000E+00
 4    2    5    .126E+02    .337E+01    .000E+00   -.126E+02   -.337E+01    .337E+03
 5    3    6   -.207E+01    .257E+01   -.131E+03    .207E+01   -.257E+01    .388E+03
 6    6    7    .226E+01   -.207E+01   -.388E+03   -.226E+01    .207E+01    .181E+03
 7    7    8    .226E+01   -.744E+01   -.181E+03   -.226E+01    .744E+01   -.563E+03
 8    5    8    .744E+01    .226E+01   -.337E+03   -.744E+01   -.226E+01    .563E+03
                            TOTAL END FORCES
 1    1    3    .111E+02    .467E+02    .296E+04   -.111E+02   -.467E+02    .171E+04
 2    3    4    .830E+01    .371E+01   -.159E+04   -.830E+01   -.371E+01    .196E+04
 3    4    5    .830E+01   -.492E+02   -.196E+04   -.830E+01    .492E+02   -.296E+04
 4    2    5    .948E+02    .486E+02    .296E+04   -.948E+02   -.486E+02    .190E+04
 5    3    6    .740E+01    .734E+01   -.117E+03   -.740E+01   -.734E+01    .851E+03
 6    6    7    .403E+02    .740E+01   -.851E+03   -.403E+02   -.740E+01    .159E+04
 7    7    8    .403E+02   -.455E+02   -.159E+04   -.403E+02    .455E+02   -.296E+04
 8    5    8    .455E+02    .403E+02    .107E+04   -.455E+02   -.403E+02    .296E+04
```

```
Plastic Hinge Locations for Load Level = 52.936
At Joint    In Member
     8            7
     8            8
```

===

```
                          INCREMENT NO.     5
                          ==================
Incr. Load Factor  =          7.700
Total Load Factor  =         60.636
Condition Number   = 1/   .396E-06
```

```
NODAL DISPLACEMENTS
-------------------
                    INCREMENTAL                           TOTAL
jt       hrz.          vrt.         rot.         hrz.         vrt.          rot.

 1     .000E+00      .000E+00     -.869E-02    .000E+00     .000E+00     -.133E-01
 2     .000E+00      .000E+00     -.821E-02    .000E+00     .000E+00     -.130E-01
 3     .780E+00      .699E-03     -.601E-02    .166E+01    -.218E-02     -.143E-01
 4     .777E+00     -.259E+00      .579E-03    .165E+01    -.590E+00      .270E-02
 5     .775E+00     -.469E-02     -.682E-02    .165E+01    -.293E-01     -.147E-01
 6     .136E+01      .794E-03     -.448E-02    .296E+01    -.400E-02     -.995E-02
 7     .136E+01     -.203E+00      .376E-03    .295E+01    -.432E+00      .209E-02
 8     .136E+01     -.678E-02      .000E+00    .294E+01    -.432E-01     -.232E-02


END FORCES
----------
                          INCREMENTAL END FORCES
Mem j1  j2      N1            V1            M1            N2           V2           M2
 1   1   3   -.270E+01     .912E+01      .000E+00      .270E+01    -.912E+01     .912E+03
 2   3   4    .948E+01    -.233E+01     -.124E+04     -.948E+01     .233E+01     .100E+04
 3   4   5    .948E+01    -.100E+02     -.100E+04     -.948E+01     .100E+02     .000E+00
 4   2   5    .181E+02     .474E+01      .000E+00     -.181E+02    -.474E+01     .474E+03
 5   3   6   -.366E+00     .117E+02      .324E+03      .366E+00    -.117E+02     .843E+03
 6   6   7   -.474E+01    -.366E+00     -.843E+03      .474E+01     .366E+00     .807E+03
 7   7   8   -.474E+01    -.807E+01     -.807E+03      .474E+01     .807E+01     .000E+00
 8   5   8    .807E+01    -.474E+01     -.474E+03     -.807E+01     .474E+01     .000E+00
                            TOTAL END FORCES
 1   1   3    .841E+01     .558E+02      .296E+04     -.841E+01    -.558E+02     .262E+04
 2   3   4    .178E+02     .138E+01     -.283E+04     -.178E+02    -.138E+01     .296E+04
 3   4   5    .178E+02    -.593E+02     -.296E+04     -.178E+02     .593E+02    -.296E+04
 4   2   5    .113E+03     .533E+02      .296E+04     -.113E+03    -.533E+02     .237E+04
 5   3   6    .703E+01     .190E+02      .207E+03     -.703E+01    -.190E+02     .169E+04
 6   6   7    .356E+02     .703E+01     -.169E+04     -.356E+02    -.703E+01     .240E+04
 7   7   8    .356E+02    -.536E+02     -.240E+04     -.356E+02     .536E+02    -.296E+04
 8   5   8    .536E+02     .356E+02      .593E+03     -.536E+02    -.356E+02     .296E+04

Plastic Hinge Locations for Load Level = 60.636
At Joint    In Member
    4           2
    4           3
================================================================================
                          INCREMENT NO.     6
                          ==================
Incr. Load Factor  =        2.406
Total Load Factor  =       63.043
Condition Number   = 1/  .358E-06


NODAL DISPLACEMENTS
-------------------
                    INCREMENTAL                           TOTAL
jt       hrz.          vrt.         rot.         hrz.         vrt.          rot.

 1     .000E+00      .000E+00     -.753E-02    .000E+00     .000E+00     -.208E-01
 2     .000E+00      .000E+00     -.781E-02    .000E+00     .000E+00     -.208E-01
 3     .748E+00      .218E-03     -.736E-02    .240E+01    -.196E-02     -.217E-01
 4     .746E+00     -.783E+00      .000E+00    .240E+01    -.137E+01      .270E-02
 5     .744E+00     -.147E-02     -.670E-02    .239E+01    -.307E-01     -.214E-01
 6     .134E+01      .106E-02     -.387E-02    .430E+01    -.294E-02     -.138E-01
 7     .134E+01     -.156E+00      .417E-03    .429E+01    -.587E+00      .250E-02
 8     .134E+01     -.293E-02      .000E+00    .428E+01    -.461E-01     -.232E-02


END FORCES
----------
```

```
                         INCREMENTAL END FORCES
Mem  j1   j2     N1           V1           M1           N2           V2           M2
 1    1    3   -.842E+00    .575E+00     .000E+00     .842E+00    -.575E+00     .575E+02
 2    3    4    .751E+01    .241E+01     .241E+03    -.751E+01    -.241E+01     .000E+00
 3    4    5    .751E+01    .000E+00     .000E+00    -.751E+01     .000E+00     .000E+00
 4    2    5    .565E+01    .376E+01     .000E+00    -.565E+01    -.376E+01     .376E+03
 5    3    6   -.325E+01    .592E+01    -.298E+03     .325E+01    -.592E+01     .890E+03
 6    6    7   -.376E+01   -.325E+01    -.890E+03     .376E+01     .325E+01     .565E+03
 7    7    8   -.376E+01   -.565E+01    -.565E+03     .376E+01     .565E+01     .000E+00
 8    5    8    .565E+01   -.376E+01    -.376E+03    -.565E+01     .376E+01     .000E+00
                           TOTAL END FORCES
 1    1    3    .757E+01    .564E+02     .296E+04    -.757E+01    -.564E+02     .268E+04
 2    3    4    .253E+02    .378E+01    -.258E+04    -.253E+02    -.378E+01     .296E+04
 3    4    5    .253E+02   -.593E+02    -.296E+04    -.253E+02     .593E+02    -.296E+04
 4    2    5    .119E+03    .571E+02     .296E+04    -.119E+03    -.571E+02     .275E+04
 5    3    6    .378E+01    .249E+02    -.912E+02    -.378E+01    -.249E+02     .258E+04
 6    6    7    .318E+02    .378E+01    -.258E+04    -.318E+02    -.378E+01     .296E+04
 7    7    8    .318E+02   -.593E+02    -.296E+04    -.318E+02     .593E+02    -.296E+04
 8    5    8    .593E+02    .318E+02     .217E+03    -.593E+02    -.318E+02     .296E+04

Plastic Hinge Locations for Load Level = 63.043
At Joint     In Member
     7            6
     7            7
===================================================================================
Incremental Displacements up to        .1510E+05
Excessive Deformations.... Must be a mechanism
===================================================================================
```

(a) Deflections: From the output, we note that

$$\lambda_c = 63.043.$$

From Example 6.8.1, we have

$$\lambda_c = 2.128 \frac{M_p}{L} = 2.128 \frac{2963}{100} = 63.05,$$

which is practically the same. Also, from the output data, we note

$$\delta_{h8} = 4.28 \text{ in.} \tag{7.8.41}$$

From Example 6.8.1, δ_{h5} (δ_{h8} in this example) can be calculated from Eq. (6.8.6) as

$$\delta_{h8} = 14.206 \frac{M_p L^2}{6EI}$$

or

$$\delta_{h8} = \frac{14.206 \times 2963 \times (100)^2}{6 \times 29{,}000 \times 586} = 4.13 \text{ in.} \tag{7.8.42}$$

Equation (6.8.6) is based on a certain assumed value of indeterminate moments in the moment check. The error between the estimated δ_{h8} and that computed from FOPA is 3.5%. A modified estimation of deflection by Eq. (6.8.6) based on the exact elastic analysis of the indeterminate moments is

$$\delta_{h5} = 14.626 \frac{M_p L^2}{6EI}$$

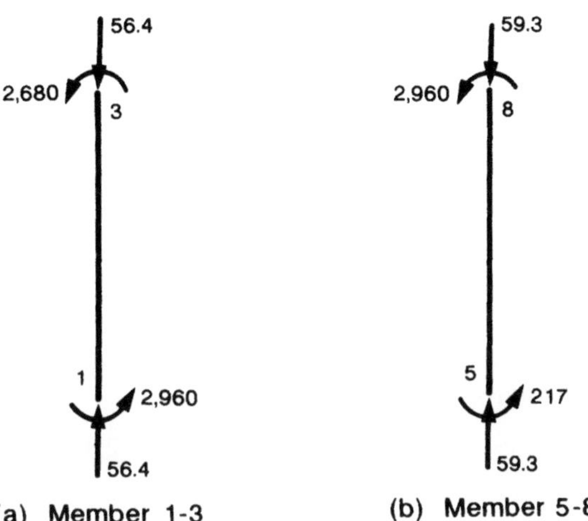

FIGURE 7.8. Member end moments of Fig. 7.7: (a) member 1-3, (b) member 5-8, and (c) member 7-8.

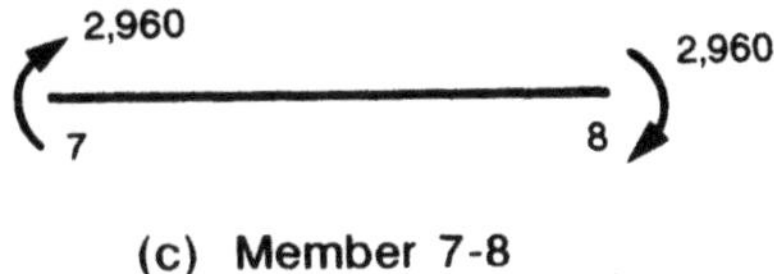

(c) Member 7-8

or

$$\delta_{h5} = \frac{14.626 \times 2963 \times (100)^2}{6 \times 29,000 \times 586} = 4.25 \text{ in.} \tag{7.8.43}$$

(b) Rotation Capacity: *(i) Point 1:* The rotation capacity required at point 1 for the formation of a mechanism can be calculated from Eq. (6.3.1) as

$$\theta_A = \frac{\Delta}{L} + \frac{L}{3EI}\left(M_{13} - \frac{1}{2}M_{31}\right)$$

in which Δ is the horizontal displacement of node 3 with respect to node 1 (displacement is zero) and M_{13} and M_{31} [Fig. 7.8(a)] are internal moments, all obtained from the output from FOPA. Note that in this equation, moments M_{13} and M_{31} are positive when clockwise and vice-versa.

$$\theta_A = \frac{2.4}{100} + \frac{100}{3 \times 29,000 \times 586}\left[-2960 + \frac{2680}{2}\right]$$

or

$$\theta_A = 0.0208.$$

(ii) Point 8: The required rotation capacity at point 8 consists of two parts: the rotation of segment 8-5 and the rotation of segment 8-7; it can be expressed as

$$\theta_8 = \theta_{85} - \theta_{87}.$$

The rotation of segment 8-5 is calculated from

$$\theta_{85} = \frac{\Delta}{L} + \frac{L}{3EI}\left(M_{85} - \frac{1}{2} M_{58} \right)$$

in which Δ is the relative horizontal displacement of nodes 8 and 5 and is calculated as

$$\Delta = 4.28 - 2.39 = 1.89 \text{ in.}$$

M_{85} and M_{58} [Fig. 7.8(b)] are internal moments in segment 8-5 and are obtained from the output of FOPA. Now, θ_{85} becomes

$$\theta_{85} = \frac{1.89}{100} + \frac{100}{3 \times 29{,}000 \times 586}\left[-2960 + \frac{217}{2} \right] = 0.01331.$$

Similarly, for segment 8-7, we have

$$\theta_{87} = \frac{\Delta_{v78}}{L} + \frac{L}{3EI}\left(M_{87} - \frac{1}{2} M_{78} \right)$$

where Δ_{v78} is the relative vertical displacement of nodes 7 and 8 and is obtained from the FOPA output as

$$\Delta_{v78} = -0.587 - (-0.0461) = -0.541.$$

The rotation of segment 8-7 has the value

$$\theta_{87} = -\frac{0.541}{100} + \frac{100}{3 \times 29{,}000 \times 586}\left[2960 - \frac{1}{2}(2960) \right] = -0.00251.$$

Now the required hinge rotation at node 8 can be expressed as

$$H_8 = 0.01331 + 0.00251 = 0.01582.$$

7.9 Summary and Conclusions

The first-order plastic hinge analysis consists of a series of elastic analyses. In each of these elastic analyses, a plastic hinge is introduced at the point of the maximum moment and a new, simpler structure is formed after replacing the plastic hinge with a real hinge. The elastic analyses are continued until a sufficient number of plastic hinges is formed to transform the structure into a failure mechanism.

In order to computerize this hinge-by-hinge procedure, we first derive the stiffness matrices for beam elements with several boundary conditions. Then we describe a numerical procedure for developing the computer program. In the latter part of the chapter, we presented three illustrative examples. In the first example, we showed the details of a step-by-step calculation procedure of the hinge-by-hinge matrix-analysis method for a fixed-ended beam. In the second example, the computer program was applied to a portal frame and the plastic limit load so obtained was compared with the plastic limit analysis solution obtained in Chapter 4. In the third example, the deflection estimated for a frame in Example 6.8.1 (Chapter 6) was checked by the computer program FOPA. The required rotation capacity of plastic hinges was also estimated. In the appendix, four examples demonstrating the capability of the computer program FOPA are presented.

References

7.1. Abdel-Ghaffar, M., White, D.W., and Chen, W.F., "Simplified Second-Order Inelastic Analysis for Steel Frame Design," *Structural Engineering Report No. CE-STR-90-34*, School of Civil Engineering, Purdue University, West Lafayette, IN, 16 pages, 1990.

Appendix

M. ABDEL-GHAFFAR

A7.1 FOPA—Program Description

The program FOPA is based on the procedure described in Section 7.7. It is useful to obtain the load-displacement history of plane rigid frames under static loads. The locations and sequence of the formation of plastic hinges are also given in the output file (OUT.DAT) and on the screen.

The input can be given to the program in two ways: either (1) by opening a file called IN.DAT and entering the necessary data to run FOPA; or (2) by using the small program DATA to interactively create the files IN.DAT and MESH.DAT. The MESH.DAT file contains the information needed for a small program called DRAW to sketch the structure on the screen so that the user can verify that the nodal coordinates and the member geometry are entered correctly. The program DRAW can also be used to sketch the deformed collapse mechanism of the structure, provided the user runs DRAW after running FOPA.

The output of FOPA comes on one major file called OUT.DAT and three other files called DISP.DAT, DISPMX.DAT, and DEF.DAT. The output file OUT.DAT has everything about the structure for each load increment as well as the input data. If the structure is relatively large with many plastic hinges to form, the file OUT.DAT will be too large to print all of it. The user has the option of printing only the conclusion of the results for each load step by entering a value of "0" (zero) for the parameter FLAG in the input data file.

The output file DISP.DAT contains four columns: the load level, the horizontal and vertical displacements, and the rotation of the selected node. This file is useful in plotting the load-displacement history of the selected node without extracting it from the file OUT.DAT. The file DISPMX.DAT is similar to DISP.DAT except that DISPMX.DAT concerns the point of maximum displacement in the elastic range, which may not be at the same point as at the collapse. The point of maximum displacement in the elastic range is chosen because displacement is of serviceability concern (at the service load level) rather than a limit state at collapse. The file DEF.DAT is used by DRAW to sketch the deformed shape of the collapse mechanism.

If the user prefers to input the structure's data without using the program DATA, i.e., by directly creating or editing a file called IN.DAT, running FOPA automatically prints the file MESH.DAT, which the DRAW program needs for sketching the structure's geometry.

A7.2 Input Data

The necessary information needed for creating the input file IN.DAT can be entered as eight read statements in a free format as follows:

1. Job name: with a maximum of 80 characters.
2. Five entries as follows:
   (1) Number of nodes: with a maximum of 99 nodes.
   (2) Number of elements: with a maximum of 99 elements.
   (3) Number of loaded nodes: with a maximum of 70 loaded nodes.
   (4) Number of supports: with a maximum of 20 supports.
   (5) Number of property groups: with a maximum of 20 groups.
3. Three entries as follows:
   (1) Modulus of elasticity.
   (2) Node of interest: where the load-displacement history is monitored and printed in the file DISP.DAT.
   (3) Flag: equals "0" (zero) if only a brief output file OUT.DAT is needed. Otherwise, enter any other number if a complete output file OUT.DAT is needed.
4. Two entries for each node:
   (1) X-coordinate
   (2) Y-coordinate
   Entered for all nodes in an ascending order (i.e., starting from node 1 until the last node).
5. Five entries for each element:
   (1) Node number for the element's first node.
   (2) Node number for the element's second node.
   (3) Connection type of first node (0 for a hinged connection; 1 for a rigid connection).
   (4) Connection type of second node.
   (5) Property group number this member belongs to.
6. Three entries for each property group:
   (1) Area.
   (2) Inertia.
   (3) Plastic moment capacity.
7. Four entries of each loaded node:
   (1) Loaded node number.
   (2) Applied horizontal force (positive $\rightarrow$).
   (3) Applied vertical force (positive upward).
   (4) Applied bending moment (positive counterclockwise).

For example, "15 -7 3 2" means that at node number 15 there is a vertical downward concentrated load of 7 units (e.g., kips or kN), a horizontal force of 3 units from left to right, and a concentrated moment of 2 units (e.g., kip-in. or kN-m.) counterclockwise.

8. Four entries for each support:
    (1) Supported node number.
    (2) Is motion restrained in the X-direction? "1" for yes and "0" for no.
    (3) Is motion restrained in the Y-direction? "1" or "0."
    (4) Is rotation restrained? "1" or "0."
    For example, "23 1 1 0" means that node number 23 is a hinged support.

Example A7.1. The same frame was solved in Example 6.8.1 with different applied loads, as shown in Fig. A7.1. The plastic moment capacity is assumed in the input file as 200 and the story height is taken as 10, as used in the previous example. Using FOPA, the load factor P is obtained as $P = 42.55$ and the horizontal displacement at node E is $\delta_E = 1.65$. These values are quite close to the exact values given in Example 6.8.1. The load-displacement history of node E is plotted in Fig. A7.2 and the last plastic hinge to form was at node D (or hinge 6).

Example A7.2. The frame shown in Fig. A7.3 is the same as that in Example 5.9.3. It is solved here using FOPA. The limit load factor is 2.22. The exact value given by mechanism (1) of Eq. (5.9.50) is 2.311, which is a little higher than the analysis result. The reason is the excessive deformations required to achieve the load factor of 2.311, which introduces ill-conditioning and singularity of the stiffness matrix. This can be noticed from Fig. A7.4, where the load-displacement curve becomes almost horizontal at load level 2.22. To improve the solution for ill-conditioned structures at or near the limit load, calculations should be done in a double-precision accuracy. The more accurate solution considering the effects of stability and plasticity on the final collapse load is based on the *second-order inelastic analysis,* which is the subject of Chapter 8.

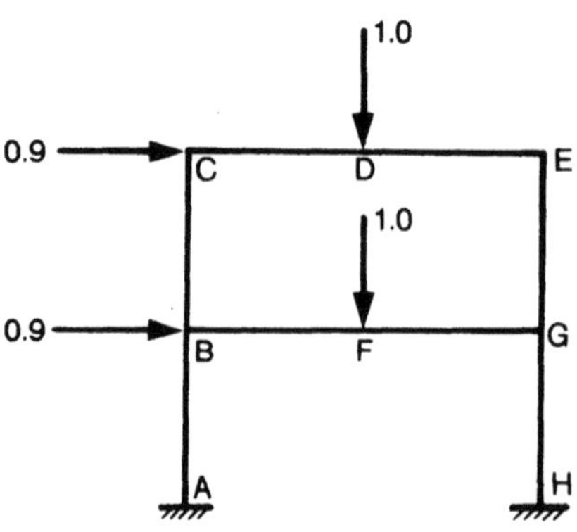

FIGURE A7.1. The finite element model for Example A7.1.

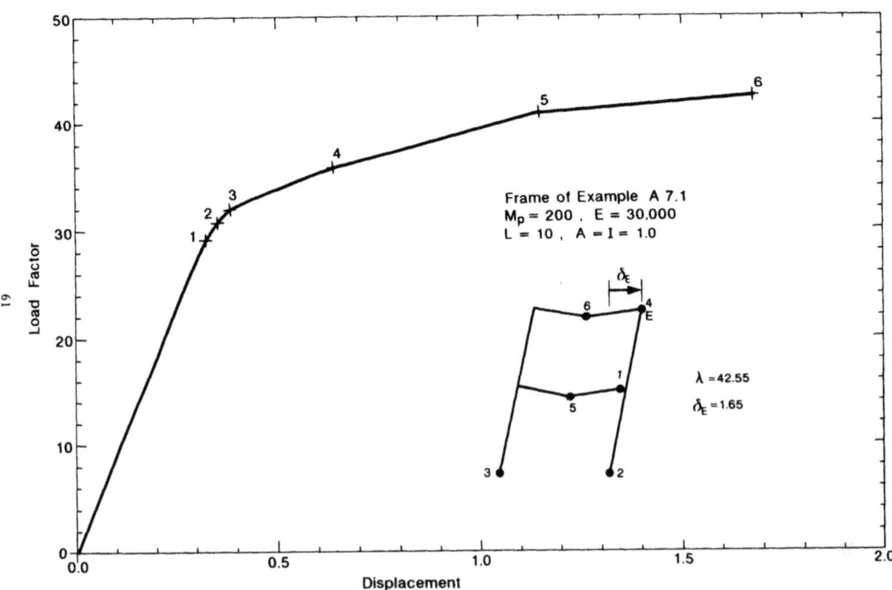

FIGURE A7.2. The load-deflection curve for the frame shown in Fig. A7.1.

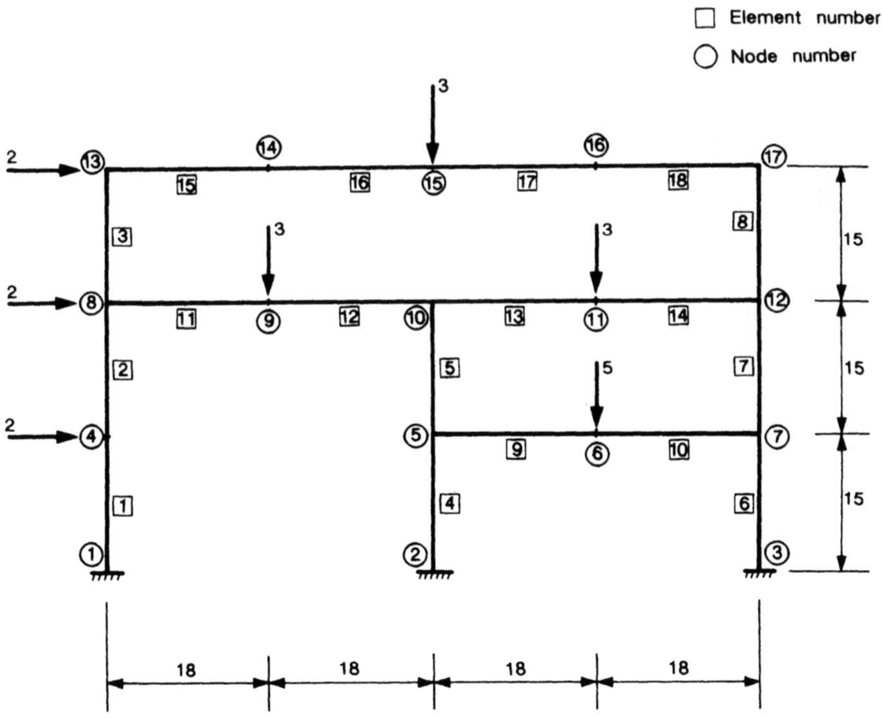

FIGURE A7.3. The finite element model for Example A7.2.

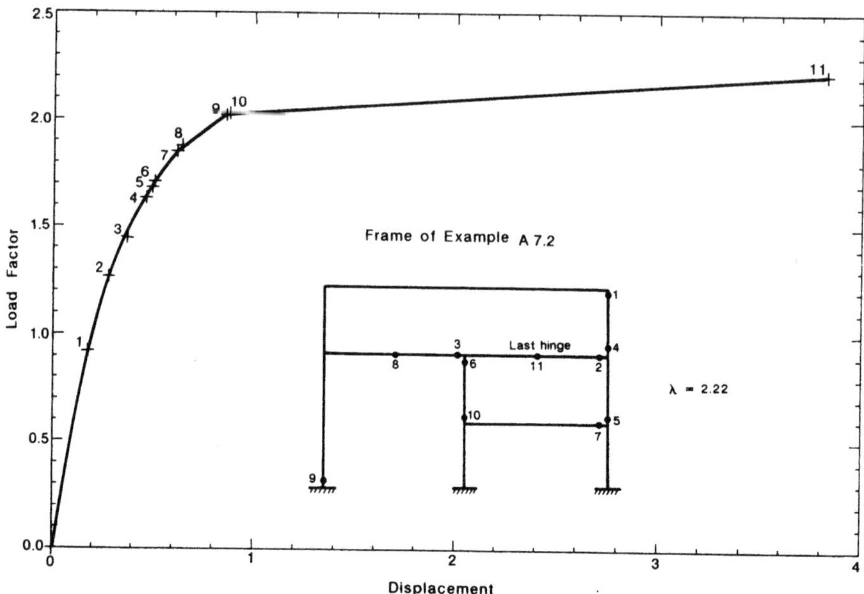

FIGURE A7.4. The load-deflection curve for the frame shown in Fig. A7.3.

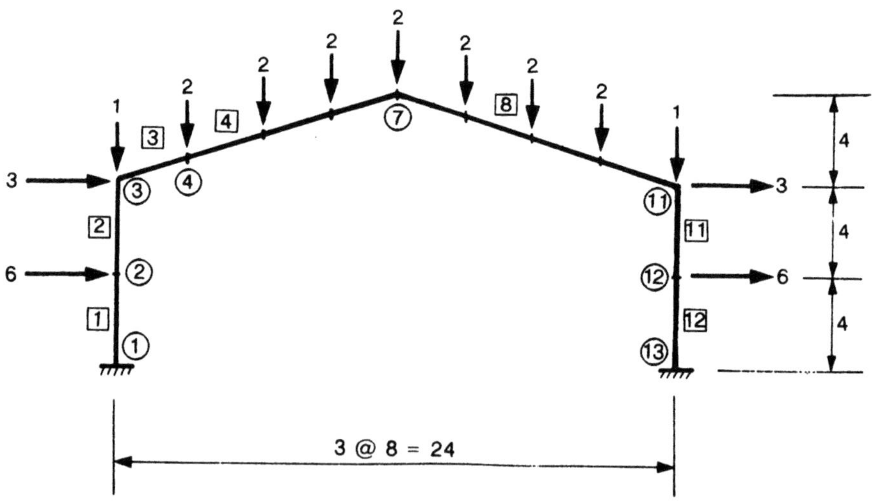

FIGURE A7.5. Finite element model for Example A7.3.

Example A7.3. In the preceding two examples, the plastic moment capacity was given and the unknown was the plastic limit load factor. In this frame, which is similar to that of Problem 5.11 in Chapter 5, we are looking for the required plastic moment for the frame to sustain the given factored loads of Fig. A7.5.

The procedure is quite simple; an assumed M_p is entered in the program (20, in this example) and the corresponding load factor is obtained, which is 0.65 for this case. The required M_p can then be calculated by scaling the assumed M_p with the inverse of the calculated load factor, i.e., the required

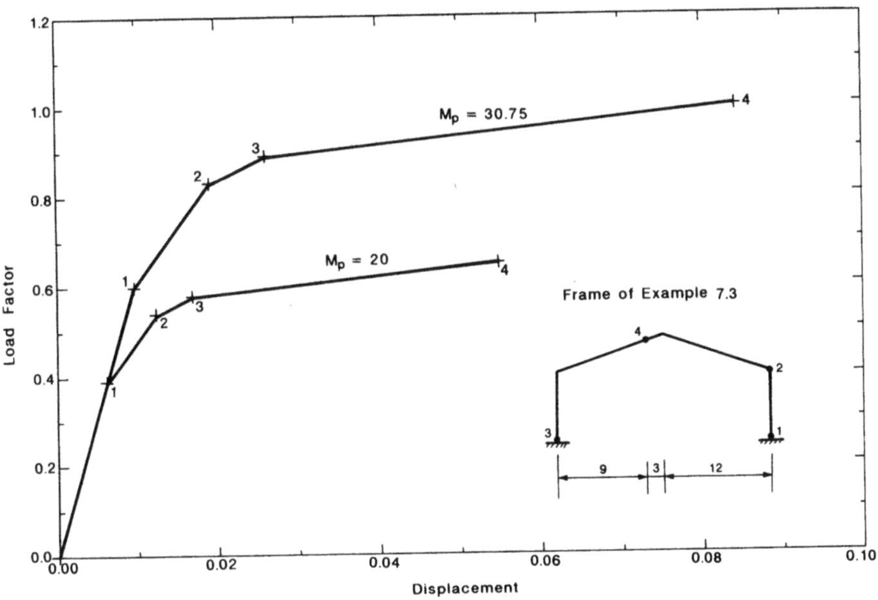

FIGURE A7.6. The load-deflection curves for the frame shown in Fig. A7.5.

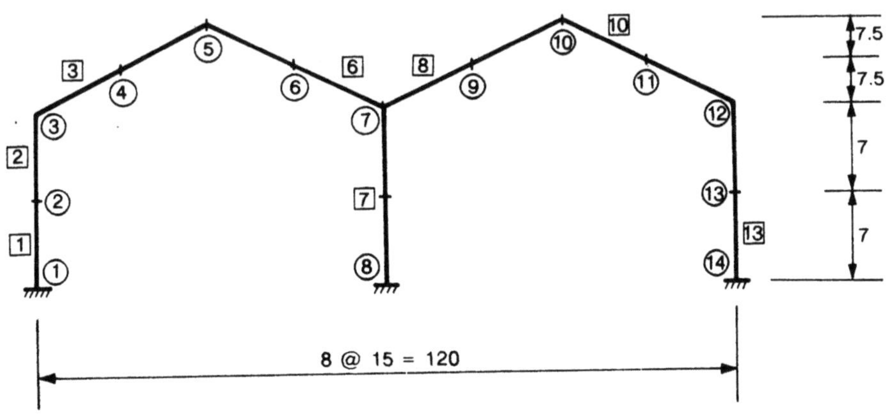

FIGURE A7.7. Finite element model for Example A7.4 with 13 elements.

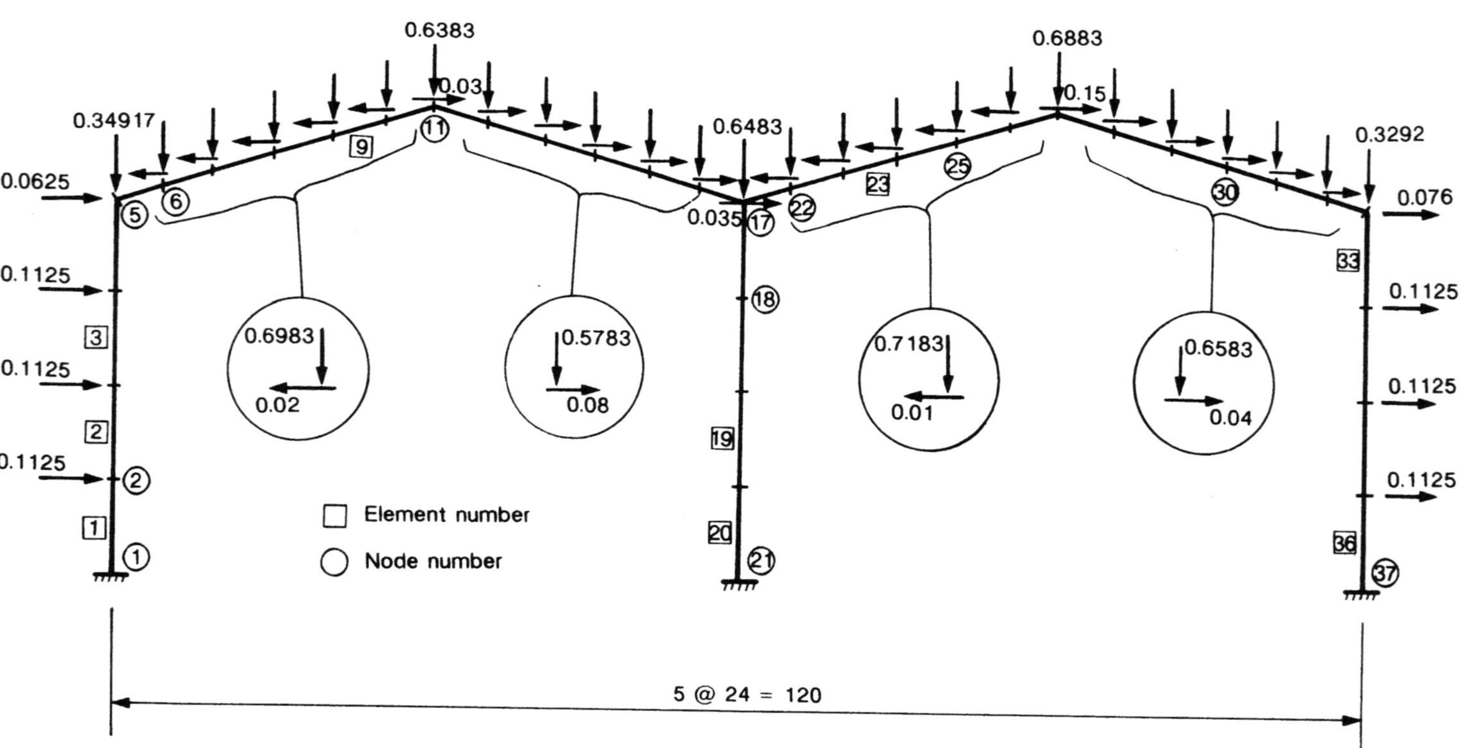

FIGURE A7.8. Finite element model for Example A7.4 with 36 elements.

M_P for this frame is

$$M_p = 20\left(\frac{1}{0.65}\right) = 30.75,$$

which is the correct answer for the problem. The response for both $M_p = 20$ and $M_p = 30.75$ is plotted in Fig. A7.6.

Several trials will be needed if the moment capacities were not the same for all members.

Example A7.4. As has been demonstrated in Example 5.11.1 in Chapter 5, the replacement of the distributed load with an equivalent concentrated load at the midspan of each member results in an overestimation of the required plastic moment capacity of the members, or alternatively the limit load factor can be underestimated. The same example is solved with the program FOPA twice, one with 13 elements (Fig. A7.7) and the other with 36 elements (Fig. A7.8). The larger number of elements is used to simulate the effect of the uniform load. The results are plotted in Fig. A7.9. The plastic moment capacity calculated from the first case is:

$$M_p = 20\left(\frac{1}{0.93}\right) = 21.50.$$

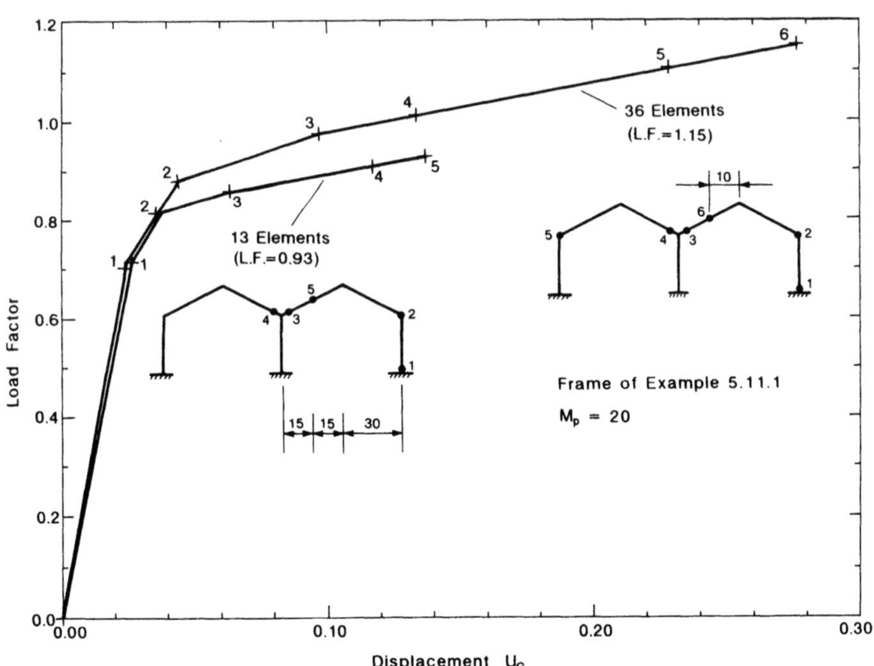

FIGURE A7.9. The load-deflection curves for the frame shown in Fig. A7.7.

Using 36 elements, a more accurate answer is obtained as:

$$M_p = 20\left(\frac{1}{1.15}\right) = 17.40.$$

The use of an element with the plastic hinge allowed to form within the member's length would give the exact answer. This approach, however, is not implemented in the present FOPA program.

8
Second-Order Plastic Hinge Analysis

J. Y. RICHARD LIEW AND W. F. CHEN

8.1 Introduction

8.1.1 Background

During the past 20 years, numerous analytical models have been developed for second-order inelastic analysis of steel frames. In general, these models may be categorized into two main types: (1) plastic zone and (2) plastic hinge. The plastic-zone model follows explicitly the gradual spread of yielding throughout the volume of the structure. Plastification in the members is modeled by discretization of members into several beam-column elements and subdivision of the cross sections into many "fibers" [8.1–8.3]. The effects of residual stresses, geometric imperfections, and material strain hardening can all be accounted for in a plastic-zone analysis model. Because of the refined discretization of the members and their cross sections, the plastic-zone analysis can accurately predict the inelastic response of the structure, and it is generally considered an "exact" method of analysis. However, this type of analysis is too computationally intensive for general design use, and because of its complexity and cost, it has not yet found application in ordinary practice. Even if such analysis methods should become generally available and reliable, a more efficient procedure to assess the structural performance and failure modes of a system would be useful. Plastic-hinge-based methods of analysis hold the promise to fulfill these requirements.

In conventional plastic-hinge-based analysis, inelasticity in frame elements is assumed to concentrate at "zero-length" plastic hinges. Regions in the frame elements other than at the plastic hinges are assumed to behave elastically. If the cross-section forces at any particular locations in an element are less than the cross-sectional plastic capacity, elastic behavior is assumed. If the section plastic capacity is reached, a plastic hinge is formed and the element stiffness matrix is adjusted to account for the presence of a plastic hinge. The cross-sectional response after the formation of a plastic hinge is usually assumed to be perfectly plastic with no strain hardening [8.4–8.6].

Plastic hinge analyses can be classified into two categories; first order and second order, depending on whether geometric second-order effects are accounted for. For the first-order elastic-plastic hinge analysis discussed in Chapter 7, equilibrium is formulated based on an undeformed geometry. Thus, only the inelasticity effects that influence the strength of the structure are included. The geometric nonlinear effects on the equilibrium of the structure are not considered. First-order elastic-plastic hinge analysis predicts the maximum load of the structure corresponding to the formation of a plastic collapse mechanism. This analysis approach essentially predicts the same maximum load as the rigid-plastic analysis approaches as discussed in Chapters 4 and 5. The elastic-plastic method is an alternative plastic design method, which in addition to finding the collapse load for a frame, gives information about the redistribution of forces prior to reaching the collapse load.

The elastic-plastic hinge analysis determines the order of plastic hinge formation, the load factor associated with each hinge formation, and member forces in the frame between each hinge formation. The main advantage of the method is that the state of the frame can be established at any load factor rather than only at the state of collapse. This allows a more accurate determination of member forces at the required design load factor.

If equilibrium is formulated based on deformed structural geometry, the analysis is normally termed *second-order*. The need for a second-order analysis of steel frames is increasing in view of the AISC LRFD specifications [8.7], which give explicit permission for the engineer to compute load effects from a direct second-order analysis. Second-order elastic analysis is also the "prefered" method in the Canadian limit states design specification [8.8] and in many other limit-states code provisions. Although there are several other approximated methods [8.9–8.12] based on the same concept as the B_1/B_2 analysis in the LRFD specification, these methods do not always give satisfactory results because they are derived based on simplified assumptions and are applicable to rectangular frames with rigid connections and small displacements. Furthermore, it is rather cumbersome and tedious to calculate the amplification factors applied to a first-order analysis. In view of this, it is more convenient and straightforward to use computer-based methods to calculate member forces directly as long as the method can be easily implemented.

Second-order elastic plastic hinge analysis considers inelasticity and stability effects at a certain level of approximation. The method often employs one element per member, and it is sufficiently accurate to capture the global behavior of the structures. The analysis method is computationally more efficient and economical than second-order plastic-zone analyses. Presently, direct second-order inelastic analysis is addressed by many specification provisions for plastic design. Given the computed load effects from this type of analysis, specifications typically provide equations that member forces must not violate if the members are deemed adequate. However, if the limit states associated with a particular design code are represented with sufficient accuracy in the analysis, then the separate checks of member equations are not required. For the discussions that follow, any methods of analysis that sufficiently capture the stability and plastic limit state behavior such that separate member capacity checks are not required are called *advanced analysis.*

At present, only plastic zone analysis has been classified as an advanced analysis technique. Australian Standard AS4100 [8.13] and EC 3 [8.14] are the only design specifications that explicitly allow engineers to disregard member capacity checks if a plastic zone analysis is employed. This chapter provides the solution and tools to advance the use of plastic-hinge-based analysis as an advanced technique for frame proportioning without the need for member capacity checks. The present work is limited to two-dimensional steel frames only. Information regarding spatial behavior of beam-columns and frames, including topics such as lateral-torsional buckling, is not considered. Also, the present work is limited to frames subjected to static loads only. Frames under dynamic and cyclic loading are not considered in the analysis.

8.1.2 Organization

This section is intended to provide an overview of the impetus behind the research in advanced analysis for steel frame design. The sequence sections are arranged to present basic theories, assumptions, and development work for implementation of second-order plastic-hinge-based analysis for the design of two-dimensional steel frames.

Sections 8.2 through 8.4 present the mathematical formulation of the conventional elastic-plastic hinge approach for modeling inelastic behavior of frame and truss elements. Section 8.5 provides assessment of the capabilities of the method and explores the limitations of the method. Benchmark studies in Section 8.5 conclude that, without modification and refinements, the elastic-plastic hinge approach is not adequate to be accepted as an advanced analysis method.

Section 8.6 presents the desirable attributes that would be considered in the development of an analysis model as an advanced technique. Section 8.7 presents an improved plastic-hinge-based method called the *refined plastic hinge model,* which satisfies the attributes outlined in Section 8.6. The refined plastic hinge approach adopts a suitable stiffness-degradation function that represents the distributed yielding behavior of beam columns. Benchmark tests are then conducted to provide confirmation of the general validity of this method.

In Section 8.8, a method of incorporating semirigid connection effects in advanced analyses of steel frames is presented. The design of semirigid frames using an advanced analysis technique is illustrated through an example.

A FORTRAN-based computer program, PHINGE, is implemented in Section 8.9, based on the theories developed in Sections 8.2 through 8.8. The computer program can perform first-order, second-order elastic, second-order plastic hinge, and second-order refined plastic hinge analyses. It is an educational software for engineers and graduate students who intend to perform second-order analysis for a more realistic estimation of a system's strength and stability. The program can be installed in personal computers or engineering workstations. Source codes, an instruction manual, and installation guides are provided for ease of implementation in a PC environment. Example problems and sample input data files are given in Sections 8.10 and 8.11. They are intended to help clarify the data-generation process.

8.2 Modeling of Elastic Beam-Column Element

8.2.1 Assumptions

The general assumptions used in the modeling of a beam-column element are:

1. all elements are initially straight and prismatic. Plane cross-sections remain plane after deformation.
2. all member cross sections are fully compact such that local buckling effects are insignificant.
3. all members are sufficiently braced such that out-of-plane flexural or lateral-torsional buckling does not influence the member response prior to failure.
4. member deformations and strains are assumed to be small, but large rigid-body displacements are allowed.
5. the member shear forces are smaller enough that the effects of shear deformation can be neglected.
6. axial shortening due to member curvature bending is neglected.

The assumptions that the member is prismatic and member distortions are small are reasonable for ordinary steel frame structures. Although the steel frame may undergo large rigid-body displacements at collapse, the distortion of each member with respect to its chord length in the displaced configuration will remain small since steel members with compact cross sections usually exhibit high bending rigidity.

Member curvature effects (bowling effect) are not considered in the present formulation because many practical frame members have small slenderness ratios for which the axial shortening is often dominated by inelastic axial deformation. For very slender beam-column members, the bowing effects may need to be considered in the stiffness formulation [8.15].

8.2.2 Stability Functions

Figure 8.1(a) shows a beam-column member subjected to end moments M_A and M_B and is acted on by an axial force P. Using the free body diagram of

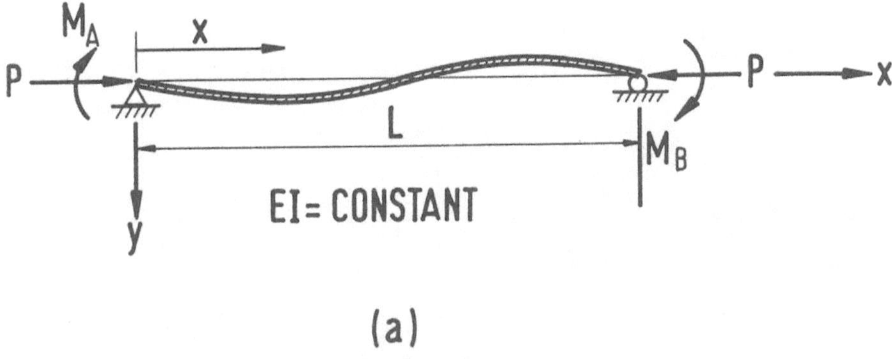

(a)

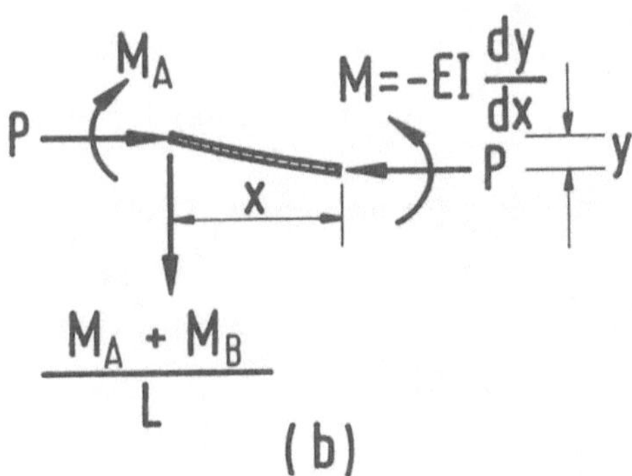

(b)

FIGURE 8.1. Beam-column subject to end moments and axial force.

a segment of beam column of length x as shown in Fig. 8.1(b), the following equilibrium equation can be derived [8.16]:

$$\frac{d^2y}{dx^2} + \rho^2 y = \frac{M_A + M_B}{LEI}x - \frac{M_A}{EI} \tag{8.2.1}$$

where E is the modulus of elasticity, I is the moment of inertia, L is the undeformed length of the element, $\rho = L\sqrt{P/EI}$, and P is taken as positive in tension.

The general solution of this ordinary differential equation is

$$y = A\sin\rho x + B\cos\rho x + \frac{M_A + M_B}{LEI\rho^2}x - \frac{M_A}{EI\rho^2}. \tag{8.2.2}$$

The constants A and B can be evaluated by using the boundary conditions

$$y(0) = 0, \; y(L) = 0 \tag{8.2.3}$$

and they are written as

$$A = -\frac{1}{EI\rho^2 \sin \rho L}(M_A \cos \rho L + M_B) \qquad (8.2.4)$$

$$B = \frac{M_A}{EI\rho^2}. \qquad (8.2.5)$$

Back-substituting A and B into Eq. (8.2.2), and rearranging terms

$$y = -\frac{1}{EI\rho^2}\left[\frac{\cos \rho L}{\sin \rho L}\sin \rho x - \cos \rho x - \frac{x}{L} + 1\right]M_A$$

$$-\frac{1}{EI\rho^2}\left[\frac{1}{\sin \rho L}\sin \rho x - \frac{x}{L}\right]M_B \qquad (8.2.6)$$

from which

$$\frac{dy}{dx} = -\frac{1}{EI\rho}\left[\frac{\cos \rho L}{\sin \rho L}\cos \rho x + \sin \rho x - \frac{1}{\rho L}\right]M_A$$

$$-\frac{1}{EI\rho}\left[\frac{1}{\sin \rho L}\cos \rho x - \frac{1}{\rho L}\right]M_B. \qquad (8.2.7)$$

The end rotation θ_A is obtained by letting $x = 0$ in Eq. (8.2.7)

$$\theta_A = \frac{dy}{dx}\bigg|_{x=0} = \frac{L}{EI}\left[\frac{\sin \rho L - \rho L \cos \rho L}{(kL)^2 \sin \rho L}\right]M_A + \frac{L}{EI}\left[\frac{\sin \rho L - \rho L}{(\rho L)^2 \sin \rho L}\right]M_B \quad (8.2.8)$$

and θ_B can be obtained from Eq. (8.27) for $x = L$

$$\theta_B = \frac{dy}{dx}\bigg|_{x=L} = \frac{L}{EI}\left[\frac{\sin \rho L - \rho L}{(\rho L)^2 \sin \rho L}\right]M_A + \frac{L}{EI}\left[\frac{\sin \rho L - \rho L \cos \rho L}{(\rho L)^2 \sin \rho L}\right]M_B. \quad (8.2.9)$$

The axial force-displacement relationship ignoring the effect of curvature shortening may be expressed as

$$P = \frac{EA}{L}e \qquad (8.2.10)$$

where e is the axial displacement of the member.

Equations (8.2.8), (8.2.9), and (8.2.10) can be expressed in matrix form as

$$\left\{\begin{array}{c} M_A \\ M_B \\ P \end{array}\right\} = \frac{EI}{L}\left[\begin{array}{ccc} S_1 & S_2 & 0 \\ S_2 & S_1 & 0 \\ 0 & 0 & A/I \end{array}\right]\left\{\begin{array}{c} \theta_A \\ \theta_B \\ e \end{array}\right\}. \qquad (8.2.11)$$

S_1 and S_2 are called the *stability functions*, which may be written as:

$$S_1 = \begin{cases} \dfrac{\rho \sin(\rho) - \rho^2 \cos(\rho)}{2 - 2\cos(\rho) - \rho \sin(\rho)} & \text{for } P < 0 \\[2ex] \dfrac{\rho^2 \cosh(\rho) - \rho \sinh(\rho)}{2 - 2\cosh(\rho) + \rho \sinh(\rho)} & \text{for } P > 0 \end{cases} \qquad (8.2.12)$$

$$S_2 = \begin{cases} \dfrac{\rho^2 - \rho\sin(\rho)}{2 - 2\cos(\rho) - \rho\sin(\rho)} & \text{for } P < 0 \\[3mm] \dfrac{\rho\sinh(\rho) - \rho^2}{2 - 2\cosh(\rho) + \rho\sinh(\rho)} & \text{for } P > 0. \end{cases} \qquad (8.2.13)$$

S_1 and S_2 account for the effect of the axial force on the bending stiffness of the member. Equations (8.2.12) and (8.2.13) are indeterminate when the axial force is equal to zero. To circumvent this problem, the following simplified equations are used to approximate the stability function when the axial force in the member falls within the range of $-2.0 \le \rho \le 2.0$ [8.5].

$$S_1 = 4 + \frac{2\pi^2\rho_e}{15} - \frac{(0.01\rho_e + 0.543)\rho_e^2}{4 + \rho_e} - \frac{(0.004\rho_e + 0.285)\rho_e^2}{8.183 + \rho_e} \qquad (8.2.14)$$

$$S_2 = 2 - \frac{\pi^2\rho_e}{30} + \frac{(0.01\rho_e + 0.543)\rho_e^2}{4 + \rho_e} - \frac{(0.004\rho_e + 0.285)\rho_e^2}{8.183 + \rho_e}. \qquad (8.2.15)$$

where $\rho_e = P/P_e = P/(\pi^2 EI/L^2) = \rho^2/\pi^2$. Equations (8.2.14) and (8.2.15) are applicable for members in tension and compression; they give excellent prediction of results compared to the "exact" solution obtained from Eqs. (8.2.12) and (8.2.13) [8.5].

8.2.3 Tangent Stiffness Relationship

In a second-order analysis, it is convenient to express the element force-displacement equations in an increment form. Denoting $\dot{M}_A$, $\dot{M}_B$, $\dot{\theta}_A$, $\dot{\theta}_B$ as the incremental end moments and joint rotations at element ends A and B, respectively, and $\dot{P}$ and $\dot{e}$ as the incremental axial force and displacement in the longitudinal direction of the element, an incremental form of Eq. (8.8.11) may be written as [8.17]:

$$\left\{ \begin{array}{c} \dot{M}_A \\ \dot{M}_B \\ \dot{P} \end{array} \right\} = \frac{EI}{L} \begin{bmatrix} S_1 & S_2 & 0 \\ S_2 & S_1 & 0 \\ 0 & 0 & A/I \end{bmatrix} \left\{ \begin{array}{c} \dot{\theta}_A \\ \dot{\theta}_B \\ \dot{e} \end{array} \right\}. \qquad (8.2.16)$$

Equation (8.2.16) may be expressed symbolically as

$$\dot{\mathbf{f}}_c = \mathbf{k}_c \dot{\mathbf{d}}_c \qquad (8.2.17)$$

in which $\dot{\mathbf{f}}_c$ and $\dot{\mathbf{d}}_c$ are the incremental force and displacement vectors, respectively, and $\mathbf{k}_c$ is the element basic tangent stiffness matrix. For a plane frame member, three additional degrees of freedom are required to describe the total displacements of the member. If d_{g1}, d_{g2}, ... and d_{g6} are defined as the global translational and rotational degrees of freedom of a frame member (see Fig. 8.2), it can be shown that the local displacements are related to the global displacements by

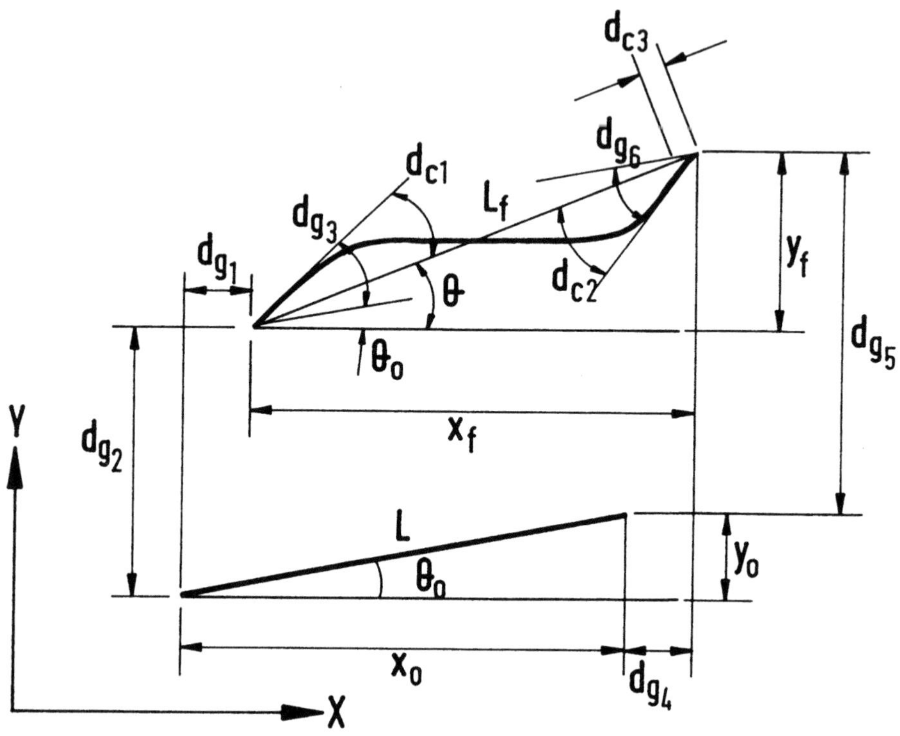

FIGURE 8.2. Global and local displacements of a beam-column element.

$$d_{c1} = \theta_A = \theta_o + d_{g3} - \tan^{-1}\frac{y_o + d_{g5} - d_{g2}}{x_o + d_{g4} - d_{g1}} \qquad (8.2.18a)$$

$$d_{c2} = \theta_B = \theta_o + d_{g6} - \tan^{-1}\frac{y_o + d_{g5} - d_{g2}}{x_o + d_{g4} - d_{g1}} \qquad (8.2.18b)$$

$$d_{c3} = \frac{(2x_o + d_{g4} - d_{g1})(d_{g4} - d_{g1}) + (2y_o + d_{g5} - d_{g2})(d_{g5} - d_{g2})}{L_f + L} \qquad (8.2.18c)$$

where L_f is the deformed chord length.

The expression for d_{c3} in Eq. (8.2.18c) is more accurate than the value calculated from $L_f - L$. This is because Eq. (8.2.18c) avoids finding the small difference between large member lengths [8.18]. This equation is obtained by writing $d_{c3} = (L_f^2 - L^2)/(L_f - L)$ and then solving for d_{c3}. In the denominator, $L_f + L \approx 2L$ may be assumed, since small displacement theory is presumed for the corotational chord element [8.19].

Upon differentiation of Eqs. (8.2.18a–c) with respect to each member end displacement variable d_{gi}, where $i = 1, 2, \ldots 6$, the incremental kinematics relationship relating the two sets of displacement vectors may be written in matrix form as

$$\begin{Bmatrix} \dot{\theta}_A \\ \dot{\theta}_B \\ \dot{e} \end{Bmatrix} = \begin{bmatrix} -\sin\theta/L & \cos\theta/L & 1 & \sin\theta/L & -\cos\theta/L & 0 \\ -\sin\theta/L & \cos\theta/L & 0 & \sin\theta/L & -\cos\theta/L & 1 \\ -\cos\theta & -\sin\theta & 0 & \cos\theta & \sin\theta & 0 \end{bmatrix} \begin{Bmatrix} \dot{d}_{g1} \\ \dot{d}_{g2} \\ \dot{d}_{g3} \\ \dot{d}_{g4} \\ \dot{d}_{g5} \\ \dot{d}_{g6} \end{Bmatrix}$$

$$(8.2.19)$$

where θ is the angle of inclination of the chord of the deformed member. Symbolically, the kinematics relationship of Eq. (8.2.19) may be written as

$$\dot{\mathbf{d}}_c = \mathbf{T}_{cg}\dot{\mathbf{d}}_g. \tag{8.2.20}$$

Based on the principle of equilibrium, the forces in the two systems shown in Fig. 8.3 are related by

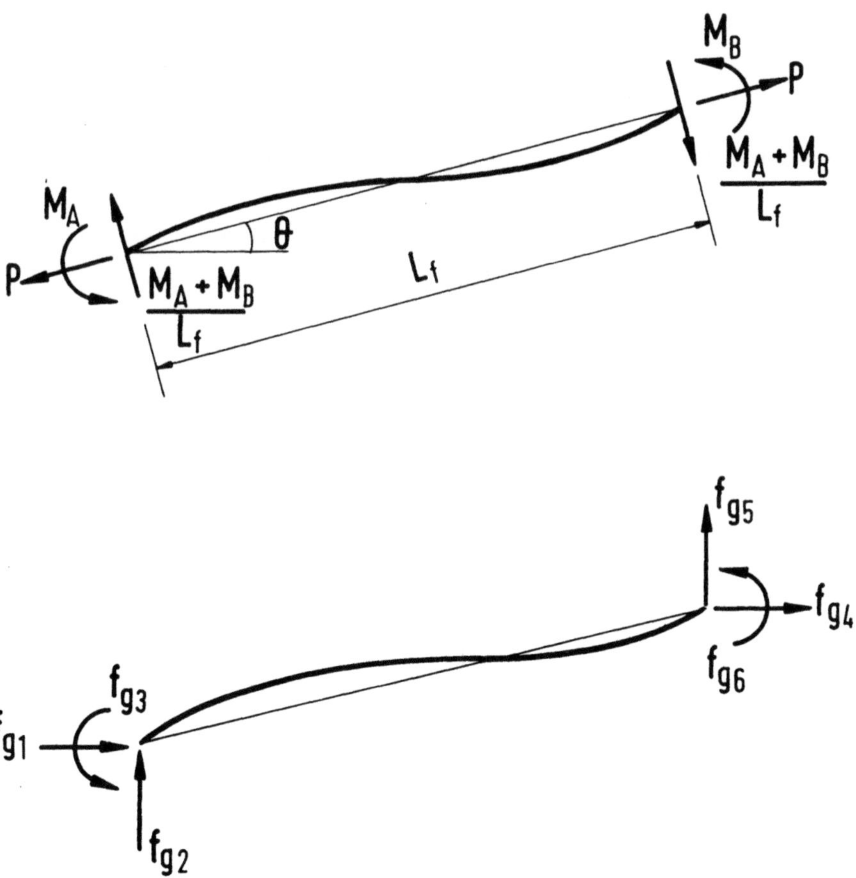

FIGURE 8.3. Equivalent force system.

$$
\begin{Bmatrix} f_{g1} \\ f_{g2} \\ f_{g3} \\ f_{g4} \\ f_{g5} \\ f_{g6} \end{Bmatrix}
=
\begin{bmatrix}
-\sin\theta/L & -\sin\theta/L & -\cos\theta \\
\cos\theta/L & \cos\theta/L & -\sin\theta \\
1 & 0 & 0 \\
\sin\theta/L & \sin\theta/L & \cos\theta \\
-\cos\theta/L & -\cos\theta/L & \sin\theta \\
0 & 1 & 0
\end{bmatrix}
\begin{Bmatrix} M_A \\ M_B \\ P \end{Bmatrix}.
\tag{8.2.21}
$$

Symbolically,

$$
\mathbf{f_g} = \mathbf{T}_{cg}^T \mathbf{f_c}.
\tag{8.2.22}
$$

Taking derivatives on both sides of Eq. (8.2.22) gives

$$
\dot{\mathbf{f}}_g = \mathbf{T}_{cg}^T \dot{\mathbf{f}}_c + \dot{\mathbf{T}}_{cg}^T \mathbf{f_c}.
\tag{8.2.23}
$$

In view of Eqs. (8.2.17) and (8.2.20), Eq. (8.2.23) may be further written as

$$
\dot{\mathbf{f}}_g = \mathbf{T}_{cg}^T \mathbf{k_c} \mathbf{T}_{cg} \dot{\mathbf{d}}_g + \dot{\mathbf{T}}_{cg}^T \mathbf{f_c}.
\tag{8.2.24}
$$

The transformation matrix $\dot{\mathbf{T}}_{cg}^T$ can be evaluated by taking the derivative of $\mathbf{T}_{cg}$ with respect to each global degree of freedom, d_{gk}, as

$$
\dot{\mathbf{T}}_{cg}^T = \left[\frac{\partial \mathbf{T}_{cg}^T}{\partial d_{gk}} \right] \dot{\mathbf{d}}_{gk}, \quad k = 1, 2, \ldots, 6.
\tag{8.2.25}
$$

Equation (8.2.25) may be written, in view of Eq. (8.2.20), as

$$
\dot{\mathbf{T}}_{cg}^T = \left[\frac{\partial^2 \mathbf{d}_{ci}}{\partial d_{gj}\, \partial d_{gk}} \right]^T \dot{\mathbf{d}}_{gk}, \quad i = 1, 2, 3;\; j = 1, 2, \ldots, 6;\; k = 1, 2, \ldots, 6
\tag{8.2.26}
$$

or

$$
\dot{\mathbf{T}}_{cg}^T = \left[\frac{\partial^2 d_{c1}}{\partial d_{gj}\, \partial d_{gk}} \,\middle|\, \frac{\partial^2 d_{c2}}{\partial d_{gj}\, \partial d_{gk}} \,\middle|\, \frac{\partial^2 d_{c3}}{\partial d_{gj}\, \partial d_{gk}} \right]^T \dot{\mathbf{d}}_{gk} = [\mathbf{T_1}|\mathbf{T_2}|\mathbf{T_3}]\dot{\mathbf{d}}_{gk}.
\tag{8.2.27}
$$

By carrying out the appropriate derivatives, the matrices $\mathbf{T_1}$, $\mathbf{T_2}$, $\mathbf{T_3}$ are given as [8.20]

$$
\mathbf{T_1} = \mathbf{T_2} = \frac{1}{L^2}
\begin{bmatrix}
-2sc & c^2 - s^2 & 0 & 2sc & -(c^2 - s^2) & 0 \\
 & 2sc & 0 & -(c^2 - s^2) & -2sc & 0 \\
 & & 0 & 0 & 0 & 0 \\
 & & & -2sc & c^2 - s^2 & 0 \\
\text{sym.} & & & & 2sc & 0 \\
 & & & & & 0
\end{bmatrix}.
\tag{8.2.28}
$$

$$
\mathbf{T_3} = \frac{1}{L}
\begin{bmatrix}
s^2 & -sc & 0 & -s^2 & sc & 0 \\
 & c^2 & 0 & sc & -c^2 & 0 \\
 & & 0 & 0 & 0 & 0 \\
 & & & s^2 & -sc & 0 \\
\text{sym.} & & & & c^2 & 0 \\
 & & & & & 0
\end{bmatrix}.
\tag{8.2.29}
$$

The complete derivation of the incremental force-displacement relationship is then obtained by substituting $\dot{\mathbf{T}}_{cg}^{T}$ from Eq. (8.2.27) into Eq. (8.2.24)

$$\dot{\mathbf{f}}_{g} = (\mathbf{T}_{cg}^{T}\mathbf{k}_{c}\mathbf{T}_{cg} + \mathbf{T}_{1}M_{A} + \mathbf{T}_{2}M_{B} + \mathbf{T}_{3}P)\dot{\mathbf{d}}_{g}. \tag{8.2.30}$$

Equation (8.2.30) is the large-displacement small-strain incremental force-displacement relationship of a beam-column element in the global coordinate system, and it may be written symbolically as

$$\dot{\mathbf{f}}_{g} = \mathbf{k}_{g}\dot{\mathbf{d}}_{g} \tag{8.2.31}$$

where $\mathbf{k}_{g}$ represents the tangent stiffness matrix of the beam-column element. It should be noted that in the derivation of the tangent stiffness matrix, $\mathbf{k}_{g}$, the joints are assumed to be rigid. If plastic hinges or connections are presented at the element ends, the tangent stiffness matrix needs to be modified. These modifications are discussed in Sections 8.4 and 8.7.

8.3 Modeling of Truss Elements

8.3.1 Assumptions

The following assumptions are made in the formulation of truss elements:

1. The element can undergo large rigid-body displacements, but the member deformation remains small.
2. The axial force-displacement relationship in the local convected coordinates is written as

$$P = \frac{EA}{L}e \tag{8.3.1}$$

where P and e are the axial force and displacement, respectively; A is the cross-sectional area; and L is the length of the truss member.
3. Failure of the truss element is said to have occurred when for compression [8.7]

$$|P| = (0.685^{\lambda_c^2})P_y \qquad \text{for } \lambda_c \leq 1.5 \tag{8.3.2a}$$

$$|P| = \frac{0.877}{\lambda_c^2}P_y \qquad \text{for } \lambda_c > 1.5 \tag{8.3.2b}$$

and when for tension

$$P = P_y = AF_y \tag{8.3.3}$$

where $\lambda_c = (L/r\pi)\sqrt{F_y/E}$ is the member slenderness parameter, r is the radius of gyration of the cross section of the member, F_y is the material yield stress, and E is the modulus of elasticity.

8.3.2 Tangent Stiffness Formulation

The tangent stiffness relationship for a bracing element can be obtained from
the tangent stiffness relationship of a frame element by deleting the appropriate rows and columns in Eq. (8.2.30) that correspond to the rotational degrees of freedom of the element. The resulting tangent stiffness relationship of
a truss element is [8.21]

$$\dot{\mathbf{f}}_g = (\mathbf{T}_{cg}^T \mathbf{k}_c \mathbf{T}_{cg} + \mathbf{T}P)\dot{\mathbf{d}}_g = \mathbf{k}_g \dot{\mathbf{d}}_g \qquad (8.3.4)$$

where, referring to Fig. 8.4,

$$\dot{\mathbf{f}}_g = [\dot{f}_{g1}, \dot{f}_{g2}, \dot{f}_{g3}, \dot{f}_{g4}]^T \qquad (8.3.5)$$

$$\dot{\mathbf{d}}_g = [\dot{d}_{g1}, \dot{d}_{g2}, \dot{d}_{g3}, \dot{d}_{g4}]^T \qquad (8.3.6)$$

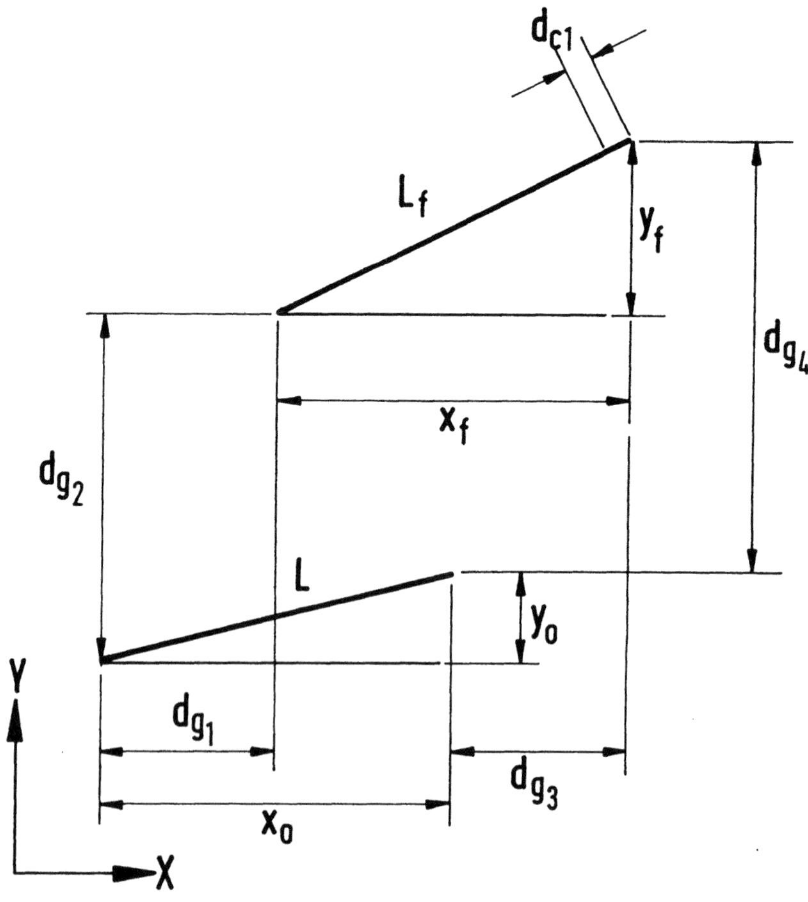

FIGURE 8.4. Global and local displacements of a truss element.

$$\mathbf{T_{cg}} = [-\cos\theta \quad -\sin\theta \quad \cos\theta \quad \sin\theta]^T \tag{8.3.7}$$

$$k_c = \frac{EA}{L} \tag{8.3.8}$$

$$\mathbf{T} = \frac{1}{L}\begin{bmatrix} \sin^2\theta & -\sin\theta\cos\theta & -\sin^2\theta & \sin\theta\cos\theta \\ & \cos^2\theta & \sin\theta\cos\theta & -\cos\theta^2 \\ \text{sym.} & & \sin^2\theta & -\sin\theta\cos\theta \\ & & & \cos^2\theta \end{bmatrix} \tag{8.3.9}$$

in which θ is the angle of inclination of the chord of the deformed element chord.

For structures subjected to severe wind or lateral earthquake loading, truss diagonal bracing may be introduced to reduce the story drifts and enhance the lateral-load resistance of the structure. In design, these braces are usually assumed to carry only axial force. Therefore, it is justifiable to use truss elements to model the bracing members.

The truss elements may also be used for modeling gravity columns that do not participate in the lateral-force resisting system. These gravity columns (leaner columns), which are commonly used in low-rise industrial buildings and tall office building frames, are usually designed to carry only gravity loads. Therefore, they can be modeled by the truss element described in this section.

8.4 Modeling of Plastic Hinges

There are two common approaches for modeling elastic-plastic hinges [8.30]. The first approach is called the beam-column stability function approach in which the plastic hinge can undergo plastic rotation only. The change in the axial force in a frame element is based solely on the element axial force-displacement relationship, with no effect from the inelasticity at the plastic hinges. The second approach is based on a force-space plasticity formulation in which an associated flow rule is used to describe the relationship between the axial and rotation plastic deformations at a fully plastified cross section. For most practical cases, the differences in results predicted by these two plastic hinge approaches are small, particularly when the axial force in the member is small. Also, both approaches can account for force-point movement on the plastic strength surface. In other words, if the axial force is increased on a fully plastified cross section, the bending moment would need to be decreased so that the cross-sectional plastic capacity is not violated. The formulation presented here is based on the beam-column stability approach.

8.4.1 Assumptions

1. Inelastic behavior in the member is assumed to be contained within a zero-length plastic hinge.

2. Plastic hinges are allowed to form only at the element ends, and plastic deformations at plastic hinges involve inelastic rotation only.
3. Once a plastic hinge is formed, the cross-sectional forces are assumed to move on the plastic strength surface. Unloading from this surface is not considered.
4. The AISC LRFD cross-sectional plastic strength equations [8.7] are adopted in the elastic-plastic hinge formulation:

$$\frac{P}{P_y} + \frac{8}{9}\frac{M}{M_p} = 1.0 \quad \text{for } \frac{P}{P_y} \geq 0.2 \tag{8.4.1}$$

$$\frac{P}{2P_y} + \frac{M}{M_p} = 1.0 \quad \text{for } \frac{P}{P_y} < 0.2 \tag{8.4.2}$$

where P_y is the squash load, M_p is the plastic moment capacity for a member under pure bending action, and P and M are the second-order axial force and bending moment at the cross section under consideration. The plastic strength equations are plotted in Fig. 8.5, and they are derived from the AISC LRFD interaction equations for beam columns of zero length, i.e., $L = 0$.

Although the present formulation is limited to modeling plastic hinges only at the element ends, the formation of a plastic hinge between both ends of the element can be detected based on a simple procedure discussed in Refs. [8.22 and 8.23]. Normally, the analysis of frame members with maximum moment within their span length, and the analysis of beam-columns with in-span loading and inclined members subjected to gravity loads would require more than one beam-column element per member to capture the plastic hinge formation in the member. Multiple beam-column elements per member are necessary to idealize the nonuniformly loaded members (both in the transverse and axial directions) as a number of finite elements with forces applied only at their ends.

Research by Chen and Atsuta [8.24] indicate that the insertion of a plastic hinge at the "exact" maximum moment point in a frame member is not required for an accurate estimate of the maximum strength of the member. As long as the "exact" location of the plastic hinge in a member is not more than $L/6$ distance away from the assumed position (L is the length of the member), the difference in strength prediction is not more than 5 percent. This observation is true for beam columns subjected to uniform patch load and concentrated lateral load [8.24].

For structures subjected to static loading only, elastic unloading at the locations of plastic hinges may be neglected. The analytical results obtained based on this assumption are generally conservative. The numerical implementation following this assumption is also straightforward in concept, and it does not have the problem of being trapped into the recurring process of loading, unloading, and reloading of plastic hinges.

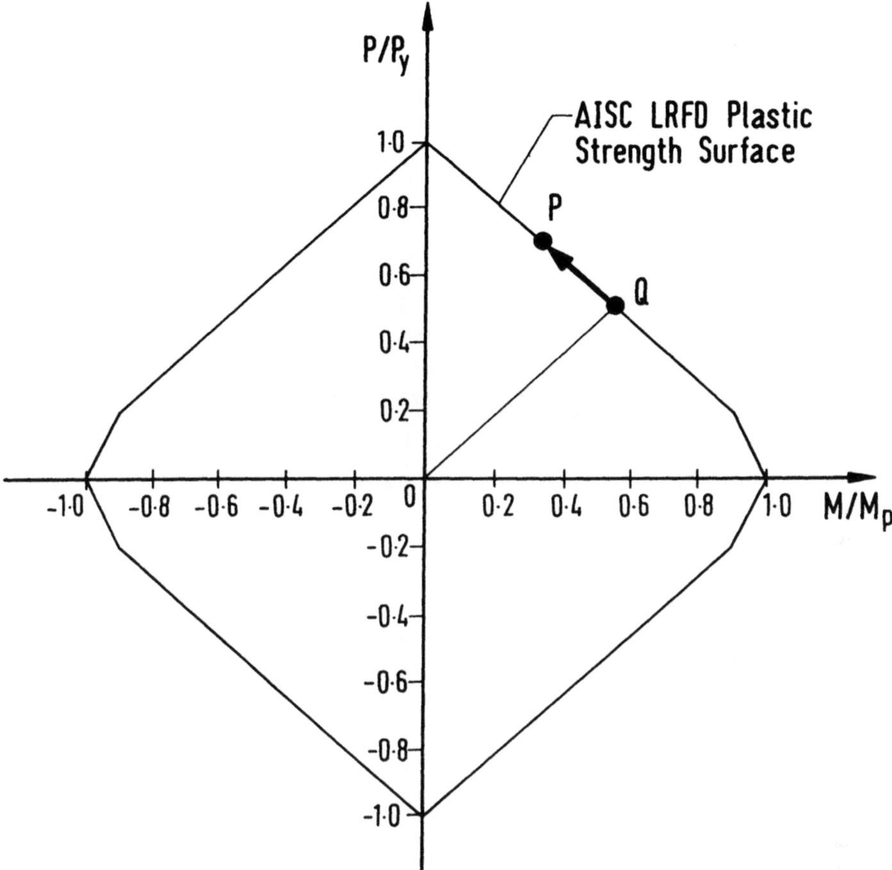

FIGURE 8.5. Cross-sectional plastic strength surface for plastic hinge analysis.

8.4.2 Modified Tangent Stiffness Relationship

If the state of forces at any cross section equals or exceeds the plastic strength, a plastic hinge is formed. The element force-displacement relationships need to be modified to reflect the change in element behavior due to formation of plastic hinge(s) at the element end(s).

If a plastic hinge is formed at element end A, the incremental force-displacement relationships from Eq. (8.2.16) may be written as:

$$\left\{ \begin{array}{c} \Delta \dot{M}_{pcA} \\ \dot{M}_B \\ \dot{P} \end{array} \right\} = \frac{EI}{L} \begin{bmatrix} S_1 & S_2 & 0 \\ S_2 & S_1 & 0 \\ 0 & 0 & A/I \end{bmatrix} \left\{ \begin{array}{c} \dot{\theta}_A \\ \dot{\theta}_B \\ \dot{e} \end{array} \right\} \qquad (8.4.3)$$

where ΔM_{pcA} is the change in plastic moment capacity at end A as P changes. $\dot{\theta}_A$ from the first row of Eq. (8.4.3) can be written as:

$$\dot{\theta}_A = \frac{L\Delta M_{pcA}}{EIS_1} - \frac{S_2}{S_1}\dot{\theta}_B. \tag{8.4.4}$$

Backsubstituting Eq. (8.4.4) into the second and third row of Eq. (8.4.3), the modified element force-displacement relationship can be obtained as

$$\left\{\begin{array}{c} \dot{M}_A \\ \dot{M}_B \\ \dot{P} \end{array}\right\} = \frac{EI}{L}\left[\begin{array}{ccc} 0 & 0 & 0 \\ 0 & (S_1 - S_2^2)/S_1 & 0 \\ 0 & 0 & A/I \end{array}\right]\left\{\begin{array}{c} \dot{\theta}_A \\ \dot{\theta}_B \\ \dot{e} \end{array}\right\} + \left\{\begin{array}{c} 1 \\ S_2/S_1 \\ 0 \end{array}\right\}\Delta M_{pcA}. \tag{8.4.5}$$

A similar approach can be followed if a plastic hinge is formed at end B. If plastic hinges are formed at both ends of the element, θ_A and θ_B can be written in terms of the change in moment at the respective end of the element. The resulting modified force-displacement relationship is

$$\left\{\begin{array}{c} \dot{M}_A \\ \dot{M}_B \\ \dot{P} \end{array}\right\} = \frac{EI}{L}\left[\begin{array}{ccc} 0 & 0 & 0 \\ 0 & 0 & 0 \\ 0 & 0 & A/I \end{array}\right]\left\{\begin{array}{c} \dot{\theta}_A \\ \dot{\theta}_B \\ \dot{e} \end{array}\right\} + \left\{\begin{array}{c} \Delta M_{pcA} \\ \Delta M_{pcB} \\ 0 \end{array}\right\} \tag{8.4.6}$$

where ΔM_{pcA} and ΔM_{pcB} are the change in the plastic moment capacity at the respective end of the member as P changes.

Equations (8.4.5) and (8.4.6) account for the presence of plastic hinge(s) at the element end(s). They may be written symbolically as

$$\dot{\mathbf{f}}_c = \mathbf{k}_{ch}\dot{\mathbf{d}}_c + \dot{\mathbf{f}}_{cp} \tag{8.4.7}$$

where $\mathbf{k}_{ch}$ is the modified tangent stiffness matrix due to the presence of plastic hinge(s). $\dot{\mathbf{f}}_{cp}$ is an equilibrium force correction vector that results from the change in moment capacity as P changes.

If Eq. (8.2.17) is replaced by Eq. (8.4.7) and the procedure in Section 8.2 is followed, the modified element force-displacement relationship in global coordinates may be written as

$$\dot{\mathbf{f}}_g = (\mathbf{T}_{cg}^T\mathbf{k}_{ch}\mathbf{T}_{cg} + \mathbf{T}_1 M_A + \mathbf{T}_2 M_B + \mathbf{T}_3 P)\dot{\mathbf{d}}_g + \mathbf{T}_{cg}^T\dot{\mathbf{f}}_{cp} \tag{8.4.8}$$

or

$$\dot{\mathbf{f}}_g = \mathbf{k}_{gh}\dot{\mathbf{d}}_g + \dot{\mathbf{f}}_{gp} \tag{8.4.9}$$

where

$$\mathbf{k}_{gh} = \mathbf{T}_{cg}^T\mathbf{k}_{ch}\mathbf{T}_{cg} + \mathbf{T}_1 M_A + \mathbf{T}_2 M_B + \mathbf{T}_3 P \tag{8.4.10}$$

is the modified tangent stiffness matrix, and

$$\dot{\mathbf{f}}_{gp} = \mathbf{T}_{cg}^T\dot{\mathbf{f}}_{cp} \tag{8.4.11}$$

is the global equilibrium force correction vector.

Once a plastic hinge is formed in a member, the subsequent change in the plastic moment capacity due to the change in axial force will affect the force-displacement relationships of the beam-column element. In other words, referring to Fig. 8.5, once the plastic strength is reached at point Q, the state of moment is stationary at Q. However, because of the presence of axial force, the state of moment will change. If the axial force is increased, the force point should move from Q to P. The force-point movement on the plastic strength surface must be considered in the tangent stiffness formulation [8.25 and 8.26].

8.5 Limitations of Elastic-Plastic Hinge Models for Advanced Analysis

Although the elastic-plastic hinge approach is capable of predicting the overall behavior of many types of frames, it cannot be used without modification to predict accurately or conservatively the capacity and load behavior of general beam-column members. Its accuracy is also suspect for frames where the failure is influenced to a great extent by the strength and stability of individual members. This section explores the limitations of the elastic-plastic hinge analysis for assessment of maximum strength behavior of planar frames. The performance of the analysis method is evaluated by comparing the results with established benchmark solutions.

8.5.1 Axially Loaded Columns

Figure 8.6 compares the AISC LRFD column strength curve [8.7] with the strength curve generated by second-order elastic-plastic hinge analysis. For comparison with the AISC LRFD column strength curve, the elastic-plastic hinge analysis is carried out using a W8 $\times$ 31 section bent about its strong axis. The column is discretized into two elements with an initial deflection δ_0 at midlength equal to $L/1500$. This initial out-of-straightness magnitude is implicitly assumed in the development of the AISC LRFD column equations [8.7]. Residual stresses and end restraints are, however, not modeled in the plastic-hinge analysis.

It is observed from Fig. 8.6 that the elastic plastic hinge analysis overpredicts the column capacity implied by the AISC LRFD column equations. The maximum error is about 23%, which corresponds to the column with a slenderness paramenter of unity, at which value residual stress and initial out-of-straightness effects interact to produce the greatest reduction in strength from the theoretical value for a perfect column. However, for columns whose slenderness parameter is less than or equal to 0.4, the maximum error from the elastic-plastic hinge analysis is not more than 5% unconservative when compared to the AISC LRFD column strength curve. These results suggest that for stocky column members with slenderness parameter λ_c less than 0.4, the column maximum strength can be predicted within 5% error using the second-order elastic-plastic hinge analysis.

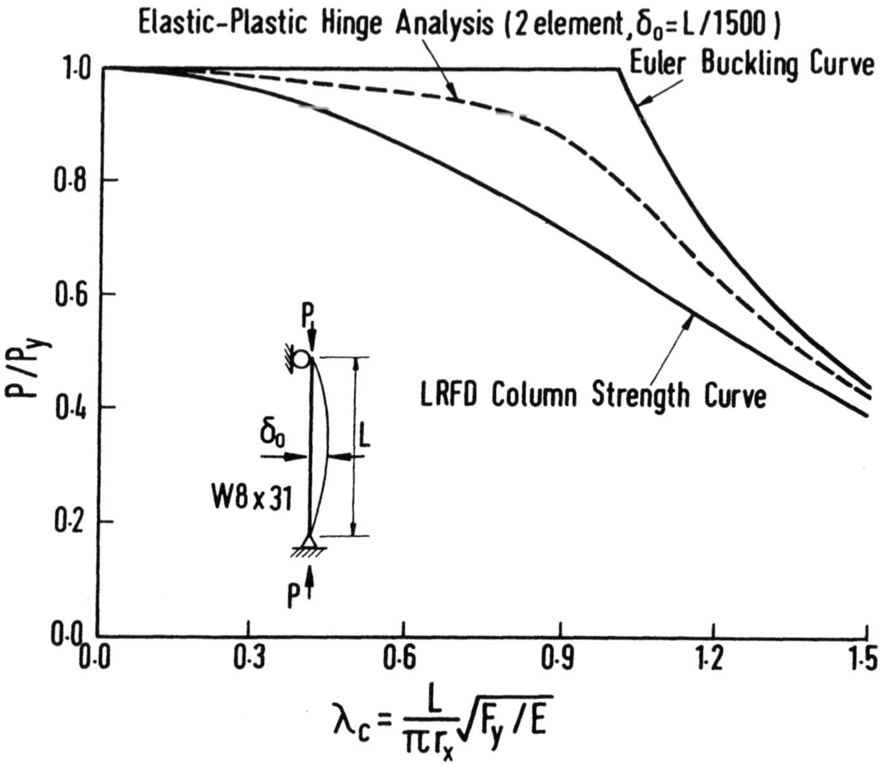

FIGURE 8.6. Comparison of strength curves for axially loaded pinned-end column.

8.5.2 Beam Columns

Zhou et al. [8.27] analyzed the beam columns shown in Fig. 8.7 using the plastic-zone method. The results are presented for beam columns subjected to equal end moments producing single curvature bending. The beam columns have residual stresses with a maximum compressive magnitude of $0.3F_y$ and an initial geometric imperfection that varied sinusoidally with a maximum in-plane deflection, $\delta_0 = L/1000$, at midlength. In Fig. 8.7, the plastic-zone strength curves are compared with the elastic-plastic hinge solutions. The plastic hinge results were generated using two elements per member. The analyses explicitly model the member initial out-of-straightness, which has a maximum magnitude of $\delta_0 = L/1000$ at the midlength.

It can be observed from Fig. 8.7 that for beam columns with L/r less than 20 and P less than $0.2P_y$, the elastic-plastic hinge analysis is sufficiently accurate for the prediction of the ultimate strengths of these beam columns. Otherwise, the elastic-plastic hinge analyses always overpredict the capacities of the beam columns by more than 5%.

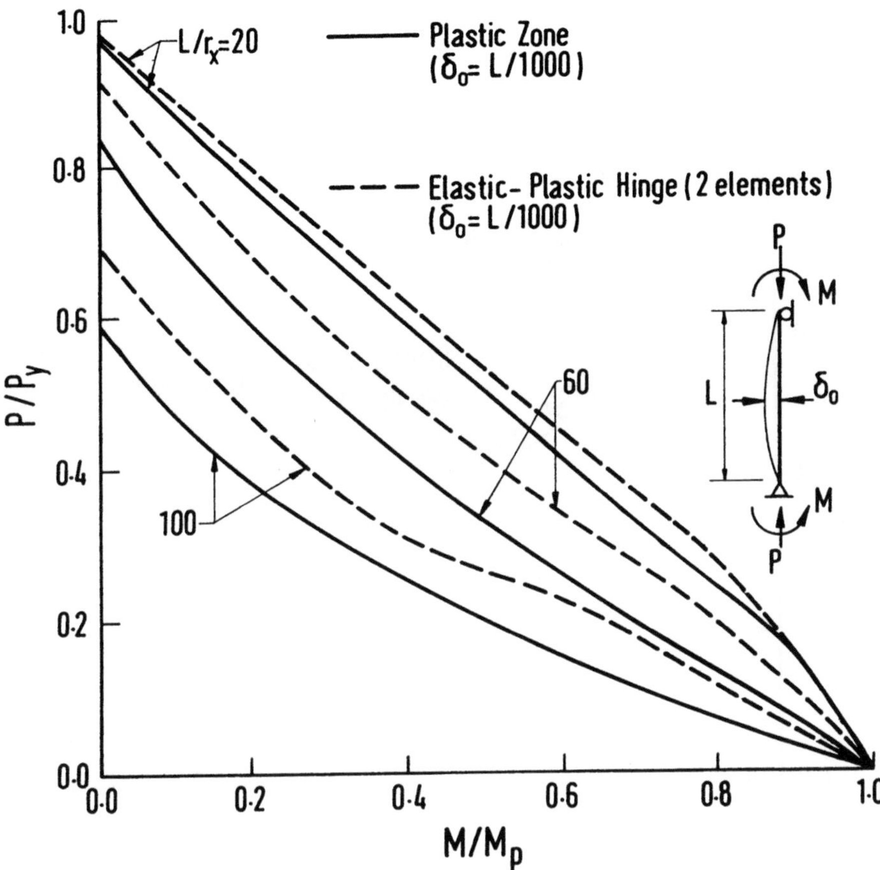

FIGURE 8.7. Comparison of beam-column strengths from plastic-hinge-based analyses with "exact" strength curves [8.27].

8.5.3 Sway Frames

Figure 8.8 through 8.10 compare the in-plane strength curves obtained by the second-order elastic-plastic hinge with the plastic-zone results from Kanchanalai [8.28]. Kanchanalai's plastic-zone solutions were used as benchmarks for the development of the AISC LRFD beam-column equations [8.22, 8.29]; any analysis methods that can match these benchmarks are expected to satisfy the requirements for two-dimensional advanced inelastic analysis, for which separate specification member capacity checks are not required.

Both the strong- and weak-axis strength curves are presented in the figures for the plastic-zone solutions. However, only the strong-axis strength curves are shown for the plastic hinge analyses. The weak-axis curves from the plastic hinge analysis are identical with the strong-axis strength curves be-

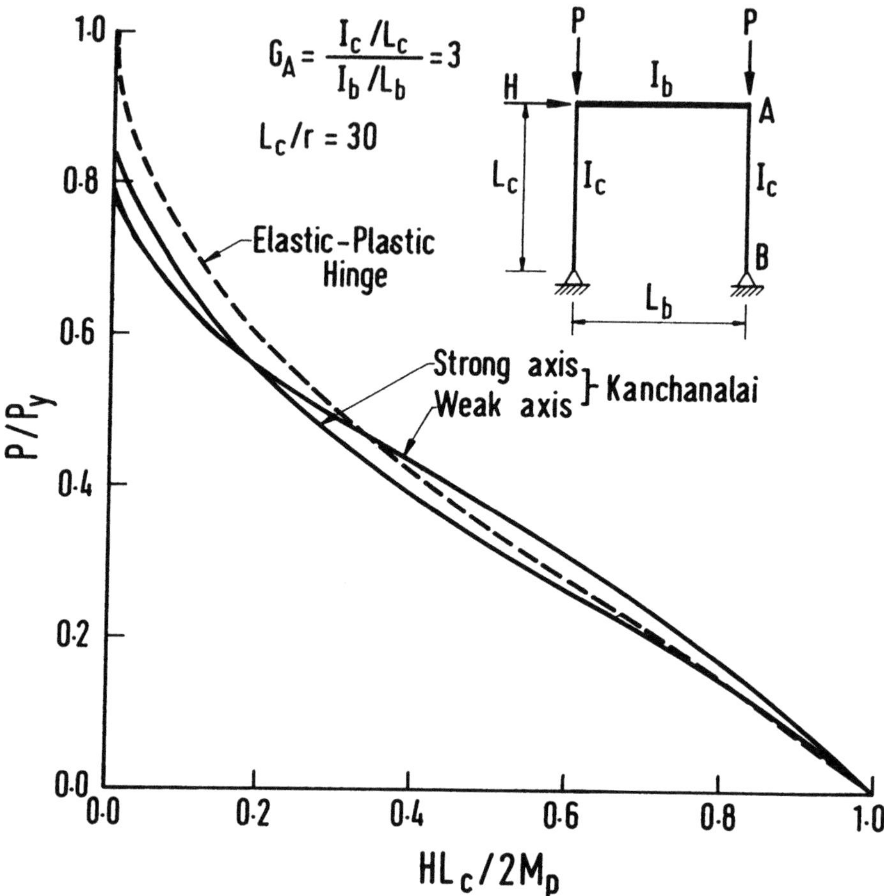

FIGURE 8.8. Comparison of strength curves for portal frame with $G_A = 3$ and $L_c/r = 30$.

cause the results are presented in a nondimensional form, and only one plastic strength curve (i.e., the LRFD beam-column interaction equations for a member of zero length) was used for both the strong- and weak-axis section strengths.

Comparisons of results show that the elastic-plastic hinge analyses overpredict the maximum strengths of the frames. For the portal frame shown in Figs. 8.8 and 8.9, the strong-axis strengths from the plastic hinge model are approximately in error by a maximum of 20 and 9%, respectively (the errors are measured radially from the origin of the plots). For the leaned-column frame shown in Fig. 8.10, the strong-axis strengths are overpredicted with a maximum error of 20%.

The weak-axis strengths are also significantly overpredicted by the elastic-

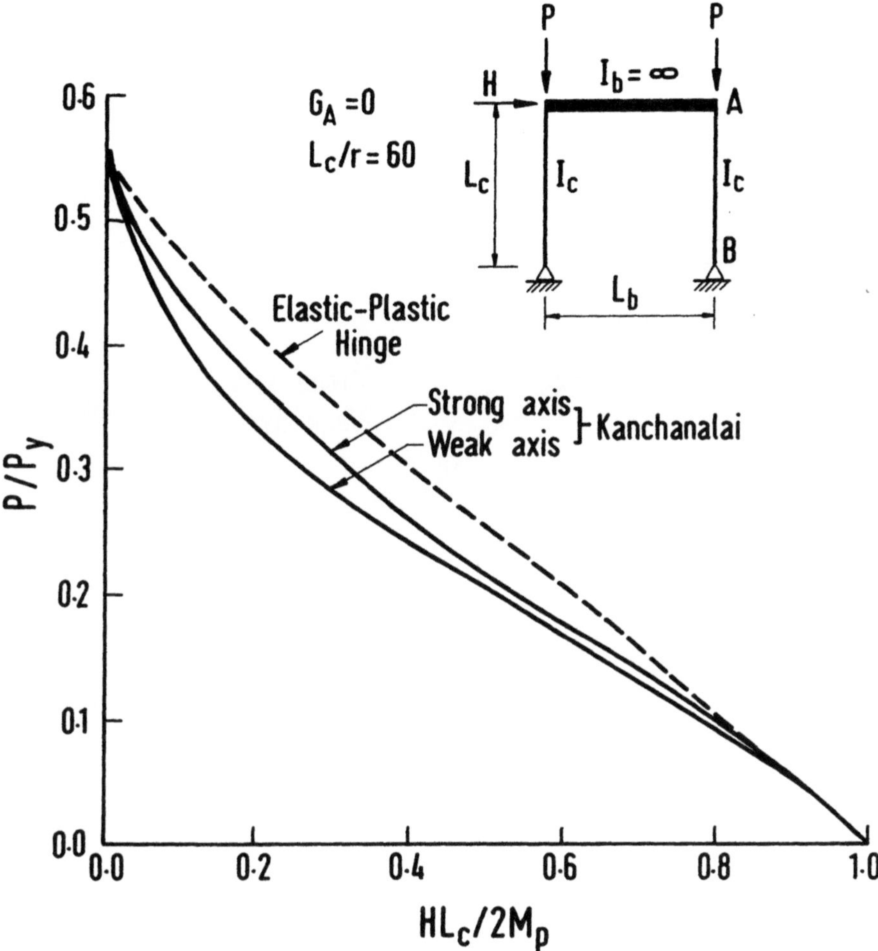

FIGURE 8.9. Comparison of strength curves for portal frame with $G_A = 0$ and $L_c/r = 60$.

plastic hinge model for the frames shown in Figs. 8.8 through 8.10. In these cases, the maximum unconservative error in the weak-axis strength predictions based on the elastic-plastic hinge analyses ranges from about 10 to 30% unconservative.

Research [8.30–8.31] shows that for very stocky or very flexible frames, the elastic-plastic hinge results are quite comparable to the strong-axis plastic-zone results for almost all ranges of axial force versus moment. This means that second-order elastic-plastic hinge analysis is sufficiently accurate in predicting the failure load either at the elastic buckling or at the collapse load associated with the formation of plastic mechanism. However, for frames such as those shown in Figs. 8.8 through 8.10, both the strong- and

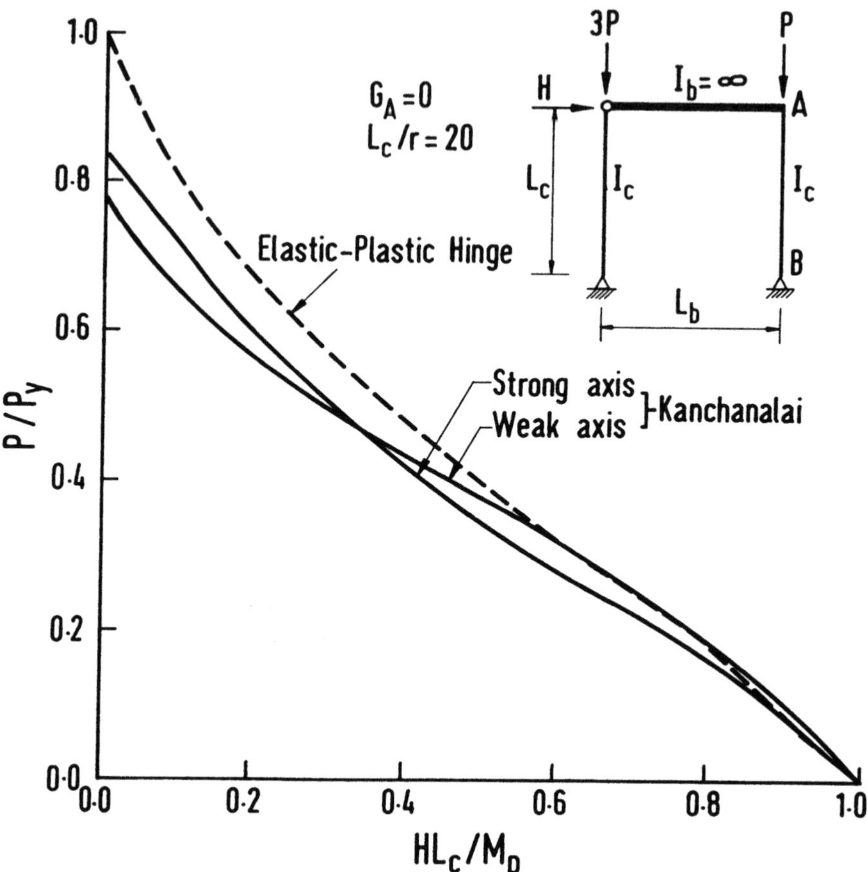

FIGURE 8.10. Comparison of strength curves for leaned column frame with $G_A = 0$ and $L_c/r = 20$.

weak-axis strengths are significantly overpredicted by the plastic hinge analyses for a wide range of moment versus axial force. Therefore, it is necessary to refine the conventional elastic-plastic hinge model to generalize its application for the analysis of a wider range of structural systems.

8.6 Desirable Attributes for Plastic-Hinge-Based Advanced Analysis

This section outlines the desirable attributes for development of an improved plastic-hinge-based-element.

1. The analysis should be capable of capturing the characteristic behavior of a wide range of structural systems and member types. For instance,

strong- and weak-axis bending action of rolled and welded beam-column sections, behavior of braced and unbraced beam columns, failure of members by predominately elastic or inelastic instability, and load-deflection and strength behavior of beam columns subjected to axial and transverse forces can all be represented with sufficient accuracy. The model should be capable of representing $P - \Delta$ and $P - \delta$ effects, including the instability effects from leaner columns. It can predict the response due to distributed plasticity associated with residual stresses, geometric imperfections at a wide range of axial force, and bending moment combinations.

2. The element model should capture, conservatively, the in-plane stability and strength of individual frame components in resisting load actions. The analysis should not more than 5% unconservative when compared to a wide range of second-order plastic-zone solutions for planar beam-column strength.

3. The response characteristics generated by the analysis model for various types of members and systems should be reasonably close to those predicted by a plastic-zone analysis.

4. The element force-displacement relationships should be derived analytically and implemented in explicit form for analysis. Residual stress effects can be accommodated implicitly using the tangent modulus model, and the effects of distributed plasticity on axial member deformations are represented.

5. The element formulation should reduce to a well-recognized behavior model in the limits of pure beam and pure column action. That is, when axial force approadches zero, the element ultimate strength behavior approaches that of the elastic-plastic hinge model. For the case of a member loaded by pure axial load, the element inelastic stiffness is close to that associated with the inelastic flexural rigidity $E_t I$ implied by the column strength equations of the AISC LRFD design specification [8.7].

6. The member forces calculated by the analysis model should not violate the cross-sectional plastic strength associated with the condition of full plasticity. Also, the model should accommodate changes in the values of the axial force and moment at a plastic hinge location. For example, if the axial force supported by a beam column increases after a plastic hinge has formed, the moment at the plastic hinge must decrease.

7. The possible benefits of strain hardening should not be relied on, since the ability of a beam column to develop significant strain hardening is dependent on many factors such as moment gradient and interaction of local and lateral-torsional buckling effects and distributed yielding along the member length. At the present time, the precise effects of yielding, strain hardening, and possible local and lateral torsional buckling on the full moment-rotation characteristics of beam-column members are still not well enough quantified for direct implementation in a practical inelastic analysis.

8. The model should be capable of achieving the requisite accuracy using

only one or a few elements per member. This attribute is necessary for analysis of moderate-size and large structural systems.

9. The model should represent the effects of distributed yielding on both the axial and flexural deformations. Framing members subjected to high axial forces and small bending moments, such as members in a truss-type system as well as in most tall building frames, may exhibit significant distributed yielding along their lengths prior to failure. The member effective axial stiffness provides a significant contribution to the response of these types of structures.

The next section presents a beam-column model that will satisfy most of the above attributes. The model is based on a fairly simple and elegant idealization of the physical behavior of beam-column response.

8.7 Approximate Effects of Distributed Yielding

As explained in Sections 8.4 and 8.5, the distributed plasticity effects associated with flexure must be captured to obtain an accurate physical model of beam-column behavior. This may be accomplished by representing the gradual degradation in stiffness that occurs at plastic hinge locations as yielding progresses under increasing moment when the cross-sectional strength is approached. Since the distributed yielding is influenced largely by member initial imperfections (member out-of-straightness and residual stresses) as well as by axial force, it is proposed that these distributed plasticity effects be modeled by an effective tangent-modulus approach for the calculation of column strength [8.30–8.31]. The detailed formulation of the model is described later.

8.7.1 Tangent-Modulus Approach

For columns subjected to pure compression, the AISC LRFD specification provides the following formula for the evaluation of axial compressive strength:

$$P = P_y(0.658^{\lambda_c^2}) \qquad \lambda_c \leq 1.5 \tag{8.7.1}$$

$$P = P_y\left(\frac{0.877}{\lambda_c^2}\right) \qquad \lambda_c > 1.5 \tag{8.7.2}$$

where P_y is the axial load at full yield, λ_c is the normalized column slenderness ratio defined by $\lambda_c = (KL/\pi r)\sqrt{F_y/E}$, F_y is the material yield stress, E is the modulus of elasticity, and KL/r is the effective slenderness ratio. The column tangent modulus, E_t, can be evaluated based on the inelastic stiffness reduction procedure given in the AISC LRFD manual for the calculation of inelastic column strength. The ratio of the tangent modulus to the elastic modulus, E_t/E, is defined as [8.30]:

$$\frac{E_t}{E} = \begin{cases} 1.0 & \text{for } P \le 0.39P_y \\ -2.7243\,\dfrac{P}{P_y}\ln\left[\dfrac{P}{P_y}\right] & \text{for } P > 0.39P_y. \end{cases} \qquad (8.7.3)$$

Since this E_t model is derived from the LRFD column strength formula, it implicitly includes the effects of residual stresses and initial out-of-straightness in modeling the member effective stiffness.

The column tangent modulus also can be evaluated based on the CRC column equations [8.32]. The E_t/E expressions may be written as:

$$\frac{E_t}{E} = \begin{cases} 1.0 & \text{for } P \le 0.5P_y \\ \dfrac{4P}{P_y}\left[1 - \dfrac{P}{P_y}\right] & \text{for } P > 0.5P_y. \end{cases} \qquad (8.7.4)$$

The axial force-displacement relationships of an axially loaded inelastic element can be derived using Eqs. (8.7.3) and (8.7.4); the results are shown in Fig. 8.11. It should be noted that the normalized axial force-displacement relationships of an element described by the tangent moduli equations are nonlinear in nature, whereas for the conventional elastic-plastic hinge model, a linear elastic–perfectly plastic normalized axial force-displacement relationship is tacitly assumed. The main difference between the CRC-E_t and the LRFD-E_t is that the former considers only the residual stress effects in mod-

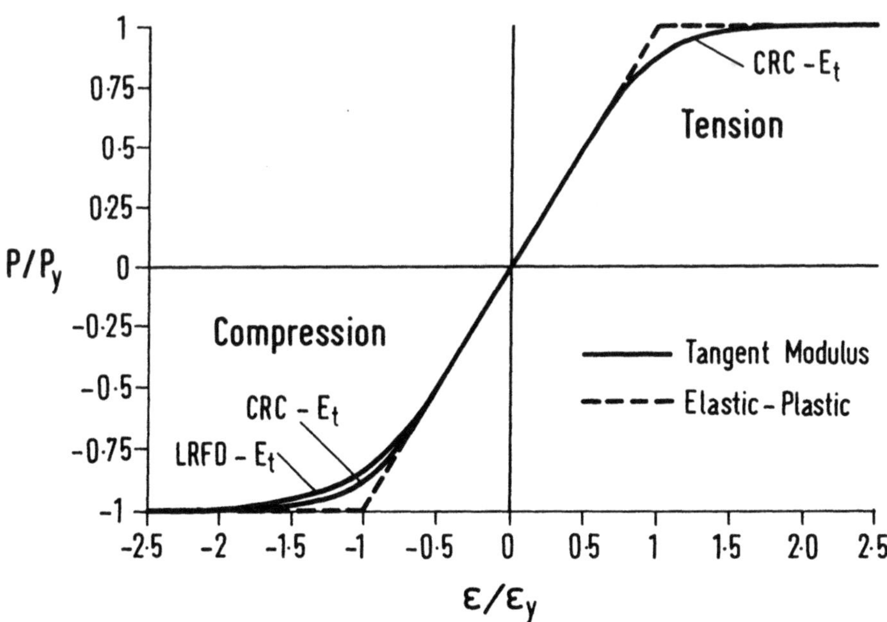

FIGURE 8.11. Axial force-strain relation.

eling the column effective stiffness, whereas the latter is based on LRFD column strength equations that account for the effects of both geometric imperfections and residual stresses.

Recent research [8.6, 8.31] found that for members subjected to significant in-plane bending, the modeling of the member stiffness equation using only the tangent-modulus approach is not sufficient to represent the gradual stiffness degradation as yielding progress through the volume of the member. Additional distributed plasticity effects in the beam-column member can be attributed to the bending action. These effects may be represented by modifying the basic elastic-plastic hinge model such that the member stiffness degrades gradually from the stiffness associated with the onset of yielding to that associated with the formation of plastic hinges. This approach, which is called the refined plastic hinge method, has been shown to be superior than the conventional elastic-plastic hinge approach in representing the inelastic behavior of many types of frame structures.

8.7.2 Effects of Plastification at Element Ends

Consider a beam-column member in which plastification may occur at both ends, the incremental force-displacement relationships may be written as [8.30]:

$$
\begin{Bmatrix} \dot{M}_A \\ \dot{M}_B \end{Bmatrix} = \frac{E_t I}{L} \begin{bmatrix} \phi_A \left(S_1 - \dfrac{S_2^2}{S_1}[1 - \phi_B] \right) & \phi_A \phi_B S_2 \\ \phi_A \phi_B S_2 & \phi_B \left(S_1 - \dfrac{S_2^2}{S_1}[1 - \phi_A] \right) \end{bmatrix} \begin{Bmatrix} \dot{\theta}_A \\ \dot{\theta}_B \end{Bmatrix}. \quad (8.7.5)
$$

The terms S_1 and S_2 are the conventional beam-column stability functions, with E_t used in place of the elastic modulus. The term ϕ (subscripts A and B denote the respective ends of the element) is a scalar parameter that allows for gradual inelastic stiffness reduction of the element associated with the effect of plastification at the element end. This term is equal to one when the element is elastic, and it is zero when a plastic hinge is formed at the end. The parameter ϕ is assumed to vary according to the following parabolic function [8.30]:

$$
\phi_i = \begin{cases} 4\alpha_i(1 - \alpha_i) & \text{for } \alpha_i > 0.5 \\ 1 & \text{for } \alpha_i \leq 0.5 \end{cases} \quad (8.7.6)
$$

where α is a force-state parameter that measures the magnitude of axial force (P) and bending moment (M) at the element end. The term α is expressed as:

$$
\alpha = \frac{P}{P_y} + \frac{8}{9}\frac{M}{M_p} \quad \text{for } \frac{P}{P_y} \geq \frac{2}{9}\frac{M}{M_p} \quad (8.7.7)
$$

$$
\alpha = \frac{P}{2P_y} + \frac{M}{M_p} \quad \text{for } \frac{P}{P_y} < \frac{2}{9}\frac{M}{M_p}. \quad (8.7.8)
$$

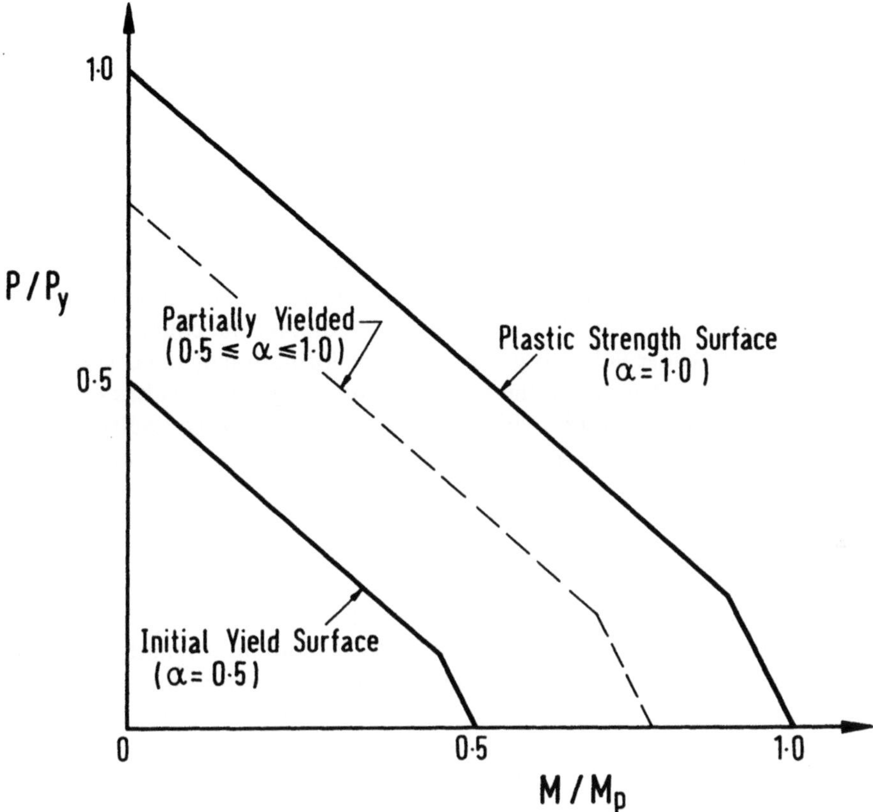

FIGURE 8.12. Two-surface stiffness degradation model for refined plastic hinge analysis.

Initial yielding is assumed to occur based on a yield surface that has the same shape as the plastic strength surface, as shown in Fig. 8.12. In this figure, the yield surface function corresponding to $\alpha = 1.0$ represents the state of forces at which the cross section is fully yielded. The yield surface function corresponding to $\alpha = 0.5$ represents the state of forces at which initial yielding starts to occur. If the state of forces is changed in such a manner that the force point moves inside or along the initial yield surface, the element is fully elastic with no stiffness reduction. If the force point moves beyond the initial yield surface, the element stiffness is reduced to account for the effect of plastification at the element end.

It should be noted that when $E_t = E$, and $\phi_A = \phi_B = 1$, the element is fully elastic. Equation (8.7.5) reduces to the conventional tangent-stiffness equations as shown here:

$$\begin{Bmatrix} \dot{M}_A \\ \dot{M}_B \end{Bmatrix} = \frac{EI}{L} \begin{bmatrix} S_1 & S_2 \\ S_2 & S_1 \end{bmatrix} \begin{Bmatrix} \dot{\theta}_A \\ \dot{\theta}_B \end{Bmatrix}. \tag{8.7.9}$$

In the refined plastic hinge formulation, the AISC LRFD beam-column interaction equations for member of zero length are adopted as the cross-sectional plastic strength equations for representing both the strong- and weak-axis in-plane strength behavior. This cross-sectional plastic strength may be reduced by the appropriate resistance factors for use in limit-states design. Different resistance factors for flexural bending and axial compression, such as those currently used in AISC LRFD for beam-column design, may also be used in the refined plastic hinge analysis, if required.

The refined hinge analysis enables the prediction of a column strength that is close to that implied by the code requirements for column design. Also, the refined plastic hinge model is successful in predicting the collapse load of a beam member whose failure is governed by the formation of a plastic collapse mechanism. The accuracy of the refined plastic hinge analysis has been verified by comparing the results with the more exact plastic-zone results reported in [8.6, 8.31]. Some results from the studies are given in the following sections.

8.7.3 *Isolated Columns and Beam Columns*

Figure 8.13 compares the inelastic column strength curve generated by the refined hinge analyses with the AISC LRFD column strength curve. In the refined plastic hinge analysis, the column is discretized into two beam-column elements. A W8 × 31 column section bending about the strong axis is assumed, and the column initial out-of-straightness with a maximum mid-length deflection equal to $L/1500$ is explicitly modeled. This out-of-straightness magnitude is equivalent to that assumed by the LRFD column strength curve.

The comparison shows that the column strength curve generated by the refined plastic hinge analysis is not more than 5% unconservative for columns with slenderness parameters up to $\lambda_c = 1.5$.

It is important to emphasize that the conventional elastic-plastic hinge method overpredicts the column strength up to a maximum error of 22.5%. This error, however, reduces to less than 5% when the refined plastic hinge method is used.

Figure 8.14 shows the beam-column strength curves generated by the refined plastic hinge analysis. The results are compared with those obtained from the elastic-plastic and plastic zone analyses. The plastic-hinge-based results in Fig. 8.14 were generated based on the use of two elements per member, and the analyses explicitly model the member initial out-of-straightness, which has a maximum magnitude of $\delta_o = L/1000$ at the mid-length. The refined plastic hinge model is based on the use of the CRC column tangent modulus since member initial out-of-straightness is modeled explicitly in the analysis. It can be observed that the refined plastic hinge approach is successful in predicting the limit strengths of the beam columns with maximum error not more than 7%. Additional studies for beam columns

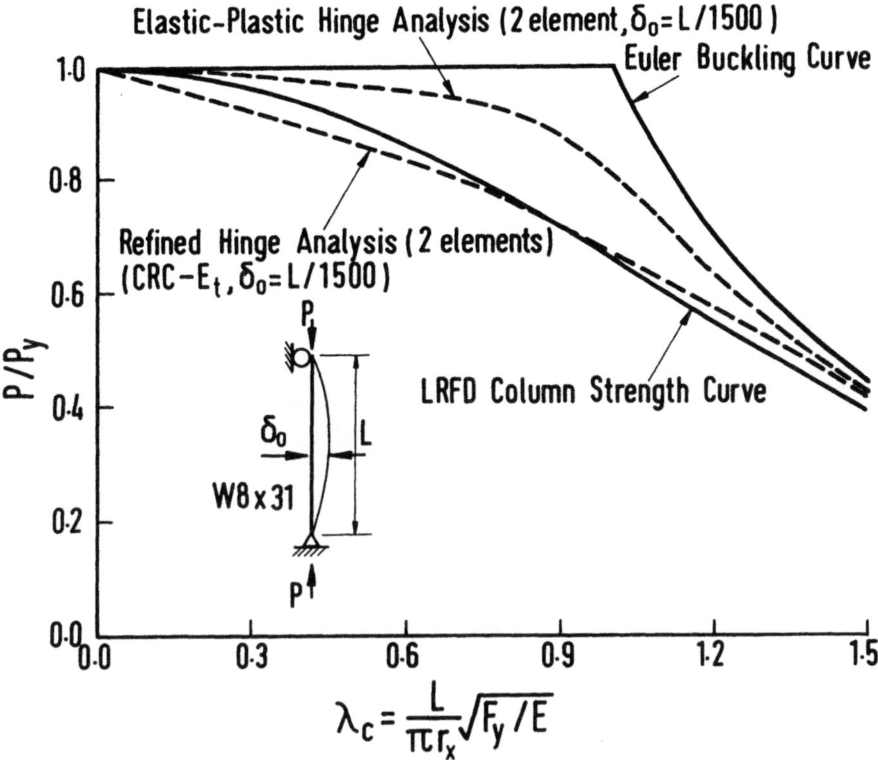

FIGURE 8.13. Comparison of column strength curves.

subjected to unequal end moments producing reverse-curvature bending are also reported in Ret. [8.6].

8.7.4 Portal Frame

A portal sway frame is presented here to show the improvements in the predicted behavior obtained using the refined plastic hinge analysis. The portal frame in Fig. 8.15 consists of a rigid girder and two columns that are pin-supported at the bases. This frame is analyzed by Kanchanalai [8.28] using a plastic-zone model. The gravity loads are applied first, followed by a lateral load H, which is increased until the failure of the frame. The results are plotted for different combinations of gravity load versus lateral load. It can be observed that the proposed model gives a closer strength prediction than the more accurate plastic-zone method. The significance of the proposed refinements is more pronounced when the column axial load is high. It should be noted that this portal frame example is rather critical in testing the accuracy of second-order inelastic analyses.

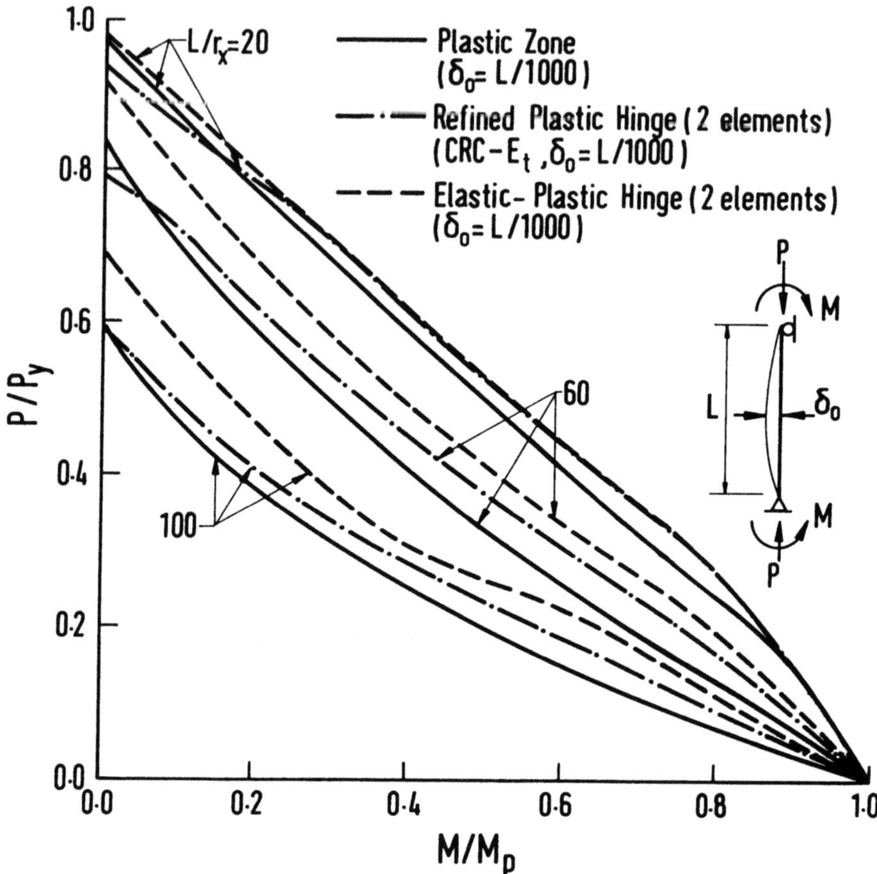

FIGURE 8.14. Comparison of beam-column strength curves.

8.7.5 Six-Story Sway Frame

A six-story three-bay frame has been proposed by Vogel [8.33] as a calibration frame for nonlinear inelastic analysis. This example is cited to provide a better comparison of the various methods in representing the yield effects. The frame (Fig. 8.16) consists of initial out-of-plumbness associated with the value recommended by EC 3 [8.14]. Both gravity and lateral loads are applied proportionally until failure occurs. The applied load versus top and fourth-story lateral displacement curves are shown in Fig. 8.16.

For this frame, all the inelastic analysis methods predict essentially the same limit load. The maximum frame resistance is reached at a load parameter of 1.111 in Vogel's plastic-zone study, at 1.118 for the refined plastic hinge analysis, and 1.124 for the elastic-plastic hinge analysis. The maximum difference between these limit loads is less than 2%. These results conclude that

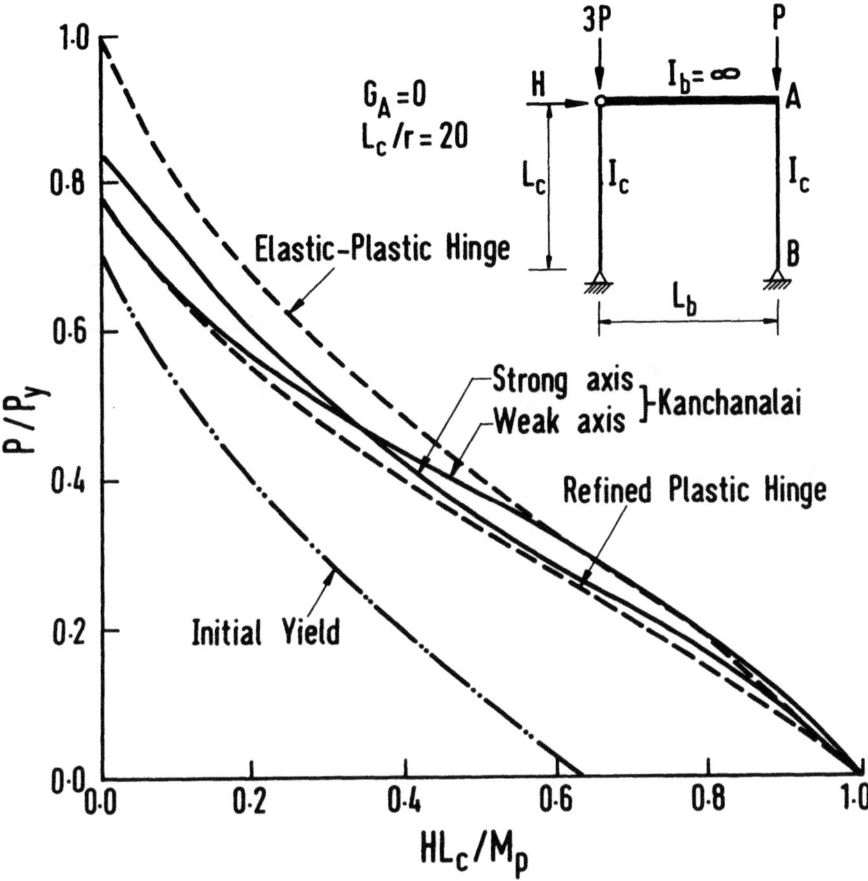

FIGURE 8.15. Comparison of strength curves for portal frame.

when the overall nonlinear behavior of the frame is dominated by inelastic action in the beams, the plastic-hinge-based models generally give sufficient representation of the overall frame behavior. For high-rise building frames, both plastic hinge analyses should compare well with the plastic-zone method as shown by the closeness of all three curves in Fig. 8.16. The inability of the conventional method to represent member strength (such as for the frame shown in Fig. 8.16) can be attributed as "local effects." That is, the conventional plastic hinge method is accurate in predicting the system response, but may not be accurate enough in predicting the strength and stability of isolated beam-column elements or subassemblies [8.31].

Bending moments at selected locations in the frame, computed based on the three inelastic analyses at the frame's limit of resistance, are compared in Fig. 8.17. The plastic-zone solutions shown are based on Ziemian's work

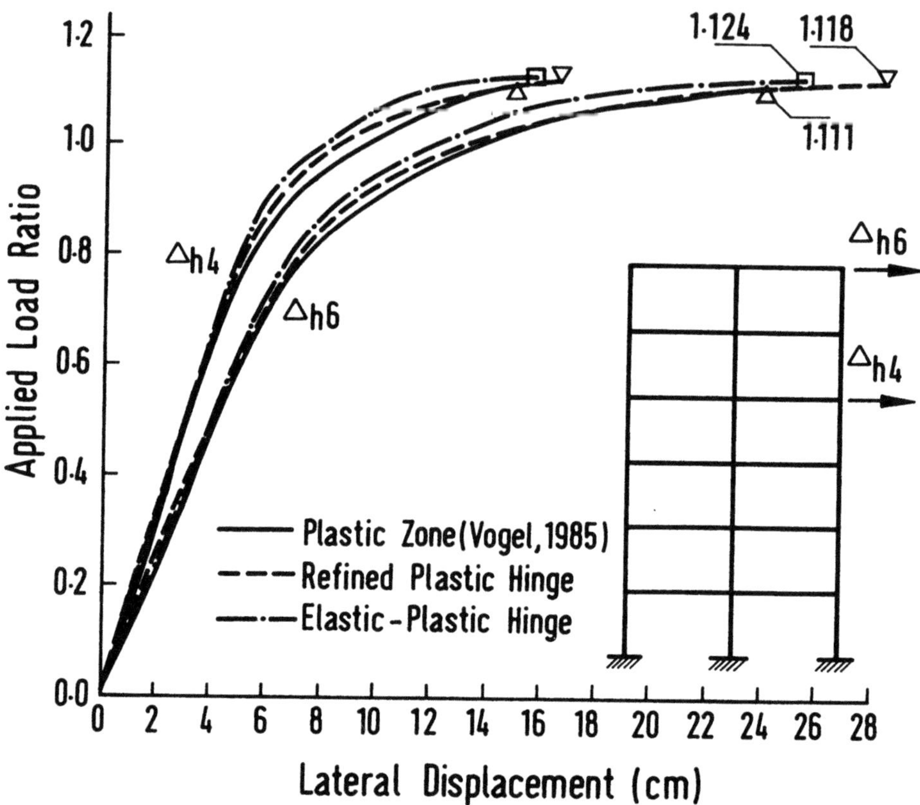

FIGURE 8.16. Comparison of load-displacement curves for Volgel six-story frame.

[8.34] in which the frame's limit of resistance is reached at a load factor of 1.180, which is slightly higher than the limit load predicted by Vogel's plastic-zone analysis. The distribution of forces predicted by all the inelastic analyses are remarkably close. Included in Fig. 8.17 are the locations of plastic hinges detected by the refined plastic hinge and elastic-plastic hinge analyses. The first plastic hinge predicted by the elastic-plastic hinge analysis occurs at an applied load ratio of 0.770, whereas in the refined plastic hinge analysis, no plastic hinge is observed until the applied load ratio reaches a value of 0.873. The elastic-plastic hinge analysis detects a total of 24 plastic hinges in comparison with the refined plastic hinge approach, which detects only 15 plastic hinges in the frame. However, several locations that are not registered as plastic hinges in the refined hinge analysis have ϕ values almost equal to zero. These ϕ values may be used to indicate the degrees of yielding at the critical locations in the frame; $\phi = 1$ indicates a fully elastic state, whereas $\phi = 0$ signifies a fully yielded state. Detailed comparisons of yielded zones predicted by the refined plastic hinge and plastic-zone analyses are given in Ref. [8.6].

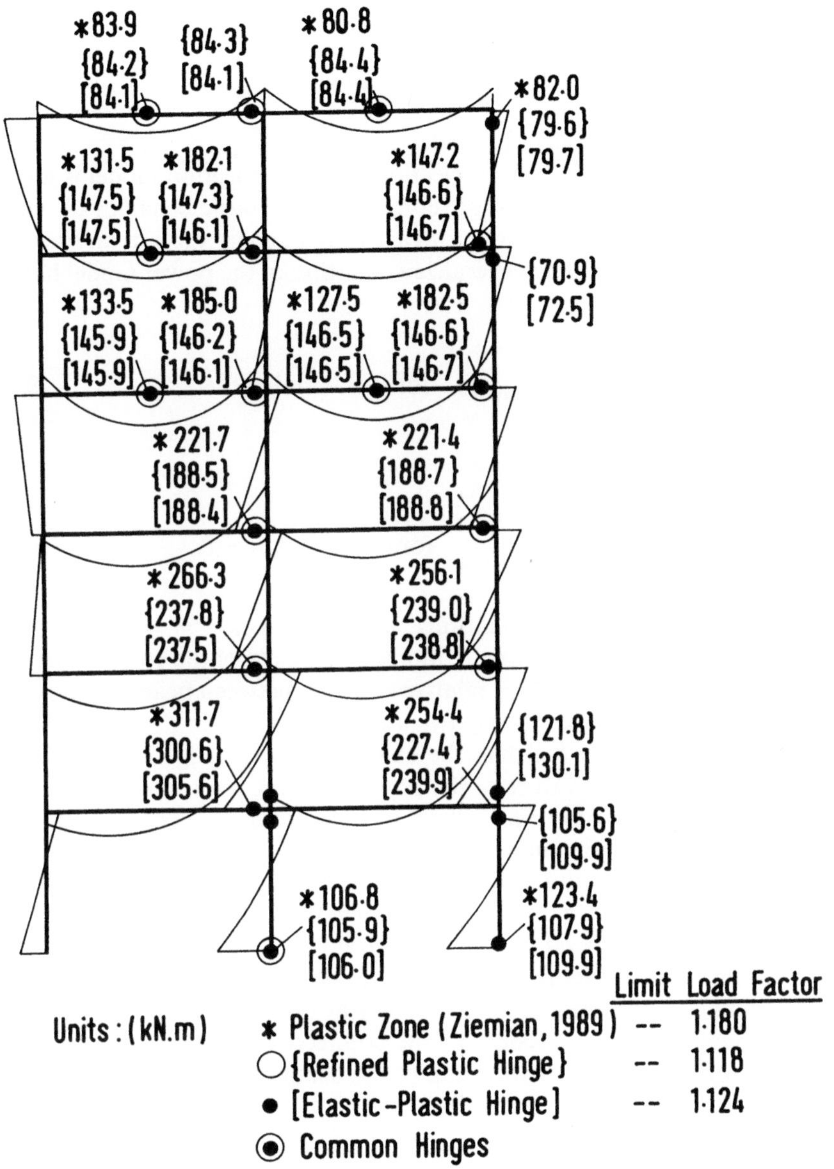

FIGURE 8.17. Comparison of internal force distribution and plastic hinge locations.

8.8 Modeling of Semirigid Frames

8.8.1 Connection Element

A three-parameter connection model [8.35] is adopted for second-order analysis of semirigid frames. The selection of the connection model is guided by its simplicity and robustness for representing the basic behavior of typical beam-to-column connections and for ease of implementation in any second-order analysis program. The generalized equation of the power model has the form:

$$m = \frac{\theta}{(1 + \theta^n)^{1/n}} \quad \text{for } \theta > 0 \text{ and } m > 0 \tag{8.8.1}$$

or equivalently

$$\theta = \frac{m}{(1 - m^n)^{1/n}} \quad \text{for } \theta > 0 \text{ and } m > 0. \tag{8.8.2}$$

The parameters in these equations are defined as: $m = M/M_u$; $M =$ connection moment; $M_u =$ ultimate moment capacity of the connection; $n =$ shape parameter; $\theta = \theta_r/\theta_o$; $\theta_r =$ relative rotation between beam and column; $\theta_o =$ reference plastic rotation, M_u/R_{ki}; and $R_{ki} =$ initial connection stiffness.

Equation (8.8.1) or (8.8.2) has the shape shown in Fig. 8.18. It can be observed that for larger values of the power index n, the transition from the initial stiffness, R_{ki}, to the final curve of maximum moment, M_u, is more abrupt. If n is infinity, the model becomes a bilinear curve that has an initial connection stiffness R_{ki} and ultimate moment capacity M_u.

In the analysis, the connections are modeled as rotational springs having the $M - \theta_r$ curves described by the three-parameter model shown in Fig. 8.18. These springs are physically tied to the ends of the beam column (Fig. 8.19) by enforcing equilibrium and compatibility at the connecting ends. Finally, the rotational degrees of freedom of the connections can be condensed out from the tangent-stiffness relationship of the beam-column element.

8.8.2 Tangent-Stiffness Formulation

The connection tangent stiffness R_{kt} at any arbitrary rotation $|\theta_r|$ can be evaluated by differentiating M with respect to $|\theta_r|$ from Eq. (8.8.1). The resulting connection tangent stiffness is [8.23, 8.36]

$$R_{kt} = \frac{dM}{d|\theta_r|_{|\theta_r|}} = \frac{M_u}{\theta_o(1 + \theta)^{1+1/n}} \tag{8.8.3}$$

when the connection is loaded, and

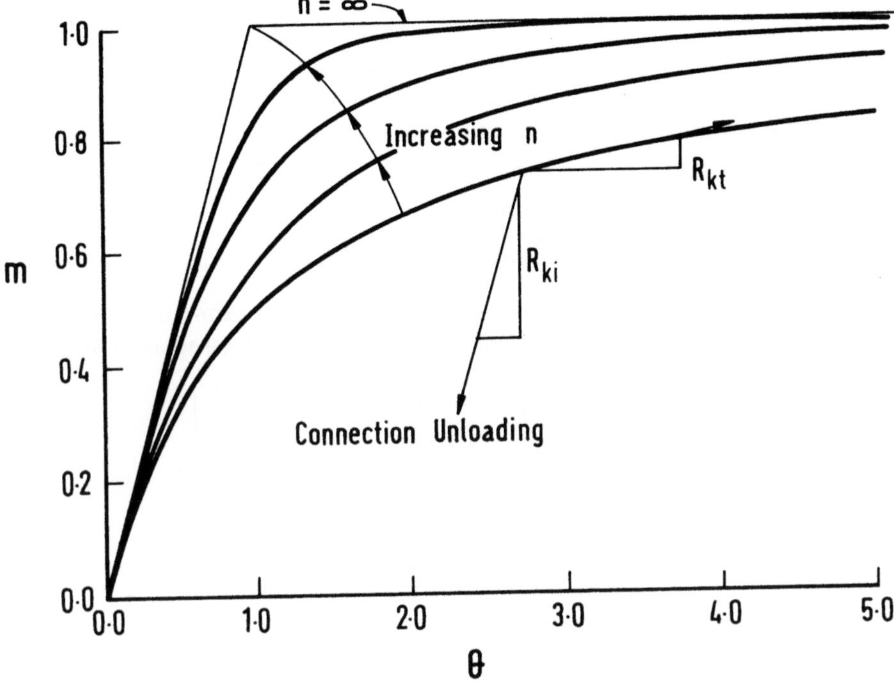

FIGURE 8.18. Moment-rotation curves of semirigid connections.

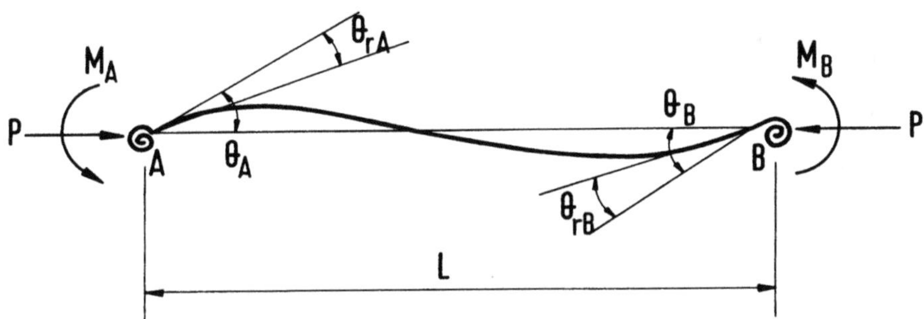

FIGURE 8.19. Beam column with semirigid connections attached at both ends.

$$R_{kt} = \frac{dM}{d|\theta_r|_{|\theta_r=0|}} = \frac{M_u}{\theta_o} = R_{ki} \qquad (8.8.4)$$

when the connection is unloaded.

The stiffness matrix of a beam-column element with semirigid connections attached at its ends may be derived based on the modified beam-column stability function approach [8.17]. Consider a beam column subjected to end moments (M_A and M_B) and axial force (P) with connections attached at both ends as shown in Fig. 8.19. The presence of connections introduces relative incremental rotations of $\dot{\theta}_{rA}$ and $\dot{\theta}_{rB}$ at the Ath and Bth end of the member, respectively. Denoting R_{ktA} and R_{ktB} as the tangent stiffness of connection A and B, respectively, the relative incremental rotations between the joints and beam ends (i.e., the incremental rotational deformations of the connections) can be expressed as:

$$\dot{\theta}_{rA} = \frac{\dot{M}_A}{R_{ktA}} \qquad \dot{\theta}_{rB} = \frac{\dot{M}_B}{R_{ktB}}. \tag{8.8.5}$$

The incremental force-displacement equations for the beam-column element modified for the presence of end connections can be expressed as [8.36]:

$$\dot{M}_A = \frac{EI}{L}\left[S_{ii}\left\{ \dot{\theta}_A - \frac{\dot{M}_A}{R_{ktA}} \right\} + S_{ij}\left\{ \dot{\theta}_B - \frac{\dot{M}_B}{R_{ktB}} \right\} \right] \tag{8.8.6}$$

$$\dot{M}_B = \frac{EI}{L}\left[S_{ij}\left\{ \dot{\theta}_A - \frac{\dot{M}_A}{R_{ktA}} \right\} + S_{jj}\left\{ \dot{\theta}_B - \frac{\dot{M}_B}{R_{ktB}} \right\} \right] \tag{8.8.7}$$

where S_{ii}, S_{ij}, S_{jj} are defined in the following manner.

1. If elastic or elastic-plastic hinge assumption is assumed, then

$$S_{ii} = S_{jj} = S_1 \quad \text{and} \quad S_{ij} = S_2 \tag{8.8.8}$$

where S_1 and S_2 are the stability functions as discussed in Section 8.2.

2. If gradual stiffness degradation due to spread-of-plasticity effects in the element is considered, then [8.23]

$$S_{ii} = \left(S_1 - \frac{S_2^2}{S_1}[1 - \phi_B] \right)\phi_A \tag{8.8.9}$$

$$S_{jj} = \left(S_1 - \frac{S_2^2}{S_1}[1 - \phi_A] \right)\phi_B \tag{8.8.10}$$

$$S_{ij} = \phi_A\phi_B S_2 \tag{8.8.11}$$

and the elastic modulus, E, in Eqs. (8.8.6) and (8.8.7) should be replaced by the tangent modulus E_t.

Solving Eqs. (8.8.6) and (8.8.7) for $\dot{M}_A$ and $\dot{M}_B$ gives

$$\dot{M}_A = \frac{E_t I}{L}[S_{ii}^*\dot{\theta}_A + S_{ij}^*\dot{\theta}_B] \tag{8.8.12}$$

$$\dot{M}_B = \frac{E_t I}{L}[S_{ij}^*\dot{\theta}_A + S_{jj}^*\dot{\theta}_B] \tag{8.8.13}$$

where

$$S_{ii}^* = \left(S_{ii} + \frac{E_t IS_{ii}S_{jj}}{LR_{ktB}} - \frac{E_t IS_{ij}^2}{LR_{ktB}}\right)\Big/ R^*$$

$$S_{jj}^* = \left(S_{jj} + \frac{E_t IS_{ii}S_{jj}}{LR_{ktA}} - \frac{E_t IS_{ij}^2}{LR_{ktA}}\right)\Big/ R^* \qquad (8.8.14)$$

$$S_{ij}^* = S_{ij}/R^*$$

and

$$R^* = \left(1 + \frac{E_t IS_{ii}}{LR_{ktA}}\right)\left(1 + \frac{E_t IS_{jj}}{LR_{ktB}}\right) - \left(\frac{E_t I}{L}\right)^2 \frac{S_{ij}^2}{R_{ktA}R_{ktB}} \qquad (8.8.15)$$

The incremental axial force-displacement equation is

$$\dot{P} = \frac{E_t A}{L}\dot{e}. \qquad (8.8.16)$$

Finally, the tangent-stiffness relationship may be written as:

$$\begin{Bmatrix} \dot{M}_A \\ \dot{M}_B \\ \dot{P} \end{Bmatrix} = \frac{EI}{L} \begin{bmatrix} S_{ii}^* & S_{ij}^* & 0 \\ S_{ij}^* & S_{jj}^* & 0 \\ 0 & 0 & A/I \end{bmatrix} \begin{Bmatrix} \dot{\theta}_A \\ \dot{\theta}_B \\ \dot{e} \end{Bmatrix}. \qquad (8.8.17)$$

Equation (8.8.17) is the modified element tangent-stiffness equation that accounts for the effects of both distributed plasticity and connection flexibility on the element. They can be transformed to global coordinates to obtain the global element tangent-stiffness relationships.

8.8.3 Analysis/Design of a Semirigid Frame

Figure 8.20 shows a frame example with prescribed configuration, dimensions, and nominal loads. The gravity loads are assumed to include the contributions from both dead and live loads. It is further assumed that the dead and live loads are of equal proportion, and the column sizes are kept the same for the top and bottom stories of the frame. For this two-story frame example, the preliminary beam sizes were chosen based on first-order plastic hinge analysis assuming rigid frame action. The preliminary column sizes were selected according to the tributary gravity load supported by each column. After a few cycles of design iterations using refined plastic hinge analyses, the least-weight members satisfying the strength limit states are obtained. The member sizes are shown in Fig. 8.20.

The details of the connection assemblage selected for the frame are shown in Fig. 8.21. The resulting geometrical dimensions of a set of possible semirigid connections are summarized in Table 8.1. The connection type is denoted in the table by the thickness of the top and seat angle as, for example, C-3/8. The initial stiffnesses and ultimate moment capacities of these connections are shown in Table 8.2.

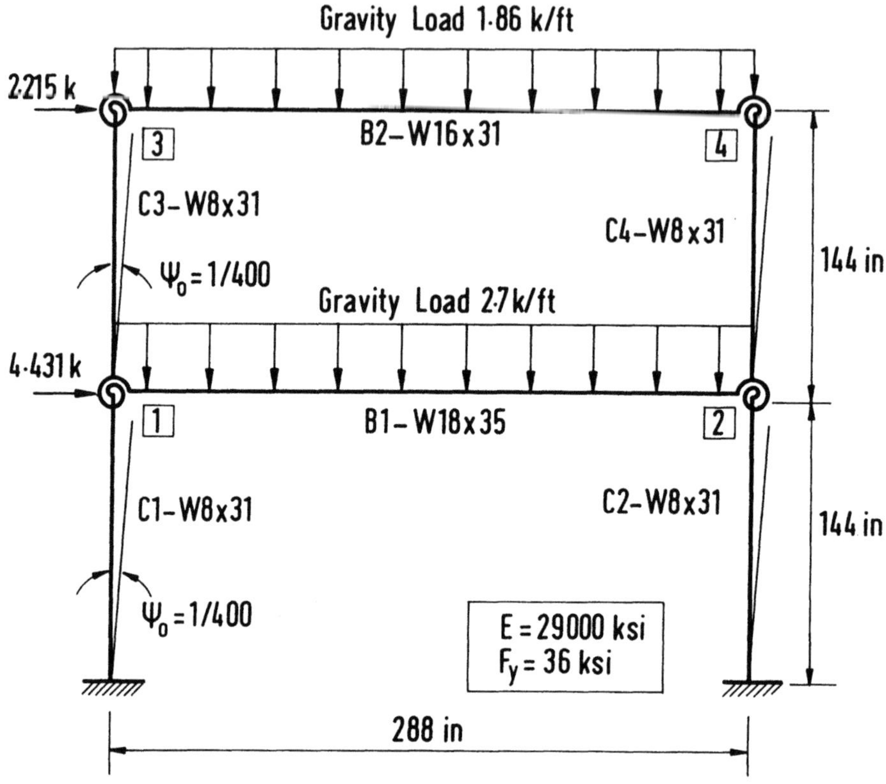

FIGURE 8.20. Dimensions and loading of a two-story frame.

Design Requirements: The frame is designed to satisfy two factored load combinations:

$$1.4 \text{ gravity} \tag{8.8.18}$$

$$1.0 \text{ gravity} + 1.3 \text{ wind.} \tag{8.8.19}$$

The refined plastic hinge analysis is carried out based on the factored cross-sectional strengths ϕP_y and ϕM_p with the LRFD-based tangent modulus [Eq. (8.7.3)] used in place of the elastic modulus. The ϕ factor used for the cross-sectional design strengths is 0.90, which is the same as that recommended by the Australian limit-states (AS4100, 1990) for use in the advanced analysis of structural steel frames. Since the LRFD tangent-modulus model is used for advanced inelastic analysis, the modeling of member initial out-of-straightness is not required. This approach simplifies the analysis/design procedures considerably, since explicit modeling of column initial out-of-straightness can be a cumbersome task for large structural systems. However, it should be noted that all the columns in the frame shown in Fig. 8.20 are not affected by the column tangent-modulus model. This is because the columns' axial forces at the factored load level are small.

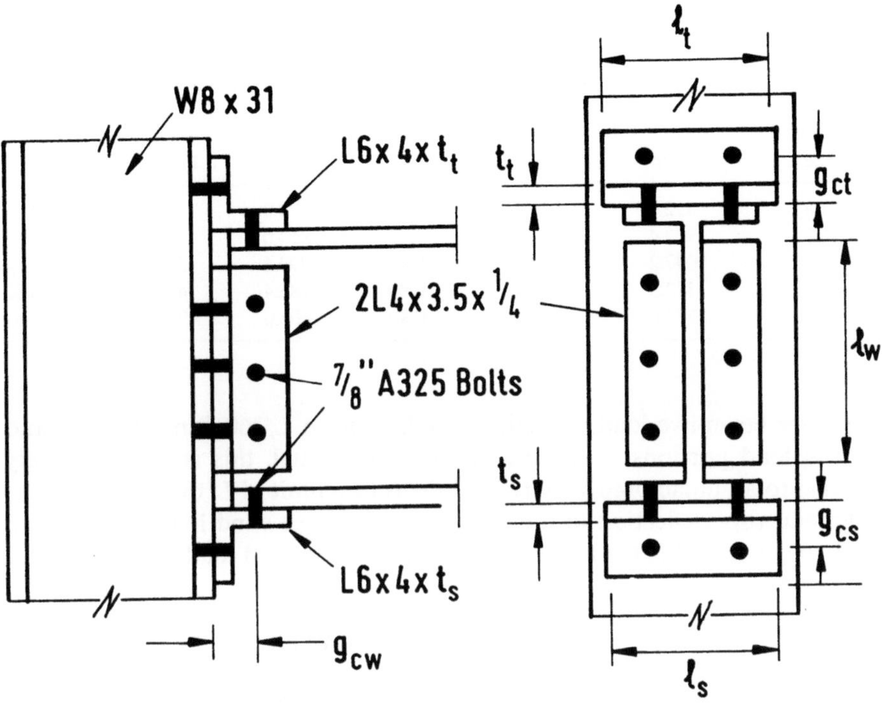

FIGURE 8.21. The connection details.

TABLE 8.1. Connection details.

Connection type	Bolt diameter, in.	Nut width, W (in.)	Top and seat angle $2L6 \times 4 \times t_t$				Web angles $L4 \times 3.5 \times 1/4$		
			t_t, in.	l_t, in.	g_{ct}, in.	k_t, in.	g_{cw}, in.	l_w, in.	k_w, in.
C-3/8	7/8	1 + 7/16	3/8	7	3	7/16	3	8	11/16
C-1/2	7/8	1 + 7/16	1/2	7	3	1	3	8	11/16
C-9/16	7/8	1 + 7/16	9/16	7	3	1 + 1/16	3	8	11/16
C-5/8	7/8	1 + 7/16	5/8	7	3	1 + 1/8	3	8	11/16
C-3/4	7/8	1 + 7/16	3/4	7	3	1 + 1/4	3	8	11/16

All columns in the frame are assumed to have an initial out-of-plumbness of $h/400$, where h is the story height. This column out-of-plumbness magnitude is obtained from ECCS provisions for use with second-order plastic-zone analysis of steel frames.

In additional to strength checks, the following serviceability requirements are also imposed:

TABLE 8.2. Connection parameters.

Connection type	Beam W18 × 35 (beam depth = 17.7 in.)			Beam W16 × 31 (beam depth = 15.8 in.)		
	M_u, kip-in	R_{ki}, kip-in/rad	n	M_u, kip-in	R_{ki}, kip-in/rad	n
C-3/8	532.0	107,548	2.10	473.0	86,980	2.23
C-1/2	903.4	254,227	1.32	814.0	205,924	1.57
C-9/16	1097.2	369,595	0.89	990.4	299,599	1.14
C-5/8	1312.5	520,528	0.80	1186.4	422,267	0.80
C-3/4	1773.0	954,013	0.80	1606.0	775,080	0.80

- Service beam live load deflections are limited to $L/360$. Since the live and dead load portions of the gravity load are equal, this is approximately equal to an unfactored gravity load deflection limit of $L/180$.
- Lateral roof deflection of the frame under service wind load is limited to $H/400$, where H is the frame height. The service wind load combination is given as

$$\text{Dead} + 0.2 \text{ Live} + 1.0 \text{ Wind.} \qquad (8.8.20)$$

Since the live and dead load portions of the gravity load are equal, the load combination in Eq. (8.8.20) may be rewritten as:

$$0.6 \text{ Gravity} + 1.0 \text{ Wind.} \qquad (8.8.21)$$

- Maximum interstory drift due to the service load combination in Eq. (8.8.21) is limited to $h/250$, where h is the story height.
- Plastic hinge formation under service loads is prohibited to prevent the possibility of excessive localized damage before the full service load is reached.

All the above serviceability requirements are checked using the refined plastic hinge method described in Section 8.7.

Summary of Result: Results showing the characteristics of the two-story frame with different connection configurations are summarized in Tables 8.3 through 8.5. For comparison, the pinned and rigid connection cases are also included in the studies.

Applied Load Ratios: Table 8.3 shows the load factor at the formation of the first plastic hinge and at the limiting strength. The load factor is expressed as a ratio of the applied load to the full factored load. For all the semirigid frames (denoted by the connection types), the first plastic hinge forms after the application of the service load. More specifically, for the gravity-loaded frames, the first plastic hinge forms shortly after the application of the service load. For the wind load combination, all the semirigid frames achieve load

TABLE 8.3. Applied load ratios for frames with semirigid connections.

| Connection type | 1.4 × Gravity | | 1.0 Gravity × 1.3 Wind |
	First plastic hinge	Limit load	Limit load
Pin	——	0.659	0.796
C-3/8	0.807	0.933	1.129
C-1/2	0.890	1.057	1.259
C-9/16	0.900	1.111	1.302
C-5/8	0.929	1.134	1.354
C-3/4	1.021	1.152	1.436
Rigid	1.143	1.157	1.464

Service load = 1.0 × gravity load = 0.714.

TABLE 8.4. Maximum beam deflections under service load (1.0 × gravity).

Connection type	Critical member	Deflection (in.)
Pin	——	——
C-3/8	B1–W18 × 36	1.031
C-1/2	B1–W18 × 36	0.88
C-9/16	B1–W18 × 36	0.87
C-5/8	B1–W18 × 36	0.83
C-3/4	B1–W18 × 36	0.73
Rigid	B2–W16 × 31	0.66

Deflection at service load is limited to $L/180 = 1.6$ in.

TABLE 8.5. Maximum lateral drifts under service load (1.0 wind + 0.6 gravity).

Connection type	Frame drift (in.)	Interstory drift (in.)
Pin	1.742	1.478
C-3/8	1.226	0.618
C-1/2	0.896	0.490
C-9/16	0.885	0.488
C-5/8	0.824	0.460
C-3/4	0.669	0.399
Rigid	0.531	0.342

1. Maximum frame drift is limited to $H/400 = 0.72$ in.
2. Maximum interstory drift is limited to $h/250 = 0.576$ in.

factors greater than 1.0 prior to the first occurrence of a plastic hinge in any of the members.

Maximum Beam Deflections: The maximum beam deflections under service gravity load (1.0 × gravity) are shown in Table 8.4. For all cases, the maximum beam deflection is significantly lower than the deflection limit criteria of $L/180$. Therefore, the beam serviceability limit is not the controlling factor for the frame.

Maximum Drifts: In the advanced analyses, proportional loading is used where the loads are applied incrementally. The maximum frame and inter-story drifts under service load combination, 0.6 gravity + 1.0 wind, are evaluated as shown in Table 8.5. The results indicate that the frames with semirigid connections have maximum frame drifts in the order of 1.26 to 2.09 times that for the frame with rigid connections. Also, the interstory drift increases as the connection thicknesses of the top and seat angles reduce from 5/8 to 3/8 inch. Finally, except for the frame with pinned connections, all other semirigid frames satisfy the interstory drift criterion of $h/250$ imposed by the design requirements. Only the frames with connection rigidity greater than or equal to that implied by C-3/4 satisfy the drift limit of $H/400 = 0.72$ inch.

Final Designd: Although the inelastic analyses show that most of the semi-rigid connections (except C-3/8) can resist the factored load effects, the frame drifts and deformation capacity of several connections violate the design

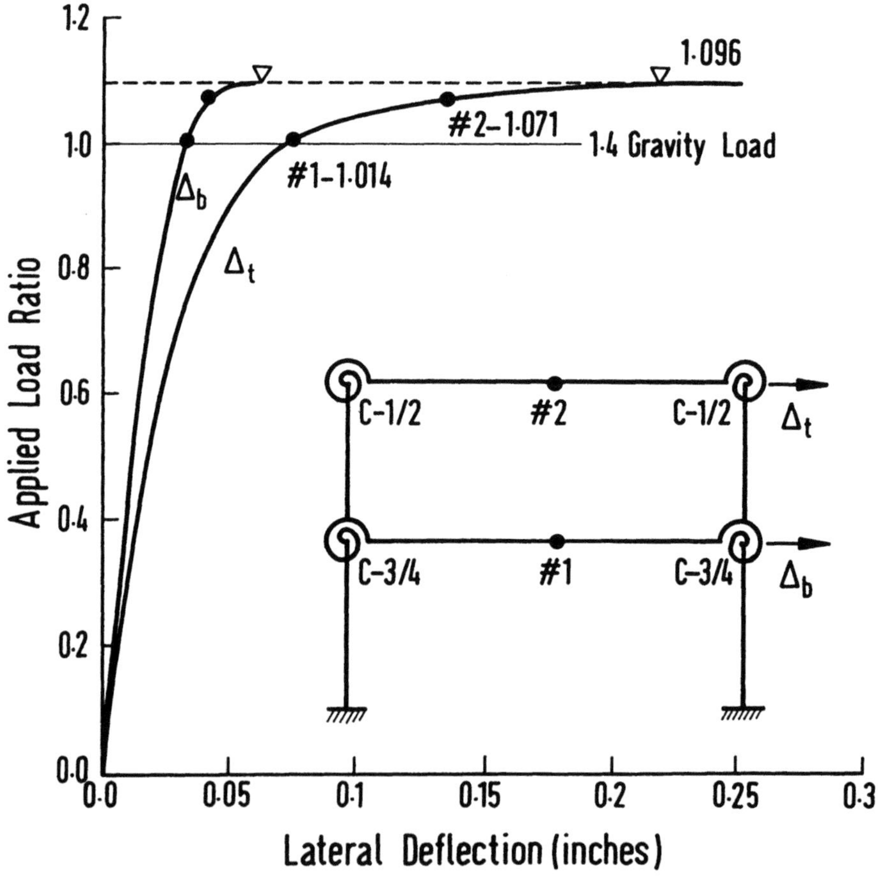

FIGURE 8.22. Lateral load-displacement traces.

requirements. To fulfill all the design requirements, it is decided to redesign the connections by satisfying, particularly, the serviceability requirements. The lightest connections in Table 8.1 that meeting the capacity requirements for the connections in beams B1 and B2 are C-3/4 and C-1/2, respectively. The final selection of connections for the two-story frame is C-3/4 connections for the lower-story girder and C-1/2 connections for the upper-story girder.

The load-displacement curves of the final frame at the top and bottom floors are shown in Fig. 8.22. The highly nonlinear response of the frame as the factored gravity load is applied is mainly due to the nonlinear response of the connections subjected to increasing end moments from the beams. The sudden increase in deflections at an applied load ratio slightly above 1.0 indicates that the frame failed by side-sway instability soon after the application of the factored gravity load. The plastic hinge formation sequence is also shown in the load-displacement diagram. Only two plastic hinges are detected by the refined plastic hinge analysis, which accounts for gradual stiffness degradation due to distributed plasticity in the members. The maximum load of this frame occurs at an applied load ratio of 1.096, which is about 10% higher than the full factored gravity load. The frame satisfies all the serviceability requirements. The first plastic hinge forms at an applied load ratio of 1.014, which is slightly above the full factored gravity load. The maximum frame and interstory drifts at service load combination (1.0 wind + 0.6 gravity) are 0.702 and 0.401 inch, respectively. These lateral drifts are slightly below the frame and interstory drift limits of $H/400 = 0.72$ inch and $h/250 = 0.576$ inch, respectively.

8.9 Computer Program—PHINGE

This section describes a FORTRAN-based computer program, PHINGE, which can be used for the analysis of two-dimensional steel frames. The computer program was jointly developed by the Structural Engineering Department at Purdue University and the Civil Engineering Department at the National University of Singapore under a research project entitled *advanced analysis for frame design*. The main objective of this project is to provide an educational type of software for engineers and graduate students to perform planar frame analysis for a more realistic prediction of the system's strength and stability. The final aim is to advance the use of second-order inelastic analysis programs for proportioning steel frame structures so that the tedious task of estimating various factors (such as effective length and moment amplification factors) for individual member capacity checks can be eliminated in the design process [8.37].

The computer program, PHINGE, has been tested in an IBM or equivalent personal computer system using Microsoft FORTRAN compiler Version 5.1. It can also be run on UNIX workstations using a standard

FORTRAN 77 compiler. The memory required to run the program depends on the size of the problem. Computer with minimum 640K of memory and a 30MB hard disk is generally required. For PC's application, the array sizes have been reduced as follow:

Maximum total degrees of freedom   MAXDOF = 150
Maximum translational degrees of freedom   MAXTOF = 100
Maximum rotational degrees of freedom   MAXROF = 50
Maximum number of truss elements   MAXTRS = 20
Maximum number of connections   MAXCNT = 20

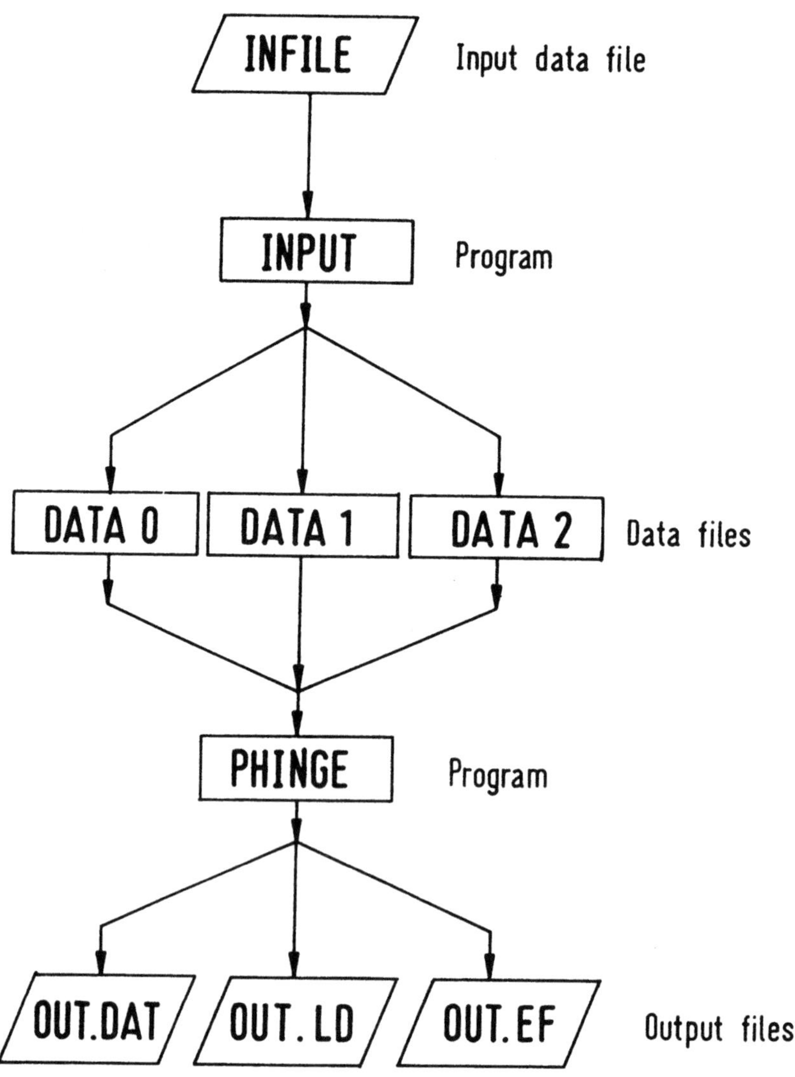

FIGURE 8.23. Operating procedure of PHINGE program.

It is possible to run a bigger job in UNIX workstations by modifying the above values in the parameter statements in the source code. The reader should read the README DOC file for more information.

The computer program is divided into two parts. The first part consists of a FORTRAN program; INPUT, which reads a input data file; INFILE, and generates three working data files, DATA0, DATA1, and DATA2. The second executable program, PHINGE, reads the working data files and generates three output files named OUT.DAT, OUT.LD, and OUT.EF. After the output files are generated, the user can view these files on the screen or print them out for detailed information on element force distribution (OUT.EF) and nodal load versus displacement results (OUT.LD). The schematic diagram shown in Fig. 8.23 explains the operation procedure of the computer program. Two executable program files named INPUT.EXE and PHINGE.EXE are supplied. These programs can be executed by issuing the commands "INPUT" and "PHINGE" separately at the terminals.

The analysis library consists of three elements, namely, a plane frame element, a plane truss element, and a connection element. The connection element is represented by a zero-length rotational spring with a user-specified nonlinear moment-rotation curve. Loading is allowed only at the nodal points with specified forces. Geometric nonlinearities can be accounted for using an iterative load-incremental scheme. Material nonlinearities are accounted for by using zero-length plastic hinges lumped at the element ends.

The complete source code listing for the computer program is provided in the attached diskette. The user is encouraged to modify or improve the program for his or her use. Program PHINGE is in its preliminary stage of development and will be under a constant state of development in the years to come.

The necessary input instructions for the user to execute PHINGE are described in Section 8.9.5. The new user is advised to read all the instructions, paying particular attention to the appended notes, to gain an initial overall view of the data required to specify an analysis task to PHINGE. Readers without prior knowledge or experience on nonlinear analysis will benefit from careful study of the example problems given in Sections 8.10 and 8.11. These examples are intended to help clarify the data-generation process. Upon examining the input instructions, the reader should recognize that no system of units is assumed by the program. It is the user's responsibility to specify data in the consistent units of his or her choice. Overlooking this requirement is a common source of erroneous results.

8.9.1 *Installation and Execution Procedure*

This section deals with the installation and execution of PHINGE on an MS-DOS-based computer system, such as an IBM-compatible PC.

A computer diskette is provided, which contains the following:

a. PHINGE program executables (INPUT.EXE and PHINGE.EXE).
b. PHINGE batch file (RUN.BAT).

c. Sample examples and their output files (EX?.*).
d. Source codes written in FORTRAN 77 (in directory SOURCE).

The programs provided must first be copied onto the hard disk, i.e., copy PHINGE.EXE, INPUT.EXE, INFILE, and RUN.BAT from the diskette to the hard disk. Before putting the program to work, the user should test the system by running the sample example provided on the same diskette.

Several input data files have been prepared and the details are explained in Sections 8.10 and 8.11. The user may test-run the program by running the PHINGE program from the directory where the executable files are resident. After the execution of the PHINGE program, three working data files and three output files will be generated. The output files produced are OUT.DAT, OUT.LD, and OUT.EF. These files should be compared to the corresponding output files, EX?.DAT, EX?LD, and EX?.EF that are provided on the diskette.

8.9.2 PHINGE Modeling Options

Program PHINGE was developed based on the theory described in Sections 8.2 through 8.8. The program can perform first-order elastic, second-order elastic, second-order elastic-plastic, and second-order refined plastic hinge analyses of planar steel frames with or without semirigid connections. The second-order elastic-plastic analysis procedure described in Section 8.4 can be reduced to first- and second-order elastic analysis by specifying appropriate load step increment and material properties in the analysis input data file. For first-order analysis, the total factored load should be applied in one load increment to suppress numerical iteration in the nonlinear analysis algorithm. For elastic analysis, section plastic moduli of arbitrary large values should be assumed for all beam-column elements to prevent the formation of plastic hinges due to bending actions.

Since second-order plastic-hinge-based analysis involves both geometric and material nonlinearities, recourse to the tangent-stiffness matrix, which is a linearized form of the nonlinear force-displacement relationships, is used to obtain solutions. In other words, instead of solving a set of nonlinear equilibrium equations, a set of linearized equations is solved at each cycle of calculation.

The simple incremental solution method is the simplest and most direct nonlinear global solution technique. This numerical procedure is straightforward in concept and implementation, and the numerical algorithm is generally well behaved and often exhibits good computational efficiency. This is especially true when the structure is loaded into the inelastic region by which a trace of hinge-by-hinge formation is required in the element stiffness formulations. However, for a finite increment size, this approach only approximates the nonlinear structural response, and equilibrium between the external applied loads and the internal element forces is not satisfied.

Iterative methods, such as the Newton-Raphson method, satisfy the equilibrium equations at a specific external load magnitude. In this method, the equilibrium out-of-balance that is present following a linear load increment is eliminated within a given tolerance by applying additional corrective steps. The Newton-Raphson method has the advantage of providing results that lie on the "exact" load-displacement trace. However, the requirement to trace the hinge-by-hinge formation in the structure may render the plastic hinge analysis slightly inefficient in the numerical iteration process.

In the PHINGE program, the simple incremental method has been implemented for second-order elastic-plastic hinge analysis. The refined plastic hinge model is implemented using an automatic load-increment procedure. To prevent plastic hinges from forming within a constant-stiffness load increment, load step sizes less than or equal to the specified increment magnitude are internally computed such that plastic hinges form only after the load increment. Thus, the subsequent element stiffness formulation will account for the stiffness reduction due to the presence of plastic hinges. For elements that are partially yielded at their ends, a limit is placed on the magnitude of the increment in the element end forces [8.30]. Increments of these force components that are relatively large, compared to their respective plastic strengths, may indicate that the path taken from the initial yield surface to the plastic strength surface may have deviated excessively from the "exact" path, upon which the calculation of element effective stiffness may lead to some error. The analysis program automatically scales an attempted load increment when the change in the element stiffness parameter exceeds a predefined tolerance. A tolerance value has been specified in the solution scheme to suppress the error that may be incurred during the computation [8.30].

The nonlinear analysis routines in PHINGE are based on the matrix stiffness method and involve the solving of the following incremental force-displacement equation:

$$[K_t]\{\dot{d}\} = \{\dot{f}\} \tag{8.9.1}$$

where $\{\dot{d}\}$ is the vector of unknown incremental nodal displacements, $\{\dot{f}\}$ is the vector of known equivalent incremental nodal loads, and $[K_t]$ is the global tangent-stiffness matrix, which reflects the current state of the deformed structure with the corresponding effective stiffness. The matrix $[K_t]$ is the assembly of the transformed element tangent-stiffness matrices discussed in Sections 8.2–8.8.

8.9.3 Convention and Terminology

This section deals with specific conventions and terminology used by PHINGE in the input preparation and the output interpretation phases of the analysis. Data preparation involves (1) describing the analysis method, size of the problems, and output format, (2) defining element properties and structural geometry, and (3) defining the loading conditions for which the structure needs to be analyzed.

8.9.3.1 Joints and Elements

The geometric dimensions of the structures are established by placing joints (or nodal points) on the structures. Each joint is given an identification number and is located in a plane associated with a global two-dimensional coordinate system.

The structural geometry is completed by connecting the predefined jonts with structural elements that are of specific types, namely, frame, truss, and connection. Each element also has an identification number.

The following are some of the factors that need to be considered in locating joints on a structure:

1. The number of joints should be sufficient to describe the geometry and the response behavior of the structures.
2. Joints need to be located at points and lines of discontinuity, e.g., at changes in material properties or section properties.
3. Joints should be located at points on the structure where forces and displacements are to be evaluated.
4. Joints should be located at points on the structure where concentrated loads are to be applied. The applied loading should be represented as concentrated loads acting on specific joints in the members.
5. Joints should be located at all support points. Support conditions are represented in the structural model by restricting the movement of the specific joints in specific directions.
6. Element mesh should be refined enough to capture the inelastic behavior and second-order effects in the regions of interest.

8.9.3.2 Global and Local Coordinate Sytems

For the generation of all the input and output data associated with the joints, a two-dimensional (x, y) global coordinate system is used. The following input data are prepared with respect to the global coordinate system:

1. joint coordinates.
2. joint restraints.
3. joint connection spring.
4. joint loading.
5. joint constraints.

The following output is also referred to in the global coordinate system:

1. joint displacements.
2. joint reactions.

8.9.3.3 Degrees of Freedom

A two-joint frame element has six displacement components as shown in Fig. 8.24. Each joint can translate in the global x- and y-directions, and it can

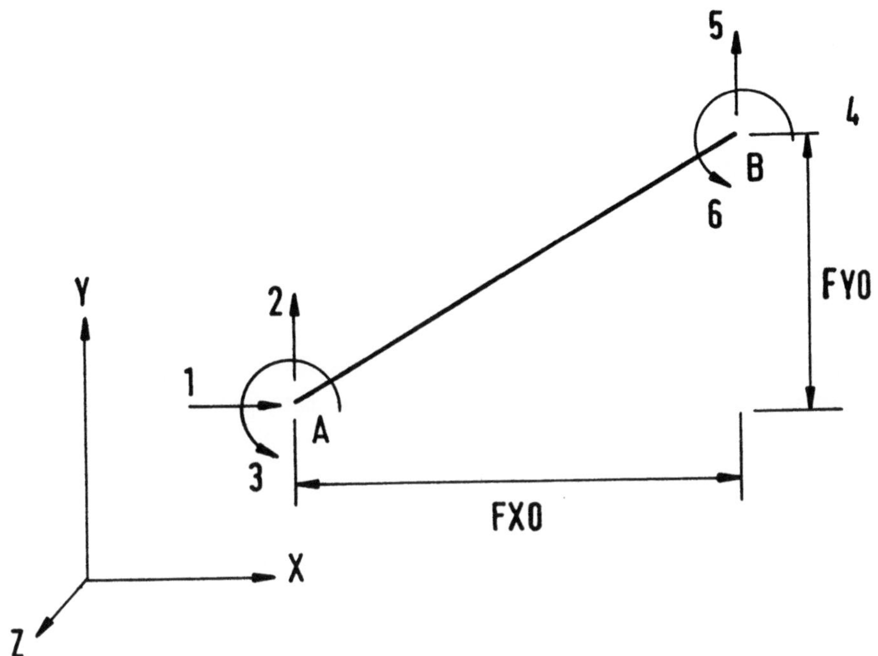

FIGURE 8.24. Degrees of freedom numbering for the beam-column element.

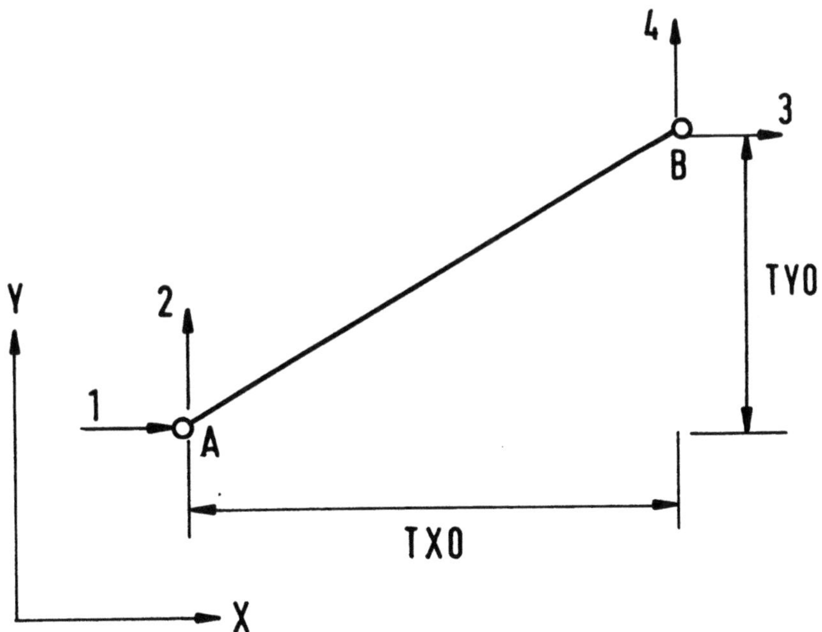

FIGURE 8.25. Degrees of freedom numbering for the truss element.

rotate about the global z-axis. The directions associated with these displacement components are known as ***degrees of freedom*** of the joint.

A two-joint truss element has four degrees of freedom as shown in Fig. 8.25. Each joint consists of two translational degrees of freedom and with no rotational components.

If the displacement of a joint corresponding to any one of its degrees of freedom is known to be zero, such as at a support point, then the degree of freedom is known as an inactive degree of freedom. Degrees of freedom at which the displacements are not known are termed active degree of freedoms. In general, the displacement of an inactive degree of freedom is usually known, and the purpose of the analysis is to find the reaction in that direction. For an inactive degree of freedom the applied load is usually known (it could be zero), and the purpose of the analysis is to find the corresponding displacement.

8.9.3.4 Units

There are no built-in units in the PHINGE computer program. The user must prepare the input in a consistent set of units. The output produced by the program will then conform to the same set of units. Therefore, if the user chooses to use kips and inches as the input units, all the dimensions of the structure must be entered in inches and all the loads in kips. The material properties should also conform to these units. The output units will then be in kips and inches, so that the frame member axial force will be in kips, bending moments will be in kip-inches and displacements will be in inches. However, joint rotations are in radians, irrespective of units.

8.9.4 Data Preparation

Described here is the input sequence and data structure to create an input file called INFILE. The analysis program, PHINGE, can analyze any frame structures with maximum degrees of freedom not more than 165. However, it is possible to recompile the source code to accommodate more degrees of freedom by changing the size of the arrays in the PARAMETER statements. This problem can be overcome by using dynamic storage allocation. This procedure is rather common in finite element programs [8.18, 8.38], and it will be used in the next release of the program.

The input data file INFILE is prepared in a specific format. The following input sequence must be followed:

1. Title line.
2. Analysis method.
3. Analysis attributes.
4. Job control.
5. Element types.
6. Element group.
7. Connection properties.
8. Frame element properties.

9. Truss element properites.
10. Connection element data.
11. Frame element data and geometry.
11. Truss element data and geometry.
12. Nodal forces.

Some of the data input are mandatory; however, the existence of some of the elements in the input data depends on the problem being analyzed. The order in which the data lines occur in the input file must be strictly followed.

The various functions and uses of these data are summarized in the following.

Line 1: Title line

Note	Columns	Variable	Description
(1)	1–8	——	A comment card

Note:

1. The comment line is skipped, without interpretation, at the execution of INPUT. This line is usually required to identity one input file from the other.

Line 2 (*I5*): Analysis method

Note	Columns	Variable	Description
(1)	1–5	ISOLVE	Specify the analysis method. Specify ISOLVE = 0 for elastic-plastic hinge analysis, and ISOLVE = 1 for refined plastic hinge analysis

Note:

1. IF ISOLVE = 0 is specified, the second-order elastic-plastic hinge analysis is activated. If ISOLVE = 1 is specified, the second-order refined plastic hinge method described in Section 8.7 is initiated.

Line 3 (2*I5*): Analysis attributes
This line is skipped if ISOLVE = 0

Note	Columns	Variable	Description
(1)	1–5	IET	Tangent-modulus model; IET = 1, CRC tangent modulus; IET = 2, LRFD tangent modulus
(2)	6–10	IPHI	Element stiffness degradation model (IPHI = 1) IPHI = 1, parabolic stiffness reduction

Note:

1. If IET = 1 is selected, the CRC tangent-modulus model given by Eq. (8.7.4) is adopted in the refined plastic hinge analysis for members subjected to compression or tension. If IET = 2 is specified, the LRFD tangent modulus from Eq. (8.7.4) is effected for members subjected to axial compression only. For members subjected to tension, the CRC tangent-modulus model is automatically selected, as discussed in [8.2].
2. A parabolic stiffness-reduction function [Eq. (8.7.6)] with initial yield surface equal to one-half the plastic strength surface is adopted in the refined plastic hinge analysis if IPHI = 1 is selected. Other possible stiffness-reduction schemes used in calibrating the refined plastic hinge model to best fit the "exact" plastic-zone solutions are given in the subroutine ESTIFF. The reader is referred to [8.6] for information pertinent to this aspect.

Line 4 (4/5): Job control card

Note	Columns	Variable	Description
(1)	1–5	NDOFS	Number of degrees of freedom of the structure (≤ 165)
(2)	6–10	NINCRE	Number of load increments
(3)	11–15	NSEQN	Number of load sequences (≤ 2)
(4)	16–20	IPRTFC	Flag for printing member end forces: IPRTFC = 0, do not print element end forces; IPRTFC = 1, print elemnt end forces

Note:

1. The structure degrees of freedom is required. This value should not be larger than the default limit of 165. If this value is exceeded, the user needs to recompile the source code by setting a larger value for "MAXDOF" in the PARAMETER statements.
2. The total load incremental steps are required. The analysis is terminated when the number of the total load incremental steps is reached. For first-order analysis, the total factored load should be applied in one load increment to suppress numerical iteration in the nonlinear analysis algorithm, i.e., let NINCRE = 1.
3. Only two load sequences are allowed in the current version of the program.
4. Member end forces are not printed if IPRTFC = 0. These forces are printed at each step of load increment if IPRTFC = 1; this means a larger storage capacity is required for OUT.EF.

Line 5 (3*I*5): Element types

Note	Columns	Variable	Description
(1)	1–5	NCTYPE	Number of connection types (≤ 10)
(1)	6–10	NFTYPE	Number of frame element types (≤ 20)
(1)	11–15	NTTYPE	Number of truss element types (≤ 10)

Note:

1. If the limits specified in the above are exceeded, the user need to recompile the program by specifying suitable values for the variables in the PARAMETER statements of the source code.

Line 6 (3*I*5): Element group

Note	Columns	Variable	Description
(1)	1–5	NUMCNT	Number of connection types (≤ 10)
(1)	6–10	NUMFRM	Number of frame elements (≤ 200)
(1)	11–15	NUMTRS	Number of truss elements (≤ 20)

Note:

1. If the limits specified in the above are exceeded, the user needs to re-compile the program by specifying suitable values for the variables in the PARAMETER statements of the source code.

Line 7 (*I*5, 3*D*10.0): Connection properties
This line should be omitted if NCTYPE = 0. Otherwise, this line must be supplied NCTYPE times.

Note	Columns	Variable	Description
(1)	1–5	ICTYPE	Connection type identifier (an integer)
(2)	6–15	MU	Connection ultimate moment capacity
(2)	16–25	RKI	Connection initial stiffness
(2)	26–35	n	Connection shape parameter

Note:

1. Connection property sets must be specified in the order as shown. Users should refer to Section 8.8 for the definition of the connection paramenters.

2. The connection property set is stored in an array CTYPE(ICTYPE, J = 1,3). Step-by-step procedures for calculating M_u and R_{ki} for semirigid connections composed of angles are given in [8.36].

Line 8 (*I5, 5D*10.0): Frame element properties

If NFTYPE $\neq$ 0, this line must be supplied NETYPE times. Otherwise, this line should be omitted.

Note	Columns	Variable	Description
	1–5	IFTYPE	Frame type identifier (an integer)
(1)	6–15	A	Cross-sectional area
(1)	16–25	I	Moment of inertia
(1)	26–35	Z	Plastic section modulus
(1)	36–45	E	Modulus of elasticity
(1), (2)	46–55	FY	Material yield strength

Notes:

1. Material property sets for the frame elements must be specified in the order shown. The material set is stored in an array FTYPE(IFTYPE, J = 1,5).
2. For elastic analysis, material yield strength of arbitrary large values should be assumed for all elements to prevent the formation of plastic hinges due to bending actions, say FY = 1E + 05.

Line 9 (*I5, 4D*10.0): Truss element properties

If NFTYPE $\neq$ 0, this line must be supplied NETYPE times. Otherwise, this line should be omitted.

Note	Columns	Variable	Description
	1–5	ITTYPE	Truss type identifier (an integer)
(1)	6–10	A	Cross-sectional area
(1)	16–25	I	Moment of inertia
(1)	26–35	E	Modulus of elasticity
(1)	36–45	FY	Yield stress

Note:

1. Material property sets for the truss elements must be specified in the order shown. For truss elements under axial compression, the analysis is terminated once the axial force in the truss element exceeds the column strength implied by the LRFD column strength equations [8.7]. Therefore, the moment of inertia for weak-axis bending should be specified for a pinned-ended truss element. For element subjected to tensile force,

the cross-sectional effective area (with allowance for bolt holes) may be specified. The material set is stored in an array TTYPE(ITTYPE, J = 1,4).

Line 10 (4I5): Connection data
This line should be omitted if NUMCNT = 0. Otherwise, this line must be supplied NUMCNT times.

Note	Columns	Variable	Description
	1–5	LCNT	Connection element number
	6–10	IFMCNT(LCNT)	Beam-column element number to which the connection is attached
(1)	11–15	IEND(LCNT)	Beam-column element end to which the connection is attached: IEND = 1 for connection at end A; IEND = 2 for connection at end B
	16–20	JDCNT(LCNT)	Connection type number

Note:

1. See Fig. 8.24 for the respective end of the beam-column element.

Line 11 (I5, 2D10.0, 7I5): Frame element data and geometry
This line should be omitted if NUMFRM = 0. Otherwise, this line must be supplied NUMFRM times.

Note	Columns	Variable	Description
	1–5	LFRM	Frame element number
(1)	6–15	FXO(LFRM)	Horizontal projected length
(1)	16–25	FYO(LFRM)	Vertical projected length
	26–30	JDFRM(LFRM)	Frame type number
(2)	31–35	NFRMCO(JDFRM,1)	Node number for degree of freedom 1
(2)	36–40	NFRMCO(JDFRM,2)	Node number for degree of freedom 2
(2)	41–45	NFRMCO(JDFRM,3)	Node number for degree of freedom 3
(2)	46–50	NFRMCO(JDFRM,4)	Node number for degree of freedom 4
(2)	51–55	NFRMCO(JDFRM,5)	Node number for degree of freedom 5
(2)	56–60	NFRMCO(JDFRM,6)	Node number for degree of freedom 6

Note:

1. The projected lengths for a frame element are shown in Fig. 8.24. The nodal numbering sequence must follow the order of the degrees of freedom for the frame element shown in Fig. 8.24.

Line 12 (*I5, 2D*10.0, *5I5*): Truss element data card

This line should be omitted if NUMTRS = 0. Otherwise, this line must be supplied NUMTRS times.

Note	Columns	Variable	Description
	1–5	LTRS	Truss element number
(1)	6–15	TXO(LTRS)	Horizontal projected length (in inches)
(1)	16–25	TYO(LTRS)	Vertical projected length (in inches)
	26–30	JDTRS(LTRS)	Truss type number
(2)	31–35	NTRSCO(JTRS,1)	Node number for degree of freedom 1
(2)	36–40	NTRSCO(JTRS,2)	Node number for degree of freedom 2
(2)	41–45	NTRSCO(JTRS,3)	Node number for degree of freedom 3
(2)	46–50	NTRSCO(JTRS,4)	Node number for degree of freedom 4

Note:

1. The projected lengths for a truss element are shown in Fig. 8.25. The nodal numbering sequence must follow the order of the degrees of freedom for the truss element shown in Fig. 8.25.

Line 13 (*I5, 3D*10.0): Nodal load

Note	Columns	Variable	Description
(1)	1–5	NLDOF	Degree of freedom at which a load is applied
(2)	6–15	FAMG(I,1)	Magnitude of load increment of load sequence 1
(3)	16–35	FAMG(I,2)	Magnitude of load at which load sequence 1 end
(3)	36–45	FAMG(I,3)	Magnitude of load increment of load sequence 2

Notes:

1. All concentrated forces must be applied at the nodal point in the x and y directions of the global coordinates by specifying he translational degrees of freedom of the node. Concentrated moment may be applied at the nodal point by specifying the rotational degree of freedom of the node.
2. If the number of load increments NINCRE $= 1$, values assigned to FAMG(I,2) and FAMG(I,3) will not affect the analysis results. In this case, the analysis is terminated when the number of load increment NINCRE is reached or when the structural stiffness matrix is nonpositive definite after attempting several load-reduction steps near the maximum load point.
3. If NINCRE $= 2$, the user must specify values for FAMG(I,2) and FAMG(I,3) so that the analysis knows when to apply load sequence 2 once the total load parameter for load sequence 1 is reached.

8.9.5 PHINGE Output Files

After a successful execution of PHINGE using the RUN command, the following files will exit on the disk:

1. DATA0: Working data file generated by INPUT.
2. DATA1: Working data file generated by INPUT.
3. DATA2: Working data file generated by INPUT.
4. OUT.DAT: Input data echo generated by PHINGE.
5. OUT.LD: Load-displacement results and sequence of plastic hinge formation at joints.
6. OUT.EF: Element forces in global coordinates.

OUT.DAT contains an echo of the information from the input data file INFILE. This file is created by PHINGE and contains the detailed summary on the input data cards. This file may be used to check for numerical and incompatibility errors in the input data preparation.

OUT.LD contains the load and displacement information at various joints in the structure. The load-displacement results are presented at the end of every load increment. The sign conventions for the loads and displacements should follow the frame degrees of freedom as shown in Figs. 8.24 and 8.25.

OUT.EF contains the element joint forces for all types of elements at every load step. The element joint forces are obtained by the summation of the product of the element incremental displacements and element tangent stiffness matrices at every load step. The element joint forces are output in the global coordinate system and are forces acting on the element at the joints and must be in equilibrium with the applied forces at the joints.

8.9.6 Executing the PHINGE Program

Say the data associated with the problem the user wishes to analyze has been created in a data file called INFILE. Entering the command RUN from the directory where the PHINGE.EXE and INPUT.EXE files are resident will activate a series of PHINGE programs in the required sequence for the analysis of the structure defined by the data file INFILE. The sequence of actions generated by the batch file, RUN.BAT, is: (1) executing INPUT, which reads the input data file INFILE and generates three working data files, DATA0, DATA1, and DATA2; (2) executing PHINGE, which reads the working data files and generates three output files, OUT.DAT, OUT.LD, and OUT.EF. The schematic diagram Fig. 8.23 explains the sequential operation procedures generated by the batch file RUN.BAT.

After the OUT.LD and OUT.EF files are generated, the user can view these files on the screen or print them out for detailed information on force distribution and load-deformation characteristics of the structure. To print an output file the MS-DOS PRINT command may be used.

It should be noted that RUN is a batch command that facilitates the execution of PHINGE and the clearing of old working files and output files. Entering the command RUN will erase the old files DATA0, DATA1, DATA2, OUT.DAT, OUT.LD, and OUT.EF before creating the new ones. Therefore, it is advisable to rename the desired output files before running a new problem.

8.10 Analysis of a One-Story Portal Frame

8.10.1 Rigid Plastic Analysis

Figure 8.26 shows a rectangular portal frame subjected to two concentrated vertical loads acting at one-third points of the beam. This example is similar to the problem given in Example 5.5.1 in which the members are proportioned based on plastic analysis.

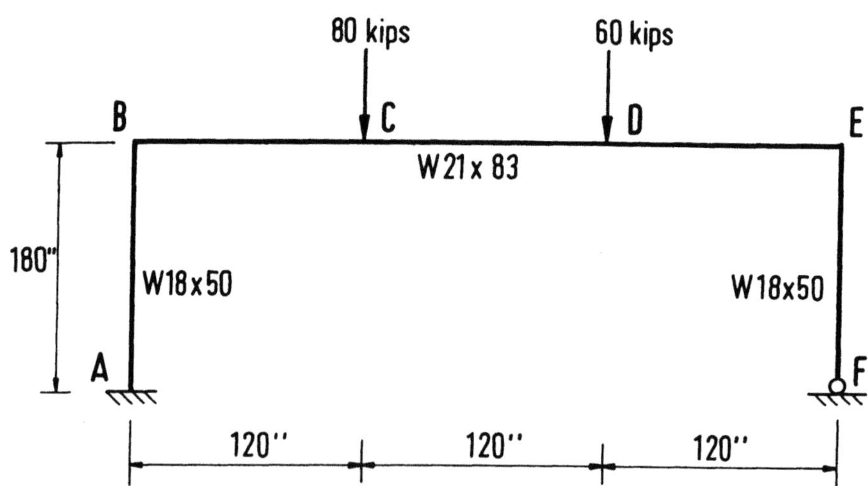

FIGURE 8.26. Portal frame subjected to two concentrated loads.

The collapse mechanism based on plastic analysis is a beam mechanism shown in Fig. 8.27. By equating the external work due to applied loads to the internal energy dissipation, the collapse load factor λ is obtained as:

$$80\lambda(20\theta) + 60\lambda(10\theta) = M_p(2\theta) + 1.94M_p(3\theta) + M_p\theta$$

$$\lambda = \frac{8.82}{2200}M_p$$

$$\text{or} \quad \lambda = \frac{8.82}{2200}(303/12) = 1.215. \tag{8.10.1}$$

From equilibrium, the bending moment diagram and reaction forces at collapse (i.e., $\lambda = 1.215$) are shown in Fig. 8.28. One of the disadvantages of plastic analysis is that the state of the frame at the factored load level cannot be determined readily. One has to perform a hinge-by-hinge analysis in order

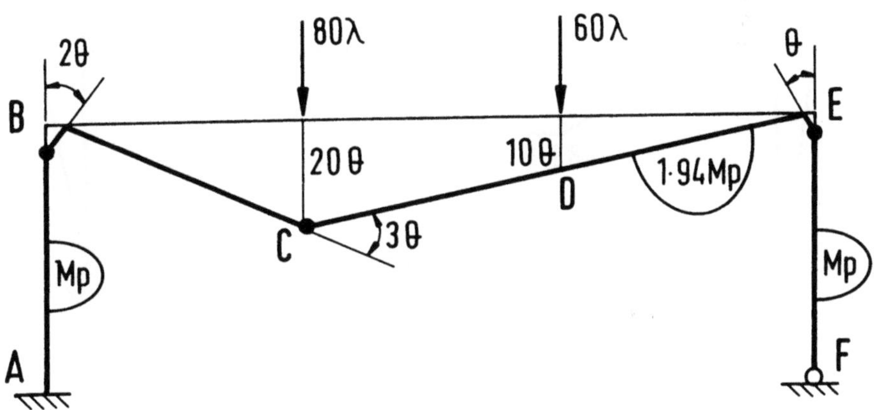

FIGURE 8.27. Collapse mechanism based on plastic analysis.

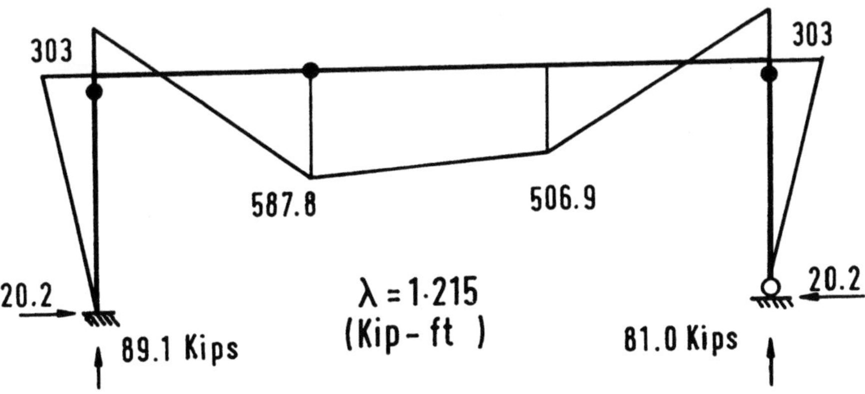

FIGURE 8.28. Force distribution at the incipient of collapse based on plastic analysis.

to obtain the "correct" force distribution in the frame for the purpose of member proportioning.

In order to perform member capacity checks, the bending moment diagram at the design load factor (i.e., at $\lambda = 1.0$) must be established. This can be obtained by factoring the bending moment diagram at collapse by a factor of 1.215 to achieve the bending moment diagram at the factored load level. The resulting bending moment diagram is shown in Fig. 8.29.

8.10.2 Second-Order Elastic-Plastic Hinge Analysis

A second-order elastic-plastic hinge analysis considers the effects of both geometric and material nonlinearities in a more exact manner. The effects of residual stress and geometric initial imperfections are, however, not modeled

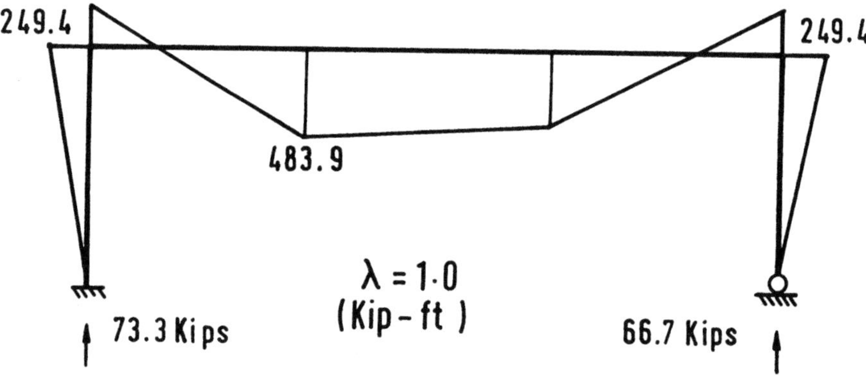

FIGURE 8.29. Force distribution at the factored load level based on plastic analysis.

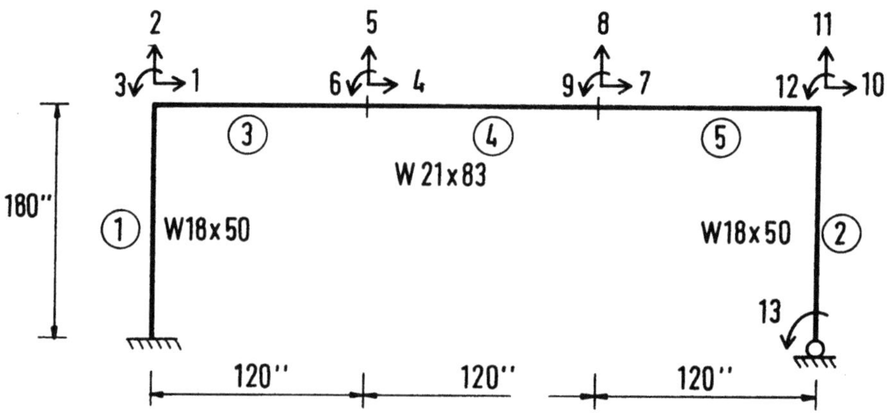

FIGURE 8.30. Structural modeling of the portal frame shown in Fig. 8.26.

in the analysis. Since the problem is nonlinear, the analysis has to be carried out using small load increments.

For the frame shown in Fig. 8.26, the beam is discretized into three elements, and the columns are represented by one element per member. The computer model is shown in Fig. 8.30. The input file, INFILE, to perform the second-order elastic-plastic analysis, is shown in Fig. 8.31. Kip-inch units are used. All members are modeled using beam-column elements, and the loads are applied proportionally to collapse.

The resulting bending moment diagram and support reaction forces at maximum load factor (at load step 31, $\lambda = 1.166$) and at design load factor (at load step 21, $\lambda = 1.0$) are shown in Figs. 8.32 and 8.33, respectively. The first

```
/*SECOND ORDER ELASTIC PLASTIC HINGE ANALYSIS
EXAMPLE 5.5.1
 0
13        50    1        1
 0         2    0
 0         5    0
 1        14.70      800.0      101.00      29000.       36.00
 2        24.30     1830.0      196.00      29000.       36.00
 1         0.0       180.0      1    0       0    0       1    2        3
 2         0.0       180.0      1    0       0   13      10   11       12
 3       120.0         0.0      2    1       2    3       4    5        6
 4       120.0         0.0      2    4       5    6       7    8        9
 5       120.0         0.0      2    7       8    9      10   11       12
 5        -4.00       -80.0      0.000
 8        -3.00       -60.0      0.000
```

FIGURE 8.31. Input data file, INFILE, for second-order elastic plastic hinge analysis of a portal frame (named ex1 in the diskette).

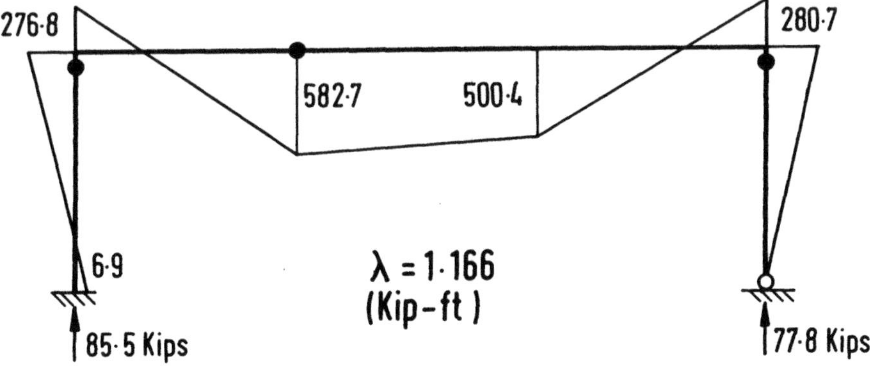

FIGURE 8.32. Force distribution at the maximum load based on second-order elastic-plastic hinge analysis.

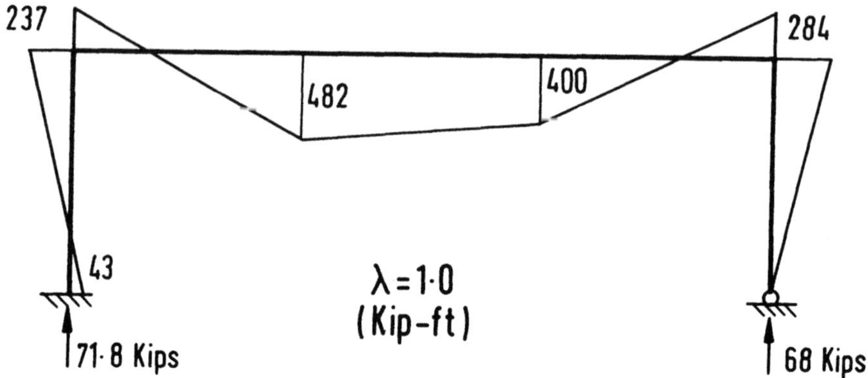

FIGURE 8.33. Force distribution at the factored load level based on second-order elastic-plastic hinge analysis.

```
/*SECOND ORDER REFINED PLASTIC HINGE ANALYSIS
EXAMPLE 5.5.1
 1
 2      1
13     40    1       1
 0      2    0
 0      5    0
 1     14.70        800.0        101.00       29000.           36.00
 2     24.30       1830.0        196.00       29000.           36.00
 1      0.0         180.0       1    0      0    0      1    2      3
 2      0.0         180.0       1    0      0   13     10   11     12
 3    120.0           0.0       2    1      2    3      4    5      6
 4    120.0           0.0       2    4      5    6      7    8      9
 5    120.0           0.0       2    7      8    9     10   11     12
 5     -4.00         -80.0       0.000
 8     -3.00         -60.0       0.000
```

FIGURE 8.34. Input data file, INFILE, for second-order refined plastic hinge analysis of a portal frame (named ex1a in the diskette).

plastic hinge is formed at the top right-hand column at $\lambda = 0.95$, and the second plastic hinge is at the larger concentrated load point at $\lambda = 1.02$. The last plastic hinge is formed at the top left-hand column at the incipient of collapse with the applied load factor $\lambda = 1.166$.

The complete input and output data files are stored in the computer diskette and are renamed as EX1 (for INFILE), EX1.DAT, EX1.LD, and EX1.EF (for the corresponding OUT.DAT, OUT.LD, and OUT.EF files).

8.10.3 Second-Order Refined Plastic Hinge Analysis

The farme model shown in Fig. 8.30 is now analyzed using the second-order refined plastic hinge analysis. The analysis adopted the LRFD tangent modulus in which explicit modeling of member initial imperfections is not required. The input file, INFILE, for this problem is shown in Fig. 8.34. A large number of load increments (NINCRE = 40) has been specified for limit-load analysis. The bending moment diagrams at design load factor (at load step 21) and at limit load factor (at load step 30) are shown in Figs. 8.36 and 8.35, respectively.

The complete input and output data files are stored in the computer diskette, and they are renamed EX1a (for INFILE), EX1a.DAT, EX1a.LD, and EX1a.EF (for the corresponding OUT.DAT, OUT.LD, and OUT.EF files).

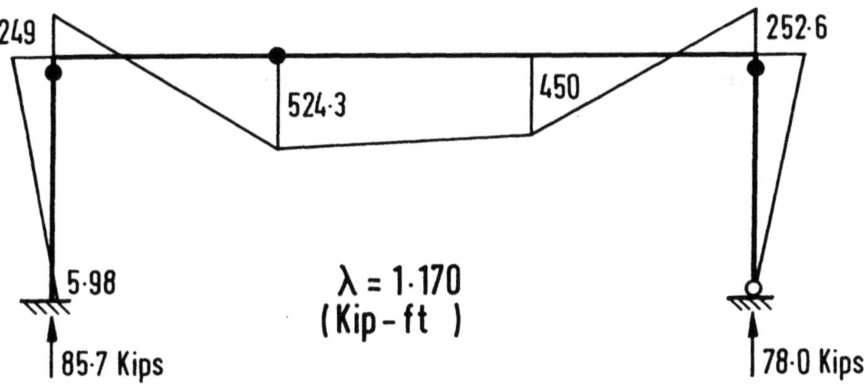

FIGURE 8.35. Force distribution at the maximum load based on second-order refined plastic hinge analysis.

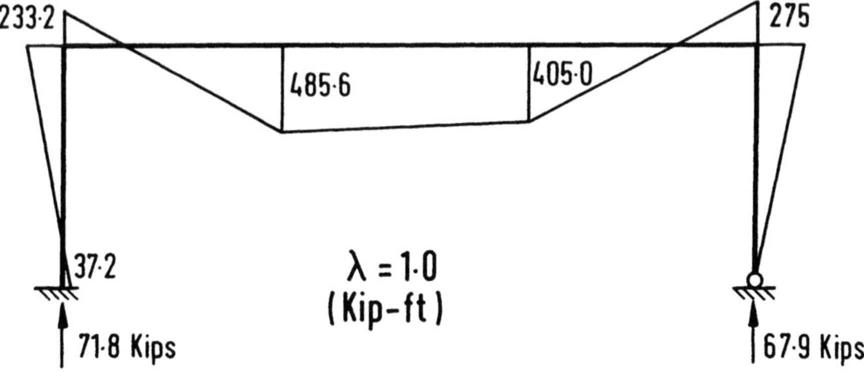

FIGURE 8.36. Force distribution at the factored load level based on second-order plastic hinge analysis.

8.10.4 *Comparison of Results*

Load-deflection results from the three analyses are compared as shown in Fig. 8.37. The plastic analysis predicts a larger collapse load at $\lambda = 1.215$, followed by the second-order elastic-plastic hinge at $\lambda = 1.166$ and the re fined plastic hinge analysis at $\lambda = 1.170$, in which the latter two results are very close to each other. In the elastic-plastic hinge analysis, the first plastic hinge is formed at the beam's right end, denoted as point E in the frame. The applied load factor corresponding to the formation of the first plastic hinge is $\lambda = 0.95$. Since the elastic-plastic hinge analysis neglects the effects of residual stresses and gradual plastification in the beam, the distribution of internal forces during the loading process is somewhat different from that predicted by the refined plastic hinge method. However, for this particular frame, the inelastic force-redistribution process does not have any influence on the prediction of the ultimate strength. This is because the second-order effect is virtually absent, and the beam failure is due to the formation of a beam mechanism.

In the refined plastic hinge analysis, the first plastic hinge is formed at point E in the frame. The load factor corresponding to the formation of the first plastic hinge is $\lambda = 0.972$, which is slightly higher than the load at first hinge predicted by the plastic hinge analysis. The formation of the first plastic hinge is delayed in the refined plastic hinge analysis because the inelastic

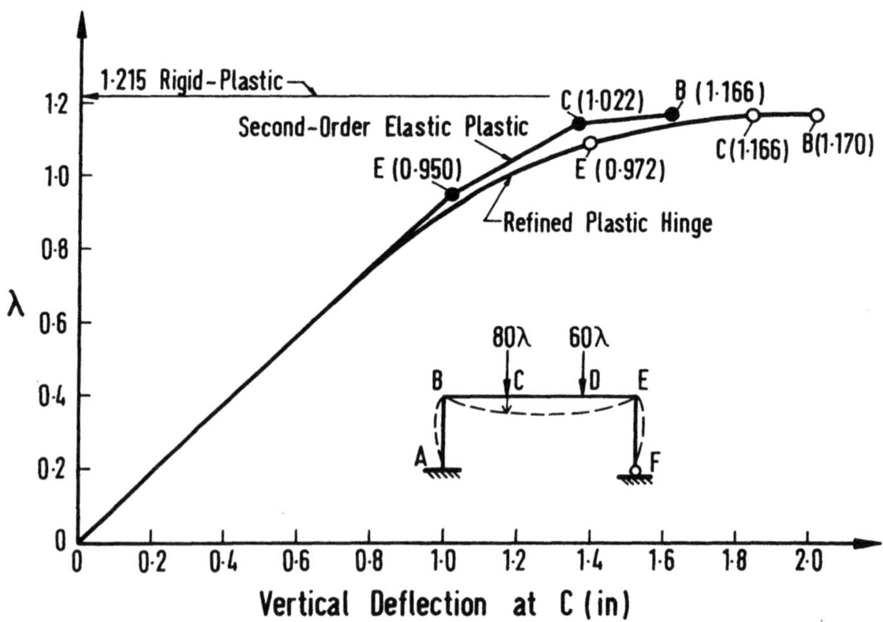

FIGURE 8.37. Comparison of load-displacement curves for the portal frame.

force redistribution occurs before the attainment of full plastic section capacity at E. The gradual plastification at E tends to shed load to other less-critical locations and therefore leads to a smooth transition of stiffness from the elastic to the inelastic range.

Analyses based on plastic, second-order elastic-plastic, and second-order refined plastic hinge methods predict different load-displacement behavior because of the different degree of accuracy in predicting the second-order and inelasticity effects in the frame. The plastic analysis does not provide any information concerning the load-deflection behavior of the frame. The elastic-plastic hinge approach predicts a piece-wise linear curve, and the refined plastic hinge approach predicts a smooth curve with gradual stiffness degradation when the limit load is approached.

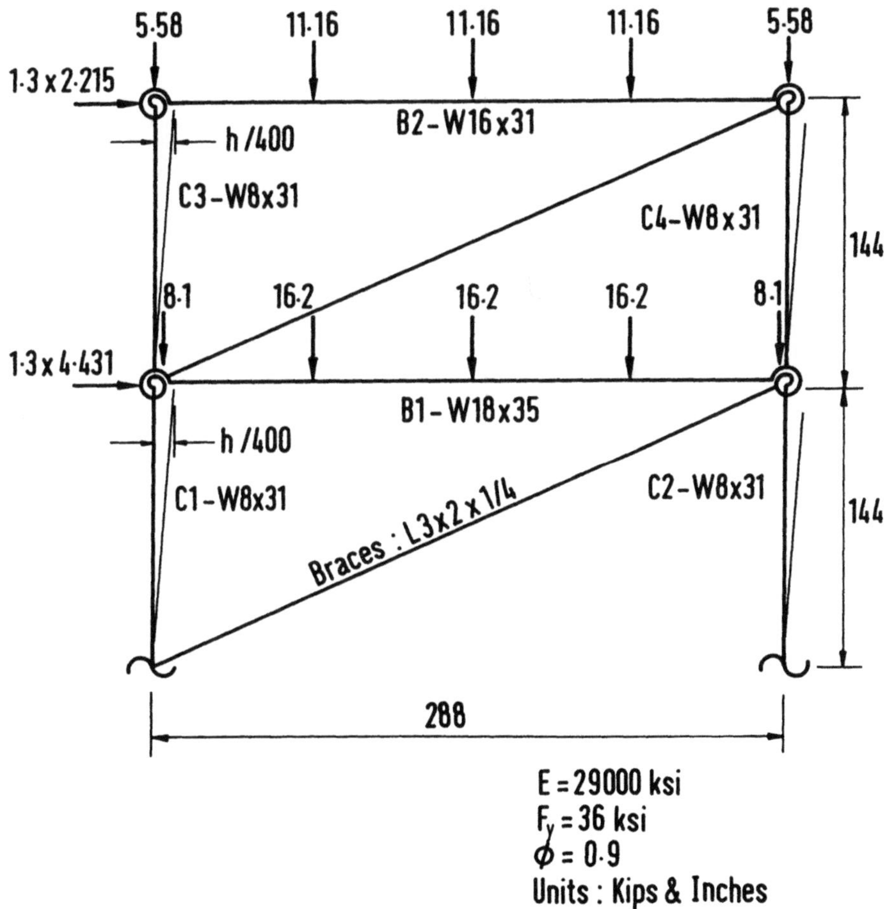

FIGURE 8.38. Semirigid frame subjected to factored loading.

8.11 Analysis of a Semirigid Braced Frame

Figure 8.38 shows a semirigid frame braced by truss diagonals and subjected to the following load combination:

$$1.0 \text{ Gravity} + 1.3 \text{ Wind.} \qquad (8.11.1)$$

The computer model for the braced semirigid frame is shown in Fig. 8.39. For illustration purposes, only the refined plastic hinge analysis is used. All beams are modeled by four discrete elements and all columns by one element. The loads are applied at the nodal points in 5% increments with respect to the full factored loads. Two load sequences are prescribed. The first load sequence applies incremental forces up to the factored load level, then is followed by the second load increment applied up to the maximum load point. Frames with column bases assumed to be fully pinned or fixed are analyzed.

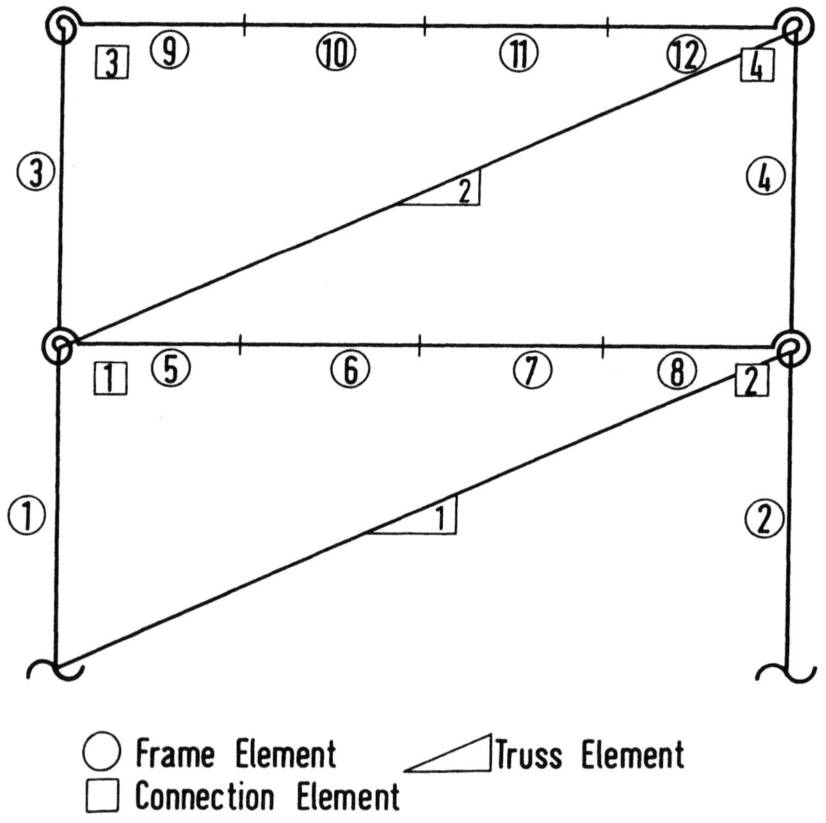

FIGURE 8.39. Structural modeling of the semirigid frame.

```
/*REFINED PLASTIC HINGE ANALYSIS OF FRAME WITH SIMPLY
SUPPORTED BASES
 1
 2     1
32    80      1          1
 2     3      1
 4    12      2
 1      1773.0      954013.0          0.80
 2       814.0      205924.0          1.57
 1         9.13         110.          30.4              29000.    32.4
 2        10.3          510.          66.5              29000.    32.4
 3         9.12         375.          54.0              29000.    32.4
 1         1.19           0.392    29000.0                 32.4
 1     5      1          1
 2     8      2          1
 3     9      1          2
 4    12      2          2
 1       0.0          144.    1      0      0      1      3      4      5
 2       0.0          144.    1      0      0      2     15     16     17
 3       0.0          144.    1      3      4      5     18     19     20
 4       0.0          144.    1     15     16     17     30     31     32
 5      72.0            0.    2      3      4      5      6      7      8
 6      72.0            0.    2      6      7      8      9     10     11
 7      72.0            0.    2      9     10     11     12     13     14
 8      72.0            0.    2     12     13     14     15     16     17
 9      72.0            0.    3     18     19     20     21     22     23
10      72.0            0.    3     21     22     23     24     25     26
11      72.0            0.    3     24     25     26     27     28     29
12      72.0            0.    3     27     28     29     30     31     32
 1     288.0          144.0   1      0      0     15     16
 2     288.0          144.0   1      3      4     30     31
 3       0.2880         5.7603        0.2880
18       0.143975       2.8795        0.143975
 4      -0.405         -8.1          -0.405
 7      -0.810        -16.2          -0.810
10      -0.810        -16.2          -0.810
13      -0.810        -16.2          -0.810
16      -0.405         -8.1          -0.405
19      -0.2790        -5.58         -0.2790
22      -0.5580       -11.16         -0.5580
25      -0.5580       -11.16         -0.5580
28      -0.5580       -11.16         -0.5580
31      -0.2790        -5.58         -0.2790
```

FIGURE 8.40. Input data file, INFILE, for second-order refined plastic hinge analysis of a semirigid braced frame with pinned based (named ex2 in the diskette).

```
/*REFINED PHINGE ANALYSIS OF A FRAME WITH FIXED BASES
 1
 2     1
30    80     2        1
 2     3     1
 4    12     2
 1      1773.0      954013.0         0.80
 2       814.0      205924.0         1.57
 1         9.13        110.          30.4              29000.    32.4
 2        10.3         510.          66.5              29000.    32.4
 3         9.12        375.          54.0              29000.    32.4
 1         0.95563       0.703    29000.0                 32.4
 1     5     1     1
 2     8     2     1
 3     9     1     2
 4    12     2     2
 1       0.360       144.    1     0     0     0     1     2     3
 2       0.360       144.    1     0     0     0    13    14    15
 3       0.360       144.    1     1     2     3    16    17    18
 4       0.360       144.    1    13    14    15    28    29    30
 5      72.0           0.    2     1     2     3     4     5     6
 6      72.0           0.    2     4     5     6     7     8     9
 7      72.0           0.    2     7     8     9    10    11    12
 8      72.0           0.    2    10    11    12    13    14    15
 9      72.0           0.    3    16    17    18    19    20    21
10      72.0           0.    3    19    20    21    22    23    24
11      72.0           0.    3    22    23    24    25    26    27
12      72.0           0.    3    25    26    27    28    29    30
 1     288.36        144.0   1     0     0    13    14
 2     288.36        144.0   1     1     2    28    29
 1       0.2880        5.76       0.2880
16       0.143975      2.8795     0.143975
 2      -0.405        -8.1       -0.405
 5      -0.810       -16.2       -0.810
 8      -0.810       -16.2       -0.810
11      -0.810       -16.2       -0.810
14      -0.405        -8.1       -0.405
17      -0.2790       -5.58      -0.2790
20      -0.5580      -11.16      -0.5580
23      -0.5580      -11.16      -0.5580
26      -0.5580      -11.16      -0.5580
29      -0.2790       -5.58      -0.2790
```

FIGURE 8.41. Input data file, INFILE, for second-order refined plastic hinge analysis of a semirigid braced frame with fixed bases (named ex2a in the diskette).

TABLE 8.6. Connection parameters.

Connections at first-story beam ends			Connections at second-story beam Ends		
M_u, kip-in.	R_{ki}, kip-in./rad	n	M_u, kip-in.	R_{ki}, kip-in./rad	n
1773	954,013	0.80	814	205,924	1.57

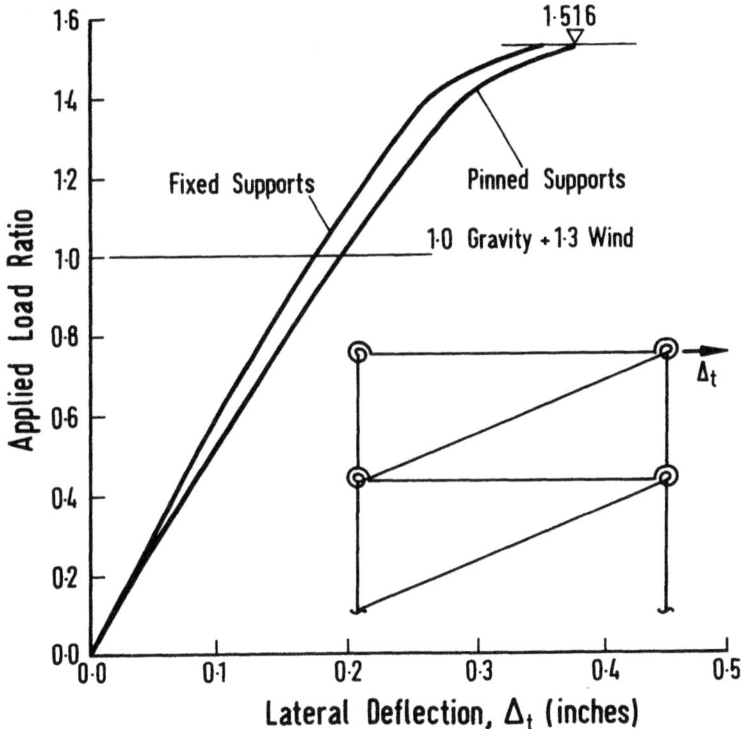

FIGURE 8.42. Load-displacement traces of semirigid braced frames with different support conditions.

The input data files for the semirigid frame with pinned and fixed bases are shown in Figs. 8.40 and 8.41, respectively. In the refined plastic hinge analysis the LRFD tangent modulus is activated by specifying IET $= 2$ as given in Figs. 8.40 and 8.41. However, the tangent-modulus model will not affect the ultimate strength results of this frame because the axial forces in the columns at factored loads are small (less than $0.3P_y$). The design yield strength is taken as 0.9 times the nominal yield strength of 36 ksi. The resistance factor of 0.9 for material yield strength is recommended by the Australian limit-states code [8.13] for use in the advanced analysis of structural steel frames.

All columns are assumed to have initial out-of-plumbness of $h/400$, where h is the story height. This column out-of-plumbness magnitude may be obtained from ECCS provisions for use with second-order plastic-zone analysis [8.39]. The column initial out-of-straightness is not modeled explicitly in the refined plastic hinge analysis since the axial force in all the columns is smaller than $0.26P_e$, where P_e is the Euler buckling load.

The semirigid connections used are top-and-seat angles with double-web angles. The evaluation of the connection's ultimate moment capacity and initial stiffness is documented in Ref. [8.36]. The connection parameters are summarized in Table 8.6.

The diagonal braces are made of angles L3 × 2 × 1/4 and are pinned at their ends by using simple bolt connections made of A325 bolt of 7/8-inch diameter. The diagonal braces are modeled using truss elements in which the net effective area of the angle section, after reduction for a single hole, is used in the analysis.

Figure 8.42 shows the load-displacement curves of the frame with two different support conditions. An applied load ratio of 1.0 corresponds to the full factored wind load combination of 1.0 gravity + 1.3 wind. The analysis shows that diagonal bracing provides an effective means to control the frame drift and enhance the lateral stiffness of the frame. The increase in load-carrying capacity is more pronounced for more flexible frames.

References

8.1. ECCS, *Ultimate Limit State Calculation of Sway Frames with Rigid Joints*, Technical Committee 8, Structural Stability Technical Working Group 8.2, System, Publication No. 33, 20 pp., 1984.

8.2. White, D.W., *Material and Geometric Nonlinear Analysis of Local Planar Behavior in Steel Frames Using Iterative Computer Graphics*, M. S. thesis., Cornell University, Ithaca, NY, 281 pp., 1985.

8.3. Clarke, M.J., Bridge, R.Q., Hancock, G.J., and Trahair, N.S., "Benchmarking and Verification of Second-Order Elastic and Inelastic Frame Analysis Programs," In *TG 29 Workshop and Nomograph on Plastic Hinge Based Methods for Advanced Analysis and Design of Steel Frames*, D.W. White and W.F. Chen, Editors, SSRC, Lehigh University, Bethlehem, PA, 1992.

8.4. Orbison, J.G., "Nonlinear Static Analysis of Three-Dimensional Steel Frames," *Report No. 82–6*, Department of Structural Engrg., Cornell University, 243 pp., 1982.

8.5. Lui, E.M., *Effect of Connection Flexibility and Panel Zone Deformation on the Behavior of Plane Steel Frames*, Ph.D. dissertation, School of Civil Engrg., Purdue University, West Lafayette, IN, 1985.

8.6. Liew, J.Y.R., *Advanced Analysis for Frame Design*, Ph.D. dissertation, School of Civil Engineering, Purdue University, West Lafayette, IN, May, 393 pp., 1992.

8.7. American Institute of Steel Construction, *Load and Resistance Factor Design Specification for Structural Steel Buildings*, 2nd. ed., Chicago, December, 1993.

8.8. CSA, *Limit States Design of Steel Structures*, CAN/CSA-S16.-M89, Canadian Standards Association, 1989.

8.9. LeMessurier, W.J., "A Practical Method of Second Order Analysis, Part 2—Rigid Frames," *Engineering J.*, AISC, 14, 2, 49–67, 1977.

8.10. Yura, J.A., "The Effective Length of Columns in Unbraced Frames," *Engineering J.*, AISC, 8, 2, 37–42, 1971.

8.11. Stevens, L.K., "Elastic Stability of Practical Multistory Frames," *Proc. Inst. Civil Engrs.*, 36, London, UK, 1967.

8.12. Cheong-Siat-Moy, F., "Consideration of Secondary Effects in Frame Design," *J. Struct. Div.*, ASCE, 103, ST10, 2005–2019, 1977.

8.13. Standards Australian, *Steel Structures*, Sydney, Australia, 1990.

8.14. EC 3, *Design of Steel Structures: Part I—General Rules and Rules for Buildings*, 1, Eurocode Edited draft, Issue 3, 1990.

8.15. Goto, Y., Suzuki, S., and Chen, W.F., "Bowing Effect on Elastic Stability of Frames Under Primary Bending Moments," *J. Struct. Engrg.*, ASCE, 117, 1, 111–127, 1991.

8.16. Chen, W.F., and Lui, E.M., *Structural Stability—Theory and Implementation*, Elsevier Science Publication Co., New York, 490 pp., 1986.

8.17. Chen, W.F., and Lui, E.M., *Stability Design of Steel Frames*, CRC Press, 380 pp., 1992.

8.18. Cook, R.D., Malkus, D.S., and Plesha, M.E., *Concepts and Applications of Finite Element Analysis*, third ed., John Wiley & Sons, 630 pp., 1989.

8.19. Belytschko, T., and Hsieh, B.J., "Non-linear Transient Finite Element Analysis with Convected Coordinates," *Int. J. Num. Meth. Engrg.*, 7, 3, 255–271, 1973.

8.20. Powell, G.H., "Theory of Nonlinear Elastic Structures," *J. Struct. Div.*, ASCE, 95, ST12, 2687–2701, 1969.

8.21. Lui, E.M., and Chen, W.F., "Behavior of Braced and Unbraced Semi-Rigid Frames," *Int. J. Solids Struct.*, 24(9), 893–913, 1988.

8.22. Liew, J.Y.R., White, D.W., and Chen, W.F., "Beam-Column Design in Steel Frameworks—Insight on Current Methods and Trends," *J. Constructional Steel Research*, 18, 269–308, 1991.

8.23. Liew, J.Y.R., White, D.W., and Chen, W.F., "Limit State Design of Semi-Rigid Frames Using Advanced Analysis: Part 2: Analysis and Design," *J. Constructional Steel Research*, Vol. 26, No. 1, 1993, pp. 29–58.

8.24. Chen, W.F., and Atsuta, T., *Theory of Beam-Column, 1, In-Plane Behavior and Design*, MacGraw-Hill Int. Pub., 513 pp., 1976.

8.25. White, D.W., Liew, J.Y.R., and Chen, W.F., *Second-Order Inelastic Analysis for Frame Design: A Report to SSRC Task Group 29 on Recent Research and the Perceived State-of-the-Art*, Structural Engineering Report, CE-STR-91-12, Purdue University, 116 pp., 1991.

8.26. White, D.W., Liew, J.Y.R., and Chen, W.F., *"Toward Advanced Analysis in LRFD"*, In *Workshop and Nomograph on Plastic Hinge Based Methods for Advanced Analysis and Design of Steel Frames*, D.W. White and W.F. Chen, Editors, SSRC, Lehigh University, Bethlehem, PA, 1992.

8.27. Zhou, S.P., Duan, L., and Chen, W.F., "Comparison of Design Equations for Steel Beam-Columns, *Structural Engineering Review*, 2(1), 45–53, 1990.

8.28. Kanchanalai, T., *The Design and Behavior of Beam-Columns in Unbraced Steel Frames*, AISI Project No. 189, Report No. 2, Civil Engineering/Structures Research Lab., University of Texas at Austin, 300 pp., 1977.

8.29. Liew, J.Y.R., White, D.W., and Chen, W.F., "Beam-Columns," *Constructional Steel Design*, an international guide, Chapter 5.1, P.J. Dowling et al., editors, Elsevier Applied Science Publishers, England, 105–132, 1992.

8.30. Liew, J.Y.R., White, D.W., and Chen, W.F., "Second-Order Refined Plastic Hinge Analysis for Frame Design: Part 1," *J. Struct. Engrg.*, ASCE, Vol. 119, No. 11, 1993, pp. 3196–3216.

8.31. Liew, J.Y.R., White, D.W., and Chen, W.F., "Second-Order Refined Plastic Hinge Analysis for Frame Design: Part 2," *J. Struct. Engrg.*, ASCE, Vol. 119, No. 11, 1993, pp. 3217–3237.

8.32. Galambos, T.V., Editor, *Guide to Stability Design Criteria for Metal Structures*, Structural Stability Research Council, fourth edition, John Wiley & Sons, New York, 786 pp., 1988.

8.33. Vogel, U., "Calibrating Frames," *Stahlbau*, 10, Oct., 295–301, 1985.

8.34. Ziemian, R.D., *Advanced Methods of Inelastic Analysis in the Limit States Design of Steel Structures*, Ph.D. dissertation, School of Civil and Environmental Engineering, Cornell University, Ithaca, NY, 265 pp., 1990.

8.35. Kishi, N., Chen, W.F., Goto, Y., and Matsuoka, K.G., "Design Aid of Semi-Rigid Connections for Frame Analysis," Engineering Journal, AISC, Vol. 30, No. 3, 3rd Quarter, 1993, pp. 90–107.

8.36. Liew, J.Y.R., White, D.W., and Chen, W.F., "Limit State Design of Semi-Rigid Frames Using Advanced Analysis: Part 1: Connection Modeling and Classification," *J. Constructional Steel Research*, Vol. 26, No. 1, 1993, pp. 1–28.

8.37. Liew, J.Y.R., and Chen, W.F., "*Trend Toward Advanced Analysis*," Chapter 1 in *Advanced Analysis of Steel Frames—Theory, Software and Applications*, W.F. Chen and S. Toma, Editors, CRC Press, Boca Raton, Florida, 1994, pp. 1–45.

8.38. Hughes, T.J.R., *The Finite Element Method: Linear Static and Dynamic Finite Element Analysis*, Prentice-Hall, Inc., Englewood Cliffs, NJ, 803 pp., 1987.

8.39. ECCS, *Essentials of Eurocode 3 Design Manual for Steel Structures in Building*, ECCS-Advisory Committee 5, No. 65, 60 pp., 1991.

Problems

8.1. The column shown in Fig. P8.1 has an initial out-of-straightness at the midspan equal to $L/1500$, and the column slenderness ratio $\lambda = (L/\pi r_x)/\sqrt{F_y/E} = 1.0$. Determine the column axial load capacity using second-order elastic-plastic hinge and refined plastic hinge analyses, assuming two frame elements per member. Compare the computer results with the solutions obtained using the AISC LRFD column strength equations, and explain why the elastic-plastic hinge analysis is not adequate for use as advanced analysis.

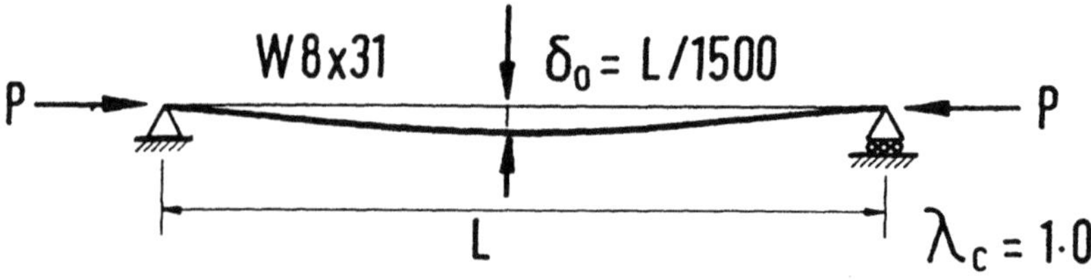

FIGURE P8.1.

8.2. The beam column shown in Fig. P8.2 is subjected to a constant axial force of $P = 0.5P_y$ and a midspan concentrated load applied incrementally to collapse. For $L/r = 80$, determine the ultimate strength of the beam column using second-order elastic-plastic hinge and refined plastic hinge analyses. Compare the results with those obtained by first-order plastic hinge analysis using the AISC LRFD beam-column interaction equations.

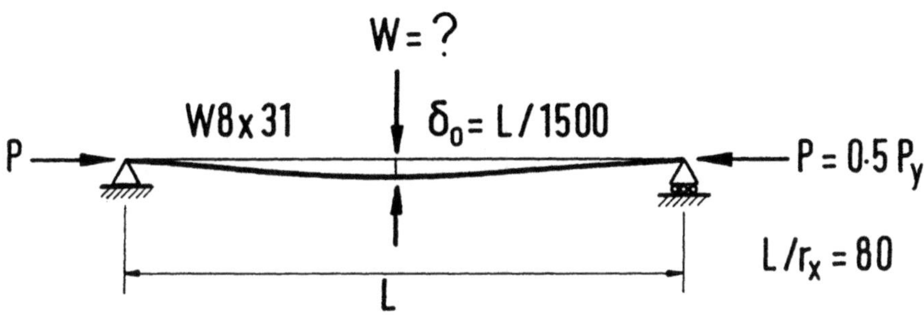

FIGURE P8.2.

8.3. The dimensions and member sizes for two portal frames in which side-sway is prevented are shown in Fig P8.3. The column slenderness ratio for type A and type B frames is $L/r = 40.3$, and the relative stiffness of the column to the beam is 0.25. The difference between the type A and type B frames is the support conditions. The type A frame has fixed supports, whereas the type B frame is simply supported at the bases of the columns. The beams and columns in these frames are rigidly connected about their strong-axis bending direction. All the member cross sections are assumed to be fully compact and the members are fully braced to prevent out-of-plane deformations.

For the load parameter $\beta = 0.34$, determine the ultimate strength of the frame using

(a) elastic analysis and the AISC LRFD beam-column equations.
(b) plastic analysis and the AISC LRFD beam-column equations.
(c) second-order elastic-plastic hinge analysis.
(d) second-order refined plastic hinge analysis.

8.4. The configurations of three calibration frames are shown in Fig. P8.4. The cross-sectional dimensions and properties of the frame members to be used in the analyses are summarized in the table. Also tabulated are the magnitudes of column sway imperfections, and the discretization of frame members to be used in the analysis of the calibration frames. For the factored loading as shown:

(a) determine the load factors at collapse using second-order plastic-hinge-based analyses.
(b) compare the load versus lateral displacement results.
(c) compare the internal force distribution and plastic hinge locations.
(d) compare the plastic-hinge-based results with the plastic-zone solutions in [8.6] and [8.33].

P_c
W_b
P_c
W16 x 50
L_c = 3556 mm
W8 x 31
Frame A
W8 x 31
L_b = 5334 mm
(a)
P_b = W_b L_b
β = P_b /(2P_c + P_b)
= 0·34
P_c
W_b
P_c
W 16x50
L_c = 3556 mm
W8x31
Frame B
W8x31
L_b = 5334 mm
(b)

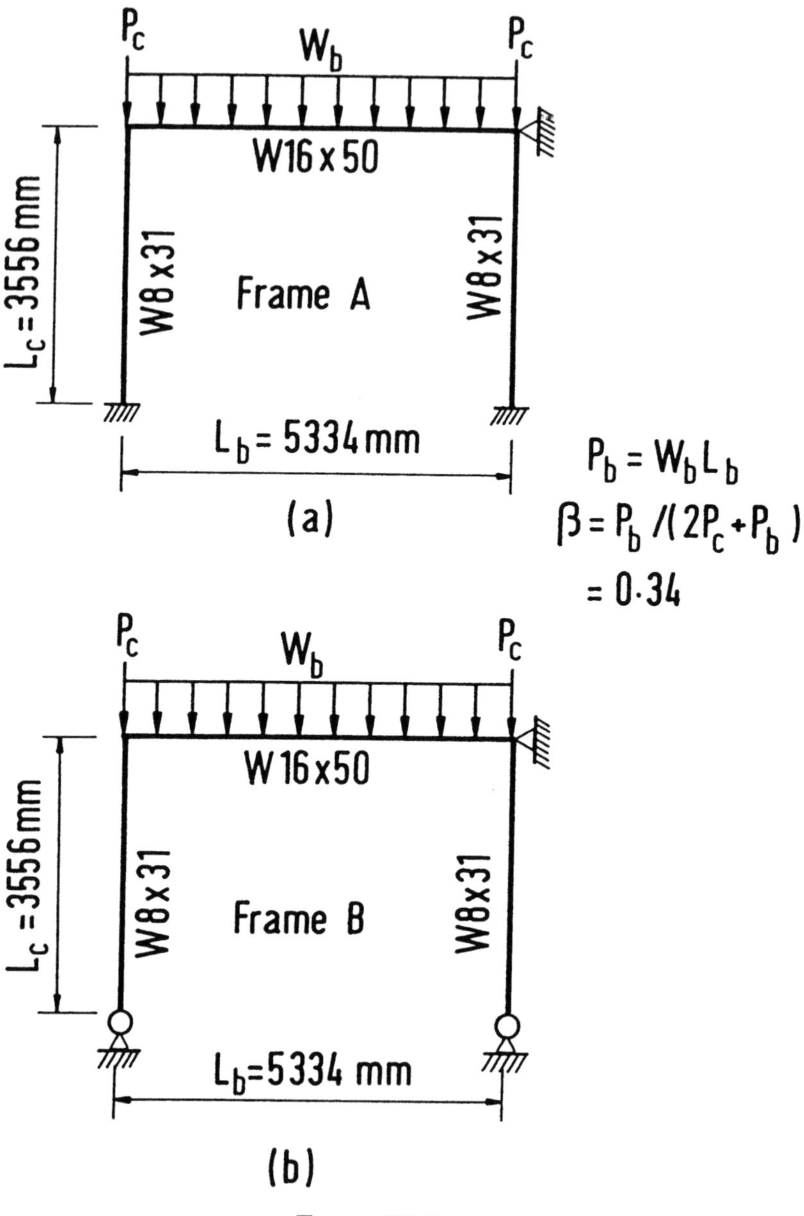

FIGURE P8.3.

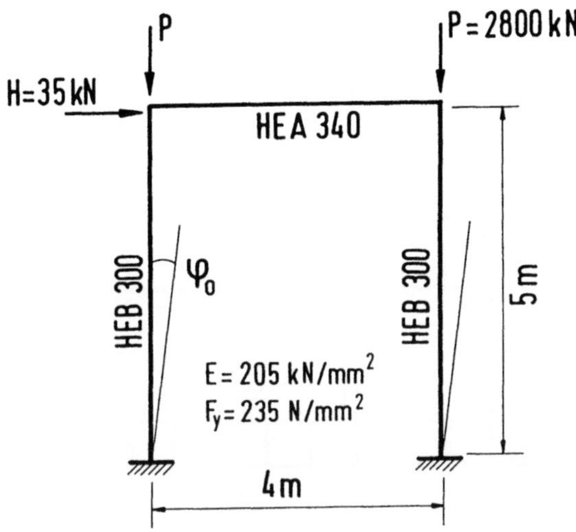

Calibration frame 1 : Portal frame

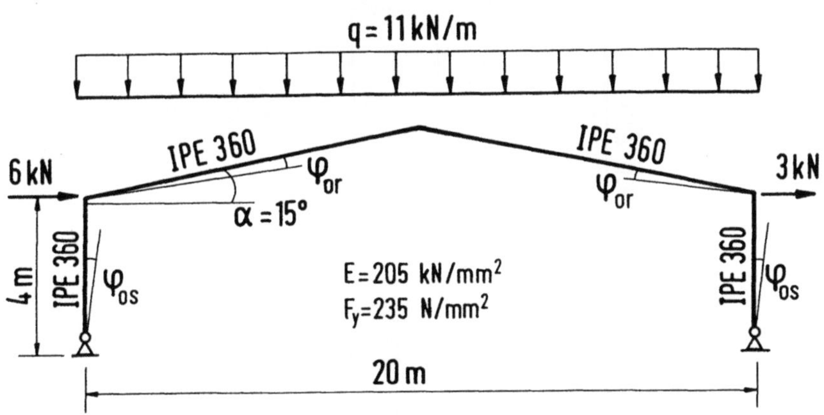

Calibration frame 2 : Gable frame

FIGURE P8.4.

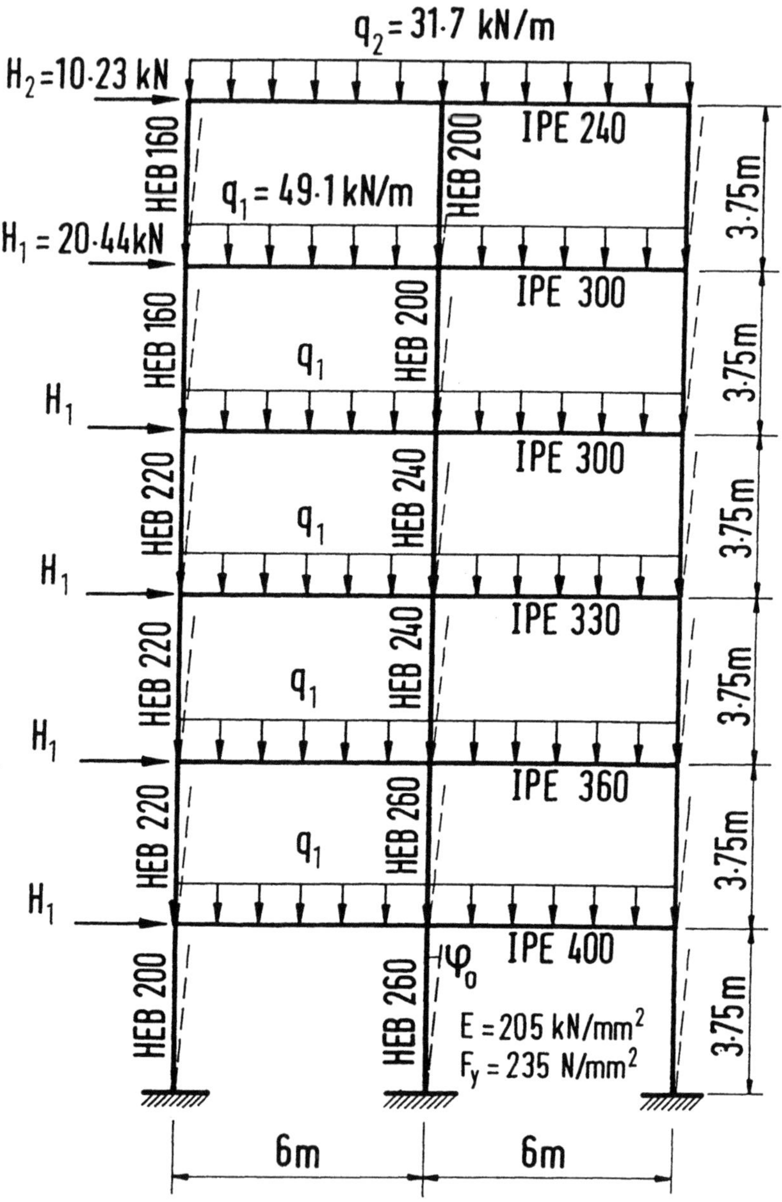

Calibration frame 3 : Six-story frame

FIGURE P8.4 (*cont.*)

TABLE FOR P.8.4: Cross-section properties of members and initial sway imperfections.

Members	A (cm²)	I_x (cm⁴)	I_y (cm⁴)	Calibration Frames		φ_0 or φ_{06} *	Sway Deflection** (mm)
IPE240	39.1	3,892	284	Portal Frame		1/400	12.5
IPE300	53.8	8,356	604				
IPE330	62.6	11,770	788				
IPE360	72.7	16,270	1,043				
IPE400	84.5	23,130	1,318	Gable Frame	Columns	1/300	13.3
HEB160	54.3	2,492	889				
HEB200	78.1	5,696	2,003		Roof Beams	1/432	24.0
HEB220	91.0	8,091	2,843				
HEB240	106.0	11,260	3,923	6-Story Frame		1/450	50.0
HEB260	118.0	14,920	5,135				
HEB300	149.0	25,170	8,563				
HEA340	133.0	27,690	7,436				

* Angle of sway imperfection $\varphi_0 = r_1 r_2/300$ (see Ref. [8.33])
** Sway deflection at the top of the frame.

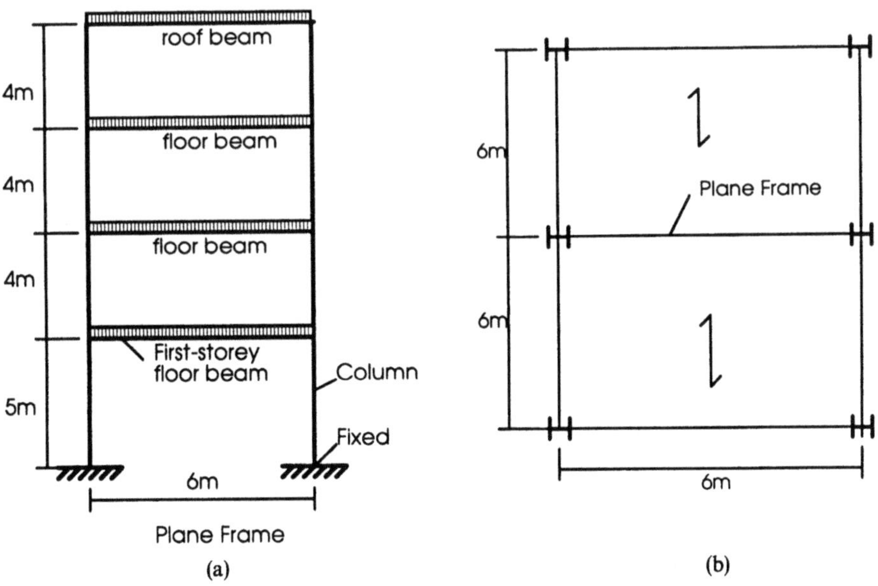

FIGURE P8.6.

8.5. For the frame configuration and loading shown in Fig. 8.20, reproportioning the frame members using AISC LRFD factored load combinations assuming full rigidity between beam-to-column connections. Using advanced analysis techniques, select a connection type from Table 1 that will satisfy ultimate strength and serviceability limit-state requiements.

8.6. Figure P8.6(a) shows an interior frame of a multistory steel building to be designed using advanced analysis methods. The frame is braced against out-of-plane sway at each story level. The floor comprises only primary beams, with flooring and roofing spanning as shown in Fig. P8.6(b). The materials used for all steel sections are A36 steel.
Based on the loading data given below:
Floor Beams
Uniformly distributed dead load = 4.5 kN/m^2.
Uniformly distributed imposed load = 5.0 kN/m^2.
Roof Beam
Uniformly distributed dead load = 4.0 kN/m^2.
Uniformly distributed imposed load = 1.5 kN/m^2.

a) Design the columns and beams using the AISC LRFD procedures. The column base may be assumed to be fixed, and same column size is to be used for the entire frame.
b) Redesign all members using advanced analysis techniques, and conduct checks to ensure that all serviceability and ultimate strength criteria are satisfied.

Index

FSC
www.fsc.org
MIX
Papier aus verantwortungsvollen Quellen
Paper from responsible sources
FSC® C105338